T0178078

Lecture Notes in Mathematics

Volume 2350

Editors-in-Chief

Jean-Michel Morel, City University of Hong Kong, Kowloon Tong, China

Bernard Teissier, IMJ-PRG, Paris, France

Series Editors

Karin Baur, University of Leeds, Leeds, UK

Michel Brion, UGA, Grenoble, France

Rupert Frank, LMU, Munich, Germany

Annette Huber, Albert Ludwig University, Freiburg, Germany

Davar Khoshnevisan, The University of Utah, Salt Lake City, UT, USA

Ioannis Kontoyiannis, University of Cambridge, Cambridge, UK

Angela Kunoth, University of Cologne, Cologne, Germany

Ariane Mézard, IMJ-PRG, Paris, France

Mark Podolskij, University of Luxembourg, Esch-sur-Alzette, Luxembourg

Mark Policott, Mathematics Institute, University of Warwick, Coventry, UK

László Székelyhidi ⓘ, MPI for Mathematics in the Sciences, Leipzig, Germany

Gabriele Vezzosi, UniFI, Florence, Italy

Anna Wienhard, MPI for Mathematics in the Sciences, Leipzig, Germany

This series reports on new developments in all areas of mathematics and their applications - quickly, informally and at a high level. Mathematical texts analysing new developments in modelling and numerical simulation are welcome. The type of material considered for publication includes:

1. Research monographs
2. Lectures on a new field or presentations of a new angle in a classical field
3. Summer schools and intensive courses on topics of current research.

Texts which are out of print but still in demand may also be considered if they fall within these categories. The timeliness of a manuscript is sometimes more important than its form, which may be preliminary or tentative. Please visit the LNM Editorial Policy (https://drive.google.com/file/d/1MOg4TbwOSokRnFJ3ZR3ciEeKs9hOnNX_/view?usp=sharing)

Titles from this series are indexed by Scopus, Web of Science, Mathematical Reviews, and zbMATH.

Yukinobu Toda

Categorical Donaldson-Thomas Theory for Local Surfaces

Springer

Yukinobu Toda
Kavli Institute for Physics and Mathematics
of the Universe
University of Tokyo
Kashiwa, Chiba, Japan

ISSN 0075-8434   ISSN 1617-9692   (electronic)
Lecture Notes in Mathematics
ISBN 978-3-031-61704-1   ISBN 978-3-031-61705-8   (eBook)
https://doi.org/10.1007/978-3-031-61705-8

Mathematics Subject Classification: 14N35, 1408, 18G80, 18N25, 14F06, 14D20, 14D23, 14E30

This Springer imprint is published by the registered company Springer Nature Switzerland AG
The registered company address is: Gewerbestrasse 11, 6330 Cham, Switzerland

If disposing of this product, please recycle the paper.

Preface

The Donaldson-Thomas invariant is an integer which counts stable coherent sheaves on a Calabi-Yau 3-fold, and was introduced by Thomas in 1998. Since its appearance, there have been significant progress of the subject and has drawn much attention in algebraic geometry (especially enumerative geometry counting curves), geometric representation theory and mathematical physics.

One of the interesting features of Donaldson-Thomas theory is its possible refinements; the notions of cohomological or motivic Donaldson-Thomas invariants were introduced by Kontsevich-Soibelman and Joyce et al., which recover the original DT invariant by taking their Euler characteristics. It is then natural to try to find a further categorification, i.e. dg-category, which recovers the Donaldson-Thomas invariant. However, the construction of such a dg-category in general has turned out to be a difficult problem, and is an open problem at this moment.

In this book, we focus on local surfaces, i.e. non-compact Calabi-Yau 3-folds given by total spaces of canonical line bundles on algebraic surfaces, and develop categorical Donaldson-Thomas theory in this case. Although the local surfaces are a particular class of Calabi-Yau 3-folds, it is an enough interesting class of Calabi-Yau 3-folds; the sheaves on them are regarded as an analogue of Higgs bundles on curves which play important role in non-abelian Hodge theory, and the categorical Donaldson-Thomas theory on local surfaces has close relation with the moduli theory of sheaves on surfaces.

One of the motivations of writing this book is to make it clear 'why categorification of Donaldson-Thomas invariants is important and interesting?' In Chap. 1, we explain that the categorical Donaldson-Thomas theory is relevant in several research fields, e.g. enumerative geometry, geometric representation theory, birational geometry, classical moduli theory or so on. Among them, we focus on the motivation from d-critical D/K equivalence conjecture and categorical wall-crossing formula. Based on these motivations, we give several conjectures on categorical Donaldson-Thomas theory for local surfaces. If the categorical Donaldson-Thomas theory in a general case will be established in a future, the conjectures we propose in this book will be also expected in general.

The work of this book is regarded as a beginning of the study of categorical Donaldson-Thomas theory. The first version of this book appeared in arXiv in 2019, and since then there have been various progress on this topic. In Sect. 1.5, we have given a updated list of the developments of this subject. We hope that this book will inspire further developments of this subject near future.

Acknowledgements

The author is grateful to Tomoyuki Abe, Yuki Hirano and Daniel Halpern-Leistner for valuable discussions, and Yalong Cao, Flancesco Sala, Mauro Porta, and Aurelio Carlucci for comments on the first version of this text. The author also thanks Kendric Schefers for pointing out an error in an earlier version of the text. The author also thanks referees for several comments. The author is supported by World Premier International Research Center Initiative (WPI initiative), MEXT, Japan, and JSPS KAKENHI Grant Numbers JP19H01779, JP24H00180.

Kashiwa, Japan   Yukinobu Toda
2024

Contents

Chapter 1
Introduction

In this chapter, we give background of Donaldson-Thomas theory, and motivations of considering categorical DT theory. The categorical DT theory is related to several branches of mathematics, e.g. enumerative geometry, geometric representation theory, birational geometry, classical moduli theory, and so on. Among them, we focus on motivations from d-critical D/K equivalence conjecture and categorical wall-crossing formula.

We then give a brief explanation of how to construct DT categories for local surfaces, based on Koszul duality equivalences and singular support quotients. We will use this construction to formulate several conjectures on d-critical D/K equivalence conjecture and categorical wall-crossing. We give three approaches toward these conjectures: linear Koszul duality, window theorem and categorified Hall product. These techniques give several results on the above conjectures.

This books is a beginning of the theory. We will also explain more recent developments of the theory.

1.1 Motivation

1.1.1 Background

On a smooth projective Calabi-Yau 3-fold, Thomas [125] introduced the integer valued invariants which virtually count stable coherent sheaves on it, and give complexifications of Casson invariants on real 3-folds. His invariants are now called *Donaldson-Thomas (DT)* invariants, and have been developed in several contexts, e.g. relation with Gromov-Witten invariants [88, 101], wall-crossing under change of stability conditions [64, 76, 129], relation with cluster algebras [32, 93], or so on.

The original DT invariant in [125] is defined via integration of the zero-dimensional virtual fundamental class on the moduli space of stable sheaves M

© The Author(s), under exclusive license to Springer Nature Switzerland AG 2024
Y. Toda, *Categorical Donaldson-Thomas Theory for Local Surfaces*, Lecture Notes
in Mathematics 2350, https://doi.org/10.1007/978-3-031-61705-8_1

on a Calabi-Yau 3-fold X (with fixed stability condition, Chern character, see Remark 1.1.1). Assuming the absence of strictly semistable sheaves (in particular M is proper), the DT invariant of X is given by

$$\text{DT}(M) := \int_{[M]^{\text{vir}}} 1 \in \mathbb{Z}. \tag{1.1.1}$$

Later, Behrend [14] showed that the DT invariant is also defined via the weighted Euler characteristics of certain constructible function χ_B (called Behrend functions) on the above moduli space M. Namely we have the identity

$$\text{DT}(M) = \int_M \chi_B \, de := \sum_{m \in \mathbb{Z}} e(\chi_B^{-1}(m)) \in \mathbb{Z}.$$

Now by the works [22, 64], the moduli space M is known to be locally written as critical loci of some function. That is for each point $p \in M$, there is an open neighborhood $p \in U$ such that

$$U \cong \text{Crit}(w), \; w \colon Y \to \mathbb{C} \tag{1.1.2}$$

where Y is a smooth scheme and w is a regular function on it. The Behrend function at p coincides with the Euler characteristics of the vanishing cycle of w. Based on this fact, Kontsevich-Soibelman [75, 76] proposed some refinements of DT invariants, e.g. motivic DT invariants and cohomological DT invariants. The cohomological ones are defined by gluing the locally defined perverse sheaves of vanishing cycles $\phi_w \in \text{Perv}(U)$, through a choice of an orientation data (see Remark 1.1.2), and taking its hypercohomology. For a glued vanishing cycle sheaves $\phi_M \in \text{Perv}(M)$, the cohomological DT invariant is given by

$$\text{CoDT}(M) := H^*(M, \phi_M) \in \text{(graded vector space)}. \tag{1.1.3}$$

The graded vector space (1.1.3) is finite dimensional, and its Euler characteristic recovers the DT invariant (1.1.1). The foundations of the above refinements of DT invariants were established in [19, 27], and for example used in [87] to give a mathematical definition of Gopakumar-Vafa invariants.

Remark 1.1.1 The moduli space M depends on a choice of a stability condition σ and a choice of a Chern character v, so should be denoted by $M^\sigma(v)$. The element v is taken in the finitely generated abelian group Γ,

$$v \in \Gamma := \text{Im}(\text{ch} \colon K(X) \to H^*(X, \mathbb{Q})).$$

In this introduction we abbreviate σ and v in the notation, and just write $M = M^\sigma(v)$, unless they are necessary.

Remark 1.1.2 An orientation data of M is a square root line bundle of its virtual canonical line bundle

$$K_M^{\text{vir}} \to M$$

whose fiber at $E \in M$ is $\det \mathbf{R}\mathrm{Hom}(E, E)$. In [65], Joyce-Upmeiler proved the existence of a canonical orientation data.

1.1.2 Categorical DT Theory

As we mentioned above, the moduli space M of stable sheaves on a CY 3-fold admits local critical chart as in (1.1.2). More strongly, it is known that M is a truncation of a derived scheme with a (-1)-shifted symplectic structure [104]. In [12, 22], it is proved that M is locally equivalent to the derived critical locus of w, and the (-1)-shifted symplectic structure induces a *d-critical structure* on M introduced by Joyce [63]. Then locally on M, we can attach the triangulated category of matrix factorizations of w, denoted by

$$\mathrm{MF}(Y, w).$$

Its objects consist of locally free sheaves $\mathcal{P}_0$, $\mathcal{P}_1$ on Y together with maps

$$\mathcal{P}_0 \underset{\alpha_1}{\overset{\alpha_0}{\rightleftarrows}} \mathcal{P}_1, \quad \alpha_0 \circ \alpha_1 = \cdot w, \quad \alpha_1 \circ \alpha_0 = \cdot w.$$

The triangulated category $\mathrm{MF}(Y, w)$ is regarded as a categorification of the vanishing cycle, as its periodic cyclic homology recovers the $\mathbb{Z}/2$-graded hypercohomology of the perverse sheaf of vanishing cycles of w (see [18, 39]). Therefore as mentioned in [62, (J)], [146, Section 6.1], it is natural to try to glue the locally defined categories $\mathrm{MF}(Y, w)$, and regard it as a categorification of the DT invariant. If such a gluing exists, say $\mathcal{DT}(M)$, then we would like to call $\mathcal{DT}(M)$ as a *DT category*, and study its properties.

However at this moment, constructing the DT category $\mathcal{DT}(M)$ is very hard and yet unsolved problem. An issue is a difficulty of gluing categories of matrix factorizations. The category $\mathcal{DT}(M)$ may be constructed by taking the homotopy limit of dg-enhancements $\mathrm{MF}_{\mathrm{dg}}(Y, w)$ of $\mathrm{MF}(Y, w)$, twisted by another dg-categories determined by a choice of an orientation data. In order to take such a limit, we need an ∞-functor from the category of open subsets $U \subset M$ written as $U = \mathrm{Crit}(w)$ to the ∞-category of dg-categories, of the form

$$U \mapsto \mathcal{DT}(U) = \mathrm{MF}_{\mathrm{dg}}(Y, w) \otimes (\text{twisting}). \tag{1.1.4}$$

Then the DT category $\mathcal{DT}(M)$ may be defined just as a limit

$$\mathcal{DT}(M) = \lim_{U \subset M} \mathcal{DT}(U).$$

A construction of an ∞-functor of the form (1.1.4) in general seems to require a substantial work on ∞-categories.

Remark 1.1.3 At the time of writing this book, there is an ongoing work on gluing matrix factorizations. The slide is available in [55].

1.1.3 The Main Player in This Book

As we mentioned above, constructing the DT category in general should require lots of technique of ∞-categories, which is beyond the scope of this book. Instead of constructing DT categories in general, we focus on the case that a CY 3-fold is the total space of the canonical line bundle on a surface, called a *local surface*. Namely for a smooth projective surface S, the local surface is given by

$$X = \mathrm{Tot}_S(\omega_S) = \mathrm{Spec}_S \left(\bigoplus_{i \geq 0} \omega_S^{\otimes(-i)} \right). \tag{1.1.5}$$

In this case $\mathbb{C}^*$ acts on fibers of the canonical bundle, and we construct $\mathbb{C}^*$-equivariant version of DT categories on local surfaces.

While this is a particular class of CY 3-folds, the moduli space M is not necessary written as a global critical locus, so the gluing problem is still not obvious. Indeed our construction is not a direct gluing of matrix factorizations, but rather indirect through the derived categories of coherent sheaves on derived moduli stacks of coherent sheaves on the surface S, together with Koszul duality. Roughly speaking, our construction is given by the Verdier quotient

$$\mathcal{DT}^{\mathbb{C}^*}(M) = D^b_{\mathrm{coh}}(\mathfrak{M}_S)/\mathcal{C}_{Z^{\mathrm{us}}}. \tag{1.1.6}$$

Here $\mathfrak{M}_S$ is the derived moduli stack of coherent sheaves on S, and $\mathcal{C}_{Z^{\mathrm{us}}}$ is its subcategory consisting of objects whose singular supports [8] are contained in the unstable locus $Z^{\mathrm{us}} \subset \mathcal{M}_X$, where $\mathcal{M}_X$ is the moduli stack of compactly supported coherent sheaves on X. A key point of this construction is the isomorphism

$$\mathcal{M}_X \cong t_0(\Omega_{\mathfrak{M}_S}[-1]) \tag{1.1.7}$$

where the right hand side is the (classical truncation of) (-1)-shifted cotangent of $\mathfrak{M}_S$

$$\Omega_{\mathfrak{M}_S}[-1] = \text{Spec Sym}(\mathbb{T}_{\mathfrak{M}_S}[1])$$

where the singular support of coherent sheaves on $\mathfrak{M}_S$ are defined. Therefore, although $D^b_{\text{coh}}(\mathfrak{M}_S)$ is defined from the surface S, the derived structure of $\mathfrak{M}_S$ and the subcategory $\mathcal{C}_{Z^{\text{us}}}$ contain information of the 3-fold geometry of X. However the above construction (1.1.6) is not precise, and we need a slight (technical) modification of it. We will explain in Sect. 1.2 that why the category (1.1.6) is related to the moduli of stable sheaves on a CY 3-fold X, why it is regarded as a gluing of categories of matrix factorizations, and what kind of modifications are required.

Although the local surfaces are special class of CY 3-folds, these are enough interesting CY 3-folds. For example, one can identify a (compactly supported) coherent sheaf on a local surface X with a pair

$$(F, \theta), \ \theta: F \to F \otimes \omega_S. \tag{1.1.8}$$

A pair as above may be regarded as a higher dimensional analogue of Higgs bundles on curves which play important role in non-abelian Hodge theory. Moreover there are two simplifications in the local surface case: the first one is that a relevant ∞-functor is a priori given from the existence of the derived moduli stack $\mathfrak{M}_S$, so we can avoid technical subtleties on ∞-categories. The second one is that there exists a canonical orientation data in this case, so the construction is canonical without taking twisting.

Then we have a well-defined categorification of DT invariants on the local surface, can formulate several conjectures on DT categories, and try to prove them. In this book, we focus on categorifications of wall-crossing formulas of DT invariants and give their relation with D/K equivalence conjecture in birational geometry, as we will discuss in Sect. 1.1.6.

1.1.4 Why Do We Want to Categorify?

Our motivation for the categorification of DT invariants comes from our ultimate desire to understand the geometry of the moduli space M. Recall that the geometry of such a moduli space on a lower dimensional variety has been a classical subject in algebraic geometry. For example, the moduli space of line bundles on a smooth projective curve is an abelian variety, called Jacobian, which is a classical geometric object in algebraic geometry. The moduli spaces of higher rank stable bundles are also classical subject, and for example studied by Narasimhan-Ramanan [95]. The moduli spaces of stable sheaves on K3 surfaces admit holomorphic symplectic structures by Mukai [92], and are known to be deformation equivalent to Hilbert schemes of points. There are many other results in classical algebraic geometry concerning the geometry of moduli spaces of stable sheaves.

However such a geometric study of the moduli spaces of sheaves has been restricted to varieties of dimensions less than or equal to two. An issue in a higher dimensional case is that the moduli spaces are usually highly singular (not irreducible, not reduced,...) which do not fit into the framework, e.g. terminal singularities, canonical singularities, in birational geometry [73].

In order to explain a relationship between geometric framework for the moduli space M and the categorification of DT invariants, let us observe the following. For a smooth projective variety Y, we have the following picture:

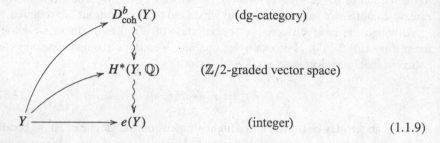

$$
\begin{array}{lll}
D^b_{\mathrm{coh}}(Y) & \text{(dg-category)} \\
\downarrow \\
H^*(Y, \mathbb{Q}) & (\mathbb{Z}/2\text{-graded vector space}) \\
\downarrow \\
Y \longrightarrow e(Y) & \text{(integer)}
\end{array}
\tag{1.1.9}
$$

Here $D^b_{\mathrm{coh}}(Y)$ is the bounded derived category of coherent sheaves, or its dg-enhancement, whose objects consist of bounded complexes of coherent sheaves

$$
\cdots \to F^0 \to F^1 \to \cdots \to F^i \to F^{i+1} \to \cdots , \quad F^i \in \mathrm{Coh}(Y).
$$

The vertical arrows mean that the source objects recover the target objects, e.g. the topological Euler number $e(Y)$ is recovered from the singular cohomology $H^*(Y, \mathbb{Q})$ by taking its Euler characteristics. There are several ways to recover $H^*(Y, \mathbb{Q})$ from the derived category $D^b_{\mathrm{coh}}(Y)$, e.g. topological K-theory, periodic cyclic homology or Hochschild homology.

The derived category of coherent sheaves was introduced by Grothendieck and Verdier in 1960s in order to give a relative version of Serre duality of cohomologies of coherent sheaves, which is called Grothendieck duality [54]. In his ICM talk in 1994, Kontsevich [74] proposed Homological mirror symmetry conjecture, which is an equivalence between the derived category of coherent sheaves on a Calabi-Yau manifold Y and derived Fukaya category of its mirror manifold $Y^\vee$,

$$
D^b_{\mathrm{coh}}(Y) \simeq D\mathrm{Fuk}(Y^\vee).
$$

Since then, the study of derived categories have been one of the main topics in algebraic geometry, and several interesting derived symmetries (which involve research fields beyond algebraic geometry) have been found, e.g. equivalences or fully-faithful functors under some birational transformations (D/K equivalence conjecture [20, 23, 68]), equivalences under variations of GIT quotients [11, 45], derived McKay correspondences [26], equivalences with non-commutative crepant resolutions [150], Homological projective duality [81], the theory of Bridgeland stability conditions [24], etc. In some sense, the derived category $D^b_{coh}(Y)$ may be regarded as a 'nice space' in non-commutative geometry, where the main players are dg-categories satisfying some nice properties.

Now for a moduli space M of stable sheaves, what we expect is the following picture analogous to (1.1.9):

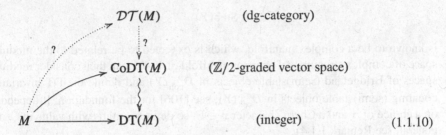

$$
\begin{array}{lll}
\mathcal{DT}(M) & \text{(dg-category)} \\
\downarrow ? \\
\text{CoDT}(M) & \text{($\mathbb{Z}/2$-graded vector space)} \\
M \longrightarrow \text{DT}(M) & \text{(integer)} & (1.1.10)
\end{array}
$$

A crucial difference of the above picture with (1.1.9) is that M is in general highly singular (again not irreducible, not reduced, ...) and $\text{DT}(M)$ is not the usual Euler number but a χ_B-weighted one which keeps track of singularities of M. Therefore if a dg-category $\mathcal{DT}(M)$ exists, then it should be a 'better' category than $D^b_{coh}(M)$, which keeps track of the singularities of M, and may be regarded as a 'virtual derived category'. Thus the category $\mathcal{DT}(M)$ may be regarded as a 'nice space' in non-commutative geometry, and we may expect that several interesting symmetries, which are known to hold for usual derived categories, also hold for DT categories.

1.1.5 Enumerative Geometry vs Category Theory

Another aspect of our motivation is to give a mutual interaction of enumerative geometry (DT invariants) which is rather an explicit research subject, and category theory (derived categories, matrix factorizations) which is an abstract research subject. The one direction (from category theory to enumerative geometry) has been already studied with lots of applications.

As we mentioned, the original DT invariants virtually count stable sheaves on CY 3-folds. Instead of stable sheaves, one can consider stable objects in $D^b_{coh}(Y)$ for a CY 3-fold Y, e.g. with respect to a Bridgeland stability condition [24]. Roughly speaking, a Bridgeland stability condition consists of data

$$\sigma = (Z, \mathcal{A}), \ \mathcal{A} \subset D^b_{coh}(Y), \ Z \colon K(Y) \to \mathbb{C} \qquad (1.1.11)$$

where $\mathcal{A}$ is the heart of a bounded t-structure (in particular abelian category) and Z is a group homomorphism (called central charge) satisfying some axioms. Given data

(1.1.11), there is the notion of (semi)stable objects in $\mathcal{A}$, where the (semi)stability of $E \in \mathcal{A}$ is given by the condition

$$\arg Z(F) < (\leq) \arg Z(E), \quad \text{for all } 0 \neq F \subsetneq E.$$

The set of stability conditions on $D^b_{\text{coh}}(Y)$

$$\text{Stab}(Y)$$

is known to be a complex manifold, which is expected to be related to the moduli space of complex structures of a mirror manifold of Y. We can then consider moduli spaces of Bridgeland (semi)stable objects in $D^b_{\text{coh}}(Y)$ and define the DT invariant counting (semi)stable objects in $D^b_{\text{coh}}(Y)$, see [106] for the foundation. It depends on a choice of σ and a Chern character $v \in \Gamma$, so defines a map (with value in $\mathbb{Q}$ in general, see Remark 1.1.4)

$$\text{DT}(v) \colon \text{Stab}(Y) \to \mathbb{Q}, \quad \sigma \mapsto \text{DT}^\sigma(v). \tag{1.1.12}$$

There is a set of real codimension one submanifolds in $\text{Stab}(Y)$, called walls, such that the moduli spaces of semistable objects (hence the associated DT invariants) are unchanged at each connected component of the complement of walls (called chamber), but may change at wall-crossing. The wall-crossing formula [64, 76] of the DT invariants (1.1.12) under change of stability conditions turned out to be important in proving several properties of the generating series of DT invariants, see [136] for a survey of recent developments in this direction.

The above wall-crossing arguments give a direction from category theory to enumerative geometry. The categoriciation problem is to give a direction in the other way, from enumerative geometry to category theory. Namely starting from DT invariants, we try to construct categories which recover them, and try to establish categorical statements (e.g. equivalences, fully-faithful functors, semiorthogonal decompositions) which reveals categorical origin of several known properties of numerical DT invariants. These possible interactions are depicted below.

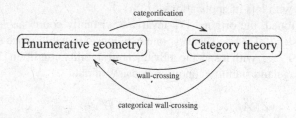

Remark 1.1.4 The map (1.1.12) depends on a choice of $v \in \Gamma$, see Remark 1.1.1. The original DT invariant (1.1.1) is an integer, as it was assumed that every semistable object with Chern character v is stable. In general there exist semistable objects which are not stable, and in that case we need to take account of contribu-

tions of strictly semisimple objects. The DT invariants which also count semistable objects are called generalized DT invariants, which was introduced and studied by Joyce-Song [64].

1.1.6 Motivation from d-Critical Birational Geometry

As already mentioned, the development of wall-crossing formulas of DT invariants [64, 76] is one of the recent important progress on the study of DT invariants. The moduli space of stable sheaves (or Bridgeland stable objects [24]) depends on a choice of a stability condition σ, so should be written as M^σ (still the Chern character v is omitted here, see Remark 1.1.1). As we mentioned in the last subsection, there is a wall-chamber structure on the space of stability conditions such that M^σ is constant if σ lies on a chamber, but may change when σ crosses a wall.

Suppose that σ lies on a wall and $\sigma_\pm$ lie on its adjacent chambers. Then we have the following diagram

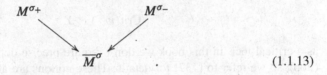

$$\tag{1.1.13}$$

where $\overline{M}^\sigma$ is the good moduli space of the moduli stack of σ-semistable objects. The above diagram resembles diagrams which appear in minimal model program in birational geometry [73]. Let us consider a diagram of birational maps

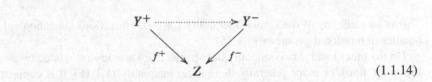

$$\tag{1.1.14}$$

where $Y^\pm$ are smooth quasi-projective varieties and $f^\pm$ are projective birational morphisms. It is called a a flip (resp. flop) in a broad sense if $-K_{Y^+}$ is f^+-ample and K_{Y^-} is f^--ample (resp. $K_{Y^\pm}$ are $f^\pm$-trivial). The flips (flops) form important class of birational transformations in minimal model program. The above flip (flop) diagram satisfies the following inequality with respect to canonical line bundles

$$Y_+ >_K (=_K) Y_-, \text{ i.e. } p_+^* K_{Y_+} - p_-^* K_{Y_-} > (=) 0 \tag{1.1.15}$$

where $p_\pm \colon W \to Y_\pm$ is a common resolution and the last inequality means the effectivity of the divisor on W.

A crucial difference between the diagram (1.1.13) and (1.1.14) is that $M^{\sigma\pm}$ are highly singular so that their canonical bundles are not defined in the usual way. Moreover even if they are smooth (by accident) the moduli spaces $M^{\sigma\pm}$ are not necessary birational, even the dimensions may be different. Nevertheless in [137], the author formulated the notion of *d-critical flips*, *d-critical flops* for diagrams of schemes with d-critical structures such as the wall-crossing diagram (1.1.13). These are d-critical analogue of usual flips and flops in birational geometry, but not honest birational maps so should be interpreted as virtual birational maps. Roughly speaking, a diagram (1.1.13) is called a d-critical flip (flop) if for each point $p \in \overline{M}^{\sigma}$, there is a flip (flop) diagram (1.1.14) and a regular function $w \colon Z \to \mathbb{C}$ such that, by setting $w^{\pm}$ by the commutative diagram

$$
\begin{array}{ccccc}
Y^+ & \xrightarrow{f^+} & Z & \xleftarrow{f^-} & Y^- \\
& w^+ \searrow & \downarrow w & \swarrow w^- & \\
& & \mathbb{C} & &
\end{array}
$$

then the moduli spaces $M^{\sigma\pm}$ are locally near p isomorphic to

$$\mathrm{Crit}(w^{\pm}) \subset Y^{\pm}$$

as d-critical loci. In this book we don't require precise definition of d-critical flip (flop), so we refer to [137] for details. These notions are also to do with the sizes of virtual canonical line bundles: for a d-critical flip (flop), we have the inequality (equality) in terms of virtual canonical line bundles (see [137, Definition 3.23])

$$M^{\sigma+} >_{K^{\mathrm{vir}}} (=_{K^{\mathrm{vir}}}) M^{\sigma-}. \tag{1.1.16}$$

This is an analogue of the similar inequality for usual flip (flop) via canonical line bundles in birational geometry (1.1.15).

On the other hand if two smooth quasi-projective varieties are related by a usual flip (resp. flop), or more generally there is an inequality (1.1.15), it is conjectured by Bondal-Orlov [20] and Kawamata [68] that there exists a fully-faithful functor (resp. equivalence) of their derived categories of coherent sheaves

$$D^b_{\mathrm{coh}}(Y_-) \hookrightarrow (\xrightarrow{\sim}) D^b_{\mathrm{coh}}(Y_+).$$

The above conjecture relates sizes of derived categories and canonical line bundles, so is called a *D/K equivalence conjecture*. Recently Halpern-Leistner [46] proves that the derived categories of coherent sheaves on Bridgeland stable objects on K3 surfaces are equivalent under the wall-crossing diagram (1.1.13). In the case of K3 surfaces, both sides of (1.1.13) are birational holomorphic symplectic manifolds, so connected by flops. Therefore his result proves D/K equivalence conjecture in this case.

Our motivation on studying DT categories is to formulate a CY 3-fold version of the above Halpern-Leistner's work for K3 surfaces. Namely under the wall-crossing diagram (1.1.13) for moduli spaces of stable objects on CY 3-folds, if it is a d-critical flip (resp. flop), or more generally if there is an inequality (1.1.16), we expect the existence of a fully-faithful functor (resp. equivalence)

$$\mathcal{DT}(M^{\sigma-}) \hookrightarrow (\overset{\sim}{\to}) \mathcal{DT}(M^{\sigma+}). \tag{1.1.17}$$

This expectation relates DT categories of $M^{\sigma\pm}$ with their virtual canonical line bundles, so is an analogue of the D/K equivalence conjecture for d-critical loci.

1.1.7 Further Motivations

Besides the motivation from d-critical birational geometry, we have several other motivations in studying DT categories:

(i) **(Categorical wall-crossing formula):** Under the wall-crossing in the previous subsection, we expect that the semiorthogonal complement of the fully-faithful functor (1.1.17) admits a semiorthogonal decomposition which recovers the wall-crossing formula of numerical DT invariants. For example wall-crossing diagram relating MNOP moduli space (moduli space of ideals sheaves of curves) and PT moduli space (moduli space of stable pairs) are related by a d-critical flip, and in that case we expect a finer semiorthogonal decomposition which recovers the MNOP/PT wall-crossing formula [25, 102]. See Theorems 1.4.2, 1.4.10, [115, 140] and Sect. 1.5.3 for results in this direction.

(ii) **(Geometric representation theory):** In the case of quiver with potential, the DT categories admit categorical Hall products and their K-groups are related to some quantum groups. See [110, 111, 151, 152] for related works. The moduli spaces of representations of quivers with potential are local model for moduli stacks of semistable objects on CY 3-folds [132]. Therefore the results for quivers with potential may be regarded as a local version of (yet to be defined in general) global DT category. In the global setting, we also expect the existence of categorical Hall products and have some connections with categorifications of geometric representation theory, see [109, 134].

(iii) **(Derived categories of classical moduli space):** In some cases the DT categories have close connections with derived categories of classical moduli spaces, e.g. symmetric products of curves, Hilbert schemes of points, Quot schemes, etc, and the approach of categorical wall-crossing formula yields interesting semiorthogonal decompositions of classical moduli spaces. See [77, 78, 135, 139, 141] for results in this direction. These may be regarded as a by-product of a categorical DT theory.

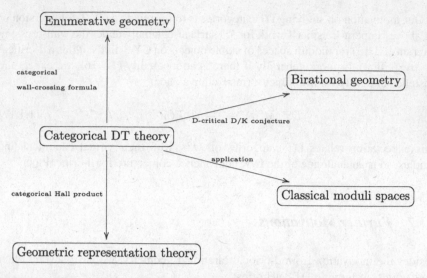

Fig. 1.1 The relation of categorical DT theory with other research subjects

From the above motivations, the study of DT categories connect several research areas: enumerative geometry, birational geometry, geometric representation theory, classical moduli spaces, etc, and it is interesting not just as a theoretical interest as a second categorification of DT invariants but also to a broad areas of research subjects, see Fig. 1.1. Indeed we have recent developments of categorical DT theory along with the motivations mentioned above, see Sect. 1.5.

1.2 Categorical DT Theory for Local Surfaces

1.2.1 (−1)-Shifted Cotangent

Here we give some more informal explanations of our idea of the construction of the DT category for local surfaces (1.1.6). Let S be a smooth projective surface and

$$\pi : X = \mathrm{Tot}_S(\omega_S) \to S$$

the local surface (1.1.6), which is a non-compact CY 3-fold. Let $\mathcal{M}_S$, $\mathcal{M}_X$ be the (classical) moduli stacks of compactly supported coherent sheaves on S, X respectively, with fixed Chern characters (but without imposing stability condition). There is a natural map

$$\pi_* : \mathcal{M}_X \to \mathcal{M}_S, \ E \mapsto \pi_* E. \tag{1.2.1}$$

It is well-known that $\mathcal{M}_X$ is the dual obstruction cone over $\mathcal{M}_S$ associated with a natural perfect obstruction theory of $\mathcal{M}_S$ (see [61] and Lemma 3.4.1). Namely for a given coherent sheaf F on S, from the spectral data (1.1.8) for sheaves on X, the fiber of the map (1.2.1) is the following vector space

$$\mathrm{Hom}(F, F \otimes \omega_S) = \mathrm{Ext}_S^2(F, F)^{\vee} \tag{1.2.2}$$

by the Serre duality. Here $\mathrm{Ext}_S^2(F, F)$ is the obstruction space of the deformation-obstruction theory for the sheaf F, so $\mathcal{M}_X$ is the cone over $\mathcal{M}_S$ with fiber the dual of the obstruction space.

The above fact is explained from derived algebraic geometry in the following way. Let $\mathcal{M}_S \subset \mathfrak{M}_S$ be the derived enhancement of $\mathcal{M}_S$ which is a quasi-smooth derived stack, and consider its (-1)-shifted cotangent derived stack $\Omega_{\mathfrak{M}_S}[-1] \to \mathfrak{M}_S$ which admits a (-1)-shifted symplectic structure [28]. Then we have

$$\mathcal{M}_X \cong t_0(\Omega_{\mathfrak{M}_S}[-1]) \tag{1.2.3}$$

where $t_0(-)$ means the classical truncation of the derived stack. This is because that the fiber of $t_0(\Omega_{\mathfrak{M}_S}[-1]) \to \mathcal{M}_S$ at F is

$$\mathcal{H}^1(\mathbb{T}_{\mathfrak{M}_S}|_F)^{\vee} = \mathrm{Ext}_S^2(F, F)^{\vee}$$

which is identified with (1.2.2). Here $\mathbb{T}_{\mathfrak{M}_S}$ is the tangent complex of $\mathfrak{M}_S$ which governs deformation-obstruction theory of coherent sheaves on S.

Locally on $\mathfrak{M}_S$, this means as follows. Let $\mathfrak{U}$ be an affine derived scheme given by the derived zero locus of a section s of a vector bundle $V \to Y$ on a smooth affine scheme Y. Namely we have the diagram

$$
\begin{array}{ccc}
 & & V \\
 & & \Big\downarrow \Big)\, s \\
\mathfrak{U} = s^{-1}(0) & \lhook\joinrel\longrightarrow & Y.
\end{array}
\tag{1.2.4}
$$

It is important to consider the derived zero locus of s in defining $\mathfrak{U}$ rather than the classical zero locus. Its structure complex is the following commutative differential graded algebra (cdga)

$$\mathcal{O}_{\mathfrak{U}} = \left(\cdots \to \overset{2}{\bigwedge} V^{\vee} \overset{s}{\to} V^{\vee} \overset{s}{\to} \mathcal{O}_Y \to 0 \right).$$

The right hand side is nothing but the Koszul complex associated with s. Recall that the Koszul complex is exact except the degree zero term if s is a regular section (i.e. the classical zero locus of s has codimension rank V), and in that case $\mathfrak{U}$ is equivalent to the classical zero locus of s. However otherwise there are cohomologies of $\mathcal{O}_{\mathfrak{U}}$ which yields non-trivial derived structure on $\mathfrak{U}$.

Then for a smooth morphism $\mathfrak{U} \to \mathfrak{M}_S$, the isomorphism (1.2.3) over $t_0(\mathfrak{U}) \to \mathcal{M}_S$ implies the isomorphism

$$t_0(\mathfrak{U}) \times_{\mathcal{M}_S} \mathcal{M}_X \xrightarrow{\cong} \operatorname{Crit}(w) \subset V^{\vee}. \tag{1.2.5}$$

Here $w \colon V^{\vee} \to \mathbb{C}$ is defined by $w(x, v) = \langle s(x), v \rangle$ for $x \in Y$ and $v \in V|_x^{\vee}$, see the following diagram

$$\operatorname{Crit}(w) \lhook\joinrel\longrightarrow V^{\vee} \xrightarrow{\ w\ } \mathbb{C}$$
$$\downarrow$$
$$Y. \tag{1.2.6}$$

1.2.2   Idea from Koszul Duality

Let us consider the above local setting (1.2.4), (1.2.6). By the isomorphism (1.2.5), what we want to construct is a gluing of categories

$$\mathcal{DT}^{\mathbb{C}^*}(\mathcal{M}_X) \stackrel{?}{=} \lim_{\mathfrak{U} \to \mathfrak{M}_S} \operatorname{MF}^{\mathbb{C}^*}(V^{\vee}, w) \tag{1.2.7}$$

where $\operatorname{MF}^{\mathbb{C}^*}(V^{\vee}, w)$ is the category of $\mathbb{C}^*$-equivariant matrix factorizations of w and $\mathfrak{U} \to \mathfrak{M}_S$ is a smooth morphism. Here $\mathbb{C}^*$ acts on fibers of $V^{\vee} \to Y$ by weight two.

Now we have the following remarkable equivalence, called Koszul duality equivalence (see Theorem 2.3.3 and [56, 58, 99, 121, 138])

$$\Phi \colon D^b_{\mathrm{coh}}(\mathfrak{U}) \xrightarrow{\sim} \operatorname{MF}^{\mathbb{C}^*}(V^{\vee}, w). \tag{1.2.8}$$

Here the left hand side is the derived category of dg-modules over $\mathcal{O}_{\mathfrak{U}}$ with bounded coherent cohomologies. Our point of view is that, via the equivalence (1.2.8), the derived category of the global moduli stack $\mathfrak{M}_S$ is regarded as a gluing of categories of $\mathbb{C}^*$-equivariant matrix factorizations. Namely, by the definition we have

$$D^b_{\mathrm{coh}}(\mathfrak{M}_S) = \lim_{\mathfrak{U} \to \mathfrak{M}_S} D^b_{\mathrm{coh}}(\mathfrak{U}),$$

so by the equivalence (1.2.8), the left hand side is regarded as a gluing of $\mathbb{C}^*$-equivariant matrix factorizations $\operatorname{MF}^{\mathbb{C}^*}(V^{\vee}, w)$. Thus it may be reasonable to set

$$\mathcal{DT}^{\mathbb{C}^*}(\mathcal{M}_X) := D^b_{\mathrm{coh}}(\mathfrak{M}_S) \tag{1.2.9}$$

and call it the $\mathbb{C}^*$-equivariant DT category for $\mathcal{M}_X$.

However $\mathcal{M}_X$ is usually not of finite type, and in order to obtain a nice moduli space we need to take its open substack of stable sheaves. As we will discuss below, we will use the notion of singular supports to extract information of (un)stable locus.

Remark 1.2.1 In [72], Kinjo proved the following dimensional reduction theorem: the cohomological DT invariant of $\mathcal{M}_X$ is isomorphic to the Borel-Moore homology of $\mathcal{M}_S$. In the local setting, the above dimension reduction is regarded as a cohomological version of the Koszul duality equivalence (1.2.8). If there is a way of defining $\mathcal{DT}^{\mathbb{C}^*}(\mathcal{M}_X)$ directly as a gluing like (1.2.7), it is expected that the resulting gluing is equivalent to $D^b_{\mathrm{coh}}(\mathfrak{M}_S)$, which should be regarded as a categorification of the above result by Kinjo.

1.2.3   DT Category for Stable Sheaves

Now we explain how to extract information of (un)stable locus from the construction (1.2.9). Let us take a stability condition σ on the abelian category of compactly supported coherent sheaves on X. For example we can take

$$\sigma = B + iH \in \mathrm{NS}(S)_{\mathbb{C}}$$

such that H is an ample class, and consider (B-twisted) H-Gieseker stability condition associated with σ (see Sect. 3.4.3). Then we have the open substack

$$\mathcal{M}_X^\sigma \subset \mathcal{M}_X$$

corresponding to σ-stable sheaves on X, which is a $\mathbb{C}^*$-gerbe over a quasi-projective scheme M_X^σ. Our purpose is to define the $\mathbb{C}^*$-equivariant DT category for M_X^σ. In general we have an open immersion

$$t_0(\Omega_{\mathfrak{M}_S^\sigma}[-1]) \subset \mathcal{M}_X^\sigma \tag{1.2.10}$$

where $\mathfrak{M}_S^\sigma \subset \mathfrak{M}_S$ is the derived open substack of σ-stable sheaves on S. If (1.2.10) is an isomorphism, then we may define

$$\mathcal{DT}^{\mathbb{C}^*}(M_X^\sigma) := D^b_{\mathrm{coh}}(\mathfrak{M}_S^{\sigma, \mathbb{C}^*\text{-rig}}) \tag{1.2.11}$$

similarly to (1.2.9). Here $\mathbb{C}^*$-rig means the $\mathbb{C}^*$-rigidification, i.e. getting rid of the trivial $\mathbb{C}^*$-autmorphisms. However (1.2.10) is not an isomorphism in many cases, as $\pi_* E$ for a stable sheaf E on X is not necessary stable on S. Therefore we need to find another way to define $\mathcal{DT}^{\mathbb{C}^*}(M_X^\sigma)$.

Remark 1.2.2 There are some cases where (1.2.10) is an isomorphism, and in that case the DT category in Definition 1.2.3 is equivalent to (1.2.11). In Sect. 6.4.4, we will give examples where (1.2.10) is an isomorphism, e.g. S is Fano, K3, the curve class is reduced. We will also discuss Conjecture 1.3.1 in that case.

The notion of singular supports appears in order to deal with the case that (1.2.10) is not an isomorphism. Let $\mathcal{Z}^{\sigma\text{-us}} \subset \mathcal{M}_X$ be the closed substack of unstable points

$$\mathcal{Z}^{\sigma\text{-us}} := \mathcal{M}_X \setminus \mathcal{M}_X^\sigma.$$

Then its pull-back under the smooth map $t_0(\mathfrak{U}) \to \mathcal{M}_S$ is a $\mathbb{C}^*$-invariant closed subscheme $Z \subset \mathrm{Crit}(w)$ via the isomorphism (1.2.5). We would like to construct the DT category for M_X^σ as a gluing of triangulated categories $\mathrm{MF}^{\mathbb{C}^*}(V^\vee \setminus Z, w)$,

$$\mathcal{DT}^{\mathbb{C}^*}(M_X^\sigma) \stackrel{?}{=} \lim_{\mathfrak{U} \to \mathfrak{M}_S} \mathrm{MF}^{\mathbb{C}^*}(V^\vee \setminus Z, w). \tag{1.2.12}$$

Our key observation is that the latter category is equivalent to the quotient category

$$\mathrm{MF}^{\mathbb{C}^*}(V^\vee, w)/\mathrm{MF}^{\mathbb{C}^*}(V^\vee, w)_Z \xrightarrow{\sim} \mathrm{MF}^{\mathbb{C}^*}(V^\vee \setminus Z, w) \tag{1.2.13}$$

by the restriction functor. Here the subcategory

$$\mathrm{MF}^{\mathbb{C}^*}(V^\vee, w)_Z \subset \mathrm{MF}^{\mathbb{C}^*}(V^\vee, w)$$

is the subcategory of matrix factorizations isomorphic to zero on $V^\vee \setminus Z$. We will see that, under the equivalence (1.2.8), the subcategory $\mathrm{MF}^{\mathbb{C}^*}(V^\vee, w)_Z$ corresponds to the subcategory

$$\mathcal{C}_Z \subset D_{\mathrm{coh}}^b(\mathfrak{U})$$

consisting of objects in $D_{\mathrm{coh}}^b(\mathfrak{U})$ whose *singular supports* are contained in Z (see [8, Section H] and Proposition 2.3.9).

The notion of singular supports for (ind)coherent sheaves on quasi-smooth derived stacks was introduced and developed by Arinkin-Gaitsgory [8] in their formulation of geometric Langlands conjecture, following an earlier work by Benson-Iyengar-Krause [16]. For an object $\mathcal{F} \in D_{\mathrm{coh}}^b(\mathfrak{U})$, its singular support is defined to be the support of the action of the Hochschild cohomology of $\mathfrak{U}$ to the graded vector space $\mathrm{Hom}^{2*}(\mathcal{F}, \mathcal{F})$, which can be shown to lie on $t_0(\Omega_{\mathfrak{U}}[-1])$. More globally, for a quasi-smooth derived stack $\mathfrak{M}$ such as $\mathfrak{M} = \mathfrak{M}_S$ and an object $\mathcal{F} \in D_{\mathrm{coh}}^b(\mathfrak{M})$, its singular support is a conical (i.e. $\mathbb{C}^*$-invariant) closed substack

$$\mathrm{Supp}^{\mathrm{sg}}(\mathcal{F}) \subset t_0(\Omega_{\mathfrak{M}}[-1]).$$

As proved in [8], the singular support is intrinsic to the derived stack, e.g. independent of presentation of $\mathfrak{U}$, preserved by pull-backs of smooth morphisms of quasi-smooth affine derived schemes, etc. Therefore combined with the isomorphism (1.2.3), we have the subcategory

$$\mathcal{C}_{\mathcal{Z}^{\sigma\text{-us}}} \subset D^b_{\mathrm{coh}}(\mathfrak{M}_S)$$

consisting of objects whose singular supports are contained in $\mathcal{Z}^{\sigma\text{-us}}$.

From the equivalences (1.2.8), (1.2.13), the Verdier quotient

$$D^b_{\mathrm{coh}}(\mathfrak{M}_S)/\mathcal{C}_{\mathcal{Z}^{\sigma\text{-us}}}$$

may be regarded as a gluing of the categories of matrix factorizations like (1.2.12). However we need two slight (technical) modifications of this construction.

- Since $\mathfrak{M}_S$ is not quasi-compact in general, we replace $\mathfrak{M}_S$ by its quasi-compact derived open substack $\mathfrak{M}_{S,\mathrm{qc}}$ whose truncation contains $\pi_*(\mathcal{M}^\sigma_X)$. This replacement avoids some technical subtleties dealing with derived stacks which are not quasi-compact.
- Since $\mathcal{M}^\sigma_X$ is a $\mathbb{C}^*$-gerbe over M^σ_X, we replace $\mathfrak{M}_{S,\mathrm{qc}}$ by its $\mathbb{C}^*$-rigidification $\mathfrak{M}^{\mathbb{C}^*\text{-rig}}_{S,\mathrm{qc}}$ to get rid of the redundant contributions of $\mathbb{C}^*$-autmorphisms.

By taking account of these modifications, we propose the following definition:

Definition 1.2.3 (Definition 3.2.2) We define the $\mathbb{C}^*$-equivariant DT category for M^σ_X to be the Verdier quotient

$$\mathcal{DT}^{\mathbb{C}^*}(M^\sigma_X) := D^b_{\mathrm{coh}}(\mathfrak{M}^{\mathbb{C}^*\text{-rig}}_{S,\mathrm{qc}})/\mathcal{C}_{\mathcal{Z}^{\sigma\text{-us},\mathbb{C}^*\text{-rig}}}.$$

We will see that the above definition does not depend (up to equivalence) on a choice of a derived open substack $\mathfrak{M}_{S,\mathrm{qc}} \subset \mathfrak{M}_S$. We will also define the λ-twisted version for $\lambda \in \mathbb{Z}$, the dg-enhancements of DT categories, and also $\widehat{\mathcal{DT}}$-version of DT categories. The relevant notations are summarized in Tables 3.1 and 3.2 in the beginning of Chap. 3.

1.2.4 De-categorification of DT Categories

As a globalization of the result in [39], we conjecture that the periodic cyclic homology of the dg-enhancement of $\mathcal{DT}^{\mathbb{C}^*}(M^\sigma_X)$ recovers the $\mathbb{Z}/2$-graded cohomological DT invariant of M^σ_X, constructed in [19] using the canonical orientation data of M^σ_X (see Conjecture 3.3.4). We only prove a numerical version of it in a very special case:

Proposition 1.2.4 (Proposition 3.3.6) *Under some technical condition on* $\mathfrak{M}_{S,\mathrm{qc}}^{\mathbb{C}^*\text{-rig}}$ *(which is automatically satisfied if* $t_0(\mathfrak{M}_{S,\mathrm{qc}}^{\mathbb{C}^*\text{-rig}})$ *is a quasi-projective scheme), we have the identity*

$$\chi(\mathrm{HP}_*(\mathcal{DT}_{\mathrm{dg}}^{\mathbb{C}^*}(M_X^\sigma))) = (-1)^{\mathrm{vdim}\,\mathcal{M}_S+1}\,\mathrm{DT}(M_X^\sigma).$$

Here dg indicates the dg-enhancement, $\mathrm{HP}_*(-)$ *is the periodic cyclic homology,* $\chi(-)$ *is the Euler characteristic,* vdim *is the virtual dimension, and* $\mathrm{DT}(-)$ *is the numerical DT invariant.*

1.3 D/K Conjectures in Categorical DT Theory

Once we have a definition of DT categories, we can formulate categorical wall-crossing based on d-critical analogue of D/K equivalence conjecture. Here we propose several conjectures on equivalences or fully-faithful functors of DT categories in the contexts which are interesting in enumerative geometry.

1.3.1 DT Categories of One-Dimensional Sheaves

Let $N_{\leq 1}(S)$ be the group of numerical classes of one or zero dimensional coherent sheaves on S. For a primitive element $v \in N_{\leq 1}(S)$ and a stability condition σ which lies on a wall, let $\sigma_\pm$ lie on its adjacent chambers. Then it is observed in [137] that the diagram (1.1.13) is a d-critical flop. Therefore we propose the following conjecture:

Conjecture 1.3.1 (Conjecture 4.1.2) There is an equivalence of DT categories

$$\mathcal{DT}^{\mathbb{C}^*}(M_X^{\sigma_+}(v)) \overset{\sim}{\to} \mathcal{DT}^{\mathbb{C}^*}(M_X^{\sigma_-}(v)).$$

The DT invariants associated with $M_X^\sigma(v)$ are known as genus zero Gopakumar-Vafa invariants, which are well-known to be wall-crossing invariant (see [129]). Therefore the above conjecture also categorifies wall-crossing invariance of genus zero GV invariants.

1.3.2 Categorical MNOP/PT Theories

The MNOP conjecture [88] is a relationship between generating series of Gromov-Witten invariants on a 3-fold and those of DT invariants counting one or zero dimensional subschemes on it. The MNOP conjecture is proved for many CY

3-folds (including quintics) by Pandharipande-Pixton [101], and a preprint by Pardon [105] appeared which proves it for any CY 3-fold. In our situation of the local surface, the relevant moduli space in the DT side is

$$I_n(X, \beta), \quad (\beta, n) \in \mathrm{NS}(S) \oplus \mathbb{Z}$$

which parametrizes ideal sheaves $I_C \subset \mathcal{O}_X$ for a compactly supported one or zero dimensional subschemes $C \subset X$ satisfying $\pi_*[C] = \beta$ and $\chi(\mathcal{O}_C) = n$.

The notion of stable pairs was introduced by Pandharipande-Thomas [102] in order to give a better formulation of MNOP conjecture. By definition a PT stable pair is a pair (F, s), where F is a compactly supported pure one dimensional sheaf on X, and $s \colon \mathcal{O}_X \to F$ is surjective in dimension one. The moduli space of stable pairs is denoted by

$$P_n(X, \beta), \quad (\beta, n) \in \mathrm{NS}(S) \oplus \mathbb{Z}$$

and it parametrizes stable pairs (F, s) such that $\pi_*[F] = \beta$ and $\chi(F) = n$. The moduli space $P_n(X, \beta)$ is regarded as a moduli space of two term complexes in the derived category

$$I^\bullet = (\mathcal{O}_X \xrightarrow{s} F) \in D^b_{\mathrm{coh}}(X). \tag{1.3.1}$$

The relationship between the corresponding DT type invariants is conjectured in [102] and proved in [25, 123, 127] for CY 3-folds using wall-crossing arguments. If we denote by $I_{n,\beta}$, $P_{n,\beta}$ the corresponding DT invariants, the formula is

$$\sum_{n \in \mathbb{Z}} I_{n,\beta} q^n = M(-q)^{e(X)} \sum_{n \in \mathbb{Z}} P_{n,\beta} q^n \tag{1.3.2}$$

where $M(q)$ is the MacMahon function

$$M(q) := \prod_{k \geq 1} \frac{1}{(1 - q^k)^k} = 1 + q + 3q^2 + 6q^3 + 13q^4 + 24q^4 + \cdots. \tag{1.3.3}$$

We would like to categorify the DT type invariants defined from $I_n(X, \beta)$ and $P_n(X, \beta)$, and try to formulate a categorical relationship of these moduli spaces. Here we note that, since an ideal sheaf I_C nor the two term complex (1.3.1) do not have compact supports, the construction in Definition 1.2.3 does not apply for this purpose. Our strategy is to realize the moduli spaces $I_n(X, \beta)$, $P_n(X, \beta)$ as open substacks of the dual obstruction cone over the moduli stack $\mathcal{M}_S^\dagger$ of pairs $(\mathcal{O}_S \to F)$, where F is a one or zero dimensional sheaf on S. Indeed we will show that the dual obstruction cone over $\mathcal{M}_S^\dagger$ is isomorphic to the moduli stack $\mathcal{M}_X^\dagger$ which parametrizes rank one objects in the extension closure

$$\mathcal{A}_X = \langle \mathcal{O}_{\overline{X}}, \mathrm{Coh}_{\leq 1}(X)[-1] \rangle_{\mathrm{ex}} \subset D^b_{\mathrm{coh}}(\overline{X}).$$

Here $X \subset \overline{X}$ is a compactification of X, and $\mathrm{Coh}_{\leq 1}(X)$ is the category of compactly supported coherent sheaves on X whose supports have dimensions less than or equal to one. The category $\mathcal{A}_X$ (called the category of *D0-D2-D6 bound states*) is an abelian category, which was introduced in [127] and studied in [129, 131, 133] to show several properties on PT invariants. As we mentioned above, we show the following:

Theorem 1.3.2 (Theorem 4.2.3) *The stack $\mathcal{M}_X^\dagger$ of rank one objects in $\mathcal{A}_X$ admits a natural morphism $\pi_*^\dagger \colon \mathcal{M}_X^\dagger \to \mathcal{M}_S^\dagger$, and is isomorphic to the dual obstruction cone over $\mathcal{M}_S^\dagger$.*

As we will mention in Remark 4.2.4, an object in $\mathcal{A}_X$ may not be written as a two term complex $(\mathcal{O}_X \to F)$ for a one or zero dimensional sheaf F, so the existence of the morphism $\pi_*^\dagger$ is a priori not obvious. Our idea for Theorem 1.3.2 is to describe rank one objects in $\mathcal{A}_X$ in terms of certain diagrams of coherent sheaves on S, and construct the map $\pi_*^\dagger$ in terms of the latter diagram (see Corollary 5.2.9). From the definition of $\mathcal{A}_X$, we have open immersions

$$I_n(X, \beta) \hookrightarrow \mathcal{M}_X^\dagger \hookleftarrow P_n(X, \beta).$$

Similarly to Definition 1.2.3, we will define the corresponding DT categories as Verdier quotients (see Definition 4.3.1)

$$\mathcal{DT}^{\mathbb{C}^*}(I_n(X, \beta)) := D^b_{\mathrm{coh}}(\mathfrak{M}^\dagger_{S,\mathrm{qc}})/\mathcal{C}_{\mathcal{Z}^{I\text{-us}}}, \qquad (1.3.4)$$

$$\mathcal{DT}^{\mathbb{C}^*}(P_n(X, \beta)) := D^b_{\mathrm{coh}}(\mathfrak{M}^\dagger_{S,\mathrm{qc}})/\mathcal{C}_{\mathcal{Z}^{P\text{-us}}},$$

where $\mathfrak{M}_S^\dagger$ is a derived enhancement of $\mathcal{M}_S^\dagger$, and $\mathcal{Z}^{I\text{-us}}$, $\mathcal{Z}^{P\text{-us}}$ are complements

$$\mathcal{Z}^{I\text{-us}} = t_0(\Omega_{\mathfrak{M}^\dagger_{S,\mathrm{qc}}}[-1]) \setminus I_n(X, \beta), \quad \mathcal{Z}^{P\text{-us}} = t_0(\Omega_{\mathfrak{M}^\dagger_{S,\mathrm{qc}}}[-1]) \setminus P_n(X, \beta).$$

We call the category (1.3.4) as $\mathbb{C}^*$-equivariant *MNOP category*, *PT category* respectively.

On the other hand, it is observed in [137] that there is a wall-crossing diagram

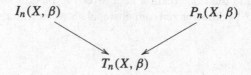

which is a d-critical flip. Therefore following d-critical analogue of D/K equivalence conjecture, we propose the following:

Conjecture 1.3.3 There exists a fully-faithful functor

$$\mathcal{DT}^{\mathbb{C}^*}(P_n(X, \beta)) \hookrightarrow \mathcal{DT}^{\mathbb{C}^*}(I_n(X, \beta)).$$

1.3.3 DT Category for Stable D0-D2-D6 Bound States

From the author's previous works [128, 129], it is known that the moduli space $P_n(X, \beta)$ is identified with the moduli space of certain stable objects in the abelian category $\mathcal{A}_X$. A stability parameter is given by $t \in \mathbb{R}$ and the stable pair moduli space corresponds to the $t \to \infty$ limit. By changing the stability parameter $t \in \mathbb{R}$, we have another moduli space of stable objects

$$P_n^t(X, \beta) \subset \mathcal{M}_X^\dagger, \quad \mathcal{Z}^{t\text{-us}} = t_0(\Omega_{\mathfrak{M}_{qc}^\dagger}[-1]) \setminus P_n^t(X, \beta)$$

and the corresponding DT category (see Sect. 4.3.2)

$$\mathcal{DT}^{\mathbb{C}^*}(P_n^t(X, \beta)) := D_{\text{coh}}^b(\mathfrak{M}_{S,qc}^\dagger)/\mathcal{C}_{\mathcal{Z}^{t\text{-us}}}.$$

As we will discuss below, we study the relationships of the above DT categories along with the arguments in Sect. 1.1.6.

In [128, 129] it is proved that, for a fixed (β, n), there exists a finite set of walls in the space of stability parameters $t \in \mathbb{R}$. By taking $t_1 > t_2 > \cdots > t_N > 0$ which do not lie on walls, we have the sequence of moduli spaces

$$P_n(X, \beta) \dashrightarrow P_n^{t_1}(X, \beta) \dashrightarrow P_n^{t_2}(X, \beta) \dashrightarrow \cdots \dashrightarrow P_n^{t_N}(X, \beta). \qquad (1.3.5)$$

If we denote by $P_{n,\beta}^t \in \mathbb{Z}$ the corresponding DT invariant for a generic t, the following wall-crossing formula for $t_i > t_j$ is proved in [128, 129]

$$\sum_{n,\beta} P_{n,\beta}^{t_i} q^n y^\beta = \prod_{t_j < n/H \cdot \beta < t_i} \exp((-1)^{n-1} n N_{n,\beta} q^n y^\beta) \sum_{n,\beta} P_{n,\beta}^{t_j} q^n y^\beta, \qquad (1.3.6)$$

where $N_{n,\beta} \in \mathbb{Q}$ is the generalized DT invariant [64] counting one dimensional semistable sheaves on X with numerical class (β, n). The above wall-crossing formula is relevant in showing the rationality of the generating series of PT invariants [128, 129].

Furthermore in [137], it is proved that the above sequence is a d-critical minimal model program, that is a d-critical analogue of usual minimal model program in birational geometry [73]. In particular we have the inequalities in terms of virtual canonical line bundles

$$P_n(X, \beta) >_{K^{\text{vir}}} P_n^{t_1}(X, \beta) >_{K^{\text{vir}}} \cdots >_{K^{\text{vir}}} P_n^{t_N}(X, \beta).$$

Then as we discussed in Sect. 1.1.6, we formulate the following conjecture:

Conjecture 1.3.4 (Conjectures 4.3.3 and 4.3.14) There exist fully-faithful functors

$$\mathcal{DT}^{\mathbb{C}^*}(P_n^{t_N}(X, \beta)) \hookrightarrow \cdots \hookrightarrow \mathcal{DT}^{\mathbb{C}^*}(P_n^{t_1}(X, \beta)) \hookrightarrow \mathcal{DT}^{\mathbb{C}^*}(P_n(X, \beta)).$$

1.4   Three Approaches

In this book, we pursue and develop three approaches toward the conjectures stated in the previous section: linear Koszul duality, window theorem for DT categories and categorified Hall products.

1.4.1   *Wall-Crossing via Linear Koszul Duality*

When the curve class β is irreducible, there is only one wall $t_0 \in \mathbb{R}$ for the moduli space $P_n^t(X, \beta)$. In this case, the Serre duality gives an isomorphism $P_n^{t_0-}(X, \beta) \cong P_{-n}(X, \beta)$. The wall-crossing diagram at $t = t_0$ is

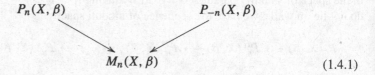

$$(1.4.1)$$

where $M_n(X, \beta)$ is the moduli space of one dimensional stable sheaves on X with numerical class (β, n). The above diagram is a d-critical flip (flop) for $n > 0$ ($n = 0$).

Our observation is that, locally on $M_n(X, \beta)$, the above diagram is described by linear Koszul dual pairs studied by Mirković-Riche [90, 91]. Let Y be a smooth affine scheme and take a two term complex of vector bundles on Y

$$\mathcal{E} = (\mathcal{E}^{-1} \to \mathcal{E}^0), \ e := \text{rank}(\mathcal{E}) \geq 0. \tag{1.4.2}$$

A Koszul dual pair in [90, 91] is given by

$$(Y^\dagger, Y^\sharp), \ Y^\dagger = \text{Spec } S(\mathcal{E}), \ Y^\sharp = \text{Spec } S(\mathcal{E}^\vee[1]).$$

The Koszul duality equivalence in [90, 91] is a derived equivalence

$$D_{\mathrm{coh}}^b([Y^\dagger/\mathbb{C}^*])^{\mathrm{op}} \xrightarrow{\sim} D_{\mathrm{coh}}^b([Y^\sharp/\mathbb{C}^*]). \qquad (1.4.3)$$

It will turn out that the diagram (1.4.1) is locally modeled by the diagram

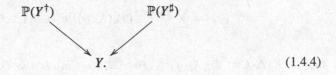

$$(1.4.4)$$

Using the equivalence (1.4.3), we show the following:

Theorem 1.4.1 (Theorem 5.3.11) *There is a semiorthogonal decomposition*

$$D^b(\mathbb{P}(Y^\dagger)) = \langle e\text{-copies of } D_{\mathrm{coh}}^b(Y), D_{\mathrm{coh}}^b(\mathbb{P}(Y^\sharp)) \rangle.$$

The above result is also proved in [60], and is called projectivization formula. Indeed the result of Theorem 5.3.11 is more general, allowing Y to be a derived stack. By applying this result to the diagram (1.4.1), we show that Conjecture 1.3.4 is true for the above wall-crossing, which holds for *any* surface S.

Theorem 1.4.2 (Theorems 5.1.7 and 5.1.9) *Suppose that β is irreducible and $n \geq 0$. Then Conjecture 1.3.4 holds. Moreover there is an equivalence*

$$\mathcal{DT}^{\mathbb{C}^*}(P_n^{t_0-}(X, \beta)) \xrightarrow{\sim} \mathcal{DT}^{\mathbb{C}^*}(P_{-n}(X, \beta))$$

together with a semiorthogonal decomposition

$$\mathcal{DT}^{\mathbb{C}^*}(P_n(X, \beta)) = \langle \Upsilon_{-n+1}, \dots, \Upsilon_0, \mathcal{DT}^{\mathbb{C}^*}(P_{-n}(X, \beta)) \rangle. \qquad (1.4.5)$$

Here each Υ_λ is equivalent to a λ-twisted DT category $\mathcal{DT}^{\mathbb{C}^*}(\mathcal{M}_n(X, \beta))_\lambda$, where $\mathcal{M}_n(X, \beta)$ is the moduli stack of one dimensional stable sheaves on X with numerical class (β, n).

The above result together with Proposition 1.2.4 recover the wall-crossing formula of numerical PT invariants for irreducible curve classes proved in [103] (which is a special case of (1.3.6)), by taking the Euler characteristics of the periodic cyclic homologies of both sides of (1.4.5) (see Lemma 5.1.10). In [135], a semiorthogonal decomposition similar to (1.4.5) is proved for derived categories of coherent sheaves on $P_n(X, \beta)$ assuming that it is non-singular. In our situation $P_n(X, \beta)$ is not necessary non-singular, and even if it is non-singular the triangulated category $\mathcal{DT}^{\mathbb{C}^*}(P_n(X, \beta))$ may not be equivalent to $D_{\mathrm{coh}}^b(P_n(X, \beta))$. Therefore the argument similar to [135] does not apply for Theorem 1.4.2.

As we mentioned above, the wall-crossing phenomena (1.3.5) was essential in showing the rationality of the PT generating series. Using Theorem 1.4.2, we have the following categorical analogue of the rationality for irreducible curve classes:

Corollary 1.4.3 (Corollary 5.1.14) *Suppose that β is irreducible. Then the generating series*

$$P_\beta^{\mathrm{cat}}(q) := \sum_{n\in\mathbb{Z}}[\mathcal{DT}^{\mathbb{C}^*}(P_n(X,\beta))]q^n \in K(\Delta\text{-}\mathrm{Cat})(\!(q)\!)$$

lies in $K(\Delta\text{-}\mathrm{Cat}) \otimes_{\mathbb{Z}} \mathbb{Q}(q)^{\mathrm{inv}}$. Here $K(\Delta\text{-}\mathrm{Cat})$ is the Grothendieck group of triangulated categories and $\mathbb{Q}(q)^{\mathrm{inv}} \subset \mathbb{Q}(\!(q)\!)$ is the subspace of rational functions invariant under $q \leftrightarrow 1/q$.

1.4.2 Window Theorem for DT Categories

Our next approach toward Conjectures 1.3.1, 1.3.3, 1.3.4 is to establish *window theorem* for DT categories and apply it. The original window theorem is developed for derived categories of GIT quotients [11, 45], and stated as follows. Let Y be a smooth affine variety with an action of a reductive algebraic group G. For a given G-equivariant line bundle l on Y, we have the G-invariant open subset $Y^{l\text{-ss}} \subset Y$ of l-semistable points. Then the window theorem implies the existence of a triangulated subcategory (called *window subcategory*)

$$\mathcal{W}([Y/G]) \subset D_{\mathrm{coh}}^b([Y/G])$$

such that the composition

$$\mathcal{W}([Y/G]) \hookrightarrow D_{\mathrm{coh}}^b([Y/G]) \twoheadrightarrow D_{\mathrm{coh}}^b([Y^{l\text{-ss}}/G])$$

is an equivalence. The window theorem has been used to show several derived equivalences of birational varieties given as variations of GIT quotients, showing many evidence for original D/K equivalence conjecture. Using the above window theorem for GIT quotients, we prove Conjectures 1.3.3 and 1.3.4 in the simplest (non-compact) CY 3-fold called resolved conifold. Note that the above conjectures also make sense for non-compact surface S, see Remark 4.3.4.

Theorem 1.4.4 (Theorems 4.4.4 and 4.4.5) *Let X be the resolved conifold defined by*

$$X = \mathrm{Tot}_{\mathbb{P}^1}(\mathcal{O}_{\mathbb{P}^1}(-1) \oplus \mathcal{O}_{\mathbb{P}^1}(-1)) = \mathrm{Tot}_S(\omega_S) \tag{1.4.6}$$

where $S \to \mathbb{C}^2$ is the blow-up at the origin. Then Conjecture 1.3.3, Conjecture 1.3.4 are true for X.

In the case of Theorem 1.4.4, the relevant moduli spaces are global critical loci so we can apply the window theorem in the setting of GIT quotients. In a more general situation, we expect that the analogy of window theorem holds for DT categories for local surfaces. Namely we expect the existence of a subcategory

$$\mathcal{W}(\mathfrak{M}_{S,qc}^{\mathbb{C}^*\text{-rig}}) \subset D_{\mathrm{coh}}^b(\mathfrak{M}_{S,qc}^{\mathbb{C}^*\text{-rig}})$$

such that the composition

$$\mathcal{W}(\mathfrak{M}_{S,qc}^{\mathbb{C}^*\text{-rig}}) \hookrightarrow D_{\mathrm{coh}}^b(\mathfrak{M}_{S,qc}^{\mathbb{C}^*\text{-rig}}) \to \mathcal{DT}^{\mathbb{C}^*}(M_X^\sigma)$$

is an equivalence. Once such a subcategory exists, we then try to compare them for different σ inside $D_{\mathrm{coh}}^b(\mathfrak{M}_{S,qc}^{\mathbb{C}^*\text{-rig}})$, which may yield the above conjectures.

We establish window theorem for DT categories, assuming the existence of good moduli space of $\mathfrak{M}_{S,qc}^{\mathbb{C}^*\text{-rig}}$. Here in general, for an Artin stack $\mathcal{M}$, its good moduli space is an algebraic space M together with a map

$$\mathcal{M} \to M$$

which generalizes GIT quotient $[Y/G] \to Y /\!\!/ G$ for an action of a reductive algebraic group G to an affine variety Y. This notion was introduced by Alper [4] and plays an important role in this book, see Sect. 3.1.4. We prove the window theorem for DT categories associated with (-1)-shifted cotangent over quasi-smooth derived stacks as follows.

Theorem 1.4.5 (Theorem 6.2.7) *For a quasi-smooth and QCA derived stack $\mathfrak{M}$, suppose that $M = t_0(\mathfrak{M})$ admits a good moduli space, a symmetric structure $\mathbb{S}$ of $\mathfrak{M}$ is given, and $\mathfrak{M}$ satisfies formal neighborhood theorem. Let us take $l, \delta \in \mathrm{Pic}(\mathcal{M})_{\mathbb{R}}$ such that δ is l-generic. Then there exists a triangulated subcategory $\mathcal{W}_\delta^{\mathrm{int}/\mathbb{S}}(\mathfrak{M}) \subset D_{\mathrm{coh}}^b(\mathfrak{M})$ such that, for the l-semistable locus $\mathcal{N}^{l\text{-ss}} \subset \mathcal{N} := t_0(\Omega_{\mathfrak{M}}[-1])$, the composition*

$$\mathcal{W}_\delta^{\mathrm{int}/\mathbb{S}}(\mathfrak{M}) \hookrightarrow D_{\mathrm{coh}}^b(\mathfrak{M}) \to \mathcal{DT}^{\mathbb{C}^*}(\mathcal{N}^{l\text{-ss}})$$

is fully-faithful, which is an equivalence if l is $\mathbb{S}$-generic and compatible with $\mathbb{S}$.

Here several notions 'symmetric structure', 'formal neighborhood theorem', '$\mathbb{S}$-generic', 'l-generic', 'compatible with $\mathbb{S}$' appear in the statement of the above theorem. We do not explain these technical notions here and refer to Sect. 6.2.1 for details. We will apply Theorem 1.4.5 to prove Conjecture 1.3.1 under some assumption.

Theorem 1.4.6 (Theorem 6.4.7) *In the setting of Conjecture 1.3.1, suppose that the following condition holds*

$$\mathcal{M}_X^\sigma(v) \subset \pi_*^{-1}(\mathcal{M}_S^\sigma(v)).$$

Then Conjecture 1.3.1 holds.

The above result also implies some derived equivalences of moduli spaces of one dimensional stable sheaves on S (not on X). For example as a corollary of Theorem 1.4.6, we show that derived moduli spaces of one dimensional stable sheaves with reduced supports are derived equivalent (see Corollary 6.4.10). The result of Theorem 1.4.6 is also used to show a derived equivalence of Sacca's Calabi-Yau manifolds [120] obtained as moduli spaces of one dimensional stable sheaves on Enriques surfaces for different polarizations (see Corollary 6.4.15), which gives an evidence of the usual D/K equivalence conjecture. Another application of Theorem 1.4.6 is the proof of Conjecture 1.3.3 for reduced curve classes.

Theorem 1.4.7 *Conjecture 1.3.3 holds if β is a reduced class.*

The results and proofs of Theorem 1.4.5 and its applications are much inspired by ideas and techniques of Halpern-Leistner [46], where he proves that Bridgeland moduli spaces of stable objects on K3 surfaces are derived equivalent under wall-crossing. Indeed Theorem 1.4.6 applied for (some rigidified version of) derived moduli stacks of semistable objects on K3 surfaces should recover the above Halpern-Leistner's result (see Remark 6.4.13), though we will not discuss this case in this book. On the other hand, our result applies other than moduli stacks of semistable objects on K3 surfaces. The latter moduli stacks are 0-shifted symplectic, and this fact is essential in the proof of derived equivalence in [46]. In our situation the derived stack $\mathfrak{M}$ is not necessary 0-shifted symplectic, but is equipped with a *symmetric structure* $\mathbb{S}$ and $l \in \text{Pic}(\mathcal{M})_\mathbb{R}$ *compatible with* $\mathbb{S}$. A symmetric structure $\mathbb{S}$ is a choice of direct sum decompositions of $\text{Aut}(x)$-representations

$$\mathcal{H}^0(\mathbb{T}_\mathfrak{M}|_x) \oplus \mathcal{H}^1(\mathbb{T}_\mathfrak{M}|_x)^\vee = \mathbb{S}_x \oplus \mathbb{U}_x \qquad (1.4.7)$$

at each closed point $x \in \mathcal{M}$ such that $\mathbb{S}_x$ is a symmetric $\text{Aut}(x)$-representation. An element $l \in \text{Pic}(\mathcal{M})_\mathbb{R}$ is compatible with $\mathbb{S}$ if the l_x-semistable locus in the LHS of (1.4.7) is the pull-back of the l_x-semistable locus in $\mathbb{S}_x$. We will use the symmetric structure $\mathbb{S}$ together with an auxiliary data $\delta \in \text{Pic}(\mathcal{M})_\mathbb{R}$ to define the *intrinsic window subcategory* (see Definition 6.3.12)

$$\mathcal{W}_\delta^{\text{int}/\mathbb{S}}(\mathfrak{M}) \subset D_{\text{coh}}^b(\mathfrak{M}) \qquad (1.4.8)$$

which gives a desired window subcategory.

Locally on the good moduli space, the intrinsic window subcategory is constructed so that it coincides with the magic window subcategory on the derived category of factorizations via Koszul duality. Here the magic window subcategory is defined by Halpern-Leistner and Sam [50] which gives stability independent descriptions of window subcategories in [11, 45], and is itself based on combinatorial arguments by Špenko and Van den Bergh [122]. So the subcategory (1.4.8) is interpreted as a gluing of magic window subcategories in the Koszul dual side, rather than those on $\mathfrak{M}$ itself.

1.4.3 Categorified Hall Products

We also give an approach toward the previous conjectures using Porta-Sala's two dimensional categorified Hall products [109]. The Porta-Sala categorified Hall product is the functor

$$D^b_{\mathrm{coh}}(\mathfrak{M}_S) \times D^b_{\mathrm{coh}}(\mathfrak{M}_S) \to D^b_{\mathrm{coh}}(\mathfrak{M}_S) \qquad (1.4.9)$$

defined through the derived moduli stack of short exact sequences of coherent sheaves on S. Namely let $\mathfrak{M}^{\mathrm{ext}}_S$ be the derived moduli stack of short exact sequences

$$0 \to F_1 \to F_2 \to F_3 \to 0, \ F_i \in \mathrm{Coh}(S). \qquad (1.4.10)$$

There exist natural evaluation maps

$$\mathfrak{M}_S \times \mathfrak{M}_S \xleftarrow{q} \mathfrak{M}^{\mathrm{ext}}_S \xrightarrow{p} \mathfrak{M}_S$$

where p sends (1.4.10) to F_2 and q sends (1.4.10) to (F_1, F_3). The functor (1.4.9) is given by

$$p_*q^*: D^b(\mathfrak{M}_S) \times D^b(\mathfrak{M}_S) \xrightarrow{q^*} D^b(\mathfrak{M}^{\mathrm{ext}}_S) \xrightarrow{p_*} D^b(\mathfrak{M}_S).$$

It gives a categorification of cohomological Hall algebras constructed by Kapranov-Vasserot [66]. There are also recent works giving categorifications of geometric representations through the correspondences similar to the categorified Hall products [96, 97, 112, 153, 154].

In [134], we proved that Porta-Sala categorified Hall products descend to products on DT categories

$$\mathcal{DT}^{\mathbb{C}^*}(\mathcal{M}^\sigma_X) \times \mathcal{DT}^{\mathbb{C}^*}(\mathcal{M}^\sigma_X) \to \mathcal{DT}^{\mathbb{C}^*}(\mathcal{M}^\sigma_X)$$

which also acts on MNOP/PT categories, and should categorify critical cohomological Hall products for local surfaces. We apply the above categorified Hall products to describe conjectural semiorthogonal decompositions associated with wall-crossing of MNOP/PT moduli spaces, which also give ansatz for d-critical analogue of D/K equivalence conjecture.

Let $\mathcal{T}_n(X, \beta)$ be the moduli stack of pairs (F, s), where F is a compactly supported one dimensional coherent sheaf on X and $s\colon \mathcal{O}_X \to F$ is surjective in dimension one. The moduli stack $\mathcal{T}_n(X, \beta)$ is nothing but the moduli stack of semistable objects at MNOP/PT wall, and its good moduli space coincides with $T_n(X, \beta)$ in the diagram (1.4.1). Also let $\mathcal{M}_i(X)$ be the moduli stack of zero dimensional sheaves on X with length i. The following is the main result by this approach specialized to the case of MNOP/PT wall, which gives another method to construct window subcategories.

Theorem 1.4.8 (Theorems 7.3.17 and 7.3.18) *Let $k_I, k_P\colon \mathbb{Z}_{\geq 1} \to \mathbb{R}$ be maps. If β is a reduced class, there exist semiorthogonal decompositions*

$$\mathcal{DT}^{\mathbb{C}^*}(\mathcal{T}_n(X, \beta)) = \langle \ldots, \Upsilon^{I,2}_{>k^I(2)}, \Upsilon^{I,1}_{>k^I(1)}, \mathcal{W}^I_{k^I}, \Upsilon^{I,1}_{\leq k^I(1)}, \Upsilon^{I,2}_{\leq k^I(2)}, \ldots \rangle$$

$$= \langle \ldots, \Upsilon^{P,2}_{<k^P(2)}, \Upsilon^{P,1}_{<k^P(1)}, \mathcal{W}^P_{k^P}, \Upsilon^{P,1}_{\geq k^P(1)}, \Upsilon^{P,2}_{\geq k^P(2)}, \ldots \rangle$$

satisfying the following:

(i) There exist semiorthogonal decompositions

$$\Upsilon^{I,i}_{>k} = \langle \ldots, \Upsilon^{I,i}_{\lfloor k \rfloor + 2}, \Upsilon^{I,i}_{\lfloor k \rfloor + 1} \rangle, \quad \Upsilon^{I,i}_{\leq k} = \langle \Upsilon^{I,i}_{\lfloor k \rfloor}, \Upsilon^{I,i}_{\lfloor k \rfloor - 1}, \ldots \rangle,$$

$$\Upsilon^{P,i}_{<k} = \langle \ldots, \Upsilon^{P,i}_{\lceil k \rceil - 2}, \Upsilon^{P,i}_{\lceil k \rceil - 1} \rangle, \quad \Upsilon^{P,i}_{\geq k} = \langle \Upsilon^{P,i}_{\lceil k \rceil}, \Upsilon^{P,i}_{\lceil k \rceil + 1}, \ldots \rangle$$

with equivalences induced by Porta-Sala categorified Hall products

$$*_j\colon \mathcal{DT}^{\mathbb{C}^*}(I_{n-i}(X, \beta)) \otimes \mathcal{DT}^{\mathbb{C}^*}(\mathcal{M}_i(X))_j \xrightarrow{\sim} \Upsilon^{I,i}_j,$$

$$*_j\colon \mathcal{DT}^{\mathbb{C}^*}(\mathcal{M}_i(X))_j \otimes \mathcal{DT}^{\mathbb{C}^*}(P_{n-i}(X, \beta)) \xrightarrow{\sim} \Upsilon^{P,i}_j.$$

(ii) The following composition functors are equivalences

$$\mathcal{W}^I_{k^I} \hookrightarrow \mathcal{DT}^{\mathbb{C}^*}(\mathcal{T}_n(X, \beta)) \twoheadrightarrow \mathcal{DT}^{\mathbb{C}^*}(I_n(X, \beta)),$$

$$\mathcal{W}^P_{k^P} \hookrightarrow \mathcal{DT}^{\mathbb{C}^*}(\mathcal{T}_n(X, \beta)) \twoheadrightarrow \mathcal{DT}^{\mathbb{C}^*}(P_n(X, \beta)).$$

(iii) By setting $k^I(i) = i/2$ and $k^P(i) = 0$, we have $\mathcal{W}^P_{k^P} \subset \mathcal{W}^I_{k^I}$. In particular we have the fully-faithful functor

$$\mathcal{DT}^{\mathbb{C}^*}(P_n(X, \beta)) \hookrightarrow \mathcal{DT}^{\mathbb{C}^*}(I_n(X, \beta)). \tag{1.4.11}$$

Note that the fully-faithful functor (1.4.11) implies Conjecture 1.3.3 for reduced curve classes, so the above result gives another proof of Theorem 1.4.7. On the other hand the result itself is conjectured to hold for any curve class, so the categorified Hall product approach may give an ansatz applied for any curve class. The above approach can be also extended to wall-crossing of D0-D2-D6 bound states. We will formulate a similar semiorthogonal decomposition for wall-crossing of D0-D2-D6 bound states, and prove it for reduced curve class β and $t > m(\beta)$ where $m(\beta)$ depends only on β.

Theorem 1.4.9 (Theorems 7.3.17 and 7.3.18) *Let β be a reduced class. Then there is $m(\beta) > 0$ which only depends on β such that for $t > m(\beta)$ there exists a fully-faithful functor*

$$\mathcal{DT}^{\mathbb{C}^*}(P_n^{t-}(X, \beta)) \hookrightarrow \mathcal{DT}^{\mathbb{C}^*}(P_n^{t+}(X, \beta)), \qquad (1.4.12)$$

i.e. Conjecture 1.3.4 holds for reduced curve classes and $t_i > m(\beta)$.

We note that $m(\beta)$ only depends on β, so for $n \gg 0$ there exist many walls in the region $t > m(\beta)$ where we can apply Theorem 1.4.9. If furthermore $t \in \mathbb{R}_{>0}$ is a simple wall, then we can describe the semiorthogonal complement of the fully-faithful functor (1.4.12), which categorifies wall-crossing formula of numerical DT invariants (1.3.6) in [128, 133].

Theorem 1.4.10 (Theorem 7.3.19) *Under the assumption of Theorem 1.4.9, suppose furthermore that $t > m(\beta)$ is a simple wall. Then there exists a semiorthogonal decomposition*

$$\mathcal{DT}^{\mathbb{C}^*}(P_n^{t+}(X, \beta))$$

$$= \left\langle \bigoplus_{\mathbf{d}} \Upsilon_{[-\frac{1}{2}\beta_1\beta_2 - \frac{1}{2}n_2, -\frac{1}{2}\beta_1\beta_2)}^{-,\mathbf{d}}, \mathcal{DT}^{\mathbb{C}^*}(P_n^{t-}(X, \beta)), \bigoplus_{\mathbf{d}} \Upsilon_{[-\frac{1}{2}\beta_1\beta_2, -\frac{1}{2}\beta_1\beta_2 + \frac{1}{2}n_2)}^{-,\mathbf{d}} \right\rangle.$$

Here $\mathbf{d}$ is a decomposition $(\beta, n) = (\beta_1, n_1) + (\beta_2, n_2)$ with $n_2/(\beta_2 \cdot H) = t$, and $\Upsilon_{[a,b)}^{-,\mathbf{d}}$ admits semiorthogonal decomposition

$$\Upsilon_{[a,b)}^{-,\mathbf{d}} = \left\langle \Upsilon_{\lceil a \rceil}^{-,\mathbf{d}}, \Upsilon_{\lceil a \rceil+1}^{-,\mathbf{d}}, \ldots, \Upsilon_{\lceil b \rceil-1}^{-,\mathbf{d}} \right\rangle$$

with equivalences induced by Porta-Sala categorified Hall products

$$*_j \colon \mathcal{DT}^{\mathbb{C}^*}(\mathcal{M}_{n_2}^H(X, \beta_2))_j \otimes \mathcal{DT}^{\mathbb{C}^*}(P_{n_2}^{t-}(X, \beta_1)) \xrightarrow{\sim} \Upsilon_j^{-,\mathbf{d}}.$$

Here $\mathcal{M}_{n_2}^H(X, \beta_2)$ is the moduli stack of one dimensional H-semistable sheaves on X with numerical class (β_2, n_2).

1.5 Updates of the Developments

The current book is a beginning of the study of categorical DT theory. Since the appearance of the first version of this text, there has been various progress on this topic:

1.5.1 The $\mathbb{Z}/2$-Periodic Version

The version of DT categories in this book is a $\mathbb{C}^*$-equivariant one. For a general CY 3-fold, a local critical chart (Y, w) in (1.1.2) does not admit a $\mathbb{C}^*$-action, and the category $\mathrm{MF}(Y, w)$ is a $\mathbb{Z}/2$-periodic triangulated category. Therefore it is better to define a $\mathbb{Z}/2$-periodic version of the DT category even for the local surface case.

In [138], we proved a $\mathbb{Z}/2$-periodic Koszul duality replacing the right hand side in (1.2.8) with the $\mathbb{Z}/2$-periodic categories of matrix factorizations, and constructed the $\mathbb{Z}/2$-periodic version of the DT category. By writing $\mathfrak{M} = \mathfrak{M}_S$ and $\mathcal{Z} = \mathcal{Z}^{\mathrm{us}}$, the $\mathbb{Z}/2$-periodic version is given by replacing (1.1.6) with

$$\mathcal{DT}^{\mathbb{Z}/2}(M) = D^b(\mathfrak{M}_\varepsilon)/\mathcal{C}_{\mathcal{Z}_\varepsilon}. \tag{1.5.1}$$

Here $\mathfrak{M}_\varepsilon := \mathfrak{M} \times \operatorname{Spec} \mathbb{C}[\varepsilon]$ with $\deg \varepsilon = -1$ and $\mathcal{Z}_\varepsilon$ is given by

$$\mathcal{Z}_\varepsilon = \mathbb{C}^*(\mathcal{Z} \times \{1\}) \cup (t_0(\Omega_{\mathfrak{M}}[-1]) \times \{0\})$$

which is a closed substack

$$\mathcal{Z}_\varepsilon \subset t_0(\Omega_{\mathfrak{M}}[-1]) \times \mathbb{A}^1 = t_0(\Omega_{\mathfrak{M}_\varepsilon}[-1])$$

where $\mathbb{C}^*$ acts on $\mathbb{A}^1$ by weight one.

In [138], it is explained that the construction (1.5.1) may be regarded as a gluing of $\mathbb{Z}/2$-periodic categories of matrix factorizations

$$\lim_{\mathfrak{U} \to \mathfrak{M}} \mathrm{MF}^{\mathbb{Z}/2}(V^\vee, w)$$

where $\mathfrak{U} \to \mathfrak{M}$ is a local chart as in (1.2.4), (1.2.6). Moreover up to idempotent completion, the $\mathbb{Z}/2$-periodic DT category is recovered from the $\mathbb{C}^*$-equivariant one developed in this book, see [138, Theorem 1.2].

1.5.2   Window Subcategories via Θ-Stacks

The categorical Hall product for surfaces is constructed by Porta-Sala [109], and it was used in [134] to give categorical Hall products for DT categories

$$DT^{\mathbb{C}^*}(\mathcal{M}_X^\sigma) \times DT^{\mathbb{C}^*}(\mathcal{M}_X^\sigma) \to DT^{\mathbb{C}^*}(\mathcal{M}_X^\sigma).$$

The arguments in Chap. 7 are generalized in [142] for more general (-1)-shifted cotangent stacks using Halpern-Leistner's theory of Θ-stacks [47]. In particular, the above categorical Hall product is interpreted in terms of the stack of maps

$$\Theta := [\mathbb{A}^1/\mathbb{C}^*] \to \mathcal{N} := t_0(\Omega_{\mathfrak{M}}[-1]),$$

where $\mathfrak{M}$ is quasi-smooth derived stack and $\mathbb{C}^*$ acts on $\mathbb{A}^1$ by weight one.

Moreover, the result of Theorem 1.4.8 is generalized to a more general quasi-smooth derived stack and its (-1)-shifted cotangent using the above Θ-stacks, as follows. Let $\mathfrak{M}$ be a quasi-smooth derived stack such that its classical truncation $\mathcal{M}$ admits a good moduli space $\mathcal{M} \to M$. By taking $l \in H^2(\mathcal{M}, \mathbb{R})$ and positive definite $b \in H^4(\mathcal{M}, \mathbb{R})$, and pulling them back to $\mathcal{N}$, we have the Θ-stratification

$$\mathcal{N} = \mathcal{S}_1^\Omega \sqcup \cdots \sqcup \mathcal{S}_N^\Omega \sqcup \mathcal{N}^{l\text{-ss}} \tag{1.5.2}$$

with center $\mathcal{Z}_i^\Omega \subset \mathcal{S}_i^\Omega$. It is proved in [142] that each $\mathcal{Z}_i^\Omega$ is an open substack of (-1)-shifted cotangent over the stack $\mathrm{Map}(B\mathbb{C}^*, \mathfrak{M})$, see [142, Subsection 6.1]. So we have the DT category $DT^{\mathbb{C}^*}(\mathcal{Z}_i^\Omega)$ defined as a singular support quotient developed in this book, together with the decomposition into $\mathbb{C}^*$-weight part with respect to the canonical $B\mathbb{C}^*$-action on $\mathcal{Z}_i^\Omega$

$$DT^{\mathbb{C}^*}(\mathcal{Z}_i^\Omega) = \bigoplus_{j \in \mathbb{Z}} DT^{\mathbb{C}^*}(\mathcal{Z}_i^\Omega)_{\mathrm{wt}=j}.$$

Theorem 1.5.1 ([142, Theorem 1.2]) *(Under some technical assumption), for each choice of $m_i \in \mathbb{R}$ for $1 \le i \le N$, there exists a semiorthogonal decomposition*

$$DT^{\mathbb{C}^*}(\mathcal{N}) = \langle \mathcal{D}_{1,<m_1}, \ldots, \mathcal{D}_{N,<m_N}, \mathcal{W}_{m_\bullet}^l, \mathcal{D}_{N,\ge m_N}, \ldots, \mathcal{D}_{1,\ge m_1} \rangle$$

satisfying the followings:

(i) There exist semiorthogonal decompositions of the form

$$\mathcal{D}_{i,<m_i} = \Big\langle \ldots, DT^{\mathbb{C}^*}(\mathcal{Z}_i^\Omega)_{\mathrm{wt}=\lceil m_i \rceil - 3}, DT^{\mathbb{C}^*}(\mathcal{Z}_i^\Omega)_{\mathrm{wt}=\lceil m_i \rceil - 2},$$

$$DT^{\mathbb{C}^*}(\mathcal{Z}_i^\Omega)_{\mathrm{wt}=\lceil m_i \rceil - 1} \Big\rangle,$$

$$\mathcal{D}_{i,\geq m_i} = \Big\langle \mathcal{DT}^{\mathbb{C}^*}(\mathcal{Z}_i^{\Omega})_{\mathrm{wt}=\lceil m_i\rceil}, \mathcal{DT}^{\mathbb{C}^*}(\mathcal{Z}_i^{\Omega})_{\mathrm{wt}=\lceil m_i\rceil+1},$$

$$\mathcal{DT}^{\mathbb{C}^*}(\mathcal{Z}_i^{\Omega})_{\mathrm{wt}=\lceil m_i\rceil+2}, \dots \Big\rangle.$$

(ii) The composition functor

$$\mathcal{W}_{m_\bullet}^l \hookrightarrow \mathcal{DT}^{\mathbb{C}^*}(\mathcal{N}) \twoheadrightarrow \mathcal{DT}^{\mathbb{C}^*}(\mathcal{N}^{l\text{-ss}})$$

is an equivalence.

The above result in [142] recovers Theorem 1.4.8 by setting $\mathcal{N} = \mathcal{T}_n(X, \beta)$, and taking $l_\pm$ to be

$$\mathcal{N}^{l_+\text{-ss}} = I_n(X, \beta), \ \mathcal{N}^{l_-\text{-ss}} = P_n(X, \beta).$$

It is also used to give further evidence of Conjecture 1.3.4, see [142, Theorem 1.4].

1.5.3 Categorical Wall-Crossing Formula

In [140], it is proved that the fully-faithful functors in Conjecture 1.3.4 for the resolved conifold (1.4.6) are further decomposed into semiorthogonal decompositions, which give categorifications of Nagao-Nakajima wall-crossing formula [94].

Theorem 1.5.2 ([140, Theorem 1.1]) *For a resolved conifold X, the category $\mathcal{DT}(P_n(X, \beta))$ admits a semiorthogonal decomposition*

$$\mathcal{DT}(P_n(X, \beta)) = \langle a_{n,\beta}\text{-copies of } \mathrm{MF}(\mathrm{Spec}\,\mathbb{C}, 0)\rangle. \tag{1.5.3}$$

Here $a_{n,\beta}$ is defined by

$$a_{n,\beta} := \sum_{\substack{l:\, \mathbb{Z}_{\geq 1}\to\mathbb{Z}_{\geq 0} \\ \sum_{m\geq 1} l(m)\cdot(m,1)=(n,\beta)}} \prod_{m\geq 1} \binom{m}{l(m)}.$$

Remark 1.5.3 The semiorthogonal decomposition (1.5.3) gives a categorification of the following formula for the PT invariants of the resolved conifold

$$\sum_{n,\beta} P_{n,\beta} q^n t^\beta = \prod_{m\geq 1}(1 - (-q)^m t)^m.$$

Let S be an arbitrary smooth projective surface. When β is a reduced class, in [115] it is proved that the fully-faithful functor in Conjecture 1.3.3 (proved in

Theorem 1.4.7) is further decomposed into a semiorthogonal decomposition, giving a categorical analogue of the formula (1.3.2):

Theorem 1.5.4 ([115, Theorem 1.1]) *Suppose that β is a reduced class. There is a semiorthogonal decomposition*

$$\mathcal{DT}^{\mathbb{C}^*}(I_n(X, \beta))$$

$$= \left\langle \bigotimes_{i=1}^{k} \mathbb{T}_X(d_i)_{v_i} \otimes \mathcal{DT}^{\mathbb{C}^*}(P_{n'}(X, \beta)) \mid -1 < \frac{v_1}{d_1} < \cdots < \frac{v_k}{d_k} \leq 0 \right\rangle.$$

Here the right hand side is after all partitions $d_1 + \cdots + d_k + n' = n$ and the subcategory

$$\mathbb{T}_X(d)_v \subset D^b_{\mathrm{coh}}(\mathfrak{M}_d(S, 0))_v$$

is the intrinsic window subcategory, which is also called quasi-BPS category in [115]. Each fully-faithful functor from a semiorthogonal summand in the right hand side to the left hand side is given by the categorical Hall product.

When $\beta = 0$, it in particular implies categorical analogue of MacMahon formula (1.3.3) for Hilbert schemes of points on 3-folds, see [114]. Also see [113, 116] for related works.

1.5.4   Applications to Derived Categories of Classical Moduli Spaces

The techniques and ideas used in this book are applied to give semiorthogonal decompositions of various classical moduli spaces. In [139], we proved the existence of semiorthogonal decompositions of some Quot schemes of relative dimension zero, proving a Quot-formula conjecture by Qingyuang Jiang [59]. It is regarded as a higher rank generalization of Theorem 1.4.1.

Let Y be a smooth quasi-projective variety. For a coherent sheaf F on Y, the relative Quot scheme of dimension d is

$$\mathrm{Quot}_{Y,d}(F) \rightarrow Y \tag{1.5.4}$$

which parametrizes rank d locally free quotients of F. For each $y \in Y$, the fiber of the above map at y is the Grassmannian parameterizing d-dimensional quotients of $F|_y$. So the map (1.5.4) is a stratified Grassmannian bundle. We consider a two term complex of vector bundles on Y and its dual

$$\phi \colon \mathcal{E}^{-1} \rightarrow \mathcal{E}^0, \ \phi^\vee \colon (\mathcal{E}^0)^\vee \rightarrow (\mathcal{E}^{-1})^\vee.$$

We set

$$\mathcal{G} := \mathrm{Cok}(\phi) \in \mathrm{Coh}(Y), \quad \mathcal{H} := \mathrm{Cok}(\phi^\vee) \in \mathrm{Coh}(Y).$$

Then we have relative Quot schemes

$$\mathrm{Quot}_{Y,d}(\mathcal{G}) \to Y \leftarrow \mathrm{Quot}_{Y,d}(\mathcal{H}). \tag{1.5.5}$$

Theorem 1.5.5 ([139, Theorem 1.1]) *Suppose that* $e := \mathrm{rank}\,\mathcal{E}^0 - \mathrm{rank}\,\mathcal{E}^{-1} \geq 0$. *There exist quasi-smooth derived schemes*

$$\mathfrak{Quot}_{Y,d}(\mathcal{G}) \to Y \leftarrow \mathfrak{Quot}_{Y,d}(\mathcal{H})$$

whose classical truncation is (1.5.5) with virtual dimensions $\dim Y + ed - d^2$, $\dim Y - ed - d^2$ *respectively, such that there is a semiorthogonal decomposition of the form*

$$D^b(\mathfrak{Quot}_{Y,d}(\mathcal{G})) = \left\langle \binom{e}{i}\text{-copies of } D^b(\mathfrak{Quot}_{Y,d-i}(\mathcal{H})) : 0 \leq i \leq \min\{d, e\} \right\rangle. \tag{1.5.6}$$

The result of Theorem 1.4.1 is recovered from Theorem 1.5.5 by setting $d = 1$. A key point of the proof is to regard that the two d-critical loci

$$t_0(\Omega_{\mathfrak{Quot}_{Y,d}(\mathcal{G})}[-1]) \dashrightarrow t_0(\Omega_{\mathfrak{Quot}_{Y,d}(\mathcal{H})}[-1])$$

are related by d-critical flip, and regard $D^b_{\mathrm{coh}}(\mathfrak{Quot}_{Y,d}(\mathcal{G}))$ as DT category

$$D^b_{\mathrm{coh}}(\mathfrak{Quot}_{Y,d}(\mathcal{G})) = \mathcal{DT}^{\mathbb{C}^*}(\Omega_{\mathfrak{Quot}_{Y,d}(\mathcal{G})}[-1]).$$

The relevant fully-faithful functors in the semiorthogonal decomposition (1.5.6) is given by categorical Hall products.

The Quot formula in Theorem 1.5.5 has lots of applications on derived categories of classical moduli spaces (see [59, Section 1.5]). One of them is a semiorthogonal decomposition of varieties associated with Brill-Noether loci for curves, as follows. Let C be a smooth projective curve over $\mathbb{C}$ with genus g. We denote by $\mathrm{Pic}^d(C)$ the Picard variety parameterizing degree d line bundles on C, which is a g-dimensional complex torus and (non-canonically) isomorphic to the Jacobian $\mathrm{Jac}(C)$ of C. The Brill-Noether locus on $\mathrm{Pic}^d(C)$ is defined by

$$W_d^r(C) := \{L \in \mathrm{Pic}^d(C) : h^0(L) \geq r + 1\}.$$

There is a scheme $G_d^r(C)$ parameterizing g_d^r's which appears in the classical study of Brill-Noether loci, which is set theoretically given by

$$G_d^r(C) = \{(L, W) : L \in W_d^r(C), W \subset H^0(C, L), \dim W = r + 1\}$$

where W is a vector subspace. As explained in [59, Section 1.5.1], for any $e \geq 0$ there is a coherent sheaf $\mathcal{G}$ on $X = \mathrm{Pic}^{g-1+e}(C)$ of rank e that has homological dimension less than or equal to 1 and such that

$$\mathrm{Quot}_{X,r+1}(\mathcal{G}) = G_{g-1+e}^r(C), \quad \mathrm{Quot}_{X,r+1}(\mathcal{H}) = G_{g-1-e}^r(C).$$

Here $\mathcal{H} = \mathcal{E}xt^1_{\mathcal{O}_X}(\mathcal{G}, \mathcal{O}_X)$. As an application of Theorem 1.5.5, we have the following:

Corollary 1.5.6 ([139, Corollary 1.6]) *Let C be a general smooth projective curve with genus g. Then for any $r \in \mathbb{Z}_{\geq 0}$ and $e \geq 0$, there is a semiorthogonal decomposition*

$$D_{\mathrm{coh}}^b(G_{g-1+\delta}^r(C)) = \left\langle \binom{e}{i}\text{-copies of } D_{\mathrm{coh}}^b(G_{g-1-e}^{r-i}(C)) : 0 \leq i \leq \min\{e, r+1\} \right\rangle.$$

Here for $i = r + 1$, we have $G_{g-1-\delta}^{-1}(C) = \mathrm{Pic}^{g-1-e}(C)$.

Remark 1.5.7 The case of $r = 0$ gives the semiorthogonal decomposition of symmetric products

$$D^b(\mathrm{Sym}^{g-1+\delta}(C)) = \left\langle D^b(\mathrm{Sym}^{g-1-\delta}(C)), \delta\text{-copies of } D^b(\mathrm{Jac}(C)) \right\rangle$$

proved in [135, Corollary 5.11].

The above result is regarded as a by-product of the study of categorical DT theory to the derived categories of classical moduli spaces. As further applications to derived categories of classical moduli spaces, we have the following:

- The result of Theorem 1.5.5 was applied by Koseki [77] to give a categorical blow-up formula of Hilbert schemes of points on surfaces, which gives a categorification of blow-up formula of the generating series of Euler numbers of Hilbert schemes of points.
- In [141], we used the similar argument to prove the existence of semiorthogonal decompositions of derived categories of Quot schemes of points on curves.

1.5.5   Quasi-BPS Categories

As we mentioned in Remark 1.1.1, the generalized DT invariant is not necessary an integer but a rational number. For $v \in \Gamma$ and a stability condition σ, if we define $\Omega^{\sigma}(v)$ by the multiple cover formula

$$\mathrm{DT}^{\sigma}(v) = \sum_{k|v} \frac{1}{k^2} \Omega^{\sigma}(v/k)$$

then it was conjectured in [64, 76] that $\Omega^{\sigma}(v)$ is an integer for a generic choice of σ. The above conjecture was proved by Davison-Meinhardt [33] in the case of quivers with potential, and combined with [132] it is now a theorem. The integer $\Omega^{\sigma}(v)$ is called *BPS invariant*. When v is primitive, then $\Omega^{\sigma}(v)$ equals to the DT invariant $\mathrm{DT}^{\sigma}(v)$. It is important and interesting to give a categorification of the BPS invariant.

In [110–112], Pădurariu studied categorical and K-theoretic Hall algebras for quivers with potential, and categorifications of BPS invariants in that case. The categorification of BPS invariants is based on Spenko-Van den Bergh's non-commutative resolutions of the GIT quotients by reductive groups [122]. Pădurariu proved categorical and K-theoretic PBW type theorem of the categories of matrix factorizations associated with moduli stacks of representations of quivers with potential. It is a semiorthogonal decomposition whose factors are categorical Hall products of the subcategories, called *quasi-BPS categories* in [118].

A quasi-BPS category approximates categorification of BPS invariants but this is not always a precise categorification. In [117], we used intrinsic window subcategory in this book as a definition of quasi-BPS category, and studied in the case of K3 surfaces in details. For a K3 surface S and an element of the Mukai lattice $v \in \Gamma$, the BPS invariant is already known

$$\Omega^{\sigma}(v) = -e(\mathrm{Hilb}^{v^2/2+1}(S)).$$

The above formula was conjectured in [130] and proved in [86].

Theorem 1.5.8 ([117, Theorem 1.1, Theorem 1.3]) *Let S be a smooth projective K3 surface, and $\sigma \in \mathrm{Stab}(S)$ be a generic stability condition. We write $v = dv_0$ with v_0 primitive, $v_0^2 = 2g - 2$. Then there exists a subcategory (called quasi-BPS category)*

$$\mathbb{T}_S^{\sigma}(v)_w \subset D_{\mathrm{coh}}^b \left(\mathfrak{M}_S^{\sigma}(v) \right)_w$$

such that there is a semiorthogonal decomposition

$$D_{\mathrm{coh}}^b \left(\mathfrak{M}_S^{\sigma}(v) \right) = \left\langle \boxtimes_{i=1}^k \mathbb{T}_S^{\sigma}(d_i v_0)_{w_i + (g-1)d_i \left(\sum_{i>j} d_j - \sum_{i<j} d_j \right)} \right\rangle.$$

The right hand side is after all partitions $(d_i)_{i=1}^k$ of d and all weights $(w_i)_{i=1}^k \in$
$\mathbb{Z}^k$ such that $w_1/d_1 < \cdots < w_k/d_k$, and each fully-faithful functor in the above
semiorthogonal decomposition is given by the categorical Hall product.

Moreover if (v, w) is coprime, we have the identity of the Euler characteristic of
the topological K-theory

$$e(K^{\text{top}}(\mathbb{T}_S^\sigma(v))) = -\Omega^\sigma(v) = e(\text{Hilb}^{v^2/2+1}(S)).$$

Davison–Hennecart–Schlegel Mejia [34, Theorem 1.5] proved that the cohomological Hall algebra of a K3 surface is generated by its BPS cohomology. The result of Theorem 1.5.8 gives a categorical analogue of their result. Surprisingly when v is not primitive, a (reduced version of) quasi-BPS category $\mathbb{T}_S^\sigma(v)^{\text{red}}$ gives a twisted categorical crepant resolution of singularities of the symplectic variety M, which does not admit a usual crepant resolution except O'Grady's 10-dimensional symplectic manifold [98], see [117, Theorem 1.2]. It is an interesting problem to relate the reduced quasi-BPS category with the derived category of holomorphic symplectic manifold. Also see [118, 119] for related works.

1.6 Organization of This Book

1.6.1 Plan of the Book

In Chap. 2, we recall the notion of singular supports for coherent sheaves on quasismooth affine derived schemes, the derived categories factorizations, Koszul duality equivalence, relation of singular supports with usual supports of factorizations under Koszul duality, and prove several functoriality under Koszul duality.

In Chap. 3, we define DT categories for (-1)-shifted cotangent stacks over quasismooth derived stacks as Verdier quotients, and prove their fundamental properties. We then define DT categories for (semi)stable sheaves on local surfaces, and formulate wall-crossing equivalence for DT categories of one dimensional stable sheaves.

In Chap. 4, we focus on D0-D2-D6 bound states and introduce MNOP category, PT category and DT categories for stable D0-D2-D6 bound states. An alternative description of D0-D2-D6 bound states plays a key role here. We then formulate categorical wall-crossing conjecture for these DT categories and prove it for local $(-1, -1)$-curve.

In Chap. 5, we also concentrate on wall-crossing for irreducible curve class, and study the conjecture in this case via linear Koszul duality.

In Chap. 6, we prove window theorem for DT categories, and use it to prove conjecture on categorical wall-crossing equivalence of DT categories for one dimensional stable sheaves, under the some assumption of preservation of semistability by push-forward.

In Chap. 7, we use categorified Hall products to describe conjectural semiorthogonal decomposition of DT categories at a wall, and prove it under some assumption, including MNOP/PT wall for reduced classes.

In Chap. 8, we prove several technical results postponed in the previous sections.

1.6.2 Notation and Convention

In this book, all the schemes or (derived) stacks are locally of finite presentation over $\mathbb{C}$, except formal fibers along with good moduli space morphisms (e.g. see Sect. 6.1.4). For a scheme or derived stack Y and a quasi-coherent sheaf $\mathcal{F}$ on it, we denote by $S_{\mathcal{O}_Y}(\mathcal{F})$ its symmetric product $\oplus_{i \geq 0} \mathrm{Sym}^i_{\mathcal{O}_Y}(\mathcal{F})$. We omit the subscript $\mathcal{O}_Y$ if it is clear from the context. We denote by $\mathbb{D}_Y$ the derived dual functor $\mathbb{D}_Y(-) := \mathbf{R}\mathcal{H}om_{\mathcal{O}_Y}(-, \mathcal{O}_Y)$. For a derived stack $\mathfrak{M}$, we always denote by $t_0(\mathfrak{M})$ the underived stack given by the truncation. For a triangulated category $\mathcal{D}$ and a set of objects $\mathcal{S} \subset \mathcal{D}$, we denote by $\langle \mathcal{S} \rangle_{\mathrm{ex}}$ the extension closure, i.e. the smallest extension closed subcategory which contains $\mathcal{S}$. For an algebraic group G which acts on Y, we denote by $[Y/G]$ the associated quotient stack.

For a triangulated category $\mathcal{D}$ and its triangulated subcategory $\mathcal{D}' \subset \mathcal{D}$, we denote by $\mathcal{D}/\mathcal{D}'$ its Verdier quotient. This is the triangulated category whose objects coincide with objects in $\mathcal{D}$, and for $E, F \in \mathcal{D}$ the set of morphisms $\mathrm{Hom}_{\mathcal{D}/\mathcal{D}'}(E, F)$ is the equivalence classes of roof diagrams in $\mathcal{D}$

$$E \xleftarrow{u} G \longrightarrow F, \quad \mathrm{Cone}(u) \in \mathcal{D}'.$$

Then the sequence of exact functors of triangulated categories $\mathcal{D}' \to \mathcal{D} \to \mathcal{D}/\mathcal{D}'$ is called a localization sequence. We denote by $\mathcal{D}^{\mathrm{cp}} \subset \mathcal{D}$ the subcategory of compact objects. A subcategory $\mathcal{D}' \subset \mathcal{D}$ is called dense if any object in $\mathcal{D}$ is a direct summand of an object in $\mathcal{D}'$.

For a smooth projective surface S, we denote by $\mathrm{NS}(S)$ the Neron-Severi group of S. An effective class $\beta \in \mathrm{NS}(S)$ is called reduced (resp. irreducible) if any effective divisor in S with class β is a reduced divisor (resp. irreducible divisor). For a coherent sheaf E on S whose supports had dimension less than or equal to one, we denote by $l(E)$ the fundamental one cycle of E.

We often take limits or ind-completions for dg-categories, and then take its homotopy categories. Here we note that a limit of triangulated categories cannot be defined in general, and in order to take it we have to replace triangulated categories with their dg-enhancements and define the limit in the ∞-category of dg-categories as in (3.1.1) (see [43, 144]). By abuse of notation, we use the following notation for the limit. For an ∞-category $\mathcal{I}$, let $\{\mathcal{D}_i^{\mathrm{dg}}\}_{i \in \mathcal{I}}$ be the $\mathcal{I}$-diagram of dg-categories $\mathcal{D}_i^{\mathrm{dg}}$

for $i \in \mathcal{I}$, and denote by $\mathcal{D}_i = \mathrm{Ho}(\mathcal{D}_i^{\mathrm{dg}})$ the homotopy category of $\mathcal{D}_i^{\mathrm{dg}}$. Then we sometimes write

$$\lim_{i \in \mathcal{I}} \mathcal{D}_i := \mathrm{Ho}\left(\lim_{i \in \mathcal{I}} \mathcal{D}_i^{\mathrm{dg}}\right).$$

Similarly for a dg-category $\mathcal{D}^{\mathrm{dg}}$ with $\mathcal{D} = \mathrm{Ho}(\mathcal{D}^{\mathrm{dg}})$, we write $\mathrm{Ind}\,\mathcal{D} := \mathrm{Ho}(\mathrm{Ind}\,\mathcal{D}^{\mathrm{dg}})$ where $\mathrm{Ind}\,\mathcal{D}^{\mathrm{dg}}$ is a dg-categorical ind-completion of $\mathcal{D}^{\mathrm{dg}}$ (denoted as $\widehat{\mathcal{D}}^{\mathrm{dg}}$ in [143, Section 7]). When we discuss induced functors on limits or ind-completions, we implicitly take these functors on dg-enhancements and finally take homotopy categories.

Chapter 2
Koszul Duality Equivalence

The notion of singular supports was introduced by Arinkin-Gaitsgory [8] (which is itself based on the earlier work by Benson-Iyengar-Krause [16]) in order to give a categorical formulation of geometric Langlands conjecture. In this chapter, we first review the singular support theory for (ind)coherent sheaves on quasi-smooth affine derived schemes. For a quasi-smooth affine derived scheme $\mathfrak{U}$ and $F \in D^b_{\mathrm{coh}}(\mathfrak{U})$, its singular support is a conical closed subspace

$$\mathrm{Supp}^{\mathrm{sg}}(F) \subset t_0(\Omega_{\mathfrak{U}}[-1])$$

where $\Omega_{\mathfrak{U}}[-1] \to \mathfrak{U}$ is the (-1)-shifted cotangent over $\mathfrak{U}$. If F is a perfect complex, then its singular support is contained in the zero section of $t_0(\Omega_{\mathfrak{U}}[-1]) \to \mathfrak{U}$, but if F is not perfect then its singular support lies beyond the zero section. So the singular support may be interpreted as a measure how a given coherent sheaf is far from a perfect complex.

Our viewpoint for singular supports is to interpret them as usual supports for factorizations via Koszul duality. The derived category of factorizations $\mathrm{MF}_{\mathrm{coh}}(Y, w)$ is determined by a pair (Y, w) for a smooth scheme Y together with a function $w \colon Y \to \mathbb{C}$, which is introduced and studied in [9, 40] and generalizes Orlov's categories of matrix factorizations [100]. Our key tool is a Koszul duality equivalence proved by several people [56, 58, 99, 121] which claims that, if $\mathfrak{U}$ is represented by a derived zero locus of a section of a vector bundle $V \to Y$ on a smooth scheme Y, then there is a function $w \colon V^\vee \to \mathbb{C}$ and an equivalence

$$\Phi \colon D^b_{\mathrm{coh}}(\mathfrak{U}) \xrightarrow{\sim} \mathrm{MF}^{\mathbb{C}^*}_{\mathrm{coh}}(V^\vee, w),$$

where the right hand side is the derived category of $\mathbb{C}^*$-equivariant factorizations for $(V^\vee, w)$. We will see that, under the above equivalence Φ, the singular support for $F \in D^b_{\mathrm{coh}}(\mathfrak{U})$ is the same as the usual support for $\Phi(F)$. This fact is briefly claimed in [8, Section H], and we will give its more details.

© The Author(s), under exclusive license to Springer Nature Switzerland AG 2024   41
Y. Toda, *Categorical Donaldson-Thomas Theory for Local Surfaces*, Lecture Notes
in Mathematics 2350, https://doi.org/10.1007/978-3-031-61705-8_2

For a conical closed subset $Z \subset t_0(\Omega_{\mathfrak{U}}[-1])$, we define $\mathcal{C}_Z \subset D^b_{\mathrm{coh}}(\mathfrak{U})$ to be the subcategory of objects whose singular supports are contained in Z. The above interpretation of singular supports yields the equivalence

$$\Phi \colon D^b_{\mathrm{coh}}(\mathfrak{U})/\mathcal{C}_Z \overset{\sim}{\to} \mathrm{MF}^{\mathbb{C}^*}_{\mathrm{coh}}(V^\vee \setminus Z, w).$$

The quotient category in the left hand side will be our local model for DT categories for local surfaces, by taking Z to be unstable locus with respect to some stability condition. The above construction also works under equivariant setting, and we will discuss with the presence of G-action for an affine algebraic group G.

In the later sections, we will use the above Koszul duality equivalence to prove several statements on singular supports using the theory of derived categories of factorizations. In particular, it will be useful to compare several functorial properties of $D^b_{\mathrm{coh}}(\mathfrak{U})$ for affine derived schemes $\mathfrak{U}$, or more generally $D^b_{\mathrm{coh}}([\mathfrak{U}/G])$ for a G-action of $\mathfrak{U}$, under Koszul duality equivalence. For example we will see that the push-forward functor associated with a morphism $\mathbf{f} \colon \mathfrak{U}_1 \to \mathfrak{U}_2$ is described by the composition of pull-back/push-forward in the Koszul dual side.

The organization of this chapter is as follows. In Sect. 2.1, we recall the notion of singular supports for quasi-smooth affine derived schemes. In Sect. 2.2, we review the theory of derived categories of factorizations and their functorial properties. In Sect. 2.3, we recall and reprove Koszul duality equivalence in equivariant setting, and compare singular supports with usual supports in the factorization category side. In Sect. 2.4, we compare several functorial properties under Koszul duality.

2.1 Singular Supports of Coherent Sheaves

2.1.1 Local Model

As a local model of our quasi-smooth derived stack, we consider the following situation. Let Y be a smooth affine $\mathbb{C}$-scheme of finite presentation and V a vector bundle on it with a section s

$$V \underset{\xleftarrow{\ s\ }}{\longrightarrow} Y.$$

By abuse of notation, we also denote by V the locally free sheaf of local sections of $V \to Y$. Then the contraction by the section $s \colon V^\vee \to \mathcal{O}_Y$ determines the Koszul complex

$$\mathcal{R}(V \to Y, s) := \left(\cdots \to \bigwedge^2 V^\vee \overset{s}{\to} V^\vee \overset{s}{\to} \mathcal{O}_Y \to 0 \right)$$

which is a dg-algebra over $\mathcal{O}_Y$. We have the quasi-smooth affine derived scheme

$$\mathfrak{U} := \operatorname{Spec} \mathcal{R}(V \to Y, s). \qquad (2.1.1)$$

Note that we have the following derived Cartesian square:

$$
\begin{array}{ccc}
\mathfrak{U} & \longrightarrow & Y \\
\downarrow & \square & \downarrow 0 \\
Y & \xrightarrow{\ s\ } & V,
\end{array}
$$

i.e. $\mathfrak{U}$ is the derived zero locus of s. The classical truncation of $\mathfrak{U}$ is the closed subscheme of Y given by

$$\mathcal{U} := t_0(\mathfrak{U}) = (s = 0) \subset Y.$$

Also the cotangent complex of $\mathfrak{U}$ is given by

$$\mathbb{L}_{\mathfrak{U}} = (V^{\vee} \xrightarrow{ds} \Omega_Y) \otimes_{\mathcal{O}_Y} \mathcal{O}_{\mathfrak{U}}.$$

Then its (-1)-shifted cotangent derived scheme of $\mathfrak{U}$ is written as

$$
\begin{aligned}
\Omega_{\mathfrak{U}}[-1] &:= \operatorname{Spec} S_{\mathcal{O}_{\mathfrak{U}}}(\mathbb{T}_{\mathfrak{U}}[1]) \\
&= \operatorname{Spec} \mathcal{R}((V \oplus \Omega_Y) \times_Y V^{\vee} \to V^{\vee}, s \oplus ds).
\end{aligned}
$$

Here $\mathbb{T}_{\mathfrak{U}}$ is the tangent complex of $\mathfrak{U}$, which is dual to its cotangent complex $\mathbb{L}_{\mathfrak{U}}$. The dg-algebra $\mathcal{R}((V \oplus \Omega_Y) \times_Y V^{\vee} \to V^{\vee}, s \oplus ds)$ is the Koszul complex over $\mathcal{O}_{V^{\vee}}$ determined by

$$s \oplus ds : V^{\vee} \oplus T_Y \to \mathcal{O}_Y \oplus V \subset \mathcal{O}_{V^{\vee}}. \qquad (2.1.2)$$

On the other hand, let $w : V^{\vee} \to \mathbb{C}$ be the function defined by

$$w = s \in \Gamma(Y, V) \subset \Gamma(Y, S(V)) = \Gamma(Y, p_* \mathcal{O}_{V^{\vee}}). \qquad (2.1.3)$$

Here $p : V^{\vee} \to Y$ is the projection. Explicitly, the function w is written as

$$w(x, v) = \langle s(x), v \rangle, \quad x \in Y, \ v \in V^{\vee}|_x.$$

Note that $0_Y^* w = 0$ where $0_Y \colon Y \to V^\vee$ is the zero section of p. From the description of the differential (2.1.2), the classical truncation of $\Omega_{\mathfrak{U}}[-1]$ is the critical locus of w,

$$t_0(\Omega_{\mathfrak{U}}[-1]) = \operatorname{Spec} S(\mathcal{H}^1(\mathbb{T}_{\mathfrak{U}})) = \operatorname{Crit}(w) \subset V^\vee. \tag{2.1.4}$$

Moreover the restriction of the projection $p \colon V^\vee \to Y$ to $\operatorname{Crit}(w)$ maps to $\mathcal{U}$. As a summary, we obtain the following diagram

$$
\begin{array}{ccccccc}
\operatorname{Crit}(w) & \hookrightarrow & \Omega_{\mathfrak{U}}[-1] & \longrightarrow & V^\vee & \overset{w}{\longrightarrow} & \mathbb{C} \\
{\scriptstyle p_0}\downarrow & & \downarrow & & p\uparrow\;\big\rangle 0_Y & & \\
\mathcal{U} & \longrightarrow & \mathfrak{U} & \longrightarrow & Y. & &
\end{array}
\tag{2.1.5}
$$

In what follows, we will consider weight two $\mathbb{C}^*$-action on the fibers of $p \colon V^\vee \to Y$, so that w is of weight two as well.

Example 2.1.1 In order to help the reader to understand (-1)-shifted cotangent, we give its explicit description when $Y = \mathbb{C}^n$ with coordinates $x = (x_1, \dots, x_n)$ and $V = \mathbb{C}^{n+r} \to Y$ is the trivial rank r vector bundle. Then the section s is given by r-tuple of functions $s = (s_1(x), \dots, s_r(x))$, and the cotangent complex is

$$\mathbb{L}_{\mathfrak{U}} = (\mathcal{O}_{\mathfrak{U}}^{\oplus r} \overset{ds}{\to} \mathcal{O}_{\mathfrak{U}}^{\oplus n}), \; ds = \left(\frac{\partial s_i(x)}{\partial x_j}\right)_{1 \le i \le r, 1 \le j \le n}.$$

By taking its shifted dual and symmetric products, we see that

$$\mathcal{H}^0(S(\mathbb{T}_{\mathfrak{U}}[1])) = \mathbb{C}[x_1, \dots, x_n, y_1, \dots, y_r]/I$$

where I is the ideal given by

$$I = \left(s_i(x), \sum_{k=1}^r y_k \frac{\partial s_k(x)}{\partial x_j} : 1 \le i \le r, 1 \le j \le n\right).$$

The above ideal I gives the defining equation of $t_0(\Omega_{\mathfrak{U}}[-1])$. It is a critical locus of the function w on $\mathbb{C}^{n+r}$

$$w(x_1, \dots, x_n, y_1, \dots, y_r) = \sum_{k=1}^r y_k \cdot s_k(x_1, \dots, s_n).$$

Remark 2.1.2 At each point $y \in \mathcal{U}$, the fiber of the map p_0 in the diagram (2.1.5) is a linear space

$$p_0^{-1}(y) = \mathcal{H}^1(\mathbb{T}_{\mathfrak{U}}|_y),$$

which is the obstruction space of the derived scheme $\mathfrak{U}$ at y. Note that $\mathcal{H}^1(\mathbb{T}_{\mathfrak{U}}|_y) = 0$ if and only if $\mathfrak{U}$ is a smooth scheme at y. Therefore the fiber of the (-1)-shifted cotangent has an information of the singularity of $\mathfrak{U}$, where the singular support lies in as in the next subsection. We also note that the (-1)-shifted cotangent is denoted by $\mathrm{Sing}(\mathfrak{U})$ in [8].

2.1.2 Definition of Singular Supports

Here we review the theory of singular supports of coherent sheaves on $\mathfrak{U}$, developed in [8] following the earlier work [16].

Let $\mathrm{HH}^*(\mathfrak{U})$ be the Hochschild cohomology

$$\mathrm{HH}^*(\mathfrak{U}) := \mathrm{Hom}^*_{\mathfrak{U} \times \mathfrak{U}}(\Delta_* \mathcal{O}_{\mathfrak{U}}, \Delta_* \mathcal{O}_{\mathfrak{U}}).$$

Here $\Delta \colon \mathfrak{U} \to \mathfrak{U} \times \mathfrak{U}$ is the diagonal. Then it is shown in [8, Section 4] that there exist canonical maps

$$\mathcal{O}_{\mathcal{U}} \to \mathrm{HH}^0(\mathfrak{U}), \ \mathcal{H}^1(\mathbb{T}_{\mathfrak{U}}) \to \mathrm{HH}^2(\mathfrak{U}), \qquad (2.1.6)$$

so the map of graded rings

$$S(\mathcal{H}^1(\mathbb{T}_{\mathfrak{U}})) \to \mathrm{HH}^{2*}(\mathfrak{U}) \to \mathrm{Nat}_{D^b_{\mathrm{coh}}(\mathfrak{U})}(\mathrm{id}, \mathrm{id}[2*]). \qquad (2.1.7)$$

Here $\mathrm{Nat}_{D^b_{\mathrm{coh}}(\mathfrak{U})}(\mathrm{id}, \mathrm{id}[2*])$ is the group of natural transformations from id to $\mathrm{id}[2*]$ on $D^b_{\mathrm{coh}}(\mathfrak{U})$, and the right arrow is defined by taking Fourier-Mukai transforms associated with morphisms $\Delta_* \mathcal{O}_{\mathfrak{U}} \to \Delta_* \mathcal{O}_{\mathfrak{U}}[2*]$. Then by (2.1.4), the above maps induce the map for each $\mathcal{F} \in D^b_{\mathrm{coh}}(\mathfrak{U})$,

$$\mathcal{O}_{\mathrm{Crit}(w)} \to \mathrm{Hom}^{2*}(\mathcal{F}, \mathcal{F}). \qquad (2.1.8)$$

The above map defines the $\mathbb{C}^*$-equivariant $\mathcal{O}_{\mathrm{Crit}(w)}$-module structure on the graded space $\mathrm{Hom}^{2*}(\mathcal{F}, \mathcal{F})$, which is finitely generated by [8, Theorem 4.1.8]. Below a closed subset $Z \subset \mathrm{Crit}(w)$ is called *conical* if it is closed under the fiberwise $\mathbb{C}^*$-action on $\mathrm{Crit}(w)$.

Definition 2.1.3 ([8, 16]) For $\mathcal{F} \in D^b_{\mathrm{coh}}(\mathfrak{U})$, its singular support is the conical closed subset

$$\mathrm{Supp}^{\mathrm{sg}}(\mathcal{F}) \subset \mathrm{Crit}(w)$$

defined to be the support of $\mathrm{Hom}^{2*}(\mathcal{F}, \mathcal{F})$ as $\mathcal{O}_{\mathrm{Crit}(w)}$-module.

Remark 2.1.4 We didn't make the maps (2.1.6) explicit, so at this moment the definition of the singular supports is not self-contained. In Sect. 2.1.3 we will give another definition using relative Hochschild cohomology, which makes the definition of singular supports explicit and self-contained.

We define the subcategory of $D^b_{\text{coh}}(\mathfrak{U})$ with fixed singular support, which is a triangulated subcategory of $D^b_{\text{coh}}(\mathfrak{U})$ (for example, this is a consequence of Proposition 2.3.9).

Definition 2.1.5 For a conical closed subset $Z \subset \text{Crit}(w)$, we define the subcategory

$$\mathcal{C}_Z \subset D^b_{\text{coh}}(\mathfrak{U})$$

to be consisting of $\mathcal{F} \in D^b_{\text{coh}}(\mathfrak{U})$ whose singular support is contained in Z.

Here we give some examples on the descriptions of $\mathcal{C}_Z$, using the notation of the diagram (2.1.5).

Example 2.1.6 Let $W \subset \mathcal{U}$ be a closed subscheme and take $Z = p_0^{-1}(W)$. Then $\mathcal{C}_Z$ is the subcategory of objects in $D^b_{\text{coh}}(\mathfrak{U})$ whose usual supports are contained in W (see [8, Corollary 4.1.2]).

Indeed if $F \in D^b_{\text{coh}}(\mathfrak{U})$ is supported on W, then $\text{Hom}^{2i}(F, F)$ is also supported on W as $\mathcal{O}_{\mathcal{U}}$-module, hence $\text{Hom}^{2*}(F, F)$ is supported over $p_0^{-1}(W)$. Conversely if $\text{Hom}^{2*}(F, F)$ is supported over $p_0^{-1}(W)$, then $\text{Hom}^0(F, F)$ is supported on W, which implies that F is supported on W.

Example 2.1.7 Let $Z = i(\mathcal{U})$. Then $\mathcal{C}_Z$ is the category of perfect complexes on $\mathfrak{U}$ (see [8, Theorem 4.2.6]).

Indeed for a perfect complex $F \in D^b_{\text{coh}}(\mathfrak{U})$, we have $\text{Hom}^{2i}(F, F) = 0$ for $i \gg 0$, which implies that $\text{Hom}^{2*}(F, F)$ is set theoretically supported over $i(\mathcal{U})$.

In order to get a feeling in the converse direction, let us consider

$$\mathfrak{U} = \text{Spec}\, \mathbb{C}[x]/x^2$$

is the classical scheme of dual numbers. A typical example of non-perfect object is the skyscraper sheaf $\mathcal{O}_0 \in D^b_{\text{coh}}(\mathfrak{U})$. We have $\text{Hom}^{2i}(\mathcal{O}_0, \mathcal{O}_0) = \mathbb{C}$ for each $i \geq 0$, so

$$\bigoplus_{i \geq 0} \text{Hom}^{2i}(\mathcal{O}_0, \mathcal{O}_0) = \mathbb{C}[y]. \tag{2.1.9}$$

The (-1)-shifted cotangent is $(x^2 = xy = 0)$ in $\mathbb{C}^2$, and $i(\mathcal{U})$ corresponds to $y = 0$. The right hand side of (2.1.9) is not supported over $y = 0$.

Example 2.1.8 Let $Z = 0_Y(W)$ for a closed subscheme $W \subset \mathcal{U}$. Since $\mathcal{C}_Z = \mathcal{C}_{p_0^{-1}(W)} \cap \mathcal{C}_{0_Y(\mathcal{U})}$, from (i) and (ii) we see that $\mathcal{C}_Z$ is the category of perfect complexes on $\mathfrak{U}$ whose usual supports are contained in W.

Let $\operatorname{Ind} D^b_{\mathrm{coh}}(\mathfrak{U})$, $\operatorname{Ind} \mathcal{C}_Z$ be the ind-completions (see [41])

$$\operatorname{Ind} D^b_{\mathrm{coh}}(\mathfrak{U}) := \operatorname{Ind}(D^b_{\mathrm{coh}}(\mathfrak{U})), \quad \operatorname{Ind} \mathcal{C}_Z := \operatorname{Ind}(\mathcal{C}_Z).$$

The notion of singular support is extended to objects in $\operatorname{Ind} D^b_{\mathrm{coh}}(\mathfrak{U})$ (see [8, Definition 4.1.4]), and the subcategory

$$\operatorname{Ind} \mathcal{C}_Z \subset \operatorname{Ind} D^b_{\mathrm{coh}}(\mathfrak{U})$$

coincides with the subcategory of objects with singular supports contained in Z (see [8, Corollary 4.3.3]).

The quotient category

$$D^b_{\mathrm{coh}}(\mathfrak{U})/\mathcal{C}_Z$$

is a local model of our $\mathbb{C}^*$-equivariant DT category. Here we describe a one simple example.

Example 2.1.9 Let $\mathfrak{U}$ be the affine derived scheme given by

$$\mathfrak{U} = \operatorname{Spec} \mathcal{R}(\mathbb{C}^3 \to \mathbb{C}^2, s = xy).$$

Then $\mathfrak{U}$ is equivalent to its classical truncation $\mathcal{U} = (xy = 0) \subset \mathbb{C}^2$. The (-1)-shifted cotangent scheme $\Omega_{\mathfrak{U}}[-1]$ is the derived critical locus of $w : \mathbb{C}^3 \to \mathbb{C}$ given by $w(x, y, z) = xyz$, so

$$t_0(\Omega_{\mathfrak{U}}[-1]) = \operatorname{Crit}(w) = \{xy = yz = zx = 0\} \subset \mathbb{C}^3.$$

Let us take $Z = \{(0, 0, 0)\} \subset \operatorname{Crit}(w)$. Then from Example 2.1.6 (iii), we see that $\mathcal{C}_Z$ is the subcategory of perfect complexes on $\mathfrak{U}$ supported on $(0, 0)$. One can check that the following natural functor is an equivalence

$$D^b_{\mathrm{coh}}(\mathfrak{U})/\mathcal{C}_Z \xrightarrow{\sim} D_{\mathrm{sg}}(\mathcal{U}) \oplus D^b_{\mathrm{coh}}(\mathcal{U}_1) \oplus D^b_{\mathrm{coh}}(\mathcal{U}_2).$$

Here $D_{\mathrm{sg}}(\mathcal{U}) = D^b_{\mathrm{coh}}(\mathcal{U})/\operatorname{Perf}(\mathcal{U})$ is the triangulated category of singularities [100], and $\mathcal{U}_1, \mathcal{U}_2$ are the connected components of $\mathcal{U} \setminus \{(0, 0)\}$, which are isomorphic to $\mathbb{C}^*$.

2.1.3 Singular Supports via Relative Hochschild Cohomology

For the later use, we also give another description of singular supports via relative Hochschild cohomology

$$\mathrm{HH}^*(\mathfrak{U}/Y) := \mathrm{Hom}^*_{\mathfrak{U} \times_Y \mathfrak{U}}(\Delta_* \mathcal{O}_{\mathfrak{U}}, \Delta_* \mathcal{O}_{\mathfrak{U}}).$$

Let us calculate $\mathrm{HH}^*(\mathfrak{U}/Y)$. The automorphism of the vector bundle

$$\cdot \quad V \oplus V \xrightarrow{\cong} V \oplus V, \ (x, y) \mapsto (x + y, x - y)/2 \tag{2.1.10}$$

induces the equivalence

$$\mathfrak{U} \times_Y \mathfrak{U} \xrightarrow{\sim} \mathfrak{U}^{\flat} \times_Y \mathfrak{U} \tag{2.1.11}$$

where $\mathfrak{U}^{\flat} = \mathrm{Spec}\, S(V^{\vee}[1])$, defined as in (2.1.1) with $s = 0$. Note that $\mathcal{O}_Y$ is a dg-$\mathcal{O}_{\mathfrak{U}^{\flat}}$-module by the projection $\mathcal{O}_{\mathfrak{U}^{\flat}} \to \mathcal{O}_Y$. Under the equivalence (2.1.11), the object $\Delta_* \mathcal{O}_{\mathfrak{U}}$ corresponds to $\mathcal{O}_Y \boxtimes \mathcal{O}_{\mathfrak{U}}$, so we have

$$\mathbf{R}\mathrm{Hom}_{\mathfrak{U} \times_Y \mathfrak{U}}(\Delta_* \mathcal{O}_{\mathfrak{U}}, \Delta_* \mathcal{O}_{\mathfrak{U}}) \cong \mathbf{R}\mathrm{Hom}_{\mathfrak{U}^{\flat} \times_Y \mathfrak{U}}(\mathcal{O}_Y \boxtimes \mathcal{O}_{\mathfrak{U}}, \mathcal{O}_Y \boxtimes \mathcal{O}_{\mathfrak{U}})$$

$$\cong \mathbf{R}\mathrm{Hom}_{\mathfrak{U}^{\flat}}(\mathcal{O}_Y, \mathcal{O}_Y) \overset{\mathbf{L}}{\otimes}_{\mathcal{O}_Y} \mathbf{R}\mathrm{Hom}_{\mathfrak{U}}(\mathcal{O}_{\mathfrak{U}}, \mathcal{O}_{\mathfrak{U}}) \tag{2.1.12}$$

$$\cong S(V[-2]) \otimes_{\mathcal{O}_Y} \mathcal{O}_{\mathfrak{U}}.$$

Here the isomorphism $\mathbf{R}\mathrm{Hom}_{\mathfrak{U}^{\flat}}(\mathcal{O}_Y, \mathcal{O}_Y) \xrightarrow{\cong} S(V[-2])$ is given by taking the Koszul resolution of $\mathcal{O}_Y$ as $S(V^{\vee}[1])$-module

$$S(V^{\vee}[2]) \otimes_{\mathcal{O}_Y} S(V^{\vee}[1])$$

$$= \left(\cdots \to \bigwedge^2 (V^{\vee}[1]) \otimes S(V^{\vee}[1]) \to V^{\vee}[1] \otimes S(V^{\vee}[1]) \to S(V^{\vee}[1]) \right) \xrightarrow{\sim} \mathcal{O}_Y.$$

In particular, the map $x \mapsto x \otimes 1$ gives a map of graded algebras

$$\cdot \quad S(V) \to \mathrm{HH}^{2*}(\mathfrak{U}/Y) \to \mathrm{Nat}_{D^b_{\mathrm{coh}}(\mathfrak{U})}(\mathrm{id}, \mathrm{id}[2*]). \tag{2.1.13}$$

Similarly to (2.1.7), the right arrow is given by taking FM transforms. Then similarly to (2.1.8), the maps (2.1.13) determine the $\mathbb{C}^*$-equivariant $S(V)$-module structure on $\mathrm{Hom}^{2*}(\mathcal{F}, \mathcal{F})$ for $\mathcal{F} \in D^b_{\mathrm{coh}}(\mathfrak{U})$.

Lemma 2.1.10 ([8, Lemma 5.3.4]) *The image of* $\mathrm{Sing}^{\mathrm{sg}}(\mathcal{F})$ *under the closed immersion* $\mathrm{Crit}(w) \hookrightarrow V^{\vee}$ *coincides with the support of* $\mathrm{Hom}^{2*}(\mathcal{F}, \mathcal{F})$ *as a* $S(V)$-*module determined by the map (2.1.13).*

2.2 The Derived Categories of Factorizations

2.2.1 Definition of Factorizations

Let Y be a smooth scheme which admits an action of an algebraic group G, and also a $\mathbb{C}^*$-action which commutes with the above G-action. Let $\tau : G \times \mathbb{C}^* \to \mathbb{C}^*$ be a character given by the second projection, and let $w \in \Gamma(Y, \mathcal{O}_Y)$ be τ-semi invariant of weight two, i.e. $g^* w = \tau(g)^2 w$ for any $g \in G \times \mathbb{C}^*$. Given data as above, the *derived category of factorizations*

$$\mathrm{MF}_{\mathrm{qcoh}}^{\mathbb{C}^*}([Y/G], w) \tag{2.2.1}$$

is defined to be the triangulated category whose objects consist of quasi-coherent $(G \times \mathbb{C}^*)$-*equivariant factorizations of* w, i.e. a pair

$$(\mathcal{P}, d_{\mathcal{P}}), \ d_{\mathcal{P}} : \mathcal{P} \to \mathcal{P}\langle 1 \rangle, \ d_{\mathcal{P}} \circ d_{\mathcal{P}} = w \tag{2.2.2}$$

where $\mathcal{P}$ is a $(G \times \mathbb{C}^*)$-equivariant quasi-coherent sheaf on Y, $d_{\mathcal{P}}$ is a $(G \times \mathbb{C}^*)$-equivariant morphism. Here $\langle n \rangle$ means the twist by the $(G \times \mathbb{C}^*)$-character τ^n. The category (2.2.1) is defined to be the localization of the homotopy category of the factorizations (2.2.2) by its subcategory of acyclic factorizations (see [9, Section 3], [40, n2.2] for details). We have the triangulated subcategory

$$\mathrm{MF}_{\mathrm{coh}}^{\mathbb{C}^*}([Y/G], w) \subset \mathrm{MF}_{\mathrm{qcoh}}^{\mathbb{C}^*}([Y/G], w)$$

consisting of factorizations (2.2.2) such that $\mathcal{P}$ is coherent. If Y is affine, the above subcategory is equivalent to the triangulated category of $(G \times \mathbb{C}^*)$-equivariant matrix factorizations of w defined by Orlov [100].

Remark 2.2.1 The derived categories of factorizations are also defined in terms of curved dg-modules over curved dg-algebras introduced in [107]. By definition, a *commutative curved dg-algebra* is a triple

$$(R^{\bullet}, d_R, c), \ d_R : R^{\bullet} \to R^{\bullet+1}, \ c \in R^2.$$

Here $(R^\bullet, d_R)$ is a commutative dg-algebra, and the element $c \in R^2$ (called *curvature*) satisfies $d_R c = 0$. A *curved dg-module* over a curved dg-algebra $(R^\bullet, d_R, c)$ is a pair

$$(M^\bullet, d_M), \ d_M \colon M^\bullet \to M^{\bullet+1}, \ d_M^2 = \cdot c$$

where $M^\bullet$ is a graded $R^\bullet$-module, and the map $R^\bullet \to \mathrm{End}(M^\bullet)$ defining the $R^\bullet$-module structure commutes with $(d_R, [d_M, -])$. By its definition, giving a $(G \times \mathbb{C}^*)$-equivariant factorization of $w \colon Y \to \mathbb{C}$ is equivalent to giving a G-equivariant curved dg-module over a G-equivariant curved dg-algebra

$$(\mathcal{O}_Y, 0, w). \tag{2.2.3}$$

Then $\mathrm{MF}^{\mathbb{C}^*}_{\mathrm{coh}}([Y/G], w)$ is naturally identified with the derived category of G-equivariant finitely generated curved dg-modules over (2.2.3).

Remark 2.2.2 The $\mathbb{Z}/2$-periodic derived category of (coherent, quasi-coherent) factorizations is also defined in a similar way. For a smooth scheme Y with G-action and a G-invariant function $w \colon Y \to \mathbb{C}$, they are denoted by

$$\mathrm{MF}^{\mathbb{Z}/2}_{\mathrm{coh}}([Y/G], w), \ \mathrm{MF}^{\mathbb{Z}/2}_{\mathrm{qcoh}}([Y/G], w)$$

respectively. They are $\mathbb{Z}/2$-periodic triangulated categories, and consists of $\mathbb{Z}/2 \times G$-equivariant factorizations

$$(\mathcal{P}, d_\mathcal{P}), \ d_\mathcal{P} \colon \mathcal{P} \to \mathcal{P}\langle 1 \rangle, \ d_\mathcal{P} \circ d_\mathcal{P} = w$$

where $\mathcal{P}$ is coherent, quasi-coherent, respectively. Here $\mathbb{Z}/2$ acts on Y trivially, and $\langle 1 \rangle$ means twist with respect to a non-trivial $\mathbb{Z}/2$-character.

For a $(G \times \mathbb{C}^*)$-invariant closed subscheme $Z \subset Y$ with $\mathcal{Z} = [Z/G]$, the subcategory

$$\mathrm{MF}^{\mathbb{C}^*}_\star([Y/G], w)_\mathcal{Z} \subset \mathrm{MF}^{\mathbb{C}^*}_\star([Y/G], w) \tag{2.2.4}$$

for $\star \in \{\mathrm{qcoh}, \mathrm{coh}\}$ is defined to be the kernel of the restriction functor

$$\mathrm{MF}^{\mathbb{C}^*}_\star([Y/G], w) \to \mathrm{MF}^{\mathbb{C}^*}_\star([(Y \setminus Z)/G], w).$$

By taking the Verdier quotients, we have an equivalence (cf. [29, Theorem 1.3])

$$\mathrm{MF}^{\mathbb{C}^*}_\star([Y/G], w)/\mathrm{MF}^{\mathbb{C}^*}_\star([Y/G], w)_\mathcal{Z} \xrightarrow{\sim} \mathrm{MF}^{\mathbb{C}^*}_\star([(Y \setminus Z)/G], w). \tag{2.2.5}$$

Since any object in $\mathrm{MF}_\star^{\mathbb{C}^*}([Y/G], w)$ is supported on $\mathrm{Crit}(w)$ (see [108, Corollary 3.18]), for $Z' = Z \cap \mathrm{Crit}(w)$ we have equivalences

$$\mathrm{MF}_\star^{\mathbb{C}^*}([Y/G], w)_{Z'} \xrightarrow{\sim} \mathrm{MF}_\star^{\mathbb{C}^*}([Y/G], w)_Z.$$

Combined with (2.2.5), we also have the equivalence

$$\mathrm{MF}_\star^{\mathbb{C}^*}([(Y \setminus Z')/G], w) \xrightarrow{\sim} \mathrm{MF}_\star^{\mathbb{C}^*}([(Y \setminus Z)/G], w). \tag{2.2.6}$$

In particular, let $Y' \subset Y$ be a $(G \times \mathbb{C}^*)$-invariant open subset which contains $\mathrm{Crit}(w)$. Then by taking $Z = Y \setminus Y'$ so that $Z' = \emptyset$, the restriction functor gives an equivalence

$$\mathrm{MF}_\star^{\mathbb{C}^*}([Y/G], w) \xrightarrow{\sim} \mathrm{MF}_\star^{\mathbb{C}^*}([Y'/G], w). \tag{2.2.7}$$

For $\star = \mathrm{qcoh}$, the embedding (2.2.4) admits a right adjoint (see [11, Proposition 2.3.9])

$$\Gamma_{\mathcal{Z}} \colon \mathrm{MF}_{\mathrm{qcoh}}^{\mathbb{C}^*}([Y/G], w) \to \mathrm{MF}_{\mathrm{qcoh}}^{\mathbb{C}^*}([Y/G], w)_{\mathcal{Z}}$$

so that for an object $\mathcal{P} \in \mathrm{MF}_{\mathrm{qcoh}}^{\mathbb{C}^*}([Y/G], w)$, there is a distinguished triangle

$$\Gamma_{\mathcal{Z}}(\mathcal{P}) \to \mathcal{P} \to \mathcal{P}|_{[Y/G]\setminus\mathcal{Z}} \tag{2.2.8}$$

where $\mathcal{P}|_{[Y/G]\setminus\mathcal{Z}} := j_* j^* \mathcal{P}$ for the open immersion $j \colon [Y/G] \setminus \mathcal{Z} \hookrightarrow [Y/G]$.

2.2.2 Functoriality of Derived Categories of Factorizations

Let Y' be another smooth scheme with $(G' \times \mathbb{C}^*)$-action and $f \colon [Y'/G'] \to [Y/G]$ a $\mathbb{C}^*$-equivariant morphism. We have the push-forward/pull-back functors (cf. [9, Section 3])

$$\mathrm{MF}_{\mathrm{qcoh}}^{\mathbb{C}^*}([Y'/G'], f^*w) \underset{f^*}{\overset{f_*}{\rightleftarrows}} \mathrm{MF}_{\mathrm{qcoh}}^{\mathbb{C}^*}([Y/G], w) \tag{2.2.9}$$

such that $f^* \dashv f_*$. If f is proper, we also have functors

$$\mathrm{MF}_{\mathrm{qcoh}}^{\mathbb{C}^*}([Y'/G'], f^*w) \underset{f^!}{\overset{f_!}{\rightleftarrows}} \mathrm{MF}_{\mathrm{qcoh}}^{\mathbb{C}^*}([Y/G], w)$$

where $f_* \dashv f^!$, $f_! \dashv f^*$. They are related to the functors (2.2.9) as

$$f^!(-) = f^*(-) \otimes \omega_f, \ f_!(-) = f_*(- \otimes \omega_f).$$

Here $\omega_f := \det \mathbb{L}_f[\dim f]$. The functor f^* preserves $\mathrm{MF}_{\mathrm{coh}}^{\mathbb{C}^*}(-)$, and if f is proper then $f^!$, f_* and $f_!$ also preserve $\mathrm{MF}_{\mathrm{coh}}^{\mathbb{C}^*}(-)$.

Let $\lambda \colon \mathbb{C}^* \to G$ be a one parameter subgroup contained in the center of G and acts on Y trivial. Then we have the decomposition for $\star \in \{\mathrm{coh}, \mathrm{qcoh}\}$

$$\mathrm{MF}_\star^{\mathbb{C}^*}([Y/G], w) = \bigoplus_{j \in \mathbb{Z}} \mathrm{MF}_\star^{\mathbb{C}^*}([Y/G], w)_{\lambda\text{-wt}=j}$$

where $\mathrm{MF}_\star^{\mathbb{C}^*}([Y/G], w)_{\mathrm{wt}=j}$ is the weight j-part with respect to λ. We define the subcategory

$$\mathrm{MF}_{\mathrm{coh}}^{\mathbb{C}^*}([Y/G], w)_{\lambda\text{-below}} \subset \mathrm{MF}_{\mathrm{qcoh}}^{\mathbb{C}^*}([Y/G], w)$$

to be consisting of objects whose λ-weights are bounded below, and each λ-weight j part is an object in $\mathrm{MF}_{\mathrm{coh}}^{\mathbb{C}^*}([Y/G], w)_{\lambda\text{-wt}=j}$.

Let $W \to Y$ be a $(G \times \mathbb{C}^*)$-equivariant vector bundle, where λ acts on fibers of $W \to Y$ by negative weights. Then we have the $\mathbb{C}^*$-equivariant morphism of stacks $f \colon [W/G] \to [Y/G]$.

Lemma 2.2.3 *The functor f_* in (2.2.9) restricts to the functor*

$$f_* \colon \mathrm{MF}_{\mathrm{coh}}^{\mathbb{C}^*}([W/G], f^*w) \to \mathrm{MF}_{\mathrm{coh}}^{\mathbb{C}^*}([Y/G], w)_{\lambda\text{-below}}.$$

Proof By push-forward f_*, a coherent sheaf on $[W/G]$ is regarded as a coherent module over the sheaf of algebras $\mathrm{Sym}(W^\vee)$ on $[Y/G]$. It has only non-negative λ-weights, and each λ-weight j part is a coherent sheaf on $[Y/G]$. Therefore the lemma holds.   □

2.3   Koszul Duality Equivalence

2.3.1   *G-equivariant Tuple*

Definition 2.3.1 Let G be an affine algebraic group. A G-equivariant tuple is a tuple

$$(Y, V, s)$$

where Y is a smooth affine $\mathbb{C}$-scheme of finite presentation with a G-action, $V \to Y$ is a G-equivariant vector bundle and s is a G-invariant section of V. Given (Y, V, s) as above, the associated derived stack is denoted by

$$[\mathfrak{U}/G] := [\operatorname{Spec} \mathcal{R}(V \to Y, s)/G].$$

Its classical truncation is denoted by $[\mathcal{U}/G]$, which is a closed substack of $\mathcal{Y} := [Y/G]$.

For a G-equivariant tuple (Y, V, s), we take the weight two $\mathbb{C}^*$-action on the fibers of $V^\vee \to Y$. Then we have the $(G \times \mathbb{C}^*)$-action on $V^\vee$ and the function w defined in (2.1.3) is τ-semi invariant of weight two, where τ is the projection $G \times \mathbb{C}^* \to \mathbb{C}^*$. For a $(G \times \mathbb{C}^*)$-invariant closed subset $Z \subset \operatorname{Crit}(w)$, we have the subcategories

$$\mathcal{C}_{\mathcal{Z}} \subset D^b_{\mathrm{coh}}([\mathfrak{U}/G]), \ \operatorname{Ind} \mathcal{C}_{\mathcal{Z}} \subset \operatorname{Ind} D^b_{\mathrm{coh}}([\mathfrak{U}/G])$$

consisting of objects whose singular supports are contained in the closed substack

$$\mathcal{Z} = [Z/G] \subset [t_0(\Omega_{\mathfrak{U}}[-1])/G].$$

More precisely these are objects whose pull-back via $\mathfrak{U} \to [\mathfrak{U}/G]$ are contained in $\mathcal{C}_Z$, $\operatorname{Ind} \mathcal{C}_Z$ respectively (see Definition 3.2.1 for a general case).

Lemma 2.3.2 *The category* $\operatorname{Ind} \mathcal{C}_{\mathcal{Z}}$ *is compactly generated such that* $(\operatorname{Ind} \mathcal{C}_{\mathcal{Z}})^{\mathrm{cp}} = \mathcal{C}_{\mathcal{Z}}$.

Proof Since the stack $[\mathfrak{U}/G]$ is a global complete intersection stack in the sense of [8, Section 9.1], the lemma follows from [8, Corollary 9.2.7, Corollary 9.2.8]. □

Note that we have the following commutative diagram, which we will often use below.

$$(2.3.1)$$

Here 0_Y is the zero section of $V^\vee \to Y$, and $\pi_{\mathcal{U}}$, π_Y are good moduli space morphisms (see Sect. 3.1.4 for good moduli spaces).

2.3.2   Koszul Duality Equivalence

The construction in the previous subsection yields the derived category of factorizations and its subcategory with fixed support

$$\mathrm{MF}_{\mathrm{coh}}^{\mathbb{C}^*}([V^\vee/G], w)_{\mathcal{Z}} \subset \mathrm{MF}_{\mathrm{coh}}^{\mathbb{C}^*}([V^\vee/G], w).$$

Let $\mathcal{K}_s$ be the following $(G \times \mathbb{C}^*)$-equivariant factorization of w (called *Koszul factorization*)

$$\mathcal{K}_s := \left(\mathcal{O}_{V^\vee} \otimes_{\mathcal{O}_Y} \mathcal{O}_{\mathfrak{U}}, d_{\mathcal{K}_s} \right). \tag{2.3.2}$$

Here G acts on $\mathcal{O}_{V^\vee} \otimes_{\mathcal{O}_Y} \mathcal{O}_{\mathfrak{U}}$ diagonally, the $\mathbb{C}^*$-action is given by the grading

$$\mathcal{O}_{V^\vee} \otimes_{\mathcal{O}_Y} \mathcal{O}_{\mathfrak{U}} = S(V[-2]) \otimes_{\mathcal{O}_Y} S(V^\vee[1]),$$

and the weight one map $d_{\mathcal{K}_s}$ is given by

$$d_{\mathcal{K}_s} = 1 \otimes d_{\mathcal{O}_{\mathfrak{U}}} + \eta \colon \mathcal{O}_{V^\vee} \otimes_{\mathcal{O}_Y} \mathcal{O}_{\mathfrak{U}} \to \mathcal{O}_{V^\vee} \otimes_{\mathcal{O}_Y} \mathcal{O}_{\mathfrak{U}}\langle 1 \rangle,$$

where $\eta \in V \otimes_{\mathcal{O}_Y} V^\vee \subset \mathcal{O}_{V^\vee} \otimes_{\mathcal{O}_Y} \mathcal{O}_{\mathfrak{U}}$ corresponds to $\mathrm{id} \in \mathrm{Hom}(V, V)$.

The object $\mathcal{K}_s$ also admits a G-equivariant dg $\mathcal{O}_{\mathfrak{U}}$-module structure by the right factor of the tensor product. In particular for a dg-$\mathcal{O}_{\mathfrak{U}}$-module $\mathcal{F}^\bullet$, we can form its tensor product $\mathcal{K}_s \otimes_{\mathcal{O}_{\mathfrak{U}}} \mathcal{F}^\bullet$ which is a factorization of w. It is explicitly described as follows. The dg-$\mathcal{O}_{\mathfrak{U}}$-module $\mathcal{F}^\bullet$ consists of a complex of $\mathcal{O}_Y$-modules $(\mathcal{F}^\bullet, d_{\mathcal{F}^\bullet})$ together with $\mathcal{O}_Y$-module homomorphisms $\eta^i \colon \mathcal{F}^i \to V \otimes \mathcal{F}^{i-1}$ satisfying the axiom of dg-modules. Then $\mathcal{K}_s \otimes_{\mathcal{O}_{\mathfrak{U}}} \mathcal{F}^\bullet$ is given by

$$\left(\bigoplus_{i \in \mathbb{Z}} \mathcal{F}^i \langle i \rangle \otimes_{\mathcal{O}_Y} \mathcal{O}_{V^\vee} \to \bigoplus_{i \in \mathbb{Z}} \mathcal{F}^i \langle i+1 \rangle \otimes_{\mathcal{O}_Y} \mathcal{O}_{V^\vee} \right)$$

where the weight one morphism is an $\mathcal{O}_{V^\vee}$-module homomorphism induced by

$$d_{\mathcal{F}^\bullet}^i + \eta^i \colon \mathcal{F}^i \to \mathcal{F}^{i+1} \oplus (V \otimes \mathcal{F}^{i-1}).$$

The correspondence $(-) \to \mathcal{K}_s \otimes_{\mathcal{O}_{\mathfrak{U}}} (-)$ is understood as a Fourier-Mukai functor (see Remark 2.3.4) so that it induces a functor on derived categories. Then the following result is proved in several references (see [56, Proposition 4.8] and also [58, 99, 121]), and called *Koszul duality equivalence*.

Theorem 2.3.3 (cf. [56, 58, 99, 121]) *The functor Φ defined by*

$$\Phi \colon D_{\mathrm{coh}}^b([\mathfrak{U}/G]) \to \mathrm{MF}_{\mathrm{coh}}^{\mathbb{C}^*}([V^\vee/G], w), \quad (-) \mapsto \mathcal{K}_s \otimes_{\mathcal{O}_{\mathfrak{U}}} (-) \tag{2.3.3}$$

is an equivalence of triangulated categories. Moreover a quasi-inverse of Φ is given by

$$\Psi \colon \mathrm{MF}^{\mathbb{C}^*}_{\mathrm{coh}}([V^\vee/G], w) \to D^b_{\mathrm{coh}}([\mathfrak{U}/G]), \qquad (2.3.4)$$

$$(-) \mapsto \mathbf{RHom}_{\mathrm{MF}^{\mathbb{C}^*}_{\mathrm{coh}}([V^\vee/G], w)}(\mathcal{K}_s, -).$$

Proof The result is essentially proved in [56, 58, 99, 121], but the currently stated version is not stated in the references so we give its proof.

We first prove the case of $G = \{1\}$. Since $\mathcal{K}_s$ consists of free $\mathcal{O}_{\mathfrak{U}}$-modules, we do not need to take $\mathbf{L}$ in the computation of Φ. In particular it takes coherent $\mathcal{O}_{\mathfrak{U}}$-modules to coherent factorizations of $w \colon V^\vee \to \mathbb{C}$, so the functor (2.3.3) is well-defined. Also for $\mathcal{P} \in \mathrm{MF}^{\mathbb{C}^*}_{\mathrm{coh}}(V^\vee, w)$, we have

$$\Psi(\mathcal{P}) = \mathbf{RHom}_{\mathrm{MF}^{\mathbb{C}^*}_{\mathrm{coh}}(V^\vee, w)}(\mathcal{K}_s, \mathcal{P}) \cong \mathcal{K}_s^\vee \otimes_{\mathcal{O}_{V^\vee}} \mathcal{P} \qquad (2.3.5)$$

where $\mathcal{K}_s^\vee$ is the $\mathcal{O}_{V^\vee}$-dual of $\mathcal{K}_s$, which is a factorization of $-w$. Below we will use the following functors of derived categories of factorizations

$$\mathrm{MF}^{\mathbb{C}^*}_{\mathrm{coh}}(V^\vee, w) \xrightarrow{0^*_Y} \mathrm{MF}^{\mathbb{C}^*}_{\mathrm{coh}}(Y, 0) \xrightarrow{0^{-w}_{Y*}} \mathrm{MF}^{\mathbb{C}^*}_{\mathrm{coh}}(V^\vee, -w).$$

Here 0^{-w}_{Y*} is the push-forward along with the zero section $0_Y \colon Y \to V^\vee$ which makes sense as $0^*_Y w = 0$. Also for a $\mathbb{C}^*$-equivariant coherent sheaf $\mathcal{F}$ on Y (resp. on $V^\vee$), we regard it as an object in $\mathrm{MF}^{\mathbb{C}^*}_{\mathrm{coh}}(Y, 0)$ (resp. $\mathrm{MF}^{\mathbb{C}^*}_{\mathrm{coh}}(V^\vee, 0)$) by taking the factorization of 0, $(\mathcal{F} \xrightarrow{0} \mathcal{F}\langle 1\rangle)$. Using the above notation, by [9, Proposition 3.20, Lemma 3.21], we have the isomorphisms as factorizations of $-w$:

$$\mathcal{K}_s^\vee \cong \mathcal{K}_{-s} \otimes_{\mathcal{O}_Y} \det V[-\mathrm{rank}(V)] \cong 0^{-w}_{Y*} \mathcal{O}_Y. \qquad (2.3.6)$$

Therefore we have the isomorphism of dg-algebras

$$\mathbf{RHom}_{\mathrm{MF}^{\mathbb{C}^*}_{\mathrm{coh}}(V^\vee, w)}(\mathcal{K}_s, \mathcal{K}_s) = \mathcal{K}_s^\vee \otimes_{\mathcal{O}_{V^\vee}} \mathcal{K}_s, \cong \mathcal{O}_{\mathfrak{U}}, \qquad (2.3.7)$$

therefore (2.3.5) is a dg-$\mathcal{O}_{\mathfrak{U}}$-module. Moreover each cohomology of (2.3.5) is a coherent $\mathcal{O}_Y$-module, so it determines an object in $D^b_{\mathrm{coh}}(\mathfrak{U})$, and the functor (2.3.4) is well-defined.

The composition $\Psi \circ \Phi$ is given by

$$\Psi \circ \Phi(-) = \mathcal{K}_s^\vee \otimes_{\mathcal{O}_{V^\vee}} \mathcal{K}_s \otimes_{\mathcal{O}_{\mathfrak{U}}} (-).$$

By (2.3.7), we conclude that $\Psi \circ \Phi$ is the identity functor. Therefore Φ is fully-faithful, and it remains to show that Φ is essentially surjective. As Ψ is a right adjoint of Φ, it is enough to show that if an object $\mathcal{P} \in \mathrm{MF}^{\mathbb{C}^*}_{\mathrm{coh}}(V^\vee, w)$ satisfies

$\Psi(\mathcal{P}) \cong 0$, then $\mathcal{P} \cong 0$. Suppose that $\mathcal{P}$ is such an object. By the above description of $\mathcal{K}_s^\vee$, the push-forward of $\Psi(\mathcal{P}) \cong 0$ along the closed immersion $\mathfrak{U} \hookrightarrow Y$ is $i^*\mathcal{P} \cong 0$. Let $I \subset \mathcal{O}_{V^\vee}$ be the ideal sheaf of the zero section $Y \subset V^\vee$, and set $Y^{(2)} = \mathrm{Spec}_Y \mathcal{O}_{V^\vee}/I^2$. By applying $\mathcal{P} \overset{L}{\otimes}_{\mathcal{O}_{V^\vee}} (-)$ to the exact sequence

$$0 \to I/I^2 \to \mathcal{O}_{Y^{(2)}} \to \mathcal{O}_Y \to 0$$

we also see that $\mathcal{P}|_{Y^{(2)}} \cong 0$ in $\mathrm{MF}_{\mathrm{coh}}^{\mathbb{C}^*}(Y^{(2)}, w|_{Y^{(2)}})$. By repeating this argument, we see that $\mathcal{P}|_{\widehat{V}^\vee} \cong 0$ in $\mathrm{MF}_{\mathrm{coh}}^{\mathbb{C}^*}(\widehat{V}^\vee, w|_{\widehat{V}^\vee})$, where $\widehat{V}^\vee$ is the formal completion of $V^\vee$ at Y. On the other hand $\mathrm{Hom}_{\mathrm{MF}_{\mathrm{coh}}^{\mathbb{C}^*}(V^\vee, w)}(\mathcal{P}, \mathcal{P})$ is a finitely generated $\mathcal{O}_{V^\vee}$-module whose support is a conical closed subset $Z \subset V^\vee$. As $\mathcal{P}|_{\widehat{V}^\vee} \cong 0$, we have $Z \cap \widehat{V}^\vee = \emptyset$. But Z is conical, so we must have $Z = \emptyset$, hence $\mathcal{P} \cong 0$ holds.

For a general G, the natural transformations

$$(-) \to \Psi \circ \Phi(-), \ \Psi \circ \Phi(-) \to (-)$$

are isomorphisms after forgetting G-equivariant structures by the above argument. Therefore they are isomorphisms since the forgetting functors

$$D_{\mathrm{coh}}^b([\mathfrak{U}/G]) \to D_{\mathrm{coh}}^b(\mathfrak{U}), \ \mathrm{MF}_{\mathrm{coh}}^{\mathbb{C}^*}([V^\vee/G], w) \to \mathrm{MF}_{\mathrm{coh}}^{\mathbb{C}^*}(V^\vee, w)$$

are conservative.   □

Remark 2.3.4 The functor Φ is explained in terms of curved dg-modules [107] as follows. The graded algebra $\mathcal{O}_{V^\vee} \otimes_{\mathcal{O}_Y} \mathcal{O}_{\mathfrak{U}}$ together with the degree one map $1 \otimes d\mathcal{O}_{\mathfrak{U}}$ and the degree two element $w \otimes 1 \in \mathcal{O}_{V^\vee} \otimes_{\mathcal{O}_Y} \mathcal{O}_{\mathfrak{U}}$ define the commutative curved dg-algebra (see Remark 2.2.1)

$$(\mathcal{O}_{V^\vee} \otimes_{\mathcal{O}_Y} \mathcal{O}_{\mathfrak{U}}, 1 \otimes d_{\mathcal{O}_{\mathfrak{U}}}, w \otimes 1)$$

with differential $1 \otimes d_{\mathcal{O}_{\mathfrak{U}}}$ and curvature $w \otimes 1$. Then the object $\mathcal{K}_s$ is a curved dg-module over it, and the functor Φ is regarded as a Fourier-Mukai functor from curved dg-modules over a curved dg-algebra $(\mathcal{O}_{\mathfrak{U}}, d_{\mathcal{O}_{\mathfrak{U}}}, 0)$ to those over $(\mathcal{O}_{V^\vee}, 0, w)$ whose kernel object is $\mathcal{K}_s$.

Remark 2.3.5 If s is a regular section, then the closed immersion $\mathcal{U} \hookrightarrow \mathfrak{U}$ is an equivalence, and the object (2.3.2) is isomorphic to the factorization $p^*\mathcal{O}_{\mathcal{U}} \overset{0}{\to} p^*\mathcal{O}_{\mathcal{U}}\langle 1 \rangle$. So it coincides with the equivalence constructed in [56, 121].

Remark 2.3.6 When $s = 0$ then $w = 0$, and we have objects $\mathcal{O}_Y \in D_{\mathrm{coh}}^b(\mathfrak{U})$ and $\mathcal{O}_{V^\vee} \in \mathrm{MF}_{\mathrm{coh}}^{\mathbb{C}^*}(V^\vee, 0)$, where $\mathcal{O}_{V^\vee} = (\mathcal{O}_{V^\vee} \overset{0}{\to} \mathcal{O}_{V^\vee}\langle 1 \rangle)$. Since $\Phi(\mathcal{O}_Y) = \mathcal{O}_{V^\vee}$ for the equivalence in Theorem 2.3.3, we have the quasi-isomorphism

$$\Phi^b: \mathrm{RHom}_{\mathfrak{U}}(\mathcal{O}_Y, \mathcal{O}_Y) \overset{\sim}{\to} \mathrm{RHom}_{\mathrm{MF}_{\mathrm{coh}}^{\mathbb{C}^*}(V^\vee, 0)}(\mathcal{O}_{V^\vee}, \mathcal{O}_{V^\vee}) = S(V[-2]).$$

The above quasi-isomorphism coincides with the one given in (2.1.12) used in defining the singular support.

Remark 2.3.7 By applying Theorem 2.3.3 for $V = Y$, we obtain the equivalence

$$D^b_{\mathrm{coh}}([Y/G]) \xrightarrow{\sim} \mathrm{MF}^{\mathbb{C}^*}_{\mathrm{coh}}([Y/G], 0).$$

The above equivalence is a tautological equivalence sending $(\mathcal{F}^{\bullet}, d_{\mathcal{F}^{\bullet}})$ to $(\mathcal{F}^{\bullet}, d_{\mathcal{F}^{\bullet}})$, where the cohomological gradings of the LHS are $\mathbb{C}^*$-weights of the RHS.

Remark 2.3.8 The result of Theorem 2.3.3 is not true for $\mathbb{Z}/2$-periodic version unless $\mathrm{Crit}(w) = \mathcal{U}$, or equivalently $\mathfrak{U} = \mathcal{U}$ and it is smooth. Namely the similarly defined functor

$$\Psi \colon \mathrm{MF}^{\mathbb{Z}/2}_{\mathrm{coh}}(V^{\vee}, w) \to D^{\mathbb{Z}/2}_{\mathrm{coh}}(\mathfrak{U})$$

is not an equivalence. Here the RHS is the $\mathbb{Z}/2$-periodic derived category of dg-modules over $\mathfrak{U}$ with coherent cohomologies. Indeed if $\mathcal{U} \subsetneq \mathrm{Crit}(w)$, there is an object in $\mathrm{MF}^{\mathbb{Z}/2}_{\mathrm{coh}}(V^{\vee}, w)$ supported away from $\mathcal{U}$, so the above functor has a non-trivial kernel. This is the reason we construct the DT category for $\mathbb{C}^*$-equivariant case. A correct formulation of $\mathbb{Z}/2$-periodic Koszul duality is available in [138].

2.3.3 Singular Supports Under Koszul Duality

Under the Koszul duality equivalence in Theorem 2.3.3, we can compare the singular supports and the usual supports in the derived categories of factorizations. The following proposition was claimed in [8, Section H], and we give its details.

Proposition 2.3.9 *Let $Z \subset \mathrm{Crit}(w)$ be a $(G \times \mathbb{C}^*)$-invariant closed subset and set $\mathcal{Z} = [Z/G]$. Then the equivalence Φ in Theorem 2.3.3 restricts to the equivalence*

$$\Phi \colon \mathcal{C}_{\mathcal{Z}} \xrightarrow{\sim} \mathrm{MF}^{\mathbb{C}^*}_{\mathrm{coh}}([V^{\vee}/G], w)_{\mathcal{Z}}. \tag{2.3.8}$$

In particular by taking Verdier quotients and using (2.2.5), we also have an equivalence

$$\Phi \colon D^b_{\mathrm{coh}}([\mathfrak{U}/G])/\mathcal{C}_{\mathcal{Z}} \xrightarrow{\sim} \mathrm{MF}^{\mathbb{C}^*}_{\mathrm{coh}}([(V^{\vee} \setminus Z)/G], w). \tag{2.3.9}$$

Proof We may assume that $G = \{1\}$. The equivalence Φ in (2.3.3) induces the isomorphism of natural transforms

$$\Phi^{\mathrm{Nat}} \colon \mathrm{Nat}_{D^b_{\mathrm{coh}}(\mathfrak{U})}(\mathrm{id}, \mathrm{id}[m]) \xrightarrow{\cong} \mathrm{Nat}_{\mathrm{MF}^{\mathbb{C}^*}_{\mathrm{coh}}(V^{\vee}, w)}(\mathrm{id}, \mathrm{id}[m])$$

given by $(-) \mapsto \Phi \circ (-) \circ \Phi^{-1}$. By Lemma 2.1.10, it is enough to show that the
following diagram commutes:

$$
\begin{array}{ccc}
V & = & V \\
\downarrow & & \downarrow \\
\mathrm{Nat}_{D^b_{\mathrm{coh}}(\mathfrak{U})}(\mathrm{id}, \mathrm{id}[2]) & \xrightarrow{\Phi^{\mathrm{Nat}}} & \mathrm{Nat}_{\mathrm{MF}^{\mathbb{C}^*}_{\mathrm{coh}}(V^\vee, w)}(\mathrm{id}, \mathrm{id}[2]).
\end{array}
\tag{2.3.10}
$$

Here the left vertical arrow is given in (2.1.13), and the right vertical arrow is given
by the multiplication via $\mathcal{O}_{V^\vee}$-module structure on each factorization.

Let Φ^{Fun} be the functor

$$
\Phi^{\mathrm{Fun}} \colon D^b_{\mathrm{coh}}(\mathfrak{U} \times_Y \mathfrak{U}) \to \mathrm{MF}^{\mathbb{C}^*}_{\mathrm{coh}}(V^\vee \times_Y V^\vee, -w \boxplus w)
$$

given by the Fourier-Mukai kernel $p^*_{13}\mathcal{K}^\vee_s \otimes p^*_{24}\mathcal{K}_s$, where p_{ij} are the projections
from $\mathfrak{U} \times_Y \mathfrak{U} \times_Y V^\vee \times_Y V^\vee$ onto the corresponding factors. Here $\mathbb{C}^*$ acts on fibers
of $V^\vee \times_Y V^\vee \to Y$ by weight two. Note that $\mathfrak{U} \times_Y \mathfrak{U}$ is equivalent to the derived
zero locus of $(-s, s) \colon Y \to V \oplus V$. Moreover we have the isomorphism by the first
isomorphism in (2.3.6)

$$
p^*_{13}\mathcal{K}^\vee_s \otimes p^*_{24}\mathcal{K}_s \cong \mathcal{K}_{(-s,s)} \otimes_{\mathcal{O}_Y} \det V[-\mathrm{rank}(V)].
$$

Therefore an argument similar to Proposition 2.3.3 shows that the functor Φ^{Fun} is an
equivalence. From a categorical point of view, the equivalence Φ^{Fun} identifies FM
transforms on $D^b_{\mathrm{coh}}(\mathfrak{U})$ over Y with those on $\mathrm{MF}^{\mathbb{C}^*}_{\mathrm{coh}}(V^\vee, w)$ via $(-) \mapsto \Phi \circ (-) \circ$
Φ^{-1}, which can be verified along with the similar argument of [30, Proposition 8.2].
Since $\Phi \circ \mathrm{id} \circ \Phi^{-1}$ is the identity functor, we should have

$$
\Phi^{\mathrm{Fun}}(\Delta_*\mathcal{O}_{\mathfrak{U}}) \cong \Delta_*\mathcal{O}_{V^\vee} := (\Delta_*\mathcal{O}_{V^\vee} \xrightarrow{0} \Delta_*\mathcal{O}_{V^\vee}\langle 1 \rangle)
\tag{2.3.11}
$$

in $\mathrm{MF}^{\mathbb{C}^*}_{\mathrm{coh}}(V^\vee \times_Y V^\vee, -w \boxplus w)$. We can check the above isomorphism directly as
follows:

$$
\begin{aligned}
(\Phi^{\mathrm{Fun}})^{-1}(\Delta_*\mathcal{O}_{V^\vee}) &= \mathbf{R}\mathrm{Hom}_{\mathrm{MF}^{\mathbb{C}^*}_{\mathrm{coh}}(V^\vee \times_Y V^\vee, -w \boxplus w)}(p^*_{13}\mathcal{K}^\vee_s \otimes p^*_{24}\mathcal{K}_s, \Delta_*\mathcal{O}_{V^\vee}) \\
&\cong \mathbf{R}\mathrm{Hom}_{\mathrm{MF}^{\mathbb{C}^*}_{\mathrm{coh}}(V^\vee, 0)}(\mathcal{K}^\vee_s \otimes_{\mathfrak{U}} \mathcal{K}_s, \mathcal{O}_{V^\vee}) \\
&\cong \mathbf{R}\mathrm{Hom}_{\mathrm{MF}^{\mathbb{C}^*}_{\mathrm{coh}}(V^\vee, w)}(\mathcal{K}_s, \mathcal{K}_s) \\
&\cong \Delta_*\mathcal{O}_{\mathfrak{U}}.
\end{aligned}
$$

Here the last isomorphism follows using (2.3.6).

By the equivalence of Φ^{Fun} together with the isomorphism (2.3.11), the functor Φ^{Fun} induces the isomorphism on relative Hochschild cohomologies

$$\Phi^{\mathrm{HH}} : \mathrm{HH}^*(\mathfrak{U}/Y) \xrightarrow{\cong} \mathrm{HH}^*((V^\vee, w)/Y)). \tag{2.3.12}$$

Here the right hand side is defined by

$$\mathrm{HH}^*((V^\vee, w)/Y)) := \mathrm{Hom}^*_{\mathrm{MF}^{\mathbb{C}^*}_{\mathrm{coh}}(V^\vee \times_Y V^\vee, -w \boxplus w)}(\Delta_* \mathcal{O}_{V^\vee}, \Delta_* \mathcal{O}_{V^\vee}).$$

The isomorphism (2.3.12) is compatible with Φ^{Nat} under the natural maps from relative Hochschild cohomologies to natural transformations $\mathrm{id} \to \mathrm{id}[*]$.

Using the automorphism of $V^{\oplus 2}$ given in (2.1.10), the equivalence Φ^{Fun} is identified with the functor

$$D^b_{\mathrm{coh}}(\mathfrak{U}^\flat \times_Y \mathfrak{U}) \xrightarrow{\sim} \mathrm{MF}^{\mathbb{C}^*}_{\mathrm{coh}}(V^\vee \times_Y V^\vee, 0 \boxplus w)$$

given by the FM transform with kernel $p^*_{13} \mathcal{K}^\vee_0 \otimes p^*_{24} \mathcal{K}_s$ on $\mathfrak{U}^\flat \times_Y \mathfrak{U} \times_Y V^\vee \times_Y V^\vee$. Moreover it takes $\mathcal{O}_Y \boxtimes \mathcal{O}_{\mathfrak{U}}$ to $\mathcal{O}_{V^\vee} \boxtimes \mathcal{O}_Y$, as they correspond to $\Delta_* \mathcal{O}_{\mathfrak{U}}, \Delta_* \mathcal{O}_{V^\vee}$ under the automorphism (2.1.10). Also we have

$$\mathbf{RHom}^*_{\mathrm{MF}^{\mathbb{C}^*}_{\mathrm{coh}}(V^\vee \times_Y V^\vee, -w \boxplus w)}(\Delta_* \mathcal{O}_{V^\vee}, \Delta_* \mathcal{O}_{V^\vee})$$

$$\cong \mathbf{RHom}_{\mathrm{MF}^{\mathbb{C}^*}_{\mathrm{coh}}(V^\vee \times_Y V^\vee, 0 \boxplus w)}(\mathcal{O}_{V^\vee} \boxtimes \mathcal{O}_Y, \mathcal{O}_{V^\vee} \boxtimes \mathcal{O}_Y)$$

$$\cong \mathbf{RHom}_{\mathrm{MF}^{\mathbb{C}^*}_{\mathrm{coh}}(V^\vee, 0)}(\mathcal{O}_{V^\vee}, \mathcal{O}_{V^\vee}) \overset{\mathbf{L}}{\otimes}_{\mathcal{O}_Y} \mathbf{RHom}_{\mathrm{MF}^{\mathbb{C}^*}_{\mathrm{coh}}(V^\vee, w)}(\mathcal{O}_Y, \mathcal{O}_Y)$$

and the map Φ^{HH} is obtained by taking the cohomologies of the following map:

$$\Phi^\flat \overset{\mathbf{L}}{\otimes} \Phi : \mathbf{RHom}_{\mathfrak{U}^\flat}(\mathcal{O}_Y, \mathcal{O}_Y) \overset{\mathbf{L}}{\otimes}_{\mathcal{O}_Y} \mathbf{RHom}_{\mathfrak{U}}(\mathcal{O}_{\mathfrak{U}}, \mathcal{O}_{\mathfrak{U}})$$

$$\to \mathbf{RHom}_{\mathrm{MF}^{\mathbb{C}^*}_{\mathrm{coh}}(V^\vee, 0)}(\mathcal{O}_{V^\vee}, \mathcal{O}_{V^\vee}) \overset{\mathbf{L}}{\otimes}_{\mathcal{O}_Y} \mathbf{RHom}_{\mathrm{MF}^{\mathbb{C}^*}_{\mathrm{coh}}(V^\vee, w)}(\mathcal{O}_Y, \mathcal{O}_Y).$$

Here $\Phi^\flat$ is given in Remark 2.3.6. Note that the first factor of the right hand side is

$$\mathbf{RHom}_{\mathrm{MF}^{\mathbb{C}^*}_{\mathrm{coh}}(V^\vee, 0)}(\mathcal{O}_{V^\vee}, \mathcal{O}_{V^\vee}) = S(V[-2])$$

and it maps to the natural transforms $\mathrm{id} \to \mathrm{id}[*]$ on $\mathrm{MF}^{\mathbb{C}^*}_{\mathrm{coh}}(V^\vee, w)$ by the multiplication via the $\mathcal{O}_{V^\vee}$-module structure on factorizations of w. Therefore together with Remark 2.3.6, the above description of Φ^{HH} immediately implies the commutativity of the diagram (2.3.10). $\qquad\square$

The following lemma gives an ind-version of Theorem 2.3.3 and Proposition 2.3.9:

Lemma 2.3.10 *The equivalences (2.3.3), (2.3.8) extend to equivalences*

$$\Phi^{\mathrm{ind}} \colon \operatorname{Ind} D^b_{\mathrm{coh}}([\mathfrak{U}/G]) \xrightarrow{\sim} \mathrm{MF}^{\mathbb{C}^*}_{\mathrm{qcoh}}([V^\vee/G], w),$$

$$\Phi^{\mathrm{ind}} \colon \operatorname{Ind} \mathcal{C}_{\mathcal{Z}} \xrightarrow{\sim} \mathrm{MF}^{\mathbb{C}^*}_{\mathrm{qcoh}}([V^\vee/G], w)_{\mathcal{Z}}.$$

Proof It is enough to show the second equivalence. By taking ind-completion of the equivalence (2.3.8), we have the equivalence

$$\Phi^{\mathrm{ind}} \colon \operatorname{Ind}(\mathcal{C}_{\mathcal{Z}}) \xrightarrow{\sim} \operatorname{Ind}(\mathrm{MF}^{\mathbb{C}^*}_{\mathrm{coh}}([V^\vee/G], w)_{\mathcal{Z}}).$$

The left hand side is $\operatorname{Ind} \mathcal{C}_{\mathcal{Z}}$ by Lemma 2.3.2. As for the right hand side, it is proved in [9, Proposition 3.15] that $\mathrm{MF}^{\mathbb{C}^*}_{\mathrm{qcoh}}([V^\vee/G], w)$ is generated by the subcategory of compact objects $\mathrm{MF}^{\mathbb{C}^*}_{\mathrm{coh}}([V^\vee/G], w)$. The same argument applies to the fixed support case, i.e. $\mathrm{MF}^{\mathbb{C}^*}_{\mathrm{qcoh}}([V^\vee/G], w)_{\mathcal{Z}}$ is generated by the subcategory of compact objects $\mathrm{MF}^{\mathbb{C}^*}_{\mathrm{coh}}([V^\vee/G], w)_{\mathcal{Z}}$. Indeed the essential point in the proof of *loc. cit.* is the fact that any G-equivariant quasi-coherent sheaf is a union of G-equivariant coherent sheaves. This obviously holds for fixed supported case: any G-equivariant quasi-coherent sheaf supported on Z is a union of G-equivariant coherent sheaves supported on Z. Therefore the naturally defined functor

$$\operatorname{Ind}(\mathrm{MF}^{\mathbb{C}^*}_{\mathrm{coh}}([V^\vee/G], w)_{\mathcal{Z}}) \to \mathrm{MF}^{\mathbb{C}^*}_{\mathrm{qcoh}}([V^\vee/G], w)_{\mathcal{Z}}$$

is an equivalence by [67, Corollary 6.3.5], so we have the desired equivalence. □

Remark 2.3.11 By [41, Proposition 1.2.4], we have the equivalence

$$\operatorname{Ind} D^b_{\mathrm{coh}}([\mathfrak{U}/G])^+ \xrightarrow{\sim} D_{\mathrm{qcoh}}([\mathfrak{U}/G])^+$$

where the subscript $+$ indicates bounded below cohomologies. Under the above equivalence, the equivalence Φ^{ind} restricts to the functor

$$\Phi \colon D_{\mathrm{qcoh}}([\mathfrak{U}/G])^+ \xrightarrow{\sim} \mathrm{MF}^{\mathbb{C}^*}_{\mathrm{qcoh}}([V^\vee/G], w)^+$$

where the RHS is the subcategory of $\mathrm{MF}^{\mathbb{C}^*}_{\mathrm{qcoh}}([V^\vee/G], w)^+$ with bounded below $\mathbb{C}^*$-weights. The above equivalence is given by $\mathcal{K}_s \otimes_{\mathcal{O}_\mathfrak{U}} (-)$ as in Theorem 2.3.3.

2.4 Some Functorial Properties of Koszul Duality Equivalence

Here we show some functorial properties of Koszul duality equivalence in Theorem 2.3.3. Below we will use some terminology of functors in Sect. 3.1.2.

2.4.1 Functoriality Under Push-Forward

We define the morphism of equivariant tuples as follows:

Definition 2.4.1 Let G_i for $i = 1, 2$ be affine algebraic groups with an algebraic group homomorphism $\phi \colon G_1 \to G_2$, and (Y_i, V_i, s_i) be G_i-equivariant tuples. A *morphism of equivariant tuples* is a commutative diagram

$$
\begin{array}{ccc}
V_1 & \xrightarrow{g} & V_2 \\
{\scriptstyle s_1}\uparrow\downarrow & & \downarrow\uparrow{\scriptstyle s_2} \\
Y_1 & \xrightarrow{f} & Y_2,
\end{array}
\qquad (2.4.1)
$$

satisfying the followings:

 (i) The bottom morphism $f \colon Y_1 \to Y_2$ is equivariant with respect to $\phi \colon G_1 \to G_2$.

 (ii) The top morphism g is a composition $V_1 \xrightarrow{g'} f^*V_2 \xrightarrow{f} V_2$, where g' is a morphism of G_1-equivariant vector bundles on Y_1. Here the G_1-equivariant structure of f^*V_2 is induced by $\phi \colon G_1 \to G_2$.

If $G = G_1 = G_2$ and $\phi = \mathrm{id}$, we call the above diagram as a morphism of G-equivariant tuples.

The diagram (2.4.1) induces the diagram for smooth stacks $\mathcal{Y}_i = [Y_i/G_i]$, $\mathcal{V}_i = [V_i/G_i]$,

$$
\begin{array}{ccc}
\mathcal{V}_1 & \xrightarrow{g} & \mathcal{V}_2 \\
{\scriptstyle s_1}\uparrow\downarrow & & \downarrow\uparrow{\scriptstyle s_2} \\
\mathcal{Y}_1 & \xrightarrow{f} & \mathcal{Y}_2,
\end{array}
\qquad (2.4.2)
$$

which also induces the morphism of derived stacks

$$\mathbf{f} \colon [\mathfrak{U}_1/G_1] \to [\mathfrak{U}_2/G_2], \quad \mathfrak{U}_i := \operatorname{Spec} \mathcal{R}(V_i \to Y_i, s_i).$$

Then we have the push-forward functor (see [38, Section 3.6])

$$\mathbf{f}_*^{\mathrm{ind}} \colon \operatorname{Ind} D_{\mathrm{coh}}^b([\mathfrak{U}_1/G_1]) \to \operatorname{Ind} D_{\mathrm{coh}}^b([\mathfrak{U}_2/G_2]).$$

Let $w_i \colon V_i^\vee \to \mathbb{C}$ be given as in (2.1.3) defined from (Y_i, V_i, s_i). The diagram (2.4.1) induces the following diagram

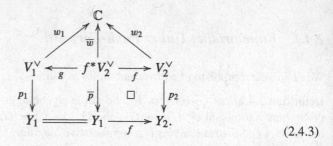

$$(2.4.3)$$

Here $\overline{w}$ is determined by

$$\overline{w} = f^* s_2 \in \Gamma(Y_1, f^* V_2) \subset \Gamma(Y_1, S(f^* V_2)).$$

The commutativity of (2.4.1) implies that the diagram (2.4.3) is also commutative. The following lemma will be used later.

Lemma 2.4.2 *In the notation of the diagram (2.4.3), we have*

$$g^{-1}(\mathrm{Crit}(w_1)) \cap f^{-1}(\mathrm{Crit}(w_2)) = t_0(\Omega_{\mathfrak{U}_2}[-1] \times_{\mathfrak{U}_2} \mathfrak{U}_1). \tag{2.4.4}$$

Proof The left hand side of (2.4.4) is defined by the ideal of $S_{\mathcal{O}_{Y_1}}(f^* V_2)$ generated by the image of the following map

$$(s_1 \oplus f^* s_2) \oplus (g \circ ds_1 \oplus f^* ds_2) \colon (V_1^\vee \oplus f^* V_2^\vee)$$
$$\oplus (T_{Y_1} \oplus f^* T_{Y_2}) \to \mathcal{O}_{Y_1} \oplus f^* V_2. \tag{2.4.5}$$

Since there exist commutative diagrams

$$f^* V_2^\vee \xrightarrow{\;\;f^* s_2\;\;} V_1^\vee \xrightarrow{\;s_1\;} \mathcal{O}_{Y_1}, \quad T_{Y_1} \xrightarrow{\;\;g \circ ds_1\;\;} f^* T_{Y_2} \xrightarrow{\;f^* ds_2\;} f^* V_2$$

the image of the map (2.4.5) coincides with the image of the map

$$s_1 \oplus f^* ds_2 \colon V_1^\vee \oplus f^* T_{Y_2} \to \mathcal{O}_{Y_1} \oplus f^* V_2. \tag{2.4.6}$$

On the other hand, we have

$$\Omega_{\mathfrak{U}_2}[-1] \times_{\mathfrak{U}_2} \mathfrak{U}_1 = \mathrm{Spec}\, S_{\mathcal{O}_{Y_1}}(V_1^\vee[1] \oplus f^*T_{Y_2}[1] \oplus f^*V_2)$$

and the differential on the RHS is the Koszul differential determined by the map (2.4.6). Therefore (2.4.4) holds. □

In the notation of the diagram (2.4.3), we define the following functor

$$f_* \circ g^* : \mathrm{MF}_{\mathrm{qcoh}}^{\mathbb{C}^*}([V_1^\vee/G_1], w_1) \xrightarrow{g^*} \mathrm{MF}_{\mathrm{qcoh}}^{\mathbb{C}^*}([f^*V_2^\vee/G_1], \overline{w})$$

$$\xrightarrow{f_*} \mathrm{MF}_{\mathrm{qcoh}}^{\mathbb{C}^*}([V_2^\vee/G_2], w_2). \tag{2.4.7}$$

The above functor is compatible with compositions of morphisms of derived stacks as follows. Let (Y_3, V_3, s_3) be another G_3-equivariant tuple, and suppose that we have morphisms of equivariant tuples with respect to algebraic group homomorphisms $G_1 \to G_2 \to G_3$

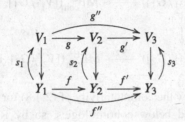

which induces the diagram of derived stacks

$$\mathbf{f}'' : [\mathfrak{U}_1/G_1] \xrightarrow{\mathbf{f}} [\mathfrak{U}_2/G_2] \xrightarrow{\mathbf{f}'} [\mathfrak{U}_3/G_3].$$

Lemma 2.4.3 *There is a natural isomorphism of functors*

$$(f'_* g'^*)(f_* g^*) \cong f''_* g''^* : \mathrm{MF}_{\mathrm{qcoh}}^{\mathbb{C}^*}([V_1^\vee/G_1], w_1) \to \mathrm{MF}_{\mathrm{qcoh}}^{\mathbb{C}^*}([V_3^\vee/G_3], w_3).$$

Proof We have the following diagram

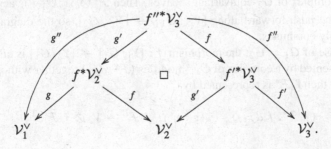

Here the middle square is a derived Cartesian. Then the lemma follows from the
derived base change.   □

We have the following lemma which describes $\mathbf{f}_*^{\mathrm{ind}}$ in the Koszul dual size.

Lemma 2.4.4 *The following diagram commutes:*

$$
\begin{array}{ccc}
\operatorname{Ind} D^b_{\mathrm{coh}}([\mathfrak{U}_1/G_1]) & \xrightarrow{\Phi_1^{\mathrm{ind}}} & \mathrm{MF}^{\mathbb{C}^*}_{\mathrm{qcoh}}([V_1^\vee/G_1], w_1) \\
{\scriptstyle \mathbf{f}_*^{\mathrm{ind}}}\Big\downarrow & & \Big\downarrow{\scriptstyle f_* \circ g^*} \\
\operatorname{Ind} D^b_{\mathrm{coh}}([\mathfrak{U}_2/G_2]) & \xrightarrow{\Phi_2^{\mathrm{ind}}} & \mathrm{MF}^{\mathbb{C}^*}_{\mathrm{qcoh}}([V_2^\vee/G_2], w_2).
\end{array}
\tag{2.4.8}
$$

*Here the horizontal arrows are equivalences in Lemma 2.3.10. In particular if the
morphism* $f \colon \mathcal{Y}_1 \to \mathcal{Y}_2$ *in the diagram (2.4.2) is proper, the above diagram restricts
to the commutative diagram*

$$
\begin{array}{ccc}
D^b_{\mathrm{coh}}([\mathfrak{U}_1/G_1]) & \xrightarrow{\Phi_1} & \mathrm{MF}^{\mathbb{C}^*}_{\mathrm{coh}}([V_1^\vee/G_1], w_1) \\
{\scriptstyle \mathbf{f}_*}\Big\downarrow & & \Big\downarrow{\scriptstyle f_* \circ g^*} \\
D^b_{\mathrm{coh}}([\mathfrak{U}_2/G_2]) & \xrightarrow{\Phi_2} & \mathrm{MF}^{\mathbb{C}^*}_{\mathrm{coh}}([V_2^\vee/G_2], w_2).
\end{array}
$$

Proof It is enough to show the commutativity of (2.4.8) for $\mathcal{F}^\bullet \in D^b_{\mathrm{coh}}([\mathfrak{U}_1/G_1])$.
Then $\mathbf{f}_*^{\mathrm{ind}}\mathcal{F}^\bullet$ has bounded below cohomologies, so by Remark 2.3.11 we can
compute $\Phi_2^{\mathrm{ind}}\mathbf{f}_*^{\mathrm{ind}}(\mathcal{F}^\bullet)$ as $\mathcal{K}_{s_2} \otimes_{\mathcal{O}_{\mathfrak{U}_2}} \mathbf{f}_*(\mathcal{F}^\bullet)$. The morphism $\mathbf{f}$ factors as

$$
\mathbf{f} \colon [\mathfrak{U}_1/G_1] \to [\mathfrak{U}_2/G_1] \to [\mathfrak{U}_2/G_2].
$$

Therefore by Lemma 2.4.3, we may assume either $G_1 = G_2$ or $(Y_1, V_1, s_1) = (Y_2, V_2, s_2)$.

In the case of $(Y_1, V_1, s_1) = (Y_2, V_2, s_2)$, the vertical arrows in (2.4.8) are push-
forwards along $[\mathfrak{U}_1/G_1] \to [\mathfrak{U}_1/G_2]$ and $[V_1^\vee/G_1] \to [V_1^\vee/G_2]$. Therefore we can
compute $\mathbf{f}_*\mathcal{F}^\bullet$ by replacing $\mathcal{F}^\bullet$ by a bounded below complex such that each term $\mathcal{F}^i$
is acyclic with respect to the push-forward along $[Y_1/G_1] \to [Y_1/G_2]$, and regard
them as a complex of G_2-equivariant sheaves. Then $\mathcal{F}^i \otimes_{\mathcal{O}_{Y_1}} \mathcal{O}_{V_1^\vee}$ is acyclic with
respect to the push-forward along $[V_1^\vee/G_1] \to [V_1^\vee/G_2]$, so the diagram (2.4.8)
tautologically commutes.

In the case of $G_1 = G_2$, the morphism $f \colon [Y_1/G_1] \to [Y_2/G_1]$ is affine. So if
$\mathcal{F}^\bullet$ is represented by a complex of $\mathcal{O}_{Y_1}$-modules $(\mathcal{F}^\bullet, d_{\mathcal{F}^\bullet})$ together with $\eta^i \colon \mathcal{F}^i \to
V_1 \otimes \mathcal{F}^{i-1}$, then $\mathbf{f}_*\mathcal{F}^\bullet$ is represented by

$$
(f_*\mathcal{F}^\bullet, f_*d_{\mathcal{F}^\bullet}), \quad f_*(g \circ \eta^i) \colon f_*\mathcal{F}^i \to V_2 \otimes f_*\mathcal{F}^{i-1}.
$$

Here $g \colon V_1 \to f^* V_2$ is a morphism in the diagram (2.4.1). Therefore $\mathcal{K}_{s_2} \otimes_{\mathcal{O}_{\mathfrak{U}_2}} \mathbf{f}_* \mathcal{F}^\bullet$ is

$$\left(\bigoplus_{i \in \mathbb{Z}} f_* \mathcal{F}^i \langle i \rangle \otimes_{\mathcal{O}_{Y_2}} \mathcal{O}_{V_2^\vee} \to \bigoplus_{i \in \mathbb{Z}} f_* \mathcal{F}^i \langle i + 1 \rangle \otimes_{\mathcal{O}_{Y_2}} \mathcal{O}_{V_2^\vee} \right) \tag{2.4.9}$$

where the morphism is a $\mathcal{O}_{V_2^\vee}$-module homomorphism determined by

$$f_* d_{\mathcal{F}}^i + f_* (g \circ \eta^i) \colon f_* \mathcal{F}^i \to f_* \mathcal{F}^{i+1} \oplus (V_2 \otimes f_* \mathcal{F}^{i-1}).$$

On the other hand, $g^*(\Phi_1(\mathcal{F}^\bullet))$ is given by

$$\left(\bigoplus_{i \in \mathbb{Z}} \mathcal{F}^i \langle i \rangle \otimes_{\mathcal{O}_{Y_1}} \mathcal{O}_{f^* V_2^\vee} \to \bigoplus_{i \in \mathbb{Z}} \mathcal{F}^i \langle i + 1 \rangle \otimes_{\mathcal{O}_{Y_1}} \mathcal{O}_{f^* V_2^\vee} \right) \tag{2.4.10}$$

where the morphism is a $\mathcal{O}_{f^* V_2^\vee}$-module homomorphism determined by

$$d_{\mathcal{F}}^i + g \circ \eta^i \colon \mathcal{F}^i \to \mathcal{F}^{i+1} \oplus (f^* V_2 \otimes_{\mathcal{O}_{Y_1}} \mathcal{F}^{i-1}).$$

Therefore by pushing forward (2.4.10) along $f \colon f^* V_2^\vee \to V_2^\vee$, we obtain the object (2.4.9). $\qquad\qquad\qquad\qquad\qquad\qquad\qquad\qquad\qquad\qquad\qquad\qquad \square$

Let (Y, V, s) be a G-equivariant tuple, and consider the diagram (2.3.1). For the zero section $0_Y \colon Y \to V^\vee$, we have $0_Y^* w = 0$. Therefore we have the functor

$$0_Y^* \colon \mathrm{MF}_{\mathrm{coh}}^{\mathbb{C}^*}([V^\vee/G], w) \to \mathrm{MF}_{\mathrm{coh}}^{\mathbb{C}^*}([Y/G], 0).$$

The following lemma is obtained from Lemma 2.4.4, which is also easy to check directly.

Lemma 2.4.5 *The following diagram commutes*

$$\begin{array}{ccc} D_{\mathrm{coh}}^b([\mathfrak{U}/G]) & \xrightarrow[\sim]{\Phi} & \mathrm{MF}_{\mathrm{coh}}^{\mathbb{C}^*}([V^\vee/G], w) \\ {\scriptstyle j_*}\big\downarrow & & \big\downarrow{\scriptstyle 0_Y^*} \\ D_{\mathrm{coh}}^b([Y/G]) & \xrightarrow[\sim]{} & \mathrm{MF}_{\mathrm{coh}}^{\mathbb{C}^*}([Y/G], 0). \end{array} \tag{2.4.11}$$

Here the bottom arrow is a tautological equivalence (see Remark 2.3.7).

Proof The lemma follows by applying Lemma 2.4.4 to the following morphism of G-equivariant tuples

$$s\left(\begin{array}{ccc} V & \xrightarrow{\;p\;} & Y \\ {\scriptstyle p}\big\downarrow & {\scriptstyle \mathrm{id}}\big\uparrow \\ Y & \xrightarrow{\;\mathrm{id}\;} & Y. \end{array}\right)\,\mathrm{id}$$

Alternatively from the construction of Φ and noting that $\mathcal{O}_Y \otimes_{\mathcal{O}_{V^\vee}} \mathcal{K}_s = \mathcal{O}_{\mathfrak{U}}$, we have

$$0_Y^* \Phi(\mathcal{F}) = \mathcal{O}_Y \otimes_{\mathcal{O}_{V^\vee}} \mathcal{K}_s \otimes_{\mathcal{O}_{\mathfrak{U}}} \mathcal{F} \cong \mathcal{F}.$$

$\square$

2.4.2   *Functoriality Under Pull-Back*

Let us consider a morphism of equivariant tuples in Definition 2.3.1. Suppose that the morphism $f \colon \mathcal{Y}_1 \to \mathcal{Y}_2$ in the diagram (2.4.2) is a proper morphism of smooth stacks, so in particular $\mathbf{f} \colon [\mathfrak{U}_1/G_1] \to [\mathfrak{U}_2/G_2]$ is proper. Then we have the continuous right adjoint of $\mathbf{f}_*^{\mathrm{ind}}$ (see [41, Section 10.1])

$$\mathbf{f}^! \colon \mathrm{Ind}\, D_{\mathrm{coh}}^b([\mathfrak{U}_2/G_2]) \to \mathrm{Ind}\, D_{\mathrm{coh}}^b([\mathfrak{U}_1/G_1]).$$

On the other hand, we also have the functor

$$g_* \circ f^! \colon \mathrm{MF}_{\mathrm{qcoh}}^{\mathbb{C}^*}([V_2^\vee/G_2], w_2) \xrightarrow{f^!} \mathrm{MF}_{\mathrm{qcoh}}^{\mathbb{C}^*}([f^* V_2^\vee/G_1], \overline{w})$$

$$\xrightarrow{g_*} \mathrm{MF}_{\mathrm{qcoh}}^{\mathbb{C}^*}([V_1^\vee/G_1], w_1)$$

which is a right adjoint of the functor (2.4.7).

Lemma 2.4.6 *Suppose that $f \colon \mathcal{Y}_1 \to \mathcal{Y}_2$ is proper. Then the following diagram commutes:*

$$\begin{array}{ccc} \mathrm{Ind}\, D_{\mathrm{coh}}^b([\mathfrak{U}_2/G_2]) & \xrightarrow{\;\Phi_2^{\mathrm{ind}}\;} & \mathrm{MF}_{\mathrm{qcoh}}^{\mathbb{C}^*}([V_2^\vee/G_2], w_2) \\ {\scriptstyle \mathbf{f}^!}\big\downarrow & & \big\downarrow{\scriptstyle g_* \circ f^!} \\ \mathrm{Ind}\, D_{\mathrm{coh}}^b([\mathfrak{U}_1/G_1]) & \xrightarrow{\;\Phi_1^{\mathrm{ind}}\;} & \mathrm{MF}_{\mathrm{qcoh}}^{\mathbb{C}^*}([V_1^\vee/G_1], w_1). \end{array}$$

Proof The lemma follows from Lemma 2.4.4 by taking the right adjoints of the vertical arrows in the diagram (2.4.8). □

Suppose that $\mathbf{f}\colon [\mathfrak{U}_1/G_1] \to [\mathfrak{U}_2/G_2]$ is quasi-smooth. Then we have the functor (see [42, Section 3.1])

$$\mathbf{f}^*\colon D^b_{\mathrm{coh}}([\mathfrak{U}_2/G_2]) \to D^b_{\mathrm{coh}}([\mathfrak{U}_1/G_1]).$$

If furthermore $g\colon f^*\mathcal{V}_2^\vee \to \mathcal{V}_1^\vee$ in the diagram (2.4.3) is proper, then we have the functor

$$g_! \circ f^*\colon \mathrm{MF}^{\mathbb{C}^*}_{\mathrm{coh}}([V_2^\vee/G_2], w_2) \xrightarrow{f^*} \mathrm{MF}^{\mathbb{C}^*}_{\mathrm{qcoh}}([f^*V_2^\vee/G_1], \overline{w})$$
$$\xrightarrow{g_!} \mathrm{MF}^{\mathbb{C}^*}_{\mathrm{qcoh}}([V_1^\vee/G_1], w_1)$$

which is a left adjoint of the functor (2.4.7).

Lemma 2.4.7 *Suppose that* $\mathbf{f}\colon [\mathfrak{U}_1/G_1] \to [\mathfrak{U}_2/G_2]$ *is quasi-smooth and the morphism* $g\colon f^*\mathcal{V}_2^\vee \to \mathcal{V}_1^\vee$ *is proper. Then the following diagram commutes:*

$$
\begin{array}{ccc}
D^b_{\mathrm{coh}}([\mathfrak{U}_2/G_2]) & \xrightarrow{\Phi_2} & \mathrm{MF}^{\mathbb{C}^*}_{\mathrm{coh}}([V_2^\vee/G_2], w_2) \\
{\scriptstyle \mathbf{f}^*}\downarrow & & \downarrow{\scriptstyle g_! \circ f^*} \\
D^b_{\mathrm{coh}}([\mathfrak{U}_1/G_1]) & \xrightarrow{\Phi_1} & \mathrm{MF}^{\mathbb{C}^*}_{\mathrm{coh}}([V_1^\vee/G_1], w_1).
\end{array}
$$

Proof The ind-completion of $\mathbf{f}^*$

$$\mathbf{f}^{\mathrm{ind}*}\colon \mathrm{Ind}\, D^b_{\mathrm{coh}}([\mathfrak{U}_2/G_2]) \to \mathrm{Ind}\, D^b_{\mathrm{coh}}([\mathfrak{U}_1/G_1])$$

is a left adjoint of $\mathbf{f}^{\mathrm{ind}}_*$ by [42, Proposition 3.1.6]. Therefore the lemma follows by taking left adjoints of the vertical arrows in the diagram (2.4.8), and restrict it to $D^b_{\mathrm{coh}}([\mathfrak{U}_2/G_2])$. □

Chapter 3
Categorical DT Theory for Local Surfaces

In this chapter, for a quasi-smooth and QCA derived stack $\mathfrak{M}$ and a conical closed substack $\mathcal{Z} \subset \mathcal{N} := t_0(\Omega_{\mathfrak{M}}[-1])$, we introduce $\mathbb{C}^*$-equivariant DT category for the complement $\mathcal{N}^{ss} := \mathcal{N} \setminus \mathcal{Z}$. A basic construction is to take the Verdier quotient

$$\mathcal{DT}^{\mathbb{C}^*}(\mathcal{N}^{ss}) := D^b_{\mathrm{coh}}(\mathfrak{M})/\mathcal{C}_{\mathcal{Z}} \tag{3.0.1}$$

where $\mathcal{C}_{\mathcal{Z}} \subset D^b_{\mathrm{coh}}(\mathfrak{M})$ is the subcategory of objects whose singular supports are contained in $\mathcal{Z}$. We call the above quotient category as *DT category*. The above construction is a global analogue of the quotient category considered in Proposition 2.3.9. So in particular if $\mathfrak{M}$ is a global derived zero locus then $\mathcal{DT}^{\mathbb{C}^*}(\mathcal{N}^{ss})$ is equivalent to the derived category of $\mathbb{C}^*$-equivariant factorizations. We will study the properties of the above quotient category in detail.

In the presence of $\mathbb{C}^*$-automorphisms in the stabilizer groups of $\mathfrak{M}$, we further introduce the variants of DT categories either by taking the $\mathbb{C}^*$-rigidifications or fix the $\mathbb{C}^*$-weight λ. The λ-twist version is an analogue of the category of twisted sheaves, twisted by a Brauer class, studied in [84]. We will also define their dg-enhancements. The relevant notation is summarized in Table 3.1.

We will also give a variant of the above construction, defined as limits of quotient categories

$$\widehat{\mathcal{DT}}^{\mathbb{C}^*}(\mathcal{N}^{ss}) := \lim_{\mathfrak{U} \xrightarrow{\alpha} \mathfrak{M}} \left(D^b_{\mathrm{coh}}(\mathfrak{U})/\mathcal{C}_{\alpha^*\mathcal{Z}} \right). \tag{3.0.2}$$

The above construction is a more direct gluing of derived factorization categories. The $\mathbb{C}^*$-rigidified version, the λ-twisted version and dg-enhancements are similarly defined, and the relevant notation is summarized in Table 3.2.

There is a natural functor from the former category (3.0.1) to the latter category (3.0.2), but in general it is not clear whether this is an equivalence or not. It will turn out that, in many cases we are interested in, both constructions give

© The Author(s), under exclusive license to Springer Nature Switzerland AG 2024
Y. Toda, *Categorical Donaldson-Thomas Theory for Local Surfaces*, Lecture Notes in Mathematics 2350, https://doi.org/10.1007/978-3-031-61705-8_3

Table 3.1 Notation of $\mathbb{C}^*$-equivariant DT categories

	$\mathbb{C}^*$-equivariant	$\mathbb{C}^*$-rigidified	λ-Twist
$\triangle$ category	$\mathcal{DT}^{\mathbb{C}^*}(\mathcal{N}^{\mathrm{ss}})$	$\mathcal{DT}^{\mathbb{C}^*}((\mathcal{N}^{\mathrm{ss}})^{\mathbb{C}^*\text{-}rig})$	$\mathcal{DT}^{\mathbb{C}^*}(\mathcal{N}^{\mathrm{ss}})_\lambda$
dg category	$\mathcal{DT}_{\mathrm{dg}}^{\mathbb{C}^*}(\mathcal{N}^{\mathrm{ss}})$	$\mathcal{DT}_{\mathrm{dg}}^{\mathbb{C}^*}((\mathcal{N}^{\mathrm{ss}})^{\mathbb{C}^*\text{-rig}})$	$\mathcal{DT}_{\mathrm{dg}}^{\mathbb{C}^*}(\mathcal{N}^{\mathrm{ss}})_\lambda$

Table 3.2 Notation of $\mathbb{C}^*$-equivariant categorical $\widehat{\mathrm{DT}}$ theories

	$\mathbb{C}^*$-equivariant	$\mathbb{C}^*$-rigidified	λ-Twist
$\triangle$ category	$\widehat{\mathcal{DT}}^{\mathbb{C}^*}(\mathcal{N}^{\mathrm{ss}})$	$\widehat{\mathcal{DT}}^{\mathbb{C}^*}((\mathcal{N}^{\mathrm{ss}})^{\mathbb{C}^*\text{-rig}})$	$\widehat{\mathcal{DT}}^{\mathbb{C}^*}(\mathcal{N}^{\mathrm{ss}})_\lambda$
dg category	$\widehat{\mathcal{DT}}_{\mathrm{dg}}^{\mathbb{C}^*}(\mathcal{N}^{\mathrm{ss}})$	$\widehat{\mathcal{DT}}_{\mathrm{dg}}^{\mathbb{C}^*}((\mathcal{N}^{\mathrm{ss}})^{\mathbb{C}^*\text{-rig}})$	$\widehat{\mathcal{DT}}_{\mathrm{dg}}^{\mathbb{C}^*}(\mathcal{N}^{\mathrm{ss}})_\lambda$

equivalent categories. Both of constructions will play important roles in this book. The former category is useful to discuss categorified Hall products for DT categories in Chap. 7, while the latter category is useful to glue window subcategories in Chap. 6. The comparisons of both constructions will be discussed in Sect. 8.1. In this chapter, we mainly focus on the former construction and develop its general theory.

It is a natural question to ask whether our DT category recovers the original DT invariant. If $\mathcal{N}^{\mathrm{ss}}$ is a scheme, then the associated DT invariant is defined by taking the integration of the Behrend function which is a point-wise Euler characteristic of the vanishing cycles. As Efimov proved in [39], there is a close relationship between vanishing cycles and derived categories of factorizations: the periodic cyclic homology of the derived category of factorizations is isomorphic to the hyper-cohomology of the perverse sheaf of vanishing cycles. Based on this result, we will give some de-categorification result for DT categories, proving that the Euler characteristic of periodic cyclic homology of the DT category recovers the DT invariant.

Based on the above construction, we will introduce DT category for local surfaces. For a smooth projective surface S, the local surface is the total space of its canonical line bundle $X = \mathrm{Tot}(\omega_S)$, which is a non-compact CY 3-fold. A key fact is that the moduli stack of compactly supported coherent sheaves on X is isomorphic to the (-1)-shifted cotangent over the derived moduli stack $\mathfrak{M}_S$ of coherent sheaves on S. If we impose stability condition, the unstable locus is a conical closed substack in $t_0(\Omega_{\mathfrak{M}_S}[-1])$. For a stability condition σ, we define the DT category for the moduli space $M_X^\sigma(v)$ of σ-stable sheaves on X with numerical class v based on the construction (3.0.1), and denoted as

$$\mathcal{DT}^{\mathbb{C}^*}(M_X^\sigma(v)).$$

The organization of this chapter is as follows. In Sect. 3.1, we recall several notions of derived stacks, quasi-smoothness, (quasi, ind)coherent sheaves on them, (-1)-shifted cotangent, and their singular supports. In Sect. 3.2, we introduce DT categories for (-1)-shifted cotangents and prove their fundamental properties. In Sect. 3.3, we prove some de-categorification result of our DT category. In Sect. 3.4, we introduce DT category for local surfaces.

3.1 Some Background on Derived Stacks

3.1.1 Quasi-Smooth Derived Stack

Below, we denote by $\mathfrak{M}$ a derived Artin stack over $\mathbb{C}$. This means that $\mathfrak{M}$ is a contravariant ∞-functor from the ∞-category of affine derived schemes over $\mathbb{C}$ to the ∞-category of simplicial sets

$$\mathfrak{M}\colon dAff^{op} \to SSets$$

satisfying some conditions (see [145, Section 3.2] for details). Here $dAff^{op}$ is defined to be the ∞-category of commutative simplicial $\mathbb{C}$-algebras, which is equivalent to the ∞-category of commutative differential graded $\mathbb{C}$-algebras with non-positive degrees (called $cdga$). The classical truncation of $\mathfrak{M}$ is denoted by

$$\mathcal{M} := t_0(\mathfrak{M})\colon Aff^{op} \hookrightarrow dAff^{op} \to SSets$$

where the first arrow is a natural functor from the category of affine schemes to affine derived schemes.

Following [145], we define the dg-category of quasi-coherent sheaves on $\mathfrak{M}$ as follows. For an affine derived scheme $\mathfrak{U} = \operatorname{Spec} R$ for a cdga R, let $\operatorname{Mod}(R)$ be the dg-category of differential graded R-modules which is equipped with a projective model category structure. Then $L_{\mathrm{qcoh}}(\mathfrak{U})$ is defined to be the dg-categorical localization of $\operatorname{Mod}(R)$ by quasi-isomorphisms in the sense of [144, Section 2.4]. The category $L_{\mathrm{qcoh}}(\mathfrak{U})$ is a dg-category, which is weakly equivalent to the dg-subcategory of $\operatorname{Mod}(R)$ consisting of cofibrant R-modules, and its homotopy category is equivalent to the derived category $D_{\mathrm{qcoh}}(\mathfrak{U})$ of dg-modules over R. Then the dg-category $L_{\mathrm{qcoh}}(\mathfrak{M})$ is defined to be the limit in the ∞-category of dg-categories

$$L_{\mathrm{qcoh}}(\mathfrak{M}) := \lim_{\mathfrak{U} \to \mathfrak{M}} L_{\mathrm{qcoh}}(\mathfrak{U}). \tag{3.1.1}$$

The limit is taken for the ∞-category of smooth morphisms $\alpha\colon \mathfrak{U} \to \mathfrak{M}$ for affine derived schemes $\mathfrak{U}$ with 1-morphisms given by commutative diagrams

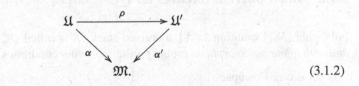

$$\tag{3.1.2}$$

Here ρ is a 0-representable smooth morphism, $\alpha' \circ \rho$ is equivalent to α, and we assign the pull-back $\rho^*\colon L_{\mathrm{qcoh}}(\mathfrak{U}') \to L_{\mathrm{qcoh}}(\mathfrak{U})$ with the above diagram. Roughly speaking an object of $L_{\mathrm{qcoh}}(\mathfrak{M})$ is a collection of objects $\mathcal{E}_{\mathfrak{U}} \in L_{\mathrm{qcoh}}(\mathfrak{U})$ with equivalences $\rho^* \mathcal{E}_\mathfrak{U} \sim \mathcal{E}_{\mathfrak{U}'}$ for each diagram (3.1.2) satisfying homotopy coherence conditions. The homotopy category of $L_{\mathrm{qcoh}}(\mathfrak{M})$ is denoted by $D_{\mathrm{qcoh}}(\mathfrak{M})$, which is a triangulated category. We have the dg and triangulated subcategories

$$L_{\mathrm{coh}}(\mathfrak{M}) \subset L_{\mathrm{qcoh}}(\mathfrak{M}), \quad D^b_{\mathrm{coh}}(\mathfrak{M}) \subset D_{\mathrm{qcoh}}(\mathfrak{M})$$

consisting of objects which have bounded coherent cohomologies. We note that there is a bounded t-structure on $D^b_{\mathrm{coh}}(\mathfrak{M})$ whose heart coincides with $\mathrm{Coh}(\mathcal{M})$. In the convention of Sect. 1.6.2, we may write $D_{\mathrm{qcoh}}(\mathfrak{M})$, $D^b_{\mathrm{coh}}(\mathfrak{M})$ as

$$D_{\mathrm{qcoh}}(\mathfrak{M}) = \lim_{\mathfrak{U} \to \mathfrak{M}} D_{\mathrm{qcoh}}(\mathfrak{U}), \quad D^b_{\mathrm{coh}}(\mathfrak{M}) = \lim_{\mathfrak{U} \to \mathfrak{M}} D^b_{\mathrm{coh}}(\mathfrak{U}).$$

A morphism of derived stacks $f \colon \mathfrak{M} \to \mathfrak{N}$ is called *quasi-smooth* if $\mathbb{L}_f$ is perfect such that for any point $x \to \mathcal{M}$ the restriction $\mathbb{L}_f|_x$ is of cohomological amplitude $[-1, 1]$. Here $\mathbb{L}_f$ is the f-relative cotangent complex. A derived stack $\mathfrak{M}$ over $\mathbb{C}$ is called *quasi-smooth* if $\mathfrak{M} \to \mathrm{Spec}\,\mathbb{C}$ is quasi-smooth. In this case, the cotangent complex $\mathbb{L}_\mathfrak{M} \in D^b_{\mathrm{coh}}(\mathfrak{M})$ of $\mathfrak{M}$ is perfect with cohomological amplitude $[-1, 1]$. By [12, Theorem 2.8], the quasi-smoothness of $\mathfrak{M}$ is equivalent to that $\mathfrak{M}$ is a 1-stack, and any point of $\mathfrak{M}$ lies in the image of a 0-representable smooth morphism

$$\alpha \colon \mathfrak{U} \to \mathfrak{M} \tag{3.1.3}$$

where $\mathfrak{U}$ is an affine derived scheme obtained as a derived zero locus as in (2.1.1). In this case, we have

$$D^b_{\mathrm{coh}}(\mathfrak{M}) = \lim_{\mathfrak{U} \xrightarrow{\alpha} \mathfrak{M}} D^b_{\mathrm{coh}}(\mathfrak{U})$$

where the limit is taken for the ∞-category $\mathcal{I}$ of smooth morphisms (3.1.3) from $\mathfrak{U}$ of the form (2.1.1) with 1-morphisms given by (3.1.2). In this book when we write $\lim_{\mathfrak{U} \xrightarrow{\alpha} \mathfrak{M}}(-)$ for a quasi-smooth $\mathfrak{M}$, the limit is always taken for the ∞-category $\mathcal{I}$ as above.

3.1.2 Ind-Coherent Sheaves on QCA Stacks

Following [38, Definition 1.1.8], a derived stack $\mathfrak{M}$ is called *QCA (quasi-compact and with affine automorphism groups)* if the following conditions hold:

(i) $\mathfrak{M}$ is quasi-compact;
(ii) The automorphism groups of its geometric points are affine;
(iii) The classical inertia stack $I_\mathcal{M} := \Delta \times_{\mathcal{M} \times \mathcal{M}} \Delta$ is of finite presentation over $\mathcal{M}$.

Let $\mathfrak{M}$ be a quasi-smooth derived stack. We denote by $\operatorname{Ind} D_{\mathrm{coh}}^b(\mathfrak{M})$ the category of its ind-coherent sheaves (see [41])

$$\operatorname{Ind} D_{\mathrm{coh}}^b(\mathfrak{M}) := \lim_{\mathfrak{U} \xrightarrow{\alpha} \mathfrak{M}} \operatorname{Ind} D_{\mathrm{coh}}^b(\mathfrak{U}), \qquad (3.1.4)$$

where α is a smooth morphism from an affine derived scheme $\mathfrak{U}$ of the form (2.1.1). The QCA condition will be useful since we have the following theorem:

Theorem 3.1.1 ([38, Theorem 3.3.5]) *If $\mathfrak{M}$ is QCA, then $\operatorname{Ind} D_{\mathrm{coh}}^b(\mathfrak{M})$ is compactly generated with $(\operatorname{Ind} D_{\mathrm{coh}}^b(\mathfrak{M}))^{\mathrm{cp}} = D_{\mathrm{coh}}^b(\mathfrak{M})$. In particular, we have*

$$\operatorname{Ind} D_{\mathrm{coh}}^b(\mathfrak{M}) = \operatorname{Ind}(D_{\mathrm{coh}}^b(\mathfrak{M})). \qquad (3.1.5)$$

Let $f : \mathfrak{M}_1 \to \mathfrak{M}_2$ be a morphism between quasi-smooth and QCA derived stacks $\mathfrak{M}_i$. Then we have the following pair of functors (see [38, Section 3.6], [41, Section 10.1])

$$\operatorname{Ind} D_{\mathrm{coh}}^b(\mathfrak{M}_1) \underset{f^!}{\overset{f_*^{\mathrm{ind}}}{\rightleftarrows}} \operatorname{Ind} D_{\mathrm{coh}}^b(\mathfrak{M}_2) .$$

If f is proper, then $f_*^{\mathrm{ind}} \dashv f^!$ and f_*^{ind} restricts to the functor (see [109, Section 4.2])

$$f_* : D_{\mathrm{coh}}^b(\mathfrak{M}_1) \to D_{\mathrm{coh}}^b(\mathfrak{M}_2). \qquad (3.1.6)$$

If f is quasi-smooth, we also have the pull-back functor (see [109, Section 4.2])

$$f^* : D_{\mathrm{coh}}^b(\mathfrak{M}_2) \to D_{\mathrm{coh}}^b(\mathfrak{M}_1). \qquad (3.1.7)$$

Its ind-completion $f^{\mathrm{ind}*}$ fits into the adjoint pair (see [42, Proposition 3.1.6], [38, Section 3.7.7])

$$\operatorname{Ind} D_{\mathrm{coh}}^b(\mathfrak{M}_1) \underset{f^{\mathrm{ind}*}}{\overset{f_*^{\mathrm{ind}}}{\rightleftarrows}} \operatorname{Ind} D_{\mathrm{coh}}^b(\mathfrak{M}_2) ,$$

such that $f^{\mathrm{ind}*} \dashv f_*^{\mathrm{ind}}$.

3.1.3 (−1)-Shifted Cotangent Derived Stack

Let $\mathfrak{M}$ be a quasi-smooth derived stack. We denote by $\Omega_{\mathfrak{M}}[-1]$ the (−1)-shifted cotangent derived stack of $\mathfrak{M}$

$$p : \Omega_{\mathfrak{M}}[-1] := \operatorname{Spec}_{\mathfrak{M}} S(\mathbb{T}_{\mathfrak{M}}[1]) \to \mathfrak{M}.$$

Here $\mathbb{T}_{\mathfrak{M}} \in D^b_{\text{coh}}(\mathfrak{M})$ is the tangent complex of $\mathfrak{M}$, which is dual to the cotangent complex $\mathbb{L}_{\mathfrak{M}}$ of $\mathfrak{M}$. The derived stack $\Omega_{\mathfrak{M}}[-1]$ admits a natural (-1)-shifted symplectic structure [28, 104], which induces the d-critical structure [63] on its classical truncation $\mathcal{N}$

$$p_0 : \mathcal{N} := t_0(\Omega_{\mathfrak{M}}[-1]) \to \mathcal{M}. \qquad (3.1.8)$$

The stack $\mathcal{N}$ is also described in terms of the dual obstruction cone introduced in [61]. Note that the quasi-smoothness of $\mathfrak{M}$ implies that the natural map of cotangent complexes

$$\mathcal{E}^\bullet := \mathbb{L}_{\mathfrak{M}}|_{\mathcal{M}} \to \tau_{\geq -1}\mathbb{L}_{\mathcal{M}} \qquad (3.1.9)$$

is a perfect obstruction theory on $\mathcal{M}$ [15]. Then we have

$$\mathcal{N} = \text{Obs}^*(\mathcal{E}^\bullet) := \text{Spec}_{\mathcal{M}} S(\mathcal{H}^1(\mathcal{E}^{\bullet\vee})).$$

The cone $\text{Obs}^*(\mathcal{E}^\bullet)$ over $\mathcal{M}$ is called the dual obstruction cone associated with the perfect obstruction theory (3.1.9).

Let $\mathfrak{M}_1$, $\mathfrak{M}_2$ be quasi-smooth derived stacks with truncations $\mathcal{M}_i = t_0(\mathfrak{M}_i)$. Let $f : \mathfrak{M}_1 \to \mathfrak{M}_2$ be a morphism. Then the morphism $f^*\mathbb{L}_{\mathfrak{M}_2} \to \mathbb{L}_{\mathfrak{M}_1}$ induces the diagram

$$
\begin{array}{ccccc}
t_0(\Omega_{\mathfrak{M}_1}[-1]) & \xleftarrow{\ f^\diamond\ } & t_0(\Omega_{\mathfrak{M}_2}[-1] \times_{\mathfrak{M}_2} \mathfrak{M}_1) & \xrightarrow{\ f^\bullet\ } & t_0(\Omega_{\mathfrak{M}_2}[-1]) \\
\downarrow & & \downarrow & \square & \downarrow \\
\mathcal{M}_1 & = \!\!=\!\!=\!\!=\!\!=\!\!= & \mathcal{M}_1 & \xrightarrow{\ \ \ f\ \ \ } & \mathcal{M}_2.
\end{array}
$$
$$(3.1.10)$$

Lemma 3.1.2 *The morphism f is quasi-smooth if and only if $f^\diamond$ is a closed immersion, f is smooth if and only if $f^\diamond$ is an isomorphism.*

Proof The lemma follows from the exact sequence

$$\cdots \to \mathcal{H}^1(\mathbb{T}_f|_{\mathcal{M}_1}) \to \mathcal{H}^1(\mathbb{T}_{\mathfrak{M}_1}|_{\mathcal{M}_1}) \to \mathcal{H}^1(f^*\mathbb{T}_{\mathfrak{M}_2}|_{\mathcal{M}_1}) \to \mathcal{H}^2(\mathbb{T}_f|_{\mathcal{M}_1}) \to \cdots .$$
□

3.1.4 Good Moduli Spaces of Artin Stacks

In general for a classical Artin stack $\mathcal{M}$, its *good moduli space* is an algebraic space M together with a quasi-compact morphism,

$$\pi_{\mathcal{M}} : \mathcal{M} \to M$$

satisfying the following conditions (cf. [4, Section 1.2]):

(i) The push-forward $\pi_{\mathcal{M}*} \colon \mathrm{QCoh}(\mathcal{M}) \to \mathrm{QCoh}(M)$ is exact.
(ii) The induced morphism $\mathcal{O}_M \to \pi_{\mathcal{M}*}\mathcal{O}_{\mathcal{M}}$ is an isomorphism.

The good moduli space morphism $\pi_{\mathcal{M}}$ is universally closed. Moreover for each closed point $y \in M$, there exists a unique closed point $x \in \pi_{\mathcal{M}}^{-1}(y)$, and its automorphism group $\mathrm{Aut}(x)$ is reductive (see [4, Theorem 4.16, Proposition 12.14]).

Let $\mathfrak{M}$ be a quasi-smooth derived stack and take its classical truncation $\mathcal{M} = t_0(\mathfrak{M})$. Suppose that it admits a good moduli space $\pi_{\mathcal{M}} \colon \mathcal{M} \to M$. We denote by $\mathcal{D}_{\text{ét}/M}$ the category of étale morphisms $\iota \colon U \to M$ for an algebraic space U, with morphisms given by commutative diagrams

$$
\begin{array}{ccc}
U' & \xrightarrow{\ \ \rho\ \ } & U \\
& \searrow{\scriptstyle \iota'} \quad \swarrow{\scriptstyle \iota} & \\
& M. &
\end{array}
\tag{3.1.11}
$$

Here ρ is an étale morphism. The following proposition will be proved in Sect. 8.3.2.

Proposition 3.1.3 *For each object $(\iota \colon U \to M) \in \mathcal{D}_{\text{ét}/M}$ there is an induced unique (up to equivalence) derived stack $\mathfrak{M}_U$ together with Cartesian diagrams*

$$
\begin{array}{ccccc}
\mathfrak{M}_U & \longleftarrow\!\!\!\supset & \mathcal{M}_U & \longrightarrow & U \\
{\scriptstyle \iota_{\mathfrak{M}}}\downarrow & \square \ \ {\scriptstyle \iota_{\mathcal{M}}}\downarrow & & \square & \downarrow{\scriptstyle \iota} \\
\mathfrak{M} & \longleftarrow\!\!\!\supset & \mathcal{M} & \xrightarrow{\ \pi_{\mathcal{M}}\ } & M
\end{array}
\tag{3.1.12}
$$

such that $\mathcal{M}_U = t_0(\mathfrak{M}_U)$ and $\iota_{\mathfrak{M}}$ is étale. Moreover given a diagram (3.1.11), there exists a unique induced morphism $\rho_{\mathfrak{M}} \colon \mathfrak{M}_{U'} \to \mathfrak{M}_U$ and Cartesian squares

$$
\begin{array}{ccccc}
\mathfrak{M}_{U'} & \longleftarrow\!\!\!\supset & \mathcal{M}_{U'} & \longrightarrow & U' \\
{\scriptstyle \rho_{\mathfrak{M}}}\downarrow & \square \ \ {\scriptstyle \rho_{\mathcal{M}}}\downarrow & & \square & \downarrow{\scriptstyle \rho} \\
\mathfrak{M}_U & \longleftarrow\!\!\!\supset & \mathcal{M}_U & \longrightarrow & U.
\end{array}
\tag{3.1.13}
$$

We will use the following étale local structure result for good moduli spaces proved in [5].

Theorem 3.1.4 ([5, Theorem 2.9]) *For any closed point $y \in M$, there exists an étale neighborhood $\iota \colon (U, 0) \to (M, y)$ and a commutative isomorphisms*

$$
\begin{array}{ccc}
[U/G] & \xrightarrow{\ \cong\ } & \mathcal{M}_U \\
\downarrow & & \downarrow \\
U /\!\!/ G & \xrightarrow{\ \cong\ } & U.
\end{array}
$$

Here $G = \mathrm{Aut}(x)$ for a unique closed point $x \in \pi_{\mathcal{M}}^{-1}(y)$, $\mathcal{U} = \mathrm{Spec}\,R$ is an affine $\mathbb{C}$-scheme with G-action, and $\mathcal{U}/\!\!/G = \mathrm{Spec}\,R^G$.

The above result can be extended to derived stacks, whose proof will be given in Sect. 8.3.3 (cf. [46, Lemma 3.2.5]).

Proposition 3.1.5 *In the situation of Theorem 3.1.4, there is an equivalence*

$$\mathfrak{M}_U \sim [\mathfrak{U}/G],$$

where $[\mathfrak{U}/G]$ is a derived stack associated with a G-equivariant tuple (Y, V, s) in Definition 2.3.1, such that $\mathcal{U} = t_0(\mathfrak{U})$ and G are given as in Theorem 3.1.4. Moreover for $l \in \mathrm{Pic}(\mathcal{M})_{\mathbb{R}}$, by replacing U with a Zariski open neighborhood of $0 \in U$, the pull-back $\iota_{\mathfrak{M}}^*(l) \in \mathrm{Pic}([\mathcal{U}/G])_{\mathbb{R}}$ is extended to a $\mathbb{R}$-line bundle on $[Y/G]$.

3.2 Categorical DT Theory via Verdier Quotients

3.2.1 Definition of DT Category

Let $\mathfrak{M}$ be a quasi-smooth and QCA derived stack, and take a conical closed substack

$$\mathcal{Z} \subset \mathcal{N} = t_0(\Omega_{\mathfrak{M}}[-1]).$$

Here similarly to before, $\mathcal{Z}$ is called conical if it is closed under the fiberwise $\mathbb{C}^*$-action on $\mathcal{N} \to \mathcal{M}$. Let $\alpha : \mathfrak{U} \to \mathfrak{M}$ be a smooth morphism as in (3.1.3). By Lemma 3.1.2, we have the associated conical closed subscheme

$$\alpha^*\mathcal{Z} := \alpha^{\diamond}(\alpha^{\spadesuit})^{-1}(\mathcal{Z}) \subset t_0(\Omega_{\mathfrak{U}}[-1]) = \mathrm{Crit}(w).$$

Here we have written $\mathfrak{U}$ as (2.1.1), and w is defined in the diagram (2.1.5). By Definition 2.1.5, we have the associated subcategory $\mathcal{C}_{\alpha^*\mathcal{Z}} \subset D_{\mathrm{coh}}^b(\mathfrak{U})$. It is proved in [8, Section 7] that the subcategory $\mathcal{C}_{\mathcal{Z}}$ satisfies functorial properties with respect to smooth morphisms. So given a diagram (3.1.2), the pull-back functors restrict to functors

$$\rho^* : \mathcal{C}_{\alpha'^*\mathcal{Z}} \to \mathcal{C}_{\alpha^*\mathcal{Z}}, \quad \rho^{\mathrm{ind}*} : \mathrm{Ind}\,\mathcal{C}_{\alpha^*\mathcal{Z}} \to \mathrm{Ind}\,\mathcal{C}_{\alpha'^*\mathcal{Z}}. \tag{3.2.1}$$

By taking the limit, we have the following definition:

Definition 3.2.1 ([8]) For a conical closed substack $\mathcal{Z} \subset \mathcal{N}$, we define

$$\mathcal{C}_{\mathcal{Z}} := \lim_{\mathfrak{U} \xrightarrow{\alpha} \mathfrak{M}} \mathcal{C}_{\alpha^*\mathcal{Z}} \subset D_{\mathrm{coh}}^b(\mathfrak{M}), \quad \mathrm{Ind}\,\mathcal{C}_{\mathcal{Z}} := \lim_{\mathfrak{U} \xrightarrow{\alpha} \mathfrak{M}} \mathrm{Ind}\,\mathcal{C}_{\alpha^*\mathcal{Z}} \subset \mathrm{Ind}\,D_{\mathrm{coh}}^b(\mathfrak{M}).$$

By [8, Corollary 8.2.6], we have adjoint pairs

$$\operatorname{Ind} \mathcal{C}_{\mathcal{Z}} \underset{\Gamma_{\mathcal{Z}}}{\overset{i}{\rightleftarrows}} \operatorname{Ind} D^b_{\mathrm{coh}}(\mathfrak{M}_2) \,, \tag{3.2.2}$$

where i is the natural inclusion and $i \dashv \Gamma_{\mathcal{Z}}$.

We denote by $\mathcal{N}^{\mathrm{ss}}$ the complement of $\mathcal{Z}$

$$\mathcal{N}^{\mathrm{ss}} := \mathcal{N} \setminus \mathcal{Z}$$

which is a $\mathbb{C}^*$-invariant open substack of $\mathcal{N}$. We define the $\mathbb{C}^*$-equivariant DT category for $\mathcal{N}^{\mathrm{ss}}$ as follows:

Definition 3.2.2 We define the triangulated category $\mathcal{DT}^{\mathbb{C}^*}(\mathcal{N}^{\mathrm{ss}})$ by the Verdier quotient

$$\mathcal{DT}^{\mathbb{C}^*}(\mathcal{N}^{\mathrm{ss}}) := D^b_{\mathrm{coh}}(\mathfrak{M})/\mathcal{C}_{\mathcal{Z}}. \tag{3.2.3}$$

The quotient category (3.2.3) admits a dg-enhancement by taking dg-quotients of dg-categories developed in [37, 70]. In general for a dg-category $\mathcal{D}$ and its dg-subcategory $\mathcal{D}' \subset \mathcal{D}$, its Drinfeld dg-quotient $\mathcal{D}/\mathcal{D}'$ is obtained from $\mathcal{D}$ by formally adding degree -1 morphisms (see [37])

$$\epsilon_U : U \to U, \ d(\epsilon_U) = \mathrm{id}_U$$

for each $U \in \mathcal{D}'$. Let

$$L\mathcal{C}_{\mathcal{Z}} \subset L_{\mathrm{coh}}(\mathfrak{M})$$

be the dg subcategory consisting of objects which are isomorphic to objects in $\mathcal{C}_{\mathcal{Z}}$ in the homotopy category.

Definition 3.2.3 We define the dg-category $\mathcal{DT}^{\mathbb{C}^*}_{\mathrm{dg}}(\mathcal{N}^{\mathrm{ss}})$ to be the Drinfeld quotient

$$\mathcal{DT}^{\mathbb{C}^*}_{\mathrm{dg}}(\mathcal{N}^{\mathrm{ss}}) := L_{\mathrm{coh}}(\mathfrak{M})/L\mathcal{C}_{\mathcal{Z}}.$$

The dg-category $\mathcal{DT}^{\mathbb{C}^*}_{\mathrm{dg}}(\mathcal{N}^{\mathrm{ss}})$ is a dg-enhancement of $\mathcal{DT}^{\mathbb{C}^*}(\mathcal{N}^{\mathrm{ss}})$, i.e. we have the canonical equivalence (see [37, Theorem 3.4])

$$\mathcal{DT}^{\mathbb{C}^*}(\mathcal{N}^{\mathrm{ss}}) \overset{\sim}{\to} \mathrm{Ho}(\mathcal{DT}^{\mathbb{C}^*}_{\mathrm{dg}}(\mathcal{N}^{\mathrm{ss}})).$$

Let $\omega_{\mathfrak{M}} := \det(\mathbb{L}_{\mathfrak{M}})[\mathrm{rank}(\mathbb{L}_{\mathfrak{M}})]$ be the dualizing line bundle of $\mathfrak{M}$. We denote by $\mathbb{D}^{\mathrm{Serre}}_{\mathfrak{M}}$ the Serre dualizing functor

$$\mathbb{D}^{\mathrm{Serre}}_{\mathfrak{M}} : D^b_{\mathrm{coh}}(\mathfrak{M}) \overset{\sim}{\to} D^b_{\mathrm{coh}}(\mathfrak{M})^{\mathrm{op}}, \ E \mapsto \mathcal{H}om(E, \omega_{\mathfrak{M}}). \tag{3.2.4}$$

Here the above functor preserves coherence as $\mathfrak{M}$ is quasi-smooth, and also preserves the singular supports by [8, Proposition 4.7.2]. Therefore we have the Serre duality equivalence

$$\mathbb{D}_{\mathcal{N}}^{\mathrm{Serre}} : \mathcal{DT}^{\mathbb{C}^*}(\mathcal{N}^{\mathrm{ss}}) \xrightarrow{\sim} \mathcal{DT}^{\mathbb{C}^*}(\mathcal{N}^{\mathrm{ss}})^{\mathrm{op}}.$$

For another quasi-smooth derived stack $\mathfrak{M}'$, a conical closed subset $\mathcal{Z}'$ and its complement $\mathcal{N}'^{\mathrm{ss}}$

$$\mathcal{Z}' \subset t_0(\Omega_{\mathfrak{M}'}[-1]), \ \mathcal{N}'^{\mathrm{ss}} = t_0(\Omega_{\mathfrak{M}'}[-1]) \setminus \mathcal{Z}'$$

we define the tensor product of DT categories as

$$\mathcal{DT}^{\mathbb{C}^*}(\mathcal{N}^{\mathrm{ss}}) \otimes \mathcal{DT}^{\mathbb{C}^*}(\mathcal{N}'^{\mathrm{ss}}) := D_{\mathrm{coh}}^b(\mathfrak{M} \times \mathfrak{M}')/\mathcal{C}_{q^*\mathcal{Z} \cup q'^*\mathcal{Z}'}$$

where q, q' are projections from $\Omega_{\mathfrak{M}}[-1] \times \Omega_{\mathfrak{M}'}[-1]$ onto corresponding factors.

3.2.2 $\widehat{\mathcal{DT}}$-Version

We can also define other kind of DT categories, defined by taking limits of quotient categories. For a diagram (3.1.2), since the pull-back $\rho^* : D_{\mathrm{coh}}^b(\mathfrak{U}) \to D_{\mathrm{coh}}^b(\mathfrak{U}')$ restrict to the functor (3.2.1), we have the induced functors

$$\rho^* : L_{\mathrm{coh}}(\mathfrak{U})/LC_{\alpha^*\mathcal{Z}} \to L_{\mathrm{coh}}(\mathfrak{U}')/LC_{\alpha'^*\mathcal{Z}'}, \ \rho^* : D_{\mathrm{coh}}^b(\mathfrak{U})/\mathcal{C}_{\alpha^*\mathcal{Z}} \to D_{\mathrm{coh}}^b(\mathfrak{U}')/\mathcal{C}_{\alpha'^*\mathcal{Z}'}.$$

By taking their limits, we have the following definition:

Definition 3.2.4 We define the dg-category $\widehat{\mathcal{DT}}_{\mathrm{dg}}^{\mathbb{C}^*}(\mathcal{N}^{\mathrm{ss}})$ and the triangulated category $\widehat{\mathcal{DT}}^{\mathbb{C}^*}(\mathcal{N}^{\mathrm{ss}})$ to be

$$\widehat{\mathcal{DT}}_{\mathrm{dg}}^{\mathbb{C}^*}(\mathcal{N}^{\mathrm{ss}}) := \lim_{\mathfrak{U} \xrightarrow{\alpha} \mathfrak{M}} (L_{\mathrm{coh}}(\mathfrak{U})/LC_{\alpha^*\mathcal{Z}}), \tag{3.2.5}$$

$$\widehat{\mathcal{DT}}^{\mathbb{C}^*}(\mathcal{N}^{\mathrm{ss}}) := \lim_{\mathfrak{U} \xrightarrow{\alpha} \mathfrak{M}} \left(D_{\mathrm{coh}}^b(\mathfrak{U})/\mathcal{C}_{\alpha^*\mathcal{Z}} \right).$$

Remark 3.2.5 By the equivalence (2.3.9), the limit (3.2.5) may be regarded as a gluing of derived categories of factorizations

$$\widehat{\mathcal{DT}}^{\mathbb{C}^*}(\mathcal{N}^{\mathrm{ss}}) = \lim_{\mathfrak{U} \xrightarrow{\alpha} \mathfrak{M}} \mathrm{MF}_{\mathrm{coh}}^{\mathbb{C}^*}(V^\vee \setminus \alpha^*\mathcal{Z}, w).$$

Remark 3.2.6 There is a natural functor

$$\mathcal{DT}^{\mathbb{C}^*}(\mathcal{N}^{ss}) \to \widehat{\mathcal{DT}}^{\mathbb{C}^*}(\mathcal{N}^{ss}),$$

but it is not clear whether this is fully-faithful nor essentially surjective in general. In Sect. 8.1 we will address the above question in the case that $\operatorname{Ind} \mathcal{C}_{\mathcal{Z}}$ is compactly generated.

For a conical closed substack $\mathcal{Z} \subset \mathcal{N} = t_0(\Omega_{\mathfrak{M}}[-1])$ and its complement $\mathcal{N}^{ss} = \mathcal{N} \setminus \mathcal{Z}$, we define the ind-completion of the DT category $\mathcal{DT}^{\mathbb{C}^*}(\mathcal{N}^{ss})$ by

$$\operatorname{Ind} \mathcal{DT}^{\mathbb{C}^*}(\mathcal{N}^{ss}) := \operatorname{Ind}(\mathcal{DT}^{\mathbb{C}^*}(\mathcal{N}^{ss})).$$

On the other hand, a similar construction in Definition 3.2.4 applies to define limits

$$\lim_{\mathfrak{U} \xrightarrow{\alpha} \mathfrak{M}} \operatorname{Ind} \left(D^b_{\mathrm{coh}}(\mathfrak{U})/\mathcal{C}_{\alpha^* \mathcal{Z}} \right), \quad \lim_{\mathfrak{U} \xrightarrow{\alpha} \mathfrak{M}} \left(\operatorname{Ind} D^b_{\mathrm{coh}}(\mathfrak{U})/ \operatorname{Ind} \mathcal{C}_{\alpha^* \mathcal{Z}} \right).$$

The following proposition will be proved in Sect. 8.1.1:

Proposition 3.2.7 *Suppose that $\operatorname{Ind} \mathcal{C}_{\mathcal{Z}}$ is compactly generated with $(\operatorname{Ind} \mathcal{C}_{\mathcal{Z}})^{\mathrm{cp}} = \mathcal{C}_{\mathcal{Z}}$. Then we have equivalences*

$$\operatorname{Ind} \mathcal{DT}^{\mathbb{C}^*}(\mathcal{N}^{ss}) \xrightarrow{\sim} \operatorname{Ind} D^b_{\mathrm{coh}}(\mathfrak{M})/ \operatorname{Ind} \mathcal{C}_{\mathcal{Z}}$$

$$\xrightarrow{\sim} \lim_{\mathfrak{U} \xrightarrow{\alpha} \mathfrak{M}} \operatorname{Ind} \left(D^b_{\mathrm{coh}}(\mathfrak{U})/\mathcal{C}_{\alpha^* \mathcal{Z}} \right)$$

$$\xrightarrow{\sim} \lim_{\mathfrak{U} \xrightarrow{\alpha} \mathfrak{M}} \left(\operatorname{Ind} D^b_{\mathrm{coh}}(\mathfrak{U})/ \operatorname{Ind} \mathcal{C}_{\alpha^* \mathcal{Z}} \right).$$

Moreover we have fully-faithful functors with dense images

$$\mathcal{DT}^{\mathbb{C}^*}(\mathcal{N}^{ss}) \hookrightarrow \widehat{\mathcal{DT}}^{\mathbb{C}^*}(\mathcal{N}^{ss}) \hookrightarrow \left(\operatorname{Ind} \mathcal{DT}^{\mathbb{C}^*}(\mathcal{N}^{ss}) \right)^{\mathrm{cp}}.$$

Using the above proposition, we also have the following lemma:

Lemma 3.2.8 *In the situation of Proposition 3.2.7, let $\mathcal{Z}_i \subset t_0(\Omega_{\mathfrak{M}}[-1])$ for $i = 1, 2$ be conical closed substacks such that $\mathcal{Z}_1 \cap \mathcal{Z}_2 = \mathcal{Z}$. Then for $\mathcal{E}_i \in \mathcal{C}_{\mathcal{Z}_i}$, we have*

$$\operatorname{Hom}_{\mathcal{DT}^{\mathbb{C}^*}(\mathcal{N}^{ss})}(\mathcal{E}_1, \mathcal{E}_2) = 0.$$

Proof By replacing $\mathcal{E}_1$ if necessary, any morphism $\mathcal{E}_1 \to \mathcal{E}_2$ in $\mathcal{DT}^{\mathbb{C}^*}(\mathcal{N}^{ss})$ is represented by a morphism $\phi \colon \mathcal{E}_1 \to \mathcal{E}_2$ in $D^b_{\mathrm{coh}}(\mathfrak{M})$. The morphism ϕ factors through $\mathcal{E}_1 \to \Gamma_{\mathcal{Z}_1}(\mathcal{E}_2) \to \mathcal{E}_2$, where $\Gamma_{\mathcal{Z}_1}$ is the right adjoint of the natural fully-faithful functor $\operatorname{Ind} \mathcal{C}_{\mathcal{Z}_1} \hookrightarrow \operatorname{Ind} D^b_{\mathrm{coh}}(\mathfrak{M})$ in (3.2.2). As $\Gamma_{\mathcal{Z}_1}(\mathcal{E}_2)$ have singular support contained in $\mathcal{Z}_1 \cap \mathcal{Z}_2 = \mathcal{Z}$ (which is obvious from the construction of the right adjoint in [8, Lemma 3.3.7]), it follows that $\phi = 0$ in $\operatorname{Ind} D^b_{\mathrm{coh}}(\mathfrak{M}) / \operatorname{Ind} \mathcal{C}_{\mathcal{Z}}$. Then the lemma follows from the first equivalence in Proposition 3.2.7. $\quad\square$

3.2.3 Replacement of the Quotient Category

The derived moduli stacks we will consider are typically not quasi-compact, and it will be useful if we can replace it by its quasi-compact open substack. In order to show that such a replacement makes sense, we need to check that the resulting quotient categories are independent of the choice of a quasi-compact open substack.

Let $\mathcal{W} \subset \mathcal{M}$ be a closed substack, and take the open derived substack

$$\mathfrak{M}_\circ \subset \mathfrak{M}$$

whose truncation is $\mathcal{M} \setminus \mathcal{W}$. Let us take a conical closed substack $\mathcal{Z} \subset \mathcal{N} = t_0(\Omega_{\mathfrak{M}}[-1])$. We have the following conical closed substack

$$\mathcal{Z}_\circ := \mathcal{Z} \setminus p_0^{-1}(\mathcal{W}) \subset \mathcal{N}_\circ := t_0(\Omega_{\mathfrak{M}_\circ}[-1]).$$

Note that $\mathcal{N}_\circ = \mathcal{N} \setminus p_0^{-1}(\mathcal{W})$, and we have the following open immersion

$$\mathcal{N}^{\mathrm{ss}}_\circ := \mathcal{N}_\circ \setminus \mathcal{Z}_\circ \hookrightarrow \mathcal{N} \setminus \mathcal{Z} = \mathcal{N}^{\mathrm{ss}}. \tag{3.2.6}$$

The following lemma will be proved in Lemmas 8.1.8 and 8.1.9.

Lemma 3.2.9 *Suppose that* $p_0^{-1}(\mathcal{W}) \subset \mathcal{Z}$. *Then the restriction functors give equivalences*

$$\mathcal{DT}^{\mathbb{C}^*}(\mathcal{N}^{\mathrm{ss}}) \xrightarrow{\sim} \mathcal{DT}^{\mathbb{C}^*}(\mathcal{N}^{\mathrm{ss}}_\circ), \ \widehat{\mathcal{DT}}^{\mathbb{C}^*}(\mathcal{N}^{\mathrm{ss}}) \xrightarrow{\sim} \widehat{\mathcal{DT}}^{\mathbb{C}^*}(\mathcal{N}^{\mathrm{ss}}_\circ).$$

Moreover if $p_0^{-1}(\mathcal{W}) = \mathcal{Z}$, *we have equivalences*

$$\mathcal{DT}^{\mathbb{C}^*}(\mathcal{N}^{\mathrm{ss}}) \xrightarrow{\sim} \widehat{\mathcal{DT}}^{\mathbb{C}^*}(\mathcal{N}^{\mathrm{ss}}) \xrightarrow{\sim} D^b_{\mathrm{coh}}(\mathfrak{M}_\circ).$$

3.2.4 $\mathbb{C}^*$-Rigidification

In some case our derived stack $\mathfrak{M}$ has $\mathbb{C}^*$-automorphisms at any point, and in this case $\mathcal{M} = t_0(\mathfrak{M})$ is never a Deligne-Mumford stack. In this case, we rigidify $\mathbb{C}^*$-automorphisms and define the DT category via the $\mathbb{C}^*$-rigidified derived stack.

Suppose that we have a closed embedding into the inertia stack

$$(\mathbb{C}^*)_{\mathcal{M}} \hookrightarrow I_{\mathcal{M}} := \Delta \times_{\mathcal{M} \times \mathcal{M}} \Delta. \qquad (3.2.7)$$

Here Δ is the diagonal. Under the presence of the embedding (3.2.7), by [1, Theorem A.1] there exists an Artin stack $\mathcal{M}^{\mathbb{C}^*\text{-rig}-\text{rig}}$ together with a map

$$\varrho_{\mathcal{M}} : \mathcal{M} \to \mathcal{M}^{\mathbb{C}^*\text{-rig}} \qquad (3.2.8)$$

uniquely characterized by the properties that $\varrho_{\mathcal{M}}$ is a $\mathbb{C}^*$-gerbe (see [84] for the notion of gerbes) and for any map $\xi : T \to \mathcal{M}$ from a $\mathbb{C}$-scheme T, the homomorphism of group schemes

$$\mathrm{Aut}_T(\xi) \to \mathrm{Aut}_T(\varrho_{\mathcal{M}} \circ \xi)$$

is surjective with kernel $(\mathbb{C}^*)_T$. In the above case, any object $\mathcal{F} \in \mathrm{Coh}(\mathcal{M})$ admits a $\mathbb{C}^*$-action induced by the inertia action (see [84, Section 2.1]).

The above $\mathbb{C}^*$-rigidification can be extended to derived stacks as follows (see [46, Lemma 3.4.9]). Suppose that there is a weak action of $B\mathbb{C}^*$ on $\mathfrak{M}$ such that the induced map on the inertia stack $(\mathbb{C}^*)_{\mathfrak{M}} \hookrightarrow I_{\mathfrak{M}}$ is a closed immersion. Then we have the $\mathbb{C}^*$-gerbe

$$\varrho_{\mathfrak{M}} : \mathfrak{M} \to \mathfrak{M}^{\mathbb{C}^*\text{-rig}} \qquad (3.2.9)$$

whose classical truncation is identified with (3.2.8).

Let Λ be the character group

$$\Lambda = \mathrm{Hom}_{\mathbb{Z}}(\mathbb{C}^*, \mathbb{C}^*) \cong \mathbb{Z}.$$

For $\lambda \in \Lambda$, we define the subcategories

$$\mathrm{Coh}(\mathcal{M})_\lambda \subset \mathrm{Coh}(\mathcal{M}), \ D^b_{\mathrm{coh}}(\mathfrak{M})_\lambda \subset D^b_{\mathrm{coh}}(\mathfrak{M})$$

to be consisting of sheaves with $\mathbb{C}^*$-weight λ, objects of cohomology sheaves with $\mathbb{C}^*$-weight λ, respectively. Then we have the decompositions

$$\mathrm{Coh}(\mathcal{M}) = \bigoplus_{\lambda \in \Lambda} \mathrm{Coh}(\mathcal{M})_\lambda, \ D^b_{\mathrm{coh}}(\mathfrak{M}) = \bigoplus_{\lambda \in \Lambda} D^b_{\mathrm{coh}}(\mathfrak{M})_\lambda. \qquad (3.2.10)$$

The above decompositions are obvious if $\varrho_{\mathfrak{M}}$ is a trivial $\mathbb{C}^*$-gerbe. In general, a generalization of the arguments in [17, Theorem 4.7, Theorem 5.4] easily implies the above decompositions. The above decompositions lift to dg-enhancements. Let

$L_{\mathrm{coh}}(\mathfrak{M})_\lambda \subset L_{\mathrm{coh}}(\mathfrak{M})$ be the dg-subcategory consisting of objects of cohomology sheaves with $\mathbb{C}^*$ weight λ. We have the weak equivalence

$$\bigoplus_{\lambda \in \Lambda} L_{\mathrm{coh}}(\mathfrak{M})_\lambda \xrightarrow{\sim} L_{\mathrm{coh}}(\mathfrak{M}).$$

Suppose that the $\mathbb{C}^*$-weight on $\mathbb{L}_{\mathfrak{M}}|_{\mathcal{M}}$ is zero. Then the stack $\mathcal{N} = t_0(\Omega_{\mathfrak{M}}[-1])$ also satisfies $(\mathbb{C}^*)_{\mathcal{N}} \subset I_{\mathcal{N}}$. Therefore we have its $\mathbb{C}^*$-rigidification $\varrho\colon \mathcal{N} \to \mathcal{N}^{\mathbb{C}^*\text{-rig}}$. Indeed we have

$$\mathcal{N}^{\mathbb{C}^*\text{-rig}} = t_0(\Omega_{\mathfrak{M}^{\mathbb{C}^*\text{-rig}}}[-1])$$

and we have the Cartesian square

$$
\begin{array}{ccc}
\mathcal{N} & \xrightarrow{\;\varrho_{\mathcal{N}}\;} & \mathcal{N}^{\mathbb{C}^*\text{-rig}} \\
\downarrow & \square & \downarrow \\
\mathcal{M} & \xrightarrow{\;\varrho_{\mathcal{M}}\;} & \mathcal{M}^{\mathbb{C}^*\text{-rig}}.
\end{array}
$$

Let us take a conical closed substack and its complement

$$\mathcal{Z}^{\mathbb{C}^*\text{-rig}} \subset \mathcal{N}^{\mathbb{C}^*\text{-rig}}, \quad (\mathcal{N}^{\mathrm{ss}})^{\mathbb{C}^*\text{-rig}} = \mathcal{N}^{\mathbb{C}^*\text{-rig}} \setminus \mathcal{Z}^{\mathbb{C}^*\text{-rig}}$$

and also set $\mathcal{Z} = \varrho_{\mathcal{N}}^{-1}(\mathcal{Z}^{\mathbb{C}^*\text{-rig}})$, $\mathcal{N}^{\mathrm{ss}} = \mathcal{N} \setminus \mathcal{Z}$. We define the λ-twisted version of DT categories as follows:

Definition 3.2.10 We define $\mathcal{DT}^{\mathbb{C}^*}(\mathcal{N}^{\mathrm{ss}})_\lambda$, $\mathcal{DT}_{\mathrm{dg}}^{\mathbb{C}^*}(\mathcal{N}^{\mathrm{ss}})_\lambda$ by the Verdier/Drinfeld quotients

$$\mathcal{DT}^{\mathbb{C}^*}(\mathcal{N}^{\mathrm{ss}})_\lambda := D_{\mathrm{coh}}^b(\mathfrak{M})_\lambda / (D_{\mathrm{coh}}^b(\mathfrak{M})_\lambda \cap \mathcal{C}_{\mathcal{Z}}),$$

$$\mathcal{DT}_{\mathrm{dg}}^{\mathbb{C}^*}(\mathcal{N}^{\mathrm{ss}})_\lambda := L_{\mathrm{coh}}(\mathfrak{M})_\lambda / (L_{\mathrm{coh}}(\mathfrak{M})_\lambda \cap L\mathcal{C}_{\mathcal{Z}}).$$

Note that $\mathcal{DT}_{\mathrm{dg}}^{\mathbb{C}^*}(\mathcal{N}^{\mathrm{ss}})_\lambda$ is a dg-enhancement of $\mathcal{DT}^{\mathbb{C}^*}(\mathcal{N}^{\mathrm{ss}})_\lambda$. We have the following proposition:

Proposition 3.2.11

(i) *We have the equivalence (weak equivalence)*

$$\bigoplus_{\lambda \in \Lambda} \mathcal{DT}^{\mathbb{C}^*}(\mathcal{N}^{\mathrm{ss}})_\lambda \xrightarrow{\sim} \mathcal{DT}^{\mathbb{C}^*}(\mathcal{N}^{\mathrm{ss}}), \quad \bigoplus_{\lambda \in \Lambda} \mathcal{DT}_{\mathrm{dg}}^{\mathbb{C}^*}(\mathcal{N}^{\mathrm{ss}})_\lambda \xrightarrow{\sim} \mathcal{DT}_{\mathrm{dg}}^{\mathbb{C}^*}(\mathcal{N}^{\mathrm{ss}}).$$

$$(3.2.11)$$

(ii) *We have the equivalence (weak equivalence)*

$$\varrho_{\mathfrak{M}}^* : \mathcal{DT}^{\mathbb{C}^*}((\mathcal{N}^{ss})^{\mathbb{C}^*\text{-rig}}) \xrightarrow{\sim} \mathcal{DT}^{\mathbb{C}^*}(\mathcal{N}^{ss})_{\lambda=0}, \qquad (3.2.12)$$

$$\varrho_{\mathfrak{M}}^* : \mathcal{DT}_{dg}^{\mathbb{C}^*}((\mathcal{N}^{ss})^{\mathbb{C}^*\text{-rig}}) \xrightarrow{\sim} \mathcal{DT}_{dg}^{\mathbb{C}^*}(\mathcal{N}^{ss})_{\lambda=0}.$$

(iii) *Suppose that (3.2.9) is a trivial $\mathbb{C}^*$-gerbe, i.e. $\mathfrak{M} = \mathfrak{M}^{\mathbb{C}^*\text{-rig}} \times B\mathbb{C}^*$. Then we have an equivalence (weak equivalence)*

$$\mathcal{DT}^{\mathbb{C}^*}(\mathcal{N}^{ss})_{\lambda=0} \xrightarrow{\sim} \mathcal{DT}^{\mathbb{C}^*}(\mathcal{N}^{ss})_{\lambda}, \ \mathcal{DT}_{dg}^{\mathbb{C}^*}(\mathcal{N}^{ss})_{\lambda=0} \xrightarrow{\sim} \mathcal{DT}_{dg}^{\mathbb{C}^*}(\mathcal{N}^{ss})_{\lambda}.$$
$$(3.2.13)$$

Proof In all of (i), (ii), (iii), it is enough to show the equivalences for triangulated DT categories.

(i) For $\mathcal{F} \in D_{coh}^b(\mathfrak{M})$, we take the decomposition $\mathcal{F} = \oplus_\lambda \mathcal{F}_\lambda$ where $\mathcal{F}_\lambda$ is $\mathbb{C}^*$-weight λ-part. Then $\mathrm{Hom}^*(\mathcal{F}, \mathcal{F})$ is the direct sum of $\mathrm{Hom}^*(\mathcal{F}_\lambda, \mathcal{F}_\lambda)$, so by the definition of singular supports $\mathcal{F}$ is an object in $\mathcal{C}_{\mathcal{Z}}$ if and only if each $\mathcal{F}_\lambda$ is an object in $\mathcal{C}_{\mathcal{Z}}$. Therefore the decomposition (3.2.10) restricts to the decomposition

$$\mathcal{C}_{\mathcal{Z}} = \bigoplus_{\lambda \in \Lambda} (D_{coh}^b(\mathfrak{M})_\lambda \cap \mathcal{C}_{\mathcal{Z}}).$$

By taking quotients, we have the equivalence (3.2.11).

(ii) We first show that the functor

$$D_{coh}^b(\mathfrak{M}^{\mathbb{C}^*\text{-rig}}) \xrightarrow{\varrho_{\mathfrak{M}}^*} D_{coh}^b(\mathfrak{M})_{\lambda=0} \qquad (3.2.14)$$

is an equivalence. We have the natural transformation $\mathrm{id} \to \varrho_{\mathfrak{M}*}\varrho_{\mathfrak{M}}^*$, which is an isomorphism as $\varrho_{\mathfrak{M}}$ is étale locally trivial $\mathbb{C}^*$-gerbe. Therefore (3.2.14) is fully-faithful. Moreover for an object $E \in D_{coh}^b(\mathfrak{M})_{\lambda=0}$, we have $E = 0$ if $\varrho_{\mathfrak{M}*}E = 0$, again because $\varrho_{\mathfrak{M}}$ is étale locally trivial $\mathbb{C}^*$-gerbe. Therefore (3.2.14) is also essentially surjective, and hence (3.2.14) is an equivalence. Then (3.2.14) obviously restricts to the equivalence between $\mathcal{C}_{\mathcal{Z}^{\mathbb{C}^*\text{-rig}}}$ and $D_{coh}^b(\mathfrak{M})_{\lambda=0} \cap \mathcal{C}_{\mathcal{Z}}$. By taking quotients, we obtain the equivalence (3.2.12).

(iii) Under the assumption, there exists a line bundle $\mathcal{O}_{\mathfrak{M}}(\lambda)$ on $\mathfrak{M}$ given by the pull-back of $\mathcal{O}_{B\mathbb{C}^*}(\lambda)$, where the latter is the one dimensional $\mathbb{C}^*$-representation with weight λ. The equivalence (3.2.13) is given by taking the tensor product with $\mathcal{O}_{\mathfrak{M}}(\lambda)$.

$\square$

Remark 3.2.12 If $\mathcal{M}^{\mathbb{C}^*\text{-rig}}$ is a scheme, then the $\mathbb{C}^*$-gerbe (3.2.8) is classified by a Brauer class $\alpha \in \mathrm{Br}(\mathcal{M}^{\mathbb{C}^*\text{-rig}})$. Then by [84, Proposition 2.1.3.3], the category $\mathrm{Coh}(\mathcal{M})_\lambda$ is equivalent to the category of α^λ-twisted coherent sheaves on $\mathcal{M}^{\mathbb{C}^*\text{-rig}}$ in the sense of Căldăraru [31]. Similarly the λ-twisted version $\mathcal{DT}^{\mathbb{C}^*}(\mathcal{N}^{\mathrm{ss}})_\lambda$ may be interpreted as α^λ-twisted analogue for $\mathcal{DT}^{\mathbb{C}^*}((\mathcal{N}^{\mathrm{ss}})^{\mathbb{C}^*\text{-rig}})$.

Similar argument also applies to define $\widehat{\mathcal{DT}}^{\mathbb{C}^*}(\mathcal{N}^{\mathrm{ss}})_\lambda$ and $\widehat{\mathcal{DT}}_{\mathrm{dg}}^{\mathbb{C}^*}(\mathcal{N}^{\mathrm{ss}})_\lambda$, with the properties as in Proposition 3.2.11. As we will not use them, we omit details.

3.2.5   Functoriality of DT Categories

Let $\mathfrak{M}_1$, $\mathfrak{M}_2$ be quasi-smooth and QCA derived stacks, and take a morphism $f \colon \mathfrak{M}_1 \to \mathfrak{M}_2$. First suppose that f is a quasi-smooth morphism. In this case, the morphism $f^\diamond$ in the diagram (3.1.10) is a closed immersion by Lemma 3.1.2. Therefore for any conical closed substack $\mathcal{Z}_2 \subset t_0(\Omega_{\mathfrak{M}_2}[-1])$, we have the conical closed substack

$$f^\diamond (f^\spadesuit)^{-1}(\mathcal{Z}_2) \subset t_0(\Omega_{\mathfrak{M}_1}[-1]).$$

By [8, Lemma 8.4.2] (also see [134, Lemma 2.2]), the functor (3.1.7) restricts to the functor

$$f^* \colon \mathcal{C}_{\mathcal{Z}_2} \to \mathcal{C}_{f^\diamond (f^\spadesuit)^{-1}(\mathcal{Z}_2)}. \tag{3.2.15}$$

Next suppose that $f \colon \mathfrak{M}_1 \to \mathfrak{M}_2$ is a proper morphism. In this case, the morphism $f^\spadesuit$ in the diagram (3.1.10) is a proper morphism of stacks. Therefore for any conical closed substack $\mathcal{Z}_1 \subset t_0(\Omega_{\mathfrak{M}_1}[-1])$, we have the conical closed substack

$$f^\spadesuit (f^\diamond)^{-1}(\mathcal{Z}_1) \subset t_0(\Omega_{\mathfrak{M}_2}[-1]).$$

Then the functor (3.1.6) restricts to the functor (see [8, Lemma 8.4.5])

$$f_* \colon \mathcal{C}_{\mathcal{Z}_1} \to \mathcal{C}_{f^\spadesuit (f^\diamond)^{-1}(\mathcal{Z}_1)}. \tag{3.2.16}$$

Let $\mathfrak{N}$ be another quasi-smooth and QCA derived stack with a diagram

$$\mathfrak{M}_1 \xleftarrow{f_1} \mathfrak{N} \xrightarrow{f_2} \mathfrak{M}_2. \tag{3.2.17}$$

Suppose that f_1 is quasi-smooth and f_2 is a proper morphism. Then we have the functor

$$F := f_{2*} \circ f_1^* \colon D^b_{\mathrm{coh}}(\mathfrak{M}_1) \xrightarrow{f_1^*} D^b_{\mathrm{coh}}(\mathfrak{N}) \xrightarrow{f_{2*}} D^b_{\mathrm{coh}}(\mathfrak{M}_2). \tag{3.2.18}$$

Let ev be the morphism

$$\mathrm{ev} := (f_1, f_2) \colon \mathfrak{N} \to \mathfrak{M}_1 \times \mathfrak{M}_2,$$

and $\Omega_{\mathrm{ev}}[-2]$ the (-2)-shifted conormal stack

$$\Omega_{\mathrm{ev}}[-2] := \mathrm{Spec}_{\mathfrak{M}_1 \times \mathfrak{M}_2}(S(\mathbb{T}_{\mathrm{ev}}[2])) \to \mathfrak{M}_1 \times \mathfrak{M}_2.$$

Then we have the diagram (see [134, Section 2.4])

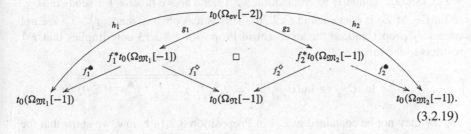

$$\tag{3.2.19}$$

Here the middle square is Cartesian. Note that $f_1^\diamond$ and g_2 are closed immersions as f_1 is quasi-smooth, and $f_2^\bullet$ is proper as f_2 is. Therefore h_2 is also proper. For a conical closed substack $\mathcal{Z}_1 \subset t_0(\Omega_{\mathfrak{M}_1}[-1])$, we have

$$f_2^\bullet (f_2^\diamond)^{-1} f_1^\diamond (f_1^\bullet)^{-1}(\mathcal{Z}_1) = h_2(h_1)^{-1}(\mathcal{Z}_1) \tag{3.2.20}$$

which is a conical closed substack in $t_0(\Omega_{\mathfrak{M}_2}[-1])$. By (3.2.15), (3.2.16), it follows that the functor F in (3.2.18) restricts to the functor

$$F \colon \mathcal{C}_{\mathcal{Z}_1} \to \mathcal{C}_{h_2(h_1)^{-1}(\mathcal{Z}_1)}.$$

Therefore we have the following:

Proposition 3.2.13 ([134, Proposition 2.4]) *Suppose that a conical closed sub-stack $\mathcal{Z}_2 \subset t_0(\Omega_{\mathfrak{M}_2}[-1])$ contains (3.2.20), or equivalently the diagram (3.2.19) restricts to the diagram*

$$\mathcal{N}_1^{\mathrm{ss}} \xleftarrow{h_1} h_2^{-1}(\mathcal{N}_2^{\mathrm{ss}}) \xrightarrow{h_2} \mathcal{N}_2^{\mathrm{ss}}.$$

Here $\mathcal{N}_i = t_0(\Omega_{\mathfrak{M}_i}[-1])$ and $\mathcal{N}_i^{ss} := \mathcal{N}_i \setminus \mathcal{Z}_i$. Then the functor (3.2.18) sends $\mathcal{C}_{\mathcal{Z}_1}$ to $\mathcal{C}_{\mathcal{Z}_2}$. In particular, the functor (3.2.18) descends to the functor

$$F : \mathcal{DT}^{\mathbb{C}^*}(\mathcal{N}_1^{ss}) \to \mathcal{DT}^{\mathbb{C}^*}(\mathcal{N}_2^{ss}). \tag{3.2.21}$$

By noting (3.1.5), the ind-completion of (3.2.18) gives rise to the functor

$$F^{ind} : := f_{2*}^{ind} \circ f_1^{ind*} : \operatorname{Ind} D^b_{coh}(\mathfrak{M}_2) \xrightarrow{f_1^{ind*}} \operatorname{Ind} D^b_{coh}(\mathfrak{N}) \xrightarrow{f_{2*}^{ind}} \operatorname{Ind} D^b_{coh}(\mathfrak{M}_2).$$

It admits the continuous right adjoint functor (see Sect. 3.1.2)

$$F^R := f_{1*}^{ind} \circ f_2^! : \operatorname{Ind} D^b_{coh}(\mathfrak{M}_2) \xrightarrow{f_2^!} \operatorname{Ind} D^b_{coh}(\mathfrak{N}) \xrightarrow{f_{1*}^{ind}} \operatorname{Ind} D^b_{coh}(\mathfrak{M}_1). \tag{3.2.22}$$

We expect that, similarly to Proposition 3.2.13, the above functor F^R sends $\operatorname{Ind} \mathcal{C}_{\mathcal{Z}_2}$ to $\operatorname{Ind} \mathcal{C}_{\mathcal{Z}_1}$ if $\mathcal{Z}_2$ is contained in (3.2.20). This is less obvious since $f_1^{\bullet}$, $f_2^{\diamond}$ are not necessary proper. Indeed the argument in Proposition 3.2.13 only implies that F^R restricts to the functor

$$F^R : \operatorname{Ind} \mathcal{C}_{\mathcal{Z}_2} \to \operatorname{Ind} \mathcal{C}_{\mathcal{Z}_1'}, \quad \mathcal{Z}_1' = \overline{f_1^{\bullet}(f_1^{\diamond})^{-1}\overline{(f_2^{\diamond}(f_2^{\bullet})^{-1}(\mathcal{Z}_2))}}$$

and $\mathcal{Z}_1'$ may not be contained in $\mathcal{Z}_1$. In Proposition 3.2.14 below, we show that the functor F^R sends $\operatorname{Ind} \mathcal{C}_{\mathcal{Z}_2}$ to $\operatorname{Ind} \mathcal{C}_{\mathcal{Z}_1}$, if the diagram (3.2.17) is represented by some diagram of equivariant tuples by discussing in the Koszul dual side.

Let (Y_i, V_i, s_i) for $i = 1, 2, 3$ be G_i-equivariant tuples, and consider morphisms of equivariant tuples with respect to some algebraic group homomorphisms $G_3 \to G_i$ for $i = 1, 2$

$$
\begin{array}{ccccc}
V_1 & \xleftarrow{g_1} & V_3 & \xrightarrow{g_2} & V_2 \\
{\scriptstyle s_1}\big\uparrow\big\downarrow & & {\scriptstyle s_3}\big\uparrow\big\downarrow & & \big\downarrow{\scriptstyle s_2} \\
Y_1 & \xleftarrow{f_1} & Y_3 & \xrightarrow{f_2} & Y_2.
\end{array}
\tag{3.2.23}
$$

The above diagram induces the diagram of derived stacks

$$[\mathfrak{U}_1/G_1] \xleftarrow{\mathbf{f}_1} [\mathfrak{U}_3/G_3] \xrightarrow{\mathbf{f}_2} [\mathfrak{U}_2/G_2]. \tag{3.2.24}$$

The diagram (3.2.23) also induces the diagram (see the construction of the diagram (2.4.3))

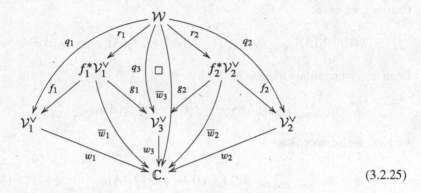

$$(3.2.25)$$

Here as before we set $\mathcal{Y}_i = [Y_i/G_i]$ and $\mathcal{V}_i = [V_i/G_i]$. We have the following proposition:

Proposition 3.2.14 *Suppose that the diagram (3.2.17) is represented by a diagram (3.2.24) induced by a diagram (3.2.23) such that $f_2\colon \mathcal{Y}_3 \to \mathcal{Y}_1$ is proper, the Cartesian in (3.2.25) is a derived Cartesian and $\mathcal{W}$ is a smooth stack. Assume that a conical closed substack $\mathcal{Z}_2 \subset t_0(\Omega_{\mathfrak{M}_2}[-1])$ is contained in (3.2.20), or equivalently the diagram (3.2.19) restricts to the diagram*

$$\mathcal{N}_1^{\mathrm{ss}} \xleftarrow{h_1} h_1^{-1}(\mathcal{N}_1^{\mathrm{ss}}) \xrightarrow{h_2} \mathcal{N}_2^{\mathrm{ss}}.$$

Then the functor F^R descends to the functor

$$F^R \colon \operatorname{Ind} \mathcal{DT}^{\mathbb{C}^*}(\mathcal{N}_2^{\mathrm{ss}}) \to \operatorname{Ind} \mathcal{DT}^{\mathbb{C}^*}(\mathcal{N}_1^{\mathrm{ss}})$$

which is a right adjoint of the ind-completion of the functor (3.2.21).

Proof We show that the functor F^R in (3.2.22) sends $\operatorname{Ind} \mathcal{C}_{\mathcal{Z}_2}$ to $\operatorname{Ind} \mathcal{C}_{\mathcal{Z}_1}$. By Lemmas 2.3.10, 2.4.4 and 2.4.6, it is enough to show that the composition

$$\operatorname{MF}_{\mathrm{qcoh}}^{\mathbb{C}^*}(\mathcal{V}_2^\vee, w_2) \xrightarrow{f_2^!} \operatorname{MF}_{\mathrm{qcoh}}^{\mathbb{C}^*}(f_2^*\mathcal{V}_2^\vee, \overline{w}_2) \xrightarrow{g_{2*}} \operatorname{MF}_{\mathrm{qcoh}}^{\mathbb{C}^*}(\mathcal{V}_3^\vee, w_3) \qquad (3.2.26)$$

$$\xrightarrow{g_1^*} \operatorname{MF}_{\mathrm{qcoh}}^{\mathbb{C}^*}(f_1^*\mathcal{V}_1^\vee, w_1) \xrightarrow{f_{1*}} \operatorname{MF}_{\mathrm{qcoh}}^{\mathbb{C}^*}(\mathcal{V}_1^\vee, w_1)$$

sends $\operatorname{MF}_{\mathrm{qcoh}}^{\mathbb{C}^*}(\mathcal{V}_2^\vee, w_2)_{\mathcal{Z}_2}$ to $\operatorname{MF}_{\mathrm{qcoh}}^{\mathbb{C}^*}(\mathcal{V}_1^\vee, w_1)_{\mathcal{Z}_1}$. By the assumption and the derived base change, the above composition is isomorphic to the composition

$$\operatorname{MF}_{\mathrm{qcoh}}^{\mathbb{C}^*}(\mathcal{V}_2^\vee, w_2) \xrightarrow{f_2^!} \operatorname{MF}_{\mathrm{qcoh}}^{\mathbb{C}^*}(f_2^*\mathcal{V}_2^\vee, \overline{w}_2) \xrightarrow{r_2^*} \operatorname{MF}_{\mathrm{qcoh}}^{\mathbb{C}^*}(\mathcal{W}, \overline{w}_3) \xrightarrow{q_{1*}} \operatorname{MF}_{\mathrm{qcoh}}^{\mathbb{C}^*}(\mathcal{V}_1^\vee, w_1).$$

Let us take an object $A \in \mathrm{MF}_{\mathrm{qcoh}}^{\mathbb{C}^*}(\mathcal{V}_2^\vee, w_2)$. Then A is supported on $\mathrm{Crit}(w_2)$, so $r_2^* f_2^!(A)$ is supported on $q_2^{-1}(\mathrm{Crit}(w_2))$. Also since $g_{2*} f_2^! A$ is supported on $\mathrm{Crit}(w_3)$, we have

$$q_{1*}(r_2^* f_2^!(A)|_{\mathcal{W} \setminus q_3^{-1}(\mathrm{Crit}(w_3))}) \cong f_{1*} g_1^*(g_{2*} f_2^!(A)|_{\mathcal{V}_3^\vee \setminus \mathrm{Crit}(w_3)}) \cong 0.$$

From the distinguished triangle in (2.2.8)

$$\Gamma_{q_3^{-1}(\mathrm{Crit}(w_3))}(r_2^* f_2^!(A)) \to r_2^* f_2^!(A) \to r_2^* f_2^!(A)|_{\mathcal{W} \setminus q_3^{-1}(\mathrm{Crit}(w_3))}$$

we have the isomorphism

$$q_{1*} \Gamma_{q_3^{-1}(\mathrm{Crit}(w_3))}(r_2^* f_2^!(A)) \xrightarrow{\cong} q_{1*} r_2^* f_2^!(A). \tag{3.2.27}$$

Since the above object is supported on $\mathrm{Crit}(w_1)$, the object

$$B := \Gamma_{q_1^{-1}(\mathrm{Crit}(w_1))} \circ \Gamma_{q_3^{-1}(\mathrm{Crit}(w_3))}(r_2^* f_2^!(A)) \in \mathrm{MF}_{\mathrm{qcoh}}^{\mathbb{C}^*}(\mathcal{W}, \overline{w}_3) \tag{3.2.28}$$

is supported on $\cap_{i=1}^3 q_i^{-1}(\mathrm{Crit}(w_i))$ such that $q_{1*} B$ is isomorphic to the object (3.2.27).

From the diagram (3.2.19) and Lemma 2.4.2, we have

$$t_0(\Omega_{\mathrm{ev}}[-2]) = \bigcap_{i=1}^3 q_i^{-1}(\mathrm{Crit}(w_i)) \subset \mathcal{W} \tag{3.2.29}$$

and h_i is the restriction of q_i to the above closed substack. Now suppose that $h_2^{-1}(\mathcal{Z}_2) \subset h_1^{-1}(\mathcal{Z}_1)$ and A is supported on $\mathcal{Z}_2$. Then the object (3.2.28) is supported on $h_2^{-1}(\mathcal{Z}_2)$, hence on $h_1^{-1}(\mathcal{Z}_1)$. Therefore the object (3.2.27) is supported on $\mathcal{Z}_1$, and hence the composition of the functors in (3.2.26) sends $\mathrm{MF}_{\mathrm{qcoh}}^{\mathbb{C}^*}(\mathcal{V}_2^\vee, w_2)_{\mathcal{Z}_2}$ to $\mathrm{MF}_{\mathrm{qcoh}}^{\mathbb{C}^*}(\mathcal{V}_1^\vee, w_1)_{\mathcal{Z}_1}$.

Since the functor (3.2.22) sends $\mathrm{Ind}\, \mathcal{C}_{\mathcal{Z}_2}$ to $\mathrm{Ind}\, \mathcal{C}_{\mathcal{Z}_1}$, it descends to the functor

$$\mathrm{Ind}\, D_{\mathrm{coh}}^b(\mathfrak{M}_2) / \mathrm{Ind}\, \mathcal{C}_{\mathcal{Z}_2} \to \mathrm{Ind}\, D_{\mathrm{coh}}^b(\mathfrak{M}_1) / \mathrm{Ind}\, \mathcal{C}_{\mathcal{Z}_1}.$$

Therefore we obtain the desired functor by Lemma 2.3.2 and Proposition 3.2.7. $\square$

3.3 Comparison with Cohomological/Numerical DT Invariants

In this section, we discuss the relationship between DT categories in the previous section with the cohomological DT invariants whose foundation is established in [19]. In [39], it is proved that the periodic cyclic homology of the category of matrix factorization is isomorphic to the $\mathbb{Z}/2$-graded hypercohomology of the perverse sheaf of the vanishing cycles, and the discussion here is based on this result. Then we show that, under some technical assumptions, we relate the Euler characteristic of the periodic cyclic homology of our DT category with the numerical DT invariant defined by the Behrend function.

3.3.1 Periodic Cyclic Homologies (Review)

Here we recall the definition of periodic cyclic homologies for small dg-categories introduced by Keller [70]. Our convention here is due to [39, Section 3.1].

Let $\mathcal{T}$ be a ($\mathbb{Z}$ or $\mathbb{Z}/2$)-graded small dg-category. Recall that the Hochschild complex $(\text{Hoch}(\mathcal{T}), b)$ is defined to be

$$
\text{Hoch}(\mathcal{T}) = \bigoplus_{E \in \mathcal{T}} \text{Hom}_{\mathcal{T}}(E, E) \oplus
$$

$$
\bigoplus_{\substack{n \geq 1, \\ E_0, \ldots, E_n \in \mathcal{T}}} \text{Hom}_{\mathcal{T}^e}(E_n, E_0) \otimes \text{Hom}_{\mathcal{T}}(E_{n-1}, E_n)[1]
$$

$$
\otimes \cdots \otimes \text{Hom}_{\mathcal{T}}(E_0, E_1)[1].
$$

Here $\mathcal{T}^e$ is the dg-category defined by formally adding a closed identity morphism e_E for each $E \in \mathcal{T}$, i.e. $\text{Ob}(\mathcal{T}^e) = \text{Ob}(\mathcal{T})$ and

$$
\text{Hom}_{\mathcal{T}^e}(E, E') = \begin{cases} \text{Hom}_{\mathcal{T}}(E, E'), & E \neq E', \\ \text{Hom}_{\mathcal{T}}(E, E) \oplus \mathbb{C} \cdot e_E, & E = E'. \end{cases}
$$

The differential b is the sum $b = b_1 + b_2$, where b_1 is given by

$$
b_1(a_n, a_{n-1}, \ldots, a_0) = \sum_{i=0}^{n} \pm(a_n \ldots, da_i, \ldots, a_0)
$$

and b_2 is given by

$$b_2(a_n, a_{n-1}, \ldots, a_0) = \pm(a_0 a_n, a_{n-1}, \ldots, a_1) + \sum_{i=0}^{n-1} \pm(a_n, \ldots, a_{i+1} a_i, \ldots, a_0).$$

Moreover there is a Connes differential B of degree -1 defined by

$$B(a_n, a_{n-1}, \ldots, a_0) = \sum_{i=0}^{n} \pm(e_{E_i}, a_{i-1}, \ldots, a_0, a_n, \ldots, a_i)$$

if $a_n \in \mathrm{Hom}_{\mathcal{T}}(E_n, E_0)$ and zero if $a_n \in \mathbb{C} \cdot e_{E_0}$. Here we refer to [39, Example, Section 3.1] for choices of signs. The triple

$$(\mathrm{Hoch}(\mathcal{T}), b, B) \tag{3.3.1}$$

is a mixed complex, i.e. dg-module over $\mathbb{C}[\epsilon]/\epsilon^2$ with $\deg \epsilon = -1$, where ϵ acts by B.

The periodic cyclic homology of $\mathcal{T}$ is defined to be

$$\mathrm{HP}_*(\mathcal{T}) := H^*(\mathrm{Hoch}(\mathcal{T})((u)), b + uB). \tag{3.3.2}$$

Here u is a formal variable of degree 2. The periodic cyclic homology (3.3.2) is also equipped with a natural connection ∇_u, which makes (3.3.2) a ($\mathbb{Z}$ or $\mathbb{Z}/2$)-graded vector bundle over the formal punctured disc $\mathrm{Spf}\,\mathbb{C}((u))$ with a connection (see [39, Example, Section 3.3] for details). Note that $\mathrm{HP}_*(\mathcal{T})$ is 2-periodic since $u : \mathrm{HP}_*(\mathcal{T}) \to \mathrm{HP}_{*+2}(\mathcal{T})$ is an isomorphism.

Below for a $\mathbb{Z}$-graded object $V^\bullet$, we set $V^{\mathbb{Z}/2}$ to be $V^\bullet$ regarded as $\mathbb{Z}/2$-graded object, i.e.

$$V^{\mathbb{Z}/2} = V^{\mathrm{even}} \oplus V^{\mathrm{odd}}. \tag{3.3.3}$$

For a $\mathbb{Z}$-graded dg-category $\mathcal{T}$, we can regard it as a $\mathbb{Z}/2$-graded dg-category $\mathcal{T}^{\mathbb{Z}/2}$ by setting

$$\mathrm{Hom}_{\mathcal{T}^{\mathbb{Z}/2}}(E, F) = \mathrm{Hom}_{\mathcal{T}}(E, F)^{\mathbb{Z}/2}.$$

Then we have the obvious identities

$$\mathrm{HP}_*(\mathcal{T}^{\mathbb{Z}/2}) = \mathrm{HP}_*(\mathcal{T})^{\mathbb{Z}/2} = (\mathrm{HP}_0(\mathcal{T}) \oplus \mathrm{HP}_1(\mathcal{T}))((u)).$$

3.3.2 Periodic Cyclic Homologies for Derived Categories of Factorizations

Let M be a smooth quasi-projective variety with a regular function $w : M \to \mathbb{C}$. Then we have the perverse sheaf of vanishing cycles (see [36, Theorem 5.2.21])

$$\phi_w := \phi_w(\mathbb{Q}_M[\dim M]) \in \mathrm{Perv}(\mathrm{Crit}(w))$$

with a monodromy automorphism $T_w \in \mathrm{Aut}(\phi_w)$.

On the other hand, let $\mathrm{MF}_{\mathrm{coh}}^{\mathbb{Z}/2}(M, w)$ be the $\mathbb{Z}/2$-periodic triangulated categorify of (non-$\mathbb{C}^*$-equivariant) coherent factorizations of $w : M \to \mathbb{C}$ (see Remark 2.2.2). We have its dg-enhancement $L_{\mathrm{coh}}^{\mathbb{Z}/2}(M, w)$, defined to be the $\mathbb{Z}/2$-graded dg-category of coherent factorizations of w, localized by acyclic factorizations. Then the following result is proved in [39]:

Theorem 3.3.1 ([39, Theorem 1.1]) *We have an isomorphism*

$$(\mathrm{HP}_*(L_{\mathrm{coh}}^{\mathbb{Z}/2}(M, w)), \nabla_u) \cong (H^{*+\dim M}(M, \phi_w) \otimes_{\mathbb{Q}} \mathbb{C}((u)), d + T_w/u)$$

as $\mathbb{Z}/2$-graded vector bundles on $\mathrm{Spf}\,\mathbb{C}((u))$ with connections.

Suppose that $\mathbb{C}^*$ acts on M which restricts to the trivial action on $\mu_2 \subset \mathbb{C}^*$, and assume that w is of weight two. Then as in Sect. 2.2.1, we have the derived category of $\mathbb{C}^*$-equivariant factorizations $\mathrm{MF}_{\mathrm{coh}}^{\mathbb{C}^*}(M, w)$, and its dg-enhancement $L_{\mathrm{coh}}^{\mathbb{C}^*}(M, w)$ defined to be the dg-category of $\mathbb{C}^*$-equivariant coherent factorizations of w localized by acyclic factorizations. Then by the following lemma together with Theorem 3.3.1, we can relate the periodic cyclic homology of $L_{\mathrm{coh}}^{\mathbb{C}^*}(M, w)$ with the vanishing cycle cohomology:

Lemma 3.3.2 *We have an isomorphism as $\mathbb{Z}/2$-graded vector spaces over $\mathbb{C}((u))$:*

$$\mathrm{HP}_*(L_{\mathrm{coh}}^{\mathbb{C}^*}(M, w))^{\mathbb{Z}/2} \cong \mathrm{HP}_*(L_{\mathrm{coh}}^{\mathbb{Z}/2}(M, w)).$$

Here $(-)^{\mathbb{Z}/2}$ means regarding $\mathbb{Z}$-graded object as a $\mathbb{Z}/2$-graded object (see (3.3.3)).

Proof Since M is smooth quasi-projective, by Sumihiro's theorem [124] M is covered by $\mathbb{C}^*$-invariant affine open subsets. Then the argument in [39, Proposition 5.1] implies that we can assume that M is affine, $M = \mathrm{Spec}\,R$ for a $\mathbb{C}$-algebra R. Note that the $\mathbb{C}^*$-action on M yields the $\mathbb{Z}$-grading $R^\bullet$ on R such that $w \in R$ is degree two, and $R^{\mathrm{odd}} = 0$.

In this case, it is proved in [39, (4.2), (4.3), (4.4)] that the Hochschild complex of $L_{\mathrm{coh}}^{\mathbb{Z}/2}(M, w)$ is, as a mixed complex, quasi-isomorphic to the Hochschild complex of the second kind (see [107, Section 2.4]) of the $\mathbb{Z}/2$-graded curved dg-algebra $(R^{\mathbb{Z}/2}, 0, w)$, where the differential is zero and w is the curvature. As we will explain below, the same arguments in [39, (4.2), (4.3), (4.4)] literally work in the $\mathbb{Z}$-graded case, and we have the following: the dg-category $L_{\mathrm{coh}}^{\mathbb{C}^*}(M, w)$ is equivalent to the dg-category of finitely generated $\mathbb{Z}$-graded curved dg-modules over $(R^\bullet, 0, w)$ and its Hochschild complex is, as a mixed complex, quasi-isomorphic to the Hochschild complex of the second kind $\mathrm{Hoch}^{\mathrm{II}}(R^\bullet, 0, w)$ of $(R^\bullet, 0, w)$. Explicitly, it is written as

$$\mathrm{Hoch}^{\mathrm{II}}(R^\bullet, 0, w) = R^\bullet \oplus \prod_{n \geq 1} \overbrace{R^\bullet \otimes R^\bullet[1] \otimes \cdots \otimes R^\bullet[1]}^{n+1}$$

with differential the sum of the Hochschild differential and the twisted differential which involves w, see [107, Section 3.1] for details. Then $\mathrm{Hoch}^{\mathrm{II}}(R^\bullet, 0, w)^{\mathbb{Z}/2}$ is nothing but the Hochschild complex of the second kind for $(R^{\mathbb{Z}/2}, 0, w)$, see [107, Section 2.5] for the change of grading. Therefore we have the desired isomorphism.

Here we explain that the arguments of [39, (4.2), (4.3), (4.4)] work in the above $\mathbb{Z}$-graded case. Indeed [39, (4.2), (4.4)] are consequences of [39, Proposition 2.13, Proposition 3.10], which refer to [107, Lemma 1.5.A, Section 2.4]. But the latter reference is formulated for both of $\mathbb{Z}$ and $\mathbb{Z}/2$ grading, so the same argument applies. Also [39, (4.3)] is a consequence of [39, Proposition (3.12)], which refers to [107, Corollary 4.8A]. The latter reference is formulated only for the $\mathbb{Z}/2$-grading case, but we can easily modify the proof for the $\mathbb{Z}$-graded case by using $\mathbb{Z}$-graded version of Orlov's equivalence between the category of graded matrix factorizations and graded singularity category (see [100, Theorem 3.10]).   □

3.3.3 Conjectural Relation with Cohomological DT Theory

Let $\mathfrak{M}$ be a quasi-smooth QCA derived stack. Below, we use notation in Sect. 3.1.3. For a smooth morphism $\alpha \colon \mathfrak{U} \to \mathfrak{M}$ with a presentation $\mathfrak{U} = \mathrm{Spec}\, \mathcal{R}(V \to Y, s)$ as in (2.1.1), we have the perverse sheaf of vanishing cycles

$$\phi_w := \phi_w(\mathbb{Q}_{V^\vee}[\dim V^\vee]) \in \mathrm{Perv}(\mathrm{Crit}(w))$$

associated with the diagram (2.1.5), equipped with a monodromy automorphism $T_w \in \mathrm{Aut}(\phi_\omega)$. By [19], we can glue these perverse sheaves to a global perverse sheaf on $\mathcal{N} = t_0(\Omega_{\mathfrak{M}}[-1])$ with an automorphism

$$\phi_{\mathcal{N}} \in \mathrm{Perv}(\mathcal{N}), \quad T_{\mathcal{N}} \in \mathrm{Aut}(\phi_{\mathcal{N}}) \tag{3.3.4}$$

once we choose an *orientation data* of the d-critical locus $\mathcal{N}$. Here an orientation data is a choice of a square root line bundle of the virtual canonical line bundle

$$K_{\mathcal{N}}^{\mathrm{vir}} = \det(\mathbb{L}_{\Omega_{\mathfrak{M}}[-1]}|_{\mathcal{N}}) \in \mathrm{Pic}(\mathcal{N}).$$

In our case, the following lemma is well-known:

Lemma 3.3.3 *We have a canonical isomorphism* $K_{\mathcal{N}}^{\mathrm{vir}} \cong p_0^* \det(\mathbb{L}_{\mathfrak{M}}|_{\mathcal{M}})^{\otimes 2}$.

Proof The lemma follows from the distinguished triangle

$$p^*\mathbb{L}_{\mathfrak{M}} \to \mathbb{L}_{\Omega_{\mathfrak{M}}[-1]} \to \mathbb{L}_{\Omega_{\mathfrak{M}}[-1]/\mathfrak{M}} = p^*\mathbb{T}_{\mathfrak{M}}[1].$$

□

By the above lemma, we can take a canonical choice of orientation data $K_{\mathcal{N}}^{\mathrm{vir},1/2} = p_0^* \det(\mathbb{L}_{\mathfrak{M}}|_{\mathcal{M}})$, and we define (3.3.4) by this choice of orientation. Let $\mathcal{Z} \subset \mathcal{N}$ be a conical closed substack such that the complement $\mathcal{N}^{\mathrm{ss}} := \mathcal{N} \setminus \mathcal{Z}$ is a scheme (not stack), which is of finite type by the QCA assumption of $\mathfrak{M}$. Then the cohomological DT invariant is defined to be the hypercohomology

$$H^*(\mathcal{N}^{\mathrm{ss}}, \phi_{\mathcal{N}}|_{\mathcal{N}^{\mathrm{ss}}}). \tag{3.3.5}$$

Based on Theorem 3.3.1 and Lemma 3.3.2, we expect the following relation of the periodic cyclic homologies of DT categories with cohomological DT invariants:

Conjecture 3.3.4 We have an isomorphism of $\mathbb{Z}/2$-graded vector bundles on $\mathrm{Spf}\,\mathbb{C}((u))$ with connections

$$(\mathrm{HP}_*(\mathcal{DT}_{\mathrm{dg}}^{\mathbb{C}^*}(\mathcal{N}^{\mathrm{ss}})), \nabla_u)^{\mathbb{Z}/2} \cong (H^{*+\mathrm{vdim}\,\mathcal{M}}(\mathcal{N}^{\mathrm{ss}}, \phi_{\mathcal{N}}|_{\mathcal{N}^{\mathrm{ss}}}) \otimes_{\mathbb{Q}} \mathbb{C}((u)), d + T_{\mathcal{N}}/u).$$

Here vdim $\mathcal{M}$ is the rank of $\mathbb{L}_{\mathfrak{M}}|_{\mathcal{M}}$.

3.3.4   Relation with Numerical DT Invariants

In the situation of the previous subsection, the hypercohomology (3.3.5) is finite dimensional and the numerical DT invariant is defined by

$$\mathrm{DT}(\mathcal{N}^{\mathrm{ss}}) := \chi(H^*(\mathcal{N}^{\mathrm{ss}}, \phi_{\mathcal{N}}|_{\mathcal{N}^{\mathrm{ss}}})) = \int_{\mathcal{N}^{\mathrm{ss}}} \nu_{\mathcal{N}}\, d\chi.$$

Here $\nu_{\mathcal{N}}\colon x \mapsto \chi(\phi_{\mathcal{N}}|_x)$ is a constructible function on $\mathcal{N}^{\mathrm{ss}}$, called *Behrend function* [14]. In this case, the isomorphism in Conjecture 3.3.4 in particular implies the numerical identity between $\mathrm{DT}(\mathcal{N}^{\mathrm{ss}})$ and the Euler characteristics of the periodic cyclic homology of $\mathcal{DT}_{\mathrm{dg}}^{\mathbb{C}^*}(\mathcal{N}^{\mathrm{ss}})$. Here we show that the numerical version of Conjecture 3.3.4 is true under some technical assumptions on $\mathfrak{M}$. We first show the following lemma:

Lemma 3.3.5 *Let* $\mathfrak{M} = \mathfrak{M}_1 \cup \mathfrak{M}_2$ *be a Zariski open cover of* $\mathfrak{M}$. *We set* $M_i = t_0(\mathfrak{M}_i)$, $M_{12} = t_0(\mathfrak{M}_{12})$ *and*

$$\mathcal{N}_i^{\mathrm{ss}} := p_0^{-1}(M_i) \cap \mathcal{N}^{\mathrm{ss}}, \quad \mathcal{N}_{12}^{\mathrm{ss}} := p_0^{-1}(M_{12}) \cap \mathcal{N}^{\mathrm{ss}}.$$

Then we have the distinguished triangle in the derived category of mixed complexes

$$\mathrm{Hoch}(\mathcal{DT}^{\mathbb{C}^*}_{\mathrm{dg}}(\mathcal{N}^{\mathrm{ss}})) \to \mathrm{Hoch}(\mathcal{DT}^{\mathbb{C}^*}_{\mathrm{dg}}(\mathcal{N}^{\mathrm{ss}}_1)) \oplus \mathrm{Hoch}(\mathcal{DT}^{\mathbb{C}^*}_{\mathrm{dg}}(\mathcal{N}^{\mathrm{ss}}_2))$$
$$\to \mathrm{Hoch}(\mathcal{DT}^{\mathbb{C}^*}_{\mathrm{dg}}(\mathcal{N}^{\mathrm{ss}}_{12})).$$

Proof The proof is similar to [39, Proposition 5.1]. We set $\mathcal{W} = \mathcal{M} \setminus \mathcal{M}_1 \subset \mathcal{M}_2$, and define $L_{\mathrm{coh}}(\mathfrak{M})_{\mathcal{W}}$ to be the dg-subcategory of $L_{\mathrm{coh}}(\mathfrak{M})$ consisting of objects whose cohomology sheaves are supported on $\mathcal{W}$. Then by Lemma 8.1.7, for $\mathcal{Z}_i = p_0^{-1}(\mathcal{M}_i) \cap \mathcal{Z}$ we have the exact sequences of dg-categories, i.e. the induced sequences of the homotopy categories are localization sequences

$$L_{\mathrm{coh}}(\mathfrak{M})_{\mathcal{W}}/LC_{p_0^{-1}(\mathcal{W}) \cap \mathcal{Z}} \to \mathcal{DT}^{\mathbb{C}^*}_{\mathrm{dg}}(\mathcal{N}^{\mathrm{ss}}) \to \mathcal{DT}^{\mathbb{C}^*}_{\mathrm{dg}}(\mathcal{N}^{\mathrm{ss}}_1),$$

$$L_{\mathrm{coh}}(\mathfrak{M}_2)_{\mathcal{W}}/LC_{p_0^{-1}(\mathcal{W}) \cap \mathcal{Z}_2} \to \mathcal{DT}^{\mathbb{C}^*}_{\mathrm{dg}}(\mathcal{N}^{\mathrm{ss}}_2) \to \mathcal{DT}^{\mathbb{C}^*}_{\mathrm{dg}}(\mathcal{N}^{\mathrm{ss}}_{12}).$$

By [69, Theorem 3.1], the assignment of dg-categories to the triples (3.3.1) takes exact sequences to the distinguished triangles of mixed complexes. Therefore we have the distinguished triangles of mixed complexes

$$\mathrm{Hoch}(L_{\mathrm{coh}}(\mathfrak{M})_{\mathcal{W}}/LC_{p_0^{-1}(\mathcal{W}) \cap \mathcal{Z}}) \to \mathrm{Hoch}(\mathcal{DT}^{\mathbb{C}^*}_{\mathrm{dg}}(\mathcal{N}^{\mathrm{ss}})) \to \mathrm{Hoch}(\mathcal{DT}^{\mathbb{C}^*}_{\mathrm{dg}}(\mathcal{N}^{\mathrm{ss}}_1)),$$

$$(3.3.6)$$

$$\mathrm{Hoch}(L_{\mathrm{coh}}(\mathfrak{M}_2)_{\mathcal{W}}/LC_{p_0^{-1}(\mathcal{W}) \cap \mathcal{Z}_2}) \to \mathrm{Hoch}(\mathcal{DT}^{\mathbb{C}^*}_{\mathrm{dg}}(\mathcal{N}^{\mathrm{ss}}_2)) \to \mathrm{Hoch}(\mathcal{DT}^{\mathbb{C}^*}_{\mathrm{dg}}(\mathcal{N}^{\mathrm{ss}}_{12})).$$

Since $\mathcal{W} \subset \mathcal{M}_2$, the restriction functor gives a weak equivalence

$$L_{\mathrm{coh}}(\mathfrak{M})_{\mathcal{W}}/LC_{p_0^{-1}(\mathcal{W}) \cap \mathcal{Z}} \xrightarrow{\sim} L_{\mathrm{coh}}(\mathfrak{M}_2)_{\mathcal{W}}/LC_{p_0^{-1}(\mathcal{W}) \cap \mathcal{Z}_2}.$$

By [69, Theorem 2.6], the mixed complexes (3.3.1) are quasi-isomorphisms under weak equivalences of dg-categories. Therefore we have the quasi-isomorphism of mixed complexes

$$\mathrm{Hoch}(L_{\mathrm{coh}}(\mathfrak{M})_{\mathcal{W}}/LC_{p_0^{-1}(\mathcal{W}) \cap \mathcal{Z}}) \xrightarrow{\sim} \mathrm{Hoch}(L_{\mathrm{coh}}(\mathfrak{M}_2)_{\mathcal{W}}/LC_{p_0^{-1}(\mathcal{W}) \cap \mathcal{Z}_2}).$$

The lemma holds from distinguished triangles (3.3.6) together with the above quasi-isomorphism.   □

For a finite dimensional $\mathbb{Z}/2$-graded vector space $V = V_0 \oplus V_1$ over $\mathbb{C}((u))$, its Euler characteristic is defined by

$$\chi(V) := \dim_{\mathbb{C}((u))} V_0 - \dim_{\mathbb{C}((u))} V_1.$$

We have the following proposition:

Proposition 3.3.6 *Let $\mathfrak{M}$ be a quasi-smooth QCA derived stack and $\mathcal{Z} \subset \mathcal{N} = t_0(\Omega_{\mathfrak{M}}[-1])$ is a conical closed substack. Suppose that for each $\mathbb{C}$-valued point $x \in \mathfrak{M}$, there is a Zariski open neighborhood $x \in \mathfrak{M}' \subset \mathfrak{M}$ satisfying the following conditions:*

(i) $\mathfrak{M}'$ is of the form $\mathfrak{M}' = [\mathfrak{U}/G]$, where $\mathfrak{U} = \mathrm{Spec}_Y \mathcal{R}(V \to Y, s)$ for a smooth quasi-projective scheme Y and a section s on a vector bundle V on it, and an affine algebraic group G acts on Y, V, such that s is G-equivariant.

(ii) Let $\mathcal{N}' = \mathcal{N} \times_{t_0(\mathfrak{M})} t_0(\mathfrak{M}')$ which is a conical closed substack in $[V^\vee/G]$, and set $\mathcal{N}'^{\mathrm{ss}} = \mathcal{N}' \cap \mathcal{N}^{\mathrm{ss}}$, $\mathcal{Z}' = \mathcal{N}' \cap \mathcal{Z}$. Then there is a $\mathbb{C}^*$-invariant quasi-projective open substack $\mathcal{W} \subset [V^\vee/G] \setminus \mathcal{Z}'$ satisfying $\mathcal{N}'^{\mathrm{ss}} = \mathcal{N}' \cap \mathcal{W}$.

Then the periodic cyclic homology $\mathrm{HP}_*(\mathcal{DT}_{\mathrm{dg}}^{\mathbb{C}^*}(\mathcal{N}^{\mathrm{ss}}))$ is finite dimensional over $\mathbb{C}((u))$, and we have the identity

$$\chi(\mathrm{HP}_*(\mathcal{DT}_{\mathrm{dg}}^{\mathbb{C}^*}(\mathcal{N}^{\mathrm{ss}}))) = (-1)^{\mathrm{vdim}\,\mathcal{M}} \mathrm{DT}(\mathcal{N}^{\mathrm{ss}}).$$

Proof By the QCA assumption on $\mathfrak{M}$, it is covered by finite number of Zariski open substacks of the form $\mathfrak{M}'$ satisfying the conditions (i), (ii). Therefore by Lemma 3.3.5 and the induction on the number of open covers, we can assume that $\mathfrak{M} = \mathfrak{M}'$. Then $\mathcal{N} = [\mathrm{Crit}(w)/G]$ where $w \colon V^\vee \to \mathbb{C}$ is given as in the diagram (2.1.5), which is G-invariant by the condition (i), and $\mathcal{Z}$ is written as $[Z/G]$ for a G-invariant conical closed subset $Z \subset \mathrm{Crit}(w)$.

By Proposition 2.3.9, we have the equivalence

$$\mathcal{DT}^{\mathbb{C}^*}(\mathcal{N}^{\mathrm{ss}}) = D_{\mathrm{coh}}^b([\mathfrak{U}/G])/\mathcal{C}_{[Z/G]} \xrightarrow{\sim} \mathrm{MF}_{\mathrm{coh}}^{\mathbb{C}^*}([(V^\vee \setminus Z)/G], w).$$

We take $\mathcal{W}$ as in the condition (ii), and write $\mathcal{W} = [W/G]$ for a G-invariant open subscheme $W \subset V^\vee \setminus Z$ on which the G-action is free. Since $\mathrm{Crit}(w) \setminus Z = \mathrm{Crit}(w|_W)$, the restriction functor gives an equivalence (see (2.2.7))

$$\mathrm{MF}_{\mathrm{coh}}^{\mathbb{C}^*}([(V^\vee \setminus Z)/G], w) \xrightarrow{\sim} \mathrm{MF}_{\mathrm{coh}}^{\mathbb{C}^*}(\mathcal{W}, w).$$

The above equivalence obviously lifts to a weak equivalence of the dg-enhancements, so by [69, Theorem 2.6] we have the quasi-isomorphism of mixed complexes

$$\mathrm{Hoch}(\mathcal{DT}_{\mathrm{dg}}^{\mathbb{C}^*}(\mathcal{N}^{\mathrm{ss}})) \xrightarrow{\sim} \mathrm{Hoch}(L_{\mathrm{coh}}^{\mathbb{C}^*}(\mathcal{W}, w)).$$

Moreover we have

$$\mathrm{vdim}\,\mathcal{M} = \dim Y - \mathrm{rank}(V) - \dim G, \quad \dim \mathcal{W} = \dim Y + \mathrm{rank}(V) - \dim G,$$

hence vdim $\mathcal{M} \equiv \dim \mathcal{W}$ (mod 2). Therefore the proposition holds by applying Theorem 3.3.1 and Lemma 3.3.2 for $w \colon \mathcal{W} \to \mathbb{C}$.   □

Remark 3.3.7 The assumption of Proposition 3.3.6 is automatically satisfied if $t_0(\mathfrak{M})$ is a quasi-projective scheme by [22, Theorem 4.1]. In this case, the algebraic group G can be taken to be trivial.

Remark 3.3.8 The existence of a local chart in Proposition 3.3.6 (i) does not follow from an argument of Proposition 3.1.5. In Proposition 3.1.5, we assumed that $\mathfrak{M}$ is assumed to have a good moduli space, while it is not assumed in Proposition 3.3.6. Also in Proposition 3.1.5, a local chart is taken étale locally while it is Zariski local in Proposition 3.3.6.

3.4 DT Categories for Local Surfaces

In this section, we introduce DT categories for compactly supported (semi)stable coherent sheaves on local surfaces, following the constructions in Sect. 3.2. We then formulate a conjecture on the wall-crossing formula of DT categories of one dimensional stable sheaves from the viewpoint of d-critical birational geometry.

3.4.1 Derived Moduli Stacks of Coherent Sheaves on Surfaces

Let S be a smooth projective surface over $\mathbb{C}$. We consider the derived Artin stack

$$\mathfrak{Perf}_S \colon dAff^{op} \to SSets \tag{3.4.1}$$

which sends an affine derived scheme T to the ∞-groupoid of perfect complexes on $T \times S$, constructed in [147]. We have the open substack

$$\mathfrak{M}_S \subset \mathfrak{Perf}_S$$

corresponding to perfect complexes on S quasi-isomorphic to coherent sheaves on S. Since any object in $\mathrm{Coh}(S)$ is perfect as S is smooth, the derived Artin stack $\mathfrak{M}_S$ is the derived moduli stack of objects in $\mathrm{Coh}(S)$.

Let $\mathcal{M}_S := t_0(\mathfrak{M}_S)$ and take the universal families

$$\mathfrak{F} \in D^b_{\mathrm{coh}}(S \times \mathfrak{M}_S), \ \mathcal{F} := \mathfrak{F}|_{S \times \mathcal{M}_S} \in D^b_{\mathrm{coh}}(S \times \mathcal{M}_S). \tag{3.4.2}$$

Then the perfect obstruction theory on $\mathcal{M}_S$ induced by the cotangent complex of $\mathfrak{M}_S$ is given by

$$\mathcal{E}^{\bullet} = \left(\mathbf{R}p_{\mathcal{M}*}\mathbf{R}\mathcal{H}om_{S \times \mathcal{M}_S}(\mathcal{F}, \mathcal{F})[1]\right)^{\vee} \to \tau_{\geq -1}\mathbb{L}_{\mathcal{M}_S}. \tag{3.4.3}$$

Here $p_\mathcal{M} \colon S \times \mathcal{M}_S \to \mathcal{M}_S$ is the projection. By the above description of the perfect obstruction theory, the derived moduli stack $\mathfrak{M}_S$ is quasi-smooth.

Let $N(S)$ be the numerical Grothendieck group of S

$$N(S) := K(S)/\equiv$$

where $F_1, F_2 \in K(S)$ satisfies $F_1 \equiv F_2$ if $\mathrm{ch}(F_1) = \mathrm{ch}(F_2)$. Then $N(S)$ is a finitely generated free abelian group. We have the decompositions into open and closed substacks

$$\mathfrak{M}_S = \coprod_{v \in N(S)} \mathfrak{M}_S(v), \ \mathcal{M}_S = \coprod_{v \in N(S)} \mathcal{M}_S(v)$$

where each component corresponds to sheaves F with $[F] = v$.

Note that the automorphism group of a sheaf F on S contains a one dimensional torus $\mathbb{C}^* \subset \mathrm{Aut}(F)$ given by the scalar multiplication. Therefore the inertia stack $I_{\mathfrak{M}_S}$ of $\mathfrak{M}_S$ admits an embedding $(\mathbb{C}^*)_{\mathfrak{M}_S} \subset I_{\mathfrak{M}_S}$. As in Sect. 3.2.4, we have the $\mathbb{C}^*$-rigidification

$$\mathfrak{M}_S(v) \to \mathfrak{M}_S^{\mathbb{C}^*\text{-rig}}(v). \tag{3.4.4}$$

3.4.2   Moduli Stacks of Compactly Supported Sheaves on Local Surfaces

For a smooth projective surface S, we consider its total space of the canonical line bundle, called *local surface*:

$$X = \mathrm{Tot}_S(\omega_S) \xrightarrow{\pi} S.$$

Here π is the projection. We denote by $\mathrm{Coh}_{\mathrm{cpt}}(X) \subset \mathrm{Coh}(X)$ the subcategory of compactly supported coherent sheaves on X. We consider the classical Artin stack

$$\mathcal{M}_X : Aff^{\mathrm{op}} \to Groupoid$$

whose T-valued points for $T \in Aff$ form the groupoid of T-flat families of objects in $\mathrm{Coh}_{\mathrm{cpt}}(X)$. We have the decomposition into open and closed substacks

$$\mathcal{M}_X = \coprod_{v \in N(S)} \mathcal{M}_X(v)$$

where $\mathcal{M}_X(v)$ corresponds to compactly supported sheaves F on X with $[\pi_* F] = v$. By pushing forward to S, we have the natural morphism

$$\pi_* \colon \mathcal{M}_X(v) \to \mathcal{M}_S(v), \quad F \mapsto \pi_* F. \tag{3.4.5}$$

The following lemma is well-known (cf. [87, Lemma 5.4]).

Lemma 3.4.1 *We have an isomorphism of stacks over* $\mathcal{M}_S(v)$

$$\eta \colon \mathcal{M}_X(v) \xrightarrow{\cong} t_0(\Omega_{\mathfrak{M}_S(v)}[-1]) = \mathrm{Obs}^*(\mathcal{E}^\bullet|_{\mathcal{M}_S(v)}). \tag{3.4.6}$$

Proof The lemma is easily proved by noting that the fiber of (3.4.5) at $[F] \in \mathcal{M}_S(v)$ is the space of $\mathcal{O}_X$-module structures on F, that is

$$(\pi_*)^{-1}([F]) = \mathrm{Hom}(F, F \otimes \omega_S) = \mathrm{Ext}^2(F, F)^\vee = \mathrm{Spec}\, S(\mathcal{H}^1(\mathcal{E}^{\bullet\vee}|_{[F]})).$$

$\square$

Similarly to (3.4.4), we have the $\mathbb{C}^*$-rigidification

$$\mathcal{M}_X(v) \to \mathcal{M}_X(v)^{\mathbb{C}^*\text{-rig}}.$$

An argument similar to Lemma 3.4.1 shows the isomorphism

$$\mathcal{M}_X(v)^{\mathbb{C}^*\text{-rig}} \xrightarrow{\cong} t_0(\Omega_{\mathfrak{M}_S(v)^{\mathbb{C}^*\text{-rig}}}[-1]). \tag{3.4.7}$$

3.4.3 Definition of DT Category for Local Surfaces

Let $A(S)_{\mathbb{C}}$ be the complexified ample cone

$$A(S)_{\mathbb{C}} := \{B + iH : H \text{ is ample}\} \subset \mathrm{NS}(S)_{\mathbb{C}}.$$

For $\sigma = B + iH \in A(S)_{\mathbb{C}}$ and an object $F \in \mathrm{Coh}_{\mathrm{cpt}}(X)$, its B-twisted reduced Hilbert polynomial is defined by

$$\overline{\chi}_\sigma(F, m) := \frac{\chi(\pi_* F \otimes \mathcal{O}_S(-B + mH))}{c} \in \mathbb{R}[m]. \tag{3.4.8}$$

Here c is the coefficient of the highest degree term of the polynomial $\chi(\pi_* F \otimes \mathcal{O}_S(-B + mH))$ in m.

Remark 3.4.2 Note that the notation $\mathcal{O}_S(-B + mH)$ makes sense only when σ is an integral class. But the right hand side of (3.4.8) is determined by the numerical class of σ, so it also makes sense for any σ.

The following definition gives the notion of B-twisted H-Gieseker stability for compactly supported sheaves on X.

Definition 3.4.3 An object $F \in \mathrm{Coh}_{\mathrm{cpt}}(X)$ is σ-(semi)stable if it is a pure dimensional sheaf, and for any subsheaf $0 \neq F' \subsetneq F$ we have $\overline{\chi}_\sigma(F', m) < (\leq)\overline{\chi}_\sigma(F, m)$ for $m \gg 0$.

We have the open substacks

$$\mathcal{M}_X^{\sigma\text{-st}}(v) \subset \mathcal{M}_X^\sigma(v) \subset \mathcal{M}_X(v)$$

corresponding to σ-stable sheaves, σ-semistable sheaves, respectively. By the GIT construction of the moduli stack $\mathcal{M}_X^\sigma(v)$, it admits a good moduli space

$$\mathcal{M}_X^\sigma(v) \to M_X^\sigma(v) \qquad (3.4.9)$$

where $M_X^\sigma(v)$ is a quasi-projective scheme whose closed points correspond to σ-polystable objects, i.e. direct sum of σ-stable objects with the same B-twisted reduced Hilbert polynomials. Suppose that the following condition holds:

$$\mathcal{M}_X^{\sigma\text{-st}}(v) = \mathcal{M}_X^\sigma(v). \qquad (3.4.10)$$

In this case, the morphism (3.4.9) is a $\mathbb{C}^*$-gerbe so that

$$M_X^\sigma(v) = \mathcal{M}_X^\sigma(v)^{\mathbb{C}^*\text{-rig}}$$

holds. Note that it admits a $\mathbb{C}^*$-action induced by the fiberwise weight two $\mathbb{C}^*$-action on $\pi \colon X \to S$.

The moduli stack of σ semistable sheaves $\mathcal{M}_X^\sigma(v)$ is of finite type, while $\mathfrak{M}_S(v)$ is not quasi-compact in general, so in particular it is not QCA. So we take a quasi-compact open derived substack $\mathfrak{M}_S(v)_{\mathrm{qc}} \subset \mathfrak{M}_S(v)$ satisfying the condition

$$\pi_*(\mathcal{M}_X^\sigma(v)) \subset t_0(\mathfrak{M}_S(v)_{\mathrm{qc}}). \qquad (3.4.11)$$

Here π_* is the morphism (3.4.5). Note that $\mathfrak{M}_S(v)_{\mathrm{qc}}$ is QCA. By the isomorphism η in Lemma 3.4.1, we have the conical closed substack

$$\mathcal{Z}^{\sigma\text{-us}} := t_0(\Omega_{\mathfrak{M}_S(v)_{\mathrm{qc}}}[-1]) \setminus \eta(\mathcal{M}_X^\sigma(v)) \subset t_0(\Omega_{\mathfrak{M}_S(v)_{\mathrm{qc}}}[-1]).$$

By taking the $\mathbb{C}^*$-rigidification and using the isomorphism (3.4.7), we also have the conical closed substack

$$(\mathcal{Z}^{\sigma\text{-us}})^{\mathbb{C}^*\text{-rig}} \subset t_0(\Omega_{\mathfrak{M}_S(v)_{\mathrm{qc}}^{\mathbb{C}^*\text{-rig}}}[-1]).$$

By Definition 3.2.2, the DT category for $M_X^\sigma(v)$ is defined as follows:

Definition 3.4.4 The $\mathbb{C}^*$-equivariant DT category for $\mathcal{M}_X^\sigma(v)$ is defined as

$$\mathcal{DT}^{\mathbb{C}^*}(\mathcal{M}_X^\sigma(v)) := D_{\mathrm{coh}}^b(\mathfrak{M}_S(v)_{\mathrm{qc}})/\mathcal{C}_{\mathcal{Z}^{\sigma\text{-us}}}.$$

When the condition (3.4.10) holds, the $\mathbb{C}^*$-equivariant DT category for the moduli space $M_X^\sigma(v)$ is defined by

$$\mathcal{DT}^{\mathbb{C}^*}(M_X^\sigma(v)) := D_{\mathrm{coh}}^b(\mathfrak{M}_S(v)_{\mathrm{qc}}^{\mathbb{C}^*\text{-rig}})/\mathcal{C}_{(\mathcal{Z}^{\sigma\text{-us}})^{\mathbb{C}^*\text{-rig}}}.$$

The dg-enhancements, λ-twisted version, $\widehat{\mathcal{DT}}$-version and ind-version are similarly defined following Sects. 3.2.1, 3.2.2, 3.2.4.

Remark 3.4.5 By Lemma 3.2.9, the DT categories in Definition 3.4.4 are independent of a choice of a quasi-compact open substack $\mathfrak{M}_S(v)_{\mathrm{qc}}$ of $\mathfrak{M}_S(v)$ satisfying (3.4.11), up to equivalence.

Remark 3.4.6 The DT category $\mathcal{DT}^{\mathbb{C}^*}(\mathcal{M}_X^\sigma(v))$ decomposes into the direct sum

$$\mathcal{DT}^{\mathbb{C}^*}(\mathcal{M}_X^\sigma(v)) = \bigoplus_{\lambda\in\mathbb{Z}}\mathcal{DT}^{\mathbb{C}^*}(\mathcal{M}_X^\sigma(v))_\lambda$$

where each summand corresponds to λ-twisted version (see Sect. 3.2.4).

Note that we have the fully-faithful functor

$$i_*\colon \mathrm{Coh}(S) \to \mathrm{Coh}_{\mathrm{cpt}}(X)$$

where i is the zero section of $\pi\colon X \to S$. The image of the above functor is closed under subobjects and quotients. Therefore an object $F \in \mathrm{Coh}(S)$ is σ-(semi)stable if and only if i_*F is σ-(semi)stable in Definition 3.4.3. Let us also take open substacks

$$\mathcal{M}_S^{\sigma\text{-st}}(v) \subset \mathcal{M}_S^\sigma(v) \subset \mathcal{M}_S(v),\ \mathfrak{M}_S^{\sigma\text{-st}}(v) \subset \mathfrak{M}_S^\sigma(v) \subset \mathfrak{M}_S(v) \qquad (3.4.12)$$

consisting of σ-(semi)stable sheaves on S.

Lemma 3.4.7 *For the map π_* in (3.4.5), we have*

$$(\pi_*)^{-1}(\mathcal{M}_S^{\sigma\text{-st}}(v)) \subset \mathcal{M}_X^{\sigma\text{-st}}(v),\ (\pi_*)^{-1}(\mathcal{M}_S^\sigma(v)) \subset \mathcal{M}_X^\sigma(v). \qquad (3.4.13)$$

Proof For a compactly supported coherent sheaf E on X, suppose that it is not σ-(semi)stable. Then a destabilizing subsheaf $F \subset E$ gives a subsheaf $\pi_*F \subset \pi_*E$ as π is affine, which destabilizes π_*E. $\qquad\Box$

Lemma 3.4.8 *Suppose that (3.4.10) holds, and the left open immersion (3.4.13) is an isomorphism. Then we have the equivalence*

$$\mathcal{DT}^{\mathbb{C}^*}(M_X^\sigma(v)) \overset{\sim}{\to} D_{\mathrm{coh}}^b(\mathfrak{M}_S^\sigma(v)^{\mathbb{C}^*\text{-rig}})$$

and the following identity

$$\mathrm{DT}(M_X^\sigma(v)) = (-1)^{\mathrm{vdim}\,\mathcal{M}_S(v)+1} \chi(\mathrm{HP}_*(\mathcal{DT}_{\mathrm{dg}}^{\mathbb{C}^*}(M_X^\sigma(v)))).$$

Proof The first statement follows from Lemma 3.2.9. The second statement follows from Proposition 3.3.6 since $\mathcal{M}_S^{\sigma\text{-st}}(v)^{\mathbb{C}^*\text{-rig}}$ is a quasi-projective scheme (see Remark 3.3.7).   □

Chapter 4
D-Critical D/K Equivalence Conjectures

The purpose of this chapter is to formulate d-critical D/K equivalence conjectures using the definition of DT categories introduced in the last chapter.

We first formulate wall-crossing equivalence of DT categories for one dimensional stable sheaves. It also categorifies wall-crossing invariance of genus zero Gopakumar-Vafa invariants.

We then construct DT categories for stable D0-D2-D6 bound states on local surfaces and formulate d-critical D/K equivalence conjectures. For an arbitrary CY three-fold X, the category of D0-D2-D6 bound states is defined by

$$\mathcal{A}_X := \langle \mathcal{O}_X, \mathrm{Coh}_{\leq 1}(X)[-1] \rangle_{\mathrm{ex}},$$

which is known to be an abelian category. The above category was first introduced in [127] in order to prove MNOP/PT correspondence conjecture. The category $\mathcal{A}_X$ contains several interesting geometric objects which have been studied in the context of curve counting DT theory, e.g. ideal sheaves of one or zero dimensional closed subschemes, Pandharipande-Thomas stable pairs, etc.

Indeed there is a stability parameter $t \in \mathbb{R}$ which defines $\mu_t^{\dagger}$-stability on $\mathcal{A}_X$, and the above MNOP/PT moduli spaces are realized as certain stable objects with respect to the $\mu_t^{\dagger}$-stability. The wall-crossing formula of the associated DT invariants was used in [128, 129] to prove the rationality conjecture of the generating series of PT invariants. For each $t \in \mathbb{R}$ and a numerical class (β, n), where β is a curve class and $n \in \mathbb{Z}$, we have the moduli space of $\mu_t^{\dagger}$-stable objects in $\mathcal{A}_X$ with numerical class (β, n), denoted by $P_n^t(X, \beta)$. We will introduce the corresponding DT category for the local surface $X = \mathrm{Tot}_S(\omega_S)$, denoted by

$$\mathcal{DT}^{\mathbb{C}^*}(P_n^t(X, \beta)).$$

© The Author(s), under exclusive license to Springer Nature Switzerland AG 2024
Y. Toda, *Categorical Donaldson-Thomas Theory for Local Surfaces*, Lecture Notes
in Mathematics 2350, https://doi.org/10.1007/978-3-031-61705-8_4

We will then formulate categorical wall-crossing for the above DT categories, based on d-critical analogue of D/K equivalence conjecture. Among them, for generic $t_1 > t_2 > 0$ we propose the existence of a fully-faithful functor

$$\mathcal{DT}^{\mathbb{C}^*}(P_n^{t_2}(X, \beta)) \hookrightarrow \mathcal{DT}^{\mathbb{C}^*}(P_n^{t_1}(X, \beta)).$$

In particular there should be a fully-faithful functor from PT category to MNOP category.

The organization of this chapter is as follows. In Sect. 4.1, we formulate categorical wall-crossing equivalence of one dimensional stable sheaves. In Sect. 4.2, we introduce moduli stacks of pairs on surfaces and moduli stacks of D0-D2-D6 bound states on local surfaces. In Sect. 4.3, we introduce DT categories for stable D0-D2-D6 bound states, and formulate our main conjecture. In Sect. 4.4, we give an example of our conjecture for a local $(-1, -1)$-curve.

4.1 Categorical DT Theory for One Dimensional Stable Sheaves

4.1.1 Moduli Stacks of One Dimensional Stable Sheaves

Let S be a smooth projective surface and $X = \mathrm{Tot}(\omega_S)$ the local surface. Here we focus on the case of moduli spaces of one dimensional stable sheaves on X. We denote by

$$\mathrm{Coh}_{\leq 1}(X) \subset \mathrm{Coh}_{\mathrm{cpt}}(X)$$

the abelian subcategory consisting of sheaves F with $\dim \mathrm{Supp}(F) \leq 1$. We also denote by $N_{\leq 1}(S)$ the subgroup of $N(S)$ spanned by sheaves $F \in \mathrm{Coh}_{\leq 1}(S)$. Note that we have an isomorphism

$$N_{\leq 1}(S) \xrightarrow{\cong} \mathrm{NS}(S) \oplus \mathbb{Z}, \ F \mapsto (l(F), \chi(F))$$

where $l(F)$ is the fundamental one cycle of F. Below we identify an element $v \in N_{\leq 1}(S)$ with $(\beta, n) \in \mathrm{NS}(S) \oplus \mathbb{Z}$ by the above isomorphism. We often write $[F] = (\beta, n)$ for $F \in \mathrm{Coh}_{\leq 1}(S)$ by the above isomorphism.

Under the above notation, we can interpret the σ-(semi)stability on $\mathrm{Coh}_{\leq 1}(X)$ for $\sigma \in A(S)_{\mathbb{C}}$ in terms of Bridgeland stability conditions [24], see Definition 3.4.3 for the definition of σ-(semi)stability. Namely for $F \in \mathrm{Coh}_{\leq 1}(X)$, it is σ-(semi)stable for $\sigma = B + iH$ if and only if it is Bridgeland (semi)stable with respect to the central charge

$$Z_\sigma : N_{\leq 1}(S) \to \mathbb{C}, \ (\beta, n) \mapsto -n + (B + iH)\beta,$$

i.e. for any subsheaf $0 \neq F' \subsetneq F$ we have $\arg Z_\sigma(F') < (\leq) \arg Z_\sigma(F)$, equivalently $\mu_\sigma(F') < (\leq)\mu_\sigma(F)$ where $\mu_\sigma(F)$ is

$$\mu_\sigma(F) := -\frac{\operatorname{Re} Z_\sigma(\pi_* F)}{\operatorname{Im} Z_\sigma(\pi_* F)} = \frac{n - B \cdot \beta}{H \cdot \beta}, \tag{4.1.1}$$

where $[\pi_* F] = (\beta, n)$. We give a description of $\mathcal{DT}^{\mathbb{C}^*}(M_X^\sigma(v))$ for $v \in N_{\leq 1}(S)$ in a simple example.

Example 4.1.1 Suppose that $C \subset S$ is a (-3)-curve, i.e.

$$C \cong \mathbb{P}^1, \ C^2 = -3, \ \omega_S|_C \cong \mathcal{O}_C(1).$$

Let us take $v = (2[C], 1) \in N_{\leq 1}(S)$ and $\sigma = iH$ for an ample divisor H on S. Then one can easily show that any compactly supported σ-stable coherent sheaf E on X with $[\pi_* E] = v$ satisfies $\pi_* E \cong F$, where $F := \mathcal{O}_C \oplus \mathcal{O}_C(-1)$. Let us set

$$V_F := \operatorname{Hom}(F, F \otimes \omega_S), \ G_F := \operatorname{Aut}(F).$$

Then we have an open immersion $\mathcal{M}_X^\sigma(v) \subset [V_F/G_F]$, so we can take $\mathfrak{M}_S(v)_{\mathrm{qc}}^{\mathbb{C}^*\text{-rig}}$ to be

$$\mathfrak{M}_S(v)_{\mathrm{qc}}^{\mathbb{C}^*\text{-rig}} = [\operatorname{Spec} S(V_F[1])/\mathbb{P}G_F].$$

Here $\mathbb{P}G_F = G_F/\mathbb{C}^*$, where $\mathbb{C}^*$ acts on F by scalar multiplication. Note that we have

$$V_F = \begin{pmatrix} H^0(\mathcal{O}_C(1)) & H^0(\mathcal{O}_C) \\ H^0(\mathcal{O}_C(2)) & H^0(\mathcal{O}_C(1)) \end{pmatrix} = \begin{pmatrix} \mathbb{C}^2 & \mathbb{C} \\ \mathbb{C}^3 & \mathbb{C}^2 \end{pmatrix}. \tag{4.1.2}$$

Let $V_F' \subset V_F$ be the codimension one linear subspace given by the zero locus of the top right component of (4.1.2). Then one can show that

$$(\mathcal{Z}^{\sigma\text{-us}})^{\mathbb{C}^*\text{-rig}} = [V_F'/\mathbb{P}G_F] = [V_F/\mathbb{P}G_F] \setminus \mathcal{M}_X^\sigma(v).$$

We have $V_F = V_F' \oplus V_F''$ for the one dimensional subspace $V_F'' \subset V_F$ generated by the top right component of (4.1.2), so

$$\operatorname{Spec} S(V_F[1]) = \mathfrak{U}' \times \mathfrak{U}'', \ \mathfrak{U}' = \operatorname{Spec} S(V_F'[1]), \ \mathfrak{U}'' = \operatorname{Spec} S(V_F''[1]).$$

Then from Example 2.1.9 (i), the subcategory

$$\mathcal{C}_{(\mathcal{Z}^{\sigma\text{-us}})^{\mathbb{C}^*\text{-rig}}} \subset D_{\mathrm{coh}}^b(\mathfrak{M}_S(v)_{\mathrm{qc}}^{\mathbb{C}^*\text{-rig}})$$

may be the subcategory of $\mathbb{P}G_F$-equivariant coherent sheaves on $\mathrm{Spec}\, S(V_F[1])$, which lie on the thick closure of $D^b_{\mathrm{coh}}(\mathfrak{U}') \boxtimes \mathrm{Perf}(\mathfrak{U}'')$. By taking the quotients, the DT category may be heuristically described as

$$\mathcal{DT}^{\mathbb{C}^*}(M^\sigma_X(v)) = \left(D^b_{\mathrm{coh}}(\mathfrak{U}') \otimes D_{\mathrm{sg}}(\mathfrak{U}'') \right)^{\mathbb{P}G_F}.$$

Here $(-)^{\mathbb{P}G_F}$ refers to the $\mathbb{P}G_F$-equivariant category.

4.1.2   Categorical Wall-Crossing for One Dimensional Sheaves

We formulate wall-crossing formula of DT categories of one dimensional stable sheaves. We fix a primitive element $v = (\beta, n) \in N_{\leq 1}(S)$ such that $\beta > 0$. Here we write $\beta > 0$ if $\beta = [C]$ for a non-zero effective divisor C on S. For each decomposition

$$v = v_1 + v_2, \ v_i = (\beta_i, n_i), \ \beta_i > 0$$

we define

$$W_{v_1, v_2} := \{\sigma \in A(S)_{\mathbb{C}} : \mu_\sigma(v_1) = \mu_\sigma(v_2)\}$$
$$= \{B + iH \in A(S)_{\mathbb{C}} : (n_1\beta_2 - n_2\beta_1) \cdot H = B\beta_1 \cdot H\beta_2 - B\beta_2 \cdot H\beta_1\}.$$

Since v is primitive, $W_{v_1, v_2} \subsetneq A(S)_{\mathbb{C}}$ and W_{v_1, v_2} is a real codimension one hypersurface in $A(S)_{\mathbb{C}}$. For a fixed v, the set of hypersurfaces of the form W_{v_1, v_2} for $v = v_1 + v_2$ are called *walls*. It is easy to see that the walls are locally finite. Also each connected component

$$\mathcal{C} \subset A(S)_{\mathbb{C}} \setminus \bigcup_{v_1 + v_2 = v} W_{v_1, v_2}$$

is called a *chamber*. From the construction of walls, the moduli stacks $\mathcal{M}^\sigma_S(v)$, $\mathcal{M}^\sigma_X(v)$ are constant if σ is contained in a chamber, but may change when σ crosses a wall. Moreover if σ lies in a chamber, they consist of σ-stable sheaves, i.e. the condition (3.4.10) holds for both of X and S.

Let us take $\sigma \in A(S)_{\mathbb{C}}$ which lies on a wall and take $\sigma_{\pm} \in A(S)_{\mathbb{C}}$ which lie on adjacent chambers. Then by Toda [137, Theorem 8.3], the wall-crossing diagram

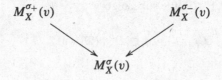

is a *d-critical flop*, which is a d-critical analogue of usual flops in birational geometry. Therefore following the discussion in Sect. 1.1.6, we propose the following conjecture:

Conjecture 4.1.2 There exists an equivalence of $\mathbb{C}^*$-equivariant DT categories:

$$\mathcal{DT}^{\mathbb{C}^*}(M_X^{\sigma_+}(v)) \xrightarrow{\sim} \mathcal{DT}^{\mathbb{C}^*}(M_X^{\sigma_-}(v)).$$

In particular $\mathcal{DT}^{\mathbb{C}^*}(M_X^\sigma(v))$ is independent of a choice of generic σ up to equivalence.

Remark 4.1.3 The DT invariants counting one dimensional (semi)stable sheaves are known as genus zero Gopakumar-Vafa invariants. It is well-known that they are independent of a choice of a stability condition (see [64, Theorem 6.16]). By Lemma 3.4.8, the equivalence in Conjecture 4.1.2 recovers this fact under the assumption in Proposition 3.3.6.

4.1.3 The Case of Irreducible Curve Class

Let us take $v = (\beta, n) \in N_{\leq 1}(S)$ such that β is irreducible (see Sect. 1.6.2). Following the notation of [103], we denote the open substacks of pure one dimensional sheaves by

$$\mathcal{M}_n(X, \beta) \subset \mathcal{M}_X(v), \; M_n(X, \beta) \subset \mathcal{M}_X(v)^{\mathbb{C}^*\text{-rig}}, \; \mathfrak{M}_n(S, \beta) \subset \mathfrak{M}_S(v),$$
$$(4.1.3)$$

respectively. As β is irreducible, the stacks (4.1.3) coincide with the substacks of σ-stable sheaves on X, S respectively, for any choice of σ. We have the following:

Lemma 4.1.4 *For $v = (\beta, n) \in N_{\leq 1}(S)$ such that β is irreducible, we have the equivalences*

$$\mathcal{DT}^{\mathbb{C}^*}(\mathcal{M}_n(X, \beta))_\lambda \xrightarrow{\sim} D_{\text{coh}}^b(\mathfrak{M}_n(S, \beta))_\lambda,$$
$$\mathcal{DT}^{\mathbb{C}^*}(M_n(X, \beta)) \xrightarrow{\sim} D_{\text{coh}}^b(\mathfrak{M}_n(S, \beta)^{\mathbb{C}^*\text{-rig}}).$$

Moreover we have the identity

$$\chi(\text{HP}_*(\mathcal{DT}_{\text{dg}}^{\mathbb{C}^*}(M_n(X, \beta)))) = (-1)^{\beta^2+1} N_{n,\beta}.$$
$$(4.1.4)$$

Here $N_{n,\beta} = \text{DT}(M_n(X, \beta))$ is the DT invariant counting one dimensional stable sheaves on X.

Proof By the condition that β is irreducible, the assumption of Lemma 3.4.8 is satisfied (see [87, Lemma 5.4]). Moreover by the Riemann-Roch theorem, we have

$$\operatorname{vdim}\mathcal{M}_S(v) = -\chi(F, F) = \beta^2$$

for a one dimensional sheaf F on S with $[F] = (\beta, n)$. Therefore the lemma holds from Lemma 3.4.8.   □

Later we will also need the following lemma:

Lemma 4.1.5 *For an irreducible class β and $n \in \mathbb{Z}$, we have the following:*

(i) *By setting $d = \gcd\{|\beta \cdot D| : D \in \mathrm{NS}(S), \beta \cdot D \neq 0\}$, there is an equivalence*

$$\mathcal{DT}^{\mathbb{C}^*}(\mathcal{M}_n(X, \beta))_\lambda \xrightarrow{\sim} \mathcal{DT}^{\mathbb{C}^*}(\mathcal{M}_{n+d}(X, \beta))_\lambda. \tag{4.1.5}$$

(ii) *Let $d \in \mathbb{Z}$ be as in (i). By setting $c = \gcd(d, n)$, there is an equivalence*

$$\mathcal{DT}^{\mathbb{C}^*}(\mathcal{M}_n(X, \beta))_\lambda \xrightarrow{\sim} \mathcal{DT}^{\mathbb{C}^*}(\mathcal{M}_n(X, \beta))_{\lambda+c}. \tag{4.1.6}$$

(iii) *There is an equivalence*

$$\mathcal{DT}^{\mathbb{C}^*}(\mathcal{M}_n(X, \beta))_\lambda \xrightarrow{\sim} \mathcal{DT}^{\mathbb{C}^*}(\mathcal{M}_{-n}(X, \beta))_{-\lambda}. \tag{4.1.7}$$

Proof

(i) Let us take $L \in \mathrm{Pic}(S)$ such that $\beta \cdot c_1(L) = d$. Then the equivalence (4.1.5) follows from the equivalence of derived stacks $\otimes L : \mathfrak{M}_n(S, \beta) \xrightarrow{\sim} \mathfrak{M}_{n+d}(S, \beta)$.

(ii) Let $L \in \mathrm{Pic}(S)$ be as above, $\mathfrak{F}$ the universal family on $S \times \mathfrak{M}_n(S, \beta)$, and take

$$\det p_{\mathfrak{M}*}(L^{\otimes k} \boxtimes \mathfrak{F}) \in \mathrm{Pic}(\mathfrak{M}_n(S, \beta)).$$

Here $p_{\mathfrak{M}}$ is the projection onto $\mathfrak{M}_n(S, \beta)$. The inertia $\mathbb{C}^*$-weight of the above line bundle is $kd + n$. Therefore there is a line bundle $\mathcal{L}$ on $\mathfrak{M}_n(S, \beta)$ whose inertia $\mathbb{C}^*$-weight is c. Indeed by writing $c = ad + bn$ for $a, b \in \mathbb{Z}$, we can take $\mathcal{L}$ to be

$$\mathcal{L} = (\det p_{\mathfrak{M}*}(L \boxtimes \mathfrak{F}))^{\otimes a} \otimes (\det p_{\mathfrak{M}*}(\mathfrak{F}))^{\otimes b-a}.$$

Then the equivalence (4.1.6) is obtained by taking the tensor product with $\mathcal{L}$.

(iii) The equivalence (4.1.7) follows from the equivalence of derived stacks $\mathfrak{M}_n(S, \beta) \xrightarrow{\sim} \mathfrak{M}_{-n}(S, \beta)$ given by $F \mapsto \mathcal{E}xt^1(F, \omega_S)$ (see Sect. 5.1.1).   □

4.2 Moduli Stacks of D0-D2-D6 Bound States

4.2.1 Moduli Stacks of Pairs

For a smooth projective surface S, let $\mathfrak{M}_S$ be the derived moduli stack of coherent sheaves on S considered in (3.4.1), and $\mathfrak{F}$ the universal object (3.4.2). We define the derived stack $\mathfrak{M}_S^\dagger$ by

$$\rho^\dagger : \mathfrak{M}_S^\dagger := \mathrm{Spec}_{\mathfrak{M}_S} S((p_{\mathfrak{M}*}\mathfrak{F})^\vee) \to \mathfrak{M}_S.$$

Here $p_\mathfrak{M} : S \times \mathfrak{M}_S \to \mathfrak{M}_S$ is the projection. Its classical truncation is a 1-stack, given by

$$\mathcal{M}_S^\dagger := t_0(\mathfrak{M}_S^\dagger) = \mathrm{Spec}_{\mathcal{M}_S}(S(\mathcal{H}^0((\mathbf{R}p_{\mathcal{M}*}\mathcal{F})^\vee))). \tag{4.2.1}$$

From the above description, the T-valued points of $\mathcal{M}_S^\dagger$ form the groupoid of pairs

$$(F_T, \xi), \quad \mathcal{O}_{S \times T} \xrightarrow{\xi} F_T$$

where $F_T \in \mathrm{Coh}(S \times T)$ is flat over T and ξ is a morphism of coherent sheaves. The isomorphisms of $\mathcal{M}_S^\dagger(T)$ are given by isomorphisms of T-flat families of coherent sheaves on S which commute with sections. In particular, we have the universal pair on $\mathcal{M}_S^\dagger$

$$\mathcal{I}^\bullet = (\mathcal{O}_{S \times \mathcal{M}_S^\dagger} \to \mathcal{F}).$$

The obstruction theory on $\mathcal{M}_S^\dagger$ induced by the cotangent complex of $\mathfrak{M}_S^\dagger$ is given by

$$\mathcal{E}^{\dagger\bullet} = \left(\mathbf{R}p_{\mathcal{M}*}\mathbf{R}\mathcal{H}om_{S \times \mathcal{M}_S^\dagger}(\mathcal{I}^\bullet, \mathcal{F})\right)^\vee \to \tau_{\geq -1}\mathbb{L}_{\mathcal{M}_S^\dagger}. \tag{4.2.2}$$

Also we have the decompositions into open and closed substacks

$$\mathfrak{M}_S^\dagger = \coprod_{v \in N(S)} \mathfrak{M}_S^\dagger(v), \quad \mathcal{M}_S^\dagger = \coprod_{v \in N(S)} \mathcal{M}_S^\dagger(v),$$

where each component corresponds to pairs $(\mathcal{O}_S \to F)$ such that $[F] = v$.

Lemma 4.2.1 *For $v \in N_{\leq 1}(S)$, the derived stack $\mathfrak{M}_S^\dagger(v)$ is quasi-smooth. Moreover the fiber of*

$$t_0(\Omega_{\mathfrak{M}_S^\dagger(v)}[-1]) = \mathrm{Obs}^*(\mathcal{E}^{\dagger\bullet}) \to \mathcal{M}_S^\dagger(v) \tag{4.2.3}$$

at $(\xi \colon \mathcal{O}_S \to F)$ is given by $\mathrm{Hom}(F \otimes \omega_S^{-1}, I^\bullet[1])$, where $I^\bullet$ is the two term complex $(\mathcal{O}_S \xrightarrow{\xi} F)$ such that $\mathcal{O}_S$ is located in degree zero.

Proof For the two term complex $I^\bullet = (\mathcal{O}_S \to F)$, we have the distinguished triangle

$$\mathbf{R}\Gamma(F) \to \mathbf{R}\mathrm{Hom}(I^\bullet, F) \to \mathbf{R}\mathrm{Hom}(F, F)[1].$$

Therefore if $\dim \mathrm{Supp}(F) \leq 1$ then $\mathcal{E}^{\dagger \bullet}|_{(\mathcal{O}_S \to F)}$ has cohomological amplitude $[-1, 1]$. The fiber of the morphism (4.2.3) is dual to the obstruction space $\mathrm{Hom}^1(I^\bullet, F)$, which is $\mathrm{Hom}(F \otimes \omega_S^{-1}, I^\bullet[1])$ by Serre duality (see Lemma 5.2.10 for T-valued fibers).   □

4.2.2 Moduli Stacks of D0-D2-D6 Bound States

Here we introduce the category of D0-D2-D6 bound states on the local surface $X = \mathrm{Tot}_S(\omega_S)$, and discuss the relationship of their moduli spaces with the dual obstruction cone over the stack of pairs $\mathcal{M}_S^\dagger$. Note that we have the compactification of X

$$X \subset \overline{X} := \mathbb{P}_S(\omega_S \oplus \mathcal{O}_S).$$

Definition 4.2.2 The category of D0-D2-D6 bound states on the non-compact CY three-fold $X = \mathrm{Tot}_S(\omega_S)$ is defined by the extension closure in $D_{\mathrm{coh}}^b(\overline{X})$

$$\mathcal{A}_X := \langle \mathcal{O}_{\overline{X}}, \mathrm{Coh}_{\leq 1}(X)[-1] \rangle_{\mathrm{ex}}.$$

Here we regard $\mathrm{Coh}_{\leq 1}(X)$ as a subcategory of $\mathrm{Coh}(\overline{X})$ by the push-forward of the open immersion $X \subset \overline{X}$.

The arguments in [127, Lemma 3.5, Proposition 3.6] show that $\mathcal{A}_X$ is the heart of a triangulated subcategory of $D_{\mathrm{coh}}^b(\overline{X})$ generated by $\mathcal{O}_{\overline{X}}$ and $\mathrm{Coh}_{\leq 1}(X)$. In particular, $\mathcal{A}_X$ is an abelian category. There is a group homomorphism

$$\mathrm{cl} \colon K(\mathcal{A}_X) \to \Gamma := \mathbb{Z} \oplus N_{\leq 1}(S)$$

characterized by the condition that $\mathrm{cl}(\mathcal{O}_X) = (1, 0)$ and $\mathrm{cl}(F[-1]) = (0, [\pi_* F])$ for $F \in \mathrm{Coh}_{\leq 1}(X)$.

Note that an object $E \in \mathcal{A}_X$ is of rank one if and only if we have an isomorphism $\mathbf{L}i_\infty^* E \cong \mathcal{O}_{S_\infty}$, where $i_\infty \colon S_\infty \hookrightarrow \overline{X}$ is the divisor at the infinity. We define the (classical) moduli stack of rank one objects in $\mathcal{A}_X$ to be the 2-functor

$$\mathcal{M}_X^\dagger \colon \mathit{Aff}^{op} \to \mathit{Groupoid}$$

whose T-valued points for $T \in Aff$ form the groupoid of data

$$\mathcal{E}_T \in D^b_{\mathrm{coh}}(\overline{X} \times T), \ \mathbf{L}(i_\infty \times \mathrm{id}_T)^* \mathcal{E}_T \overset{\cong}{\to} \mathcal{O}_{S_\infty \times T}$$

such that for any closed point $x \in T$, we have

$$\mathcal{E}_x := \mathbf{L}i_x^* \mathcal{E} \in A_X, \ i_x \colon \overline{X} \times \{x\} \hookrightarrow \overline{X} \times T.$$

The isomorphisms of the groupoid $\mathcal{M}_X^\dagger(T)$ are given by isomorphisms of objects $\mathcal{E}_T$ which commute with the trivializations at the infinity.

We have the decomposition of $\mathcal{M}_X^\dagger$ into open and closed substacks

$$\mathcal{M}_X^\dagger = \coprod_{v \subset N_{\leq 1}(S)} \mathcal{M}_X^\dagger(v)$$

where $\mathcal{M}_X^\dagger(v)$ corresponds to $E \in A_X$ with $\mathrm{cl}(E) = (1, v)$. The following result will be proved in Sect. 5.2 (see Corollary 5.2.9 and Proposition 5.2.11):

Theorem 4.2.3

(i) *For $v \in N_{\leq 1}(S)$, the stack $\mathcal{M}_X^\dagger(v)$ is isomorphic to the stack of diagrams of coherent sheaves on S*

$$
\begin{array}{ccccccccc}
0 & \longrightarrow & \mathcal{O}_S & \longrightarrow & \mathcal{U} & \longrightarrow & F \otimes \omega_S^{-1} & \longrightarrow & 0 \\
& & & {\scriptstyle \xi} \searrow & \downarrow {\scriptstyle \phi} & & & & \\
& & & & F & & & &
\end{array}
$$

$$(4.2.4)$$

where the top sequence is an exact sequence and F has numerical class v. In particular, there exists a natural morphism

$$\pi_*^\dagger \colon \mathcal{M}_X^\dagger(v) \to \mathcal{M}_S^\dagger(v) \qquad (4.2.5)$$

sending a diagram (4.2.4) to the pair $(\mathcal{O}_S \overset{\xi}{\to} F)$.

(ii) *There is an isomorphism of stacks over $\mathcal{M}_S^\dagger(v)$*

$$\eta^\dagger \colon \mathcal{M}_X^\dagger(v) \overset{\cong}{\to} t_0(\Omega_{\mathfrak{M}_S^\dagger(v)}[-1]) = \mathrm{Obs}^*(\mathcal{E}^{\dagger \bullet}|_{\mathcal{M}_S^\dagger(v)}) \qquad (4.2.6)$$

which, over the pair $(\mathcal{O}_S \overset{\xi}{\to} F)$, sends a diagram (4.2.4) to the morphism $F \otimes \omega_S^{-1} \to I^\bullet[1]$ given by (see Lemma 4.2.1)

$$F \otimes \omega_S^{-1}[-1] \overset{\sim}{\leftarrow} (\mathcal{O}_S \to \mathcal{U}) \overset{(id,\phi)}{\longrightarrow} (\mathcal{O}_S \overset{\xi}{\to} F) = I^\bullet.$$

Remark 4.2.4 Note that for a pair $(\mathcal{O}_X \to F)$ with $F \in \mathrm{Coh}_{\leq 1}(X)$, we have the associated object (see Lemma 5.2.8)

$$(\mathcal{O}_{\overline{X}} \to F) \in \mathcal{A}_X \qquad (4.2.7)$$

where $\mathcal{O}_{\overline{X}}$ is located in degree zero. The map $\pi_*^{\dagger}$ sends such an object to the associated pair $(\mathcal{O}_S \to \pi_* F)$. However a rank one object in $\mathcal{A}_X$ may not be necessary written of the form (4.2.7). As we will see in Lemma 5.2.8, a diagram (4.2.4) corresponds to an object of the form (4.2.7) if and only if the top sequence of (4.2.4) splits.

4.3 DT Category for D0-D2-D6 Bound States

4.3.1 Categorical MNOP/PT Theories

For $(\beta, n) \in N_{\leq 1}(S) = \mathrm{NS}(S) \oplus \mathbb{Z}$, we denote by

$$I_n(X, \beta)$$

the moduli space of closed subschemes $C \subset X$, where C is compactly supported with $\dim C \leq 1$ satisfying $[\pi_* \mathcal{O}_C] = (\beta, n)$. The moduli space $I_n(X, \beta)$ is a quasi-projective scheme, and considered by Maulik-Nekrasov-Okounkov-Pandharipande [88] in their formulation of GW/DT correspondence conjecture.

On the other hand, a PT stable pair consists of a pair [102]

$$(F, s), \ F \in \mathrm{Coh}_{\leq 1}(X), \ s : \mathcal{O}_X \to F \qquad (4.3.1)$$

such that F is pure one dimensional and s is surjective in dimension one. For $(\beta, n) \in N_{\leq 1}(S)$, we denote by

$$P_n(X, \beta)$$

the moduli space of PT stable pairs (4.3.1) satisfying $[\pi_* F] = (\beta, n)$. The moduli space of stable pairs $P_n(X, \beta)$ is known to be a quasi-projective scheme.

We have the open immersions

$$I_n(X, \beta) \subset \mathcal{M}_X^{\dagger}(\beta, n), \ P_n(X, \beta) \subset \mathcal{M}_X^{\dagger}(\beta, n)$$

sending a subscheme $C \subset X$, a pair (F, s) to two term complexes $(\mathcal{O}_{\overline{X}} \twoheadrightarrow \mathcal{O}_C)$, $(\mathcal{O}_{\overline{X}} \xrightarrow{s} F)$ respectively. They are quasi-projective schemes, so there is a quasi-compact derived open substack $\mathfrak{M}_S^{\dagger}(\beta, n)_{\mathrm{qc}}$ in $\mathfrak{M}_S^{\dagger}(\beta, n)$ such that

$$\pi_*^{\dagger}(I_n(X, \beta)) \subset t_0(\mathfrak{M}_S^{\dagger}(\beta, n)_{\mathrm{qc}}), \ \pi_*^{\dagger}(P_n(X, \beta)) \subset t_0(\mathfrak{M}_S^{\dagger}(\beta, n)_{\mathrm{qc}}).$$

Then by the isomorphism (4.2.6), we have the following conical closed substacks in
$t_0(\Omega_{\mathfrak{M}_S^\dagger(\beta,n)_{qc}}[-1])$

$$\mathcal{Z}^{I\text{-us}} := t_0(\Omega_{\mathfrak{M}_S^\dagger(\beta,n)_{qc}}[-1]) \setminus \eta^\dagger(I_n(X,\beta)),$$

$$\mathcal{Z}^{P\text{-us}} := t_0(\Omega_{\mathfrak{M}_S^\dagger(\beta,n)_{qc}}[-1]) \setminus \eta^\dagger(P_n(X,\beta)).$$

By Definition 3.2.2, the corresponding DT categories are defined as follows:

Definition 4.3.1 The $\mathbb{C}^*$-equivariant MNOP/PT categories are defined to be

$$\mathcal{DT}^{\mathbb{C}^*}(I_n(X,\beta)) := D^b_{\mathrm{coh}}(\mathfrak{M}_S^\dagger(\beta,n)_{qc})/\mathcal{C}_{\mathcal{Z}^{I\text{-us}}},$$

$$\mathcal{DT}^{\mathbb{C}^*}(P_n(X,\beta)) := D^b_{\mathrm{coh}}(\mathfrak{M}_S^\dagger(\beta,n)_{qc})/\mathcal{C}_{\mathcal{Z}^{P\text{-us}}}.$$

The dg-enhancements, $\widehat{\mathcal{DT}}$-version and ind-version are similarly defined following Sects. 3.2.1 and 3.2.2.

Similarly to Remark 3.4.5, the above definition is independent of a choice of $\mathfrak{M}_S^\dagger(\beta,n)_{qc}$. As proved in [127], two moduli space $I_n(X,\beta)$, $P_n(X,\beta)$ are related by wall-crossing in $\mathcal{A}_X$. Let $\mathcal{T}_n(X,\beta)$ be the open substack

$$\mathcal{T}_n(X,\beta) \subset \mathcal{M}_X^\dagger(\beta,n) \tag{4.3.2}$$

corresponding to $\mathcal{E} \in \mathcal{A}_X$ such that $\mathcal{H}^1(\mathcal{E})$ is zero dimensional. We have the following lemma proved in [127].

Lemma 4.3.2 ([127, Lemma 3.11 (ii)]) *For a rank one object $\mathcal{E} \in \mathcal{A}_X$, $\mathcal{H}^1(\mathcal{E})$ is zero dimensional if and only if it is isomorphic to a two term complex $(s : \mathcal{O}_{\overline{X}} \to F)$ where $F \in \mathrm{Coh}_{\leq 1}(X)$ and s is surjective in dimension one.*

The above lemma implies that the stack $\mathcal{T}_n(X,\beta)$ is isomorphic to the moduli stack of pairs (F,s) for $F \in \mathrm{Coh}_{\leq 1}(X)$ and $(s : \mathcal{O}_{\overline{X}} \to F)$ such that s is surjective in dimension one, and $[\pi_* F] = (\beta,n)$. As discussed in [137, Theorem B.1], it admits a good moduli space

$$\pi_{\mathcal{T}} : \mathcal{T}_n(X,\beta) \to T_n(X,\beta)$$

together with a wall-crossing diagram

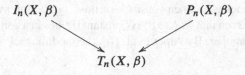

which is a d-critical flip. Therefore following the discussion in Sect. 1.1.6, we propose the following conjecture:

Conjecture 4.3.3 There exists a fully-faithful functor

$$\mathcal{DT}^{\mathbb{C}^*}(P_n(X, \beta)) \hookrightarrow \mathcal{DT}^{\mathbb{C}^*}(I_n(X, \beta)).$$

Remark 4.3.4 We can similarly formulate MNOP/PT categories for a non-compact surface S, by replacing $\mathfrak{M}_S$ by the derived moduli stack of compactly supported coherent sheaves on S. Then we can similarly formulate Conjecture 4.3.3 and study it (see Sect. 4.4). The same also applies to Conjecture 4.3.14 below.

4.3.2   Moduli Stacks of Stable D0-D2-D6 Bound States

Let S be a smooth projective surface, and $X = \text{Tot}_S(\omega_S)$ as before. Following [128, 129, 137], we introduce a one parameter family of weak stability conditions on the abelian category $\mathcal{A}_X$ given in Definition 4.2.2. Below, we fix an element $\sigma = iH \in A(S)_{\mathbb{C}}$ for an ample divisor H on S. For each $t \in \mathbb{R}$, we define the map

$$\mu_t^{\dagger} \colon \Gamma = \mathbb{Z} \oplus \text{NS}(S) \oplus \mathbb{Z} \to \mathbb{R} \cup \{\infty\}, \ (r, \beta, n) \mapsto \begin{cases} t, & r \neq 0, \\ n/(\beta \cdot H), & r = 0. \end{cases}$$

Note that we have

$$\mu_t^{\dagger}(0, \beta, n) = \mu_H(\beta, n) := \mu_{\sigma = iH}(\beta, n)$$

where the latter is defined in (4.1.1). For an object $E \in \mathcal{A}_X$, we set $\mu_t^{\dagger}(E) := \mu_t^{\dagger}(\text{cl}(E))$.

Definition 4.3.5 An object $E \in \mathcal{A}_X$ is $\mu_t^{\dagger}$-(semi) stable if for any exact sequence $0 \to E' \to E \to E'' \to 0$ in $\mathcal{A}_X$ we have the inequality $\mu_t^{\dagger}(E') < (\leq)\mu_t^{\dagger}(E'')$.

We have the substack

$$\mathcal{P}_n^t(X, \beta) \subset \mathcal{M}_X^{\dagger}(\beta, n)$$

corresponding to $\mu_t^{\dagger}$-semistable objects. The result of [128, Proposition 3.17] shows that the above substack is an open substack of finite type. Moreover there is a finite set of *walls* $W \subset \mathbb{Q}$ such that $\mathcal{P}_n^t(X, \beta)$ is constant if t lies in a connected component of $\mathbb{R} \setminus W$, called *chamber*. By Alper et al. [6], the moduli stack $\mathcal{P}_n^t(X, \beta)$ admits a good moduli space

$$\pi_{\mathcal{P}} \colon \mathcal{P}_n^t(X, \beta) \to P_n^t(X, \beta)$$

where closed points of $P_n^t(X, \beta)$ correspond to $\mu_t^\dagger$-polystable objects, i.e. direct sums of $\mu_t^\dagger$-stable objects. If $t \notin W$, then $\mathcal{P}_n^t(X, \beta)$ consists of $\mu_t^\dagger$-stable objects and the good moduli space morphism is an isomorphism

$$\pi_\mathcal{P}: \mathcal{P}_n^t(X, \beta) \xrightarrow{\cong} P_n^t(X, \beta), \ t \notin W. \tag{4.3.3}$$

Remark 4.3.6 Contrary to the case of moduli stacks of semistable sheaves in (3.4.9), the trivial $\mathbb{C}^*$-autmorphisms are rigidified in the definition of the stack $\mathcal{P}_n^t(X, \beta)$ by the trivialization at S_∞. Therefore the morphism (4.3.3) for $t \notin W$ is an isomorphism, not $\mathbb{C}^*$-gerbe.

For each $\beta \in NS(S)$, we set

$$n(\beta) := \min \left\{ \chi(\mathcal{O}_C) : \begin{array}{l} C \subset X \text{ is a compactly supported} \\ \text{closed subscheme with } \pi_*[C] \leq \beta \end{array} \right\}. \tag{4.3.4}$$

The following lemma will be used later.

Lemma 4.3.7 *If $\mathcal{P}_n^t(X, \beta) \neq \emptyset$ for $t > 0$ then $n \geq n(\beta)$.*

Proof Let $\mathcal{E} \in \mathcal{A}_X$ correspond to a point in $\mathcal{P}_n^t(X, \beta)$. Then there is an exact sequence

$$0 \to \mathcal{H}^0(\mathcal{E}) \to \mathcal{E} \to \mathcal{H}^1(\mathcal{E})[-1] \to 0.$$

Since $\mathcal{H}^0(\mathcal{E})$ is a rank one torsion free sheaf which is trivial on S_∞, we have $\mathcal{H}^0(\mathcal{E}) = I_C$ for an ideal sheaf of a compactly supported closed subscheme $C \subset X$. Then as $\pi_*[C] \leq \beta$, we have $\chi(\mathcal{O}_C) \geq n(\beta)$. Moreover $\mathcal{H}^1(\mathcal{E}) \in \mathrm{Coh}_{\leq 1}(X)$ and it satisfies $\mu_H(\mathcal{H}^1(\mathcal{E})) \geq t > 0$ by the $\mu_t^\dagger$-stability. Therefore $n = \chi(\mathcal{O}_C) + \chi(\mathcal{H}^1(\mathcal{E})) \geq n(\beta)$. $\square$

Remark 4.3.8 We have $n(\beta) > -\infty$ by Toda [126, Lemma 3.10]. Moreover if β is reduced then

$$n(\beta) = -\frac{1}{2}\beta(\beta + K_S).$$

Indeed for a closed subscheme $C \subset X$ with class β, the morphism $\mathcal{O}_S \to \pi_*\mathcal{O}_C$ is generically surjective. Since any Cohen-Macaulay curve in S with class β has Euler characteristic $-\beta(\beta + K_S)/2$, we have $\chi(\mathcal{O}_C) \geq -\beta(\beta + K_S)/2$.

4.3.3 DT Categories for Semistable D0-D2-D6 Bound States

Similarly to Sect. 4.3, there is a quasi-compact derived open substack $\mathfrak{M}_S^\dagger(\beta, n)_{qc}$ in $\mathfrak{M}_S^\dagger(\beta, n)$ such that

$$\pi_*^\dagger(P_n^t(X, \beta)) \subset t_0(\mathfrak{M}_S^\dagger(\beta, n)_{qc}). \tag{4.3.5}$$

Then by the isomorphism (4.2.6), we have the conical closed substack in the (-1)-shifted cotangent $t_0(\Omega_{\mathfrak{M}_S^\dagger(\beta,n)_{qc}}[-1])$

$$\mathcal{Z}^{t\text{-us}} := t_0(\Omega_{\mathfrak{M}_S^\dagger(\beta,n)_{qc}}[-1]) \setminus \eta^\dagger(P_n^t(X, \beta)) \subset t_0(\Omega_{\mathfrak{M}_S^\dagger(\beta,n)_{qc}}[-1]).$$

By Definition 3.2.2, we have the following definition of the DT category for $P_n^t(X, \beta)$:

Definition 4.3.9 For $(\beta, n) \in N_{\leq 1}(S)$, we define the $\mathbb{C}^*$-equivariant DT category for $\mathcal{P}_n^t(X, \beta)$ by

$$\mathcal{DT}^{\mathbb{C}^*}(\mathcal{P}_n^t(X, \beta)) := D_{coh}^b(\mathfrak{M}_S^\dagger(\beta, n)_{qc})/\mathcal{C}_{\mathcal{Z}^{t\text{-us}}}.$$

If $t \in \mathbb{R}$ lies in a chamber, then we also write it as $\mathcal{DT}^{\mathbb{C}^*}(P_n^t(X, \beta))$ by the isomorphism (4.3.3). The dg-enhancements, $\widehat{\mathcal{DT}}$-version and ind-version are similarly defined following Sects. 3.2.1, 3.2.2, and 3.2.4.

Below we will see the relation of the above DT categories with PT categories introduced in Definition 4.3.1. We recall that for $|t| \gg 0$, the moduli space $P_n^t(X, \beta)$ is related to the moduli space of PT stable pairs as follows:

Theorem 4.3.10 ([128, Theorem 3.21]) For $(\beta, n) \in N_{\leq 1}(S)$, we have the following:

(i) For $t \gg 0$, we have the isomorphism

$$P_n(X, \beta) \xrightarrow{\cong} P_n^t(X, \beta), \ (F, s) \mapsto (\mathcal{O}_{\overline{X}} \xrightarrow{s} F).$$

(ii) For $t \ll 0$, we have the isomorphism

$$P_{-n}(X, \beta) \xrightarrow{\cong} P_n^t(X, \beta), \ (F, s) \mapsto \mathbb{D}_{\overline{X}}(\mathcal{O}_{\overline{X}} \xrightarrow{s} F).$$

By Theorem 4.3.10 (i), for $v = (\beta, n)$ we have the obvious identity

$$\mathcal{DT}^{\mathbb{C}^*}(P_n^t(X, \beta)) = \mathcal{DT}^{\mathbb{C}^*}(P_n(X, \beta)), \ t \gg 0. \tag{4.3.6}$$

On the other hand, we also have an equivalence

$$\mathcal{DT}^{\mathbb{C}^*}(P_n^t(X, \beta)) \xrightarrow{\sim} \mathcal{DT}^{\mathbb{C}^*}(P_{-n}(X, \beta)), \ t \ll 0. \tag{4.3.7}$$

The equivalence (4.3.7) is less obvious than (4.3.6) since the isomorphism in Theorem 4.3.10 (ii) is given through the derived dual $\mathbb{D}_{\overline{X}}$. Indeed the equivalence (4.3.7) is a consequence of a more general duality statement in Theorem 5.1.7.

Remark 4.3.11 By formally extending the $\mu_t^\dagger$-stability in Definition 4.3.5 for $t = \infty$ and $t = \infty + 0$, we have

$$\mathcal{P}_n^{t=\infty\pm 0}(X, \beta) = P_n^{t=\infty\pm 0}(X, \beta) = \begin{cases} I_n(X, \beta), & t = \infty + 0, \\ P_n(X, \beta), & t = \infty - 0. \end{cases}$$

Moreover the moduli stack at the wall $t = \infty$ coincides with the stack in (4.3.2)

$$\mathcal{T}_n(X, \beta) = \mathcal{P}_n^{t=\infty}(X, \beta) \subset \mathcal{M}_X^\dagger(\beta, n).$$

4.3.4   Moduli Stacks of Semistable Pairs

In some case, the DT category for D0-D2-D6 bound states on X is related to the derived category of derived moduli stack of some stable pairs on S. Let $\mathcal{A}_S$ be the abelian category of pairs on S

$$W \otimes \mathcal{O}_S \to F$$

where W is a finite dimensional vector space and $F \in \mathrm{Coh}_{\leq 1}(S)$. The set of morphisms is given by commutative diagrams

$$\begin{array}{ccc} W \otimes \mathcal{O}_S & \longrightarrow & F \\ \downarrow & & \downarrow \\ W' \otimes \mathcal{O}_S & \longrightarrow & F'. \end{array}$$

Here the left arrow is induced by a linear map $W \to W'$. We have the group homomorphism

$$\mathrm{cl} \colon K(\mathcal{A}_S) \to \mathbb{Z} \oplus N_{\leq 1}(S)$$

determined by $\mathrm{cl}(\mathcal{O}_S \to 0) = (1, 0)$ and $\mathrm{cl}(0 \to F) = (0, [F])$. For an object $E \in \mathcal{A}_S$, we define $\mu_t^\dagger(E) := \mu_t^\dagger(\mathrm{cl}(E))$.

Definition 4.3.12 An object $E \in \mathcal{A}_S$ is $\mu_t^\dagger$-(semi)stable if for any exact sequence $0 \to E' \to E \to E'' \to 0$ in $\mathcal{A}_S$ we have $\mu_t^\dagger(E') < (\leq)\mu_t^\dagger(E'')$.

For each $(\beta, n) \in N_{\leq 1}(S)$, we denote by

$$\mathfrak{P}_n^t(S, \beta) \subset \mathfrak{M}_n^\dagger(S, \beta), \quad \mathcal{P}_n^t(S, \beta) \subset \mathcal{M}_n^\dagger(S, \beta) \tag{4.3.8}$$

the open substacks of $\mu_t^\dagger$-semistable objects $E \in \mathcal{A}_S$ with $\mathrm{cl}(E) = (1, \beta, n)$. By Le Potier's GIT construction of moduli spaces of semistable coherent systems [83, Theorem 4.11], we have the good moduli space

$$\pi_{\mathcal{P}} \colon \mathcal{P}_n^t(S, \beta) \to P_n^t(S, \beta)$$

where $P_n^t(S, \beta)$ is a projective scheme whose closed points correspond to $\mu_t^\dagger$-polystable objects in $\mathcal{A}_S$. Similarly to Sect. 4.3.2, there is a wall-chamber structure on $\mathbb{R}$ such that the above good moduli space morphism is an isomorphism if t lies in a chamber.

Lemma 4.3.13 *We have the open immersion*

$$(\pi_*^\dagger)^{-1}(\mathcal{P}_n^t(S, \beta)) \subset \mathcal{P}_n^t(X, \beta). \tag{4.3.9}$$

If the above inclusion is the identity, we have the equivalence

$$\mathcal{DT}^{\mathbb{C}^*}(\mathcal{P}_n^t(X, \beta)) \xrightarrow{\sim} D_{\mathrm{coh}}^b(\mathfrak{P}_n^t(S, \beta)). \tag{4.3.10}$$

Proof The inclusion (4.3.9) is obvious from the correspondence between rank objects in $\mathcal{A}_X$ and a diagram (4.2.4). Indeed if there is a destabilizing sequence of a diagram in (4.2.4), it also destabilizes the associated pair $(\mathcal{O}_S \to F)$. Then the equivalence (4.3.10) follows from Lemma 3.2.9.   □

4.3.5   Conjectural Wall-Crossing Phenomena of DT Categories

Below we discuss a conjectural wall-crossing phenomena of DT categories in Definition 4.3.9 under change of weak stability conditions. We fix $\sigma = iH$ for an ample divisor H and $(\beta, n) \in N_{\leq 1}(S)$. Let us take $t \in \mathbb{R}$ which lies on a wall and real numbers $t_\pm$ on adjacent chambers,

$$t \in W \cap \mathbb{Q}_{>0}, \quad t_\pm := t \pm \varepsilon, \ 0 < \varepsilon \ll 1.$$

Since $\mathcal{P}_n^{t\pm}(X, \beta) \subset \mathcal{P}_n^t(X, \beta)$, we have the induced diagram on good moduli spaces

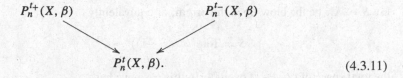

$$P_n^t(X, \beta). \tag{4.3.11}$$

By Toda [137, Theorem 9.13], the above diagram is a *d-critical flip*. Therefore following the discussion in Sect. 1.1.6, we propose the following conjecture:

Conjecture 4.3.14 There exists a fully-faithful functor

$$\mathcal{DT}^{\mathbb{C}^*}(P_n^{t-}(X, \beta)) \hookrightarrow \mathcal{DT}^{\mathbb{C}^*}(P_n^{t+}(X, \beta)).$$

The above conjecture implies that, for $t_1 > t_2 > \cdots > t_N > 0$ which lie on chambers we have a chain of fully-faithful functors:

$$\mathcal{DT}^{\mathbb{C}^*}(P_n(X, \beta)) \hookleftarrow \mathcal{DT}^{\mathbb{C}^*}(P_n^{t_1}(X, \beta)) \hookleftarrow \cdots\cdots \hookleftarrow \mathcal{DT}^{\mathbb{C}^*}(P_n^{t_N}(X, \beta)). \tag{4.3.12}$$

As for the wall-crossing at $t = 0$, the diagram (4.3.11) is a d-critical flop (see [137, Corollary 9.18]), so we conjecture the following:

Conjecture 4.3.15 There exists an equivalence

$$\mathcal{DT}^{\mathbb{C}^*}(P_n^{\varepsilon}(X, \beta)) \xrightarrow{\sim} \mathcal{DT}^{\mathbb{C}^*}(P_n^{-\varepsilon}(X, \beta)), \ 0 < \varepsilon \ll 1.$$

4.4 Example: Local $(-1, -1)$-Curve

In this section, we prove Conjectures 4.3.3, 4.3.14 in the case that S is the blow-up of $\mathbb{C}^2$ at the origin. Although S is non-compact, Conjecture 4.3.3 still makes sense (see Remark 4.3.4). In this case, X is the total space of $\mathcal{O}_{\mathbb{P}^1}(-1)^{\oplus 2}$ over $\mathbb{P}^1$, and we can explicitly describe our derived moduli stacks in terms of non-commutative crepant resolution [149]. In particular we have a global critical locus description of our moduli spaces on X, and we can reduce our problem to the comparison of factorization categories under variation of GIT quotients. Then using window theorem developed in [11, 45], we prove our assertion.

4.4.1 Local $(-1, -1)$-Curve

Let $S \to \mathbb{C}^2$ be the blow-up at the origin, or equivalently

$$S = \mathrm{Tot}_{\mathbb{P}^1}(\mathcal{O}_{\mathbb{P}^1}(-1)) \to \mathbb{P}^1 \tag{4.4.1}$$

the total space of $\mathcal{O}_{\mathbb{P}^1}(-1)$ over $\mathbb{P}^1$. In this case, we have

$$X = \mathrm{Tot}_S(\omega_S) = \mathrm{Tot}_{\mathbb{P}^1}(\mathcal{O}_{\mathbb{P}^1}(-1)^{\oplus 2}).$$

We denote by $C \subset S$ the zero section of the projection (4.4.1), which is a (-1)-curve. By setting $\mathcal{E}_S = \mathcal{O}_S \oplus \mathcal{O}_S(-C)$ and $A_S = \mathrm{End}(\mathcal{E}_S)$, we have the derived equivalence [149]

$$\Phi_S := \mathbf{RHom}_S(\mathcal{E}_S, -) \colon D^b_{\mathrm{coh}}(S) \xrightarrow{\sim} D^b(\mathrm{mod} A_S).$$

It is easy to see that A_S is the path algebra of the following quiver Q_S with relation

$$\bullet^0 \underset{\underset{b_2}{b_1}}{\overset{a_1}{\rightleftarrows}} \bullet^1 \ , \quad b_1 a_1 b_2 = b_2 a_1 b_1.$$

Here the objects $\mathcal{O}_C$, $\mathcal{O}_C(-1)[1]$ are sent to the simple A_S-modules corresponding to the vertices $\bullet^0$, $\bullet^1$ respectively.

Similarly by setting $\mathcal{E}_X = \pi^* \mathcal{E}_S$ and $A_X = \mathrm{End}(\mathcal{E}_X)$, we have the derived equivalence [149]

$$\Phi_X := \mathbf{RHom}_X(\mathcal{E}_X, -) \colon D^b_{\mathrm{coh}}(X) \xrightarrow{\sim} D^b(\mathrm{mod} A_X). \tag{4.4.2}$$

Then A_X is the path algebra of the following quiver Q_X with relation ∂W for the super-potential W (see [94])

$$\bullet^0 \underset{\underset{b_2}{b_1}}{\overset{\overset{a_2}{a_1}}{\rightrightarrows}} \bullet^1 \ , \quad W = a_2(b_1 a_1 b_2 - b_2 a_1 b_1).$$

We will also consider framed quivers

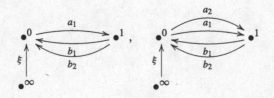

denoted by $Q_S^\dagger$, $Q_X^\dagger$ respectively. By setting $\mathcal{P}_S = \Phi_S(\mathcal{O}_S)$, giving a $Q_S^\dagger$-representation with dimension vector 1 at $\bullet^\infty$ and relation $b_1 a_1 b_2 = b_2 a_1 b_1$ is equivalent to giving a pair $(\mathcal{P}_S \to M)$ where $M \in \mathrm{mod}A_S$. Similarly for $\mathcal{P}_X = \Phi_X(\mathcal{O}_X)$, giving a $Q_X^\dagger$-representation with dimension vector 1 at $\bullet^\infty$ and relation ∂W is equivalent to giving a pair $(\mathcal{P}_X \to N)$ for $N \in \mathrm{mod}A_X$.

4.4.2   Moduli Stacks of Quiver Representations

We prepare some notation on moduli stack of quiver representations. For $\vec{v} = (v_0, v_1) \in \mathbb{Z}_{\geq 0}^2$, let V_0, V_1 be vector spaces with $\dim V_i = v_i$. We set

$$R_{Q_S^\dagger}(\vec{v}) = V_0 \oplus \mathrm{Hom}(V_0, V_1) \oplus \mathrm{Hom}(V_1, V_0)^{\oplus 2},$$

$$R_{Q_X^\dagger}(\vec{v}) = V_0 \oplus \mathrm{Hom}(V_0, V_1)^{\oplus 2} \oplus \mathrm{Hom}(V_1, V_0)^{\oplus 2}.$$

The algebraic group $G = \mathrm{GL}(V_0) \times \mathrm{GL}(V_1)$ naturally acts on $R_{Q_S^\dagger}$, $R_{Q_X^\dagger}$, and the quotient stacks

$$\mathcal{M}_{Q_S^\dagger}(\vec{v}) = [R_{Q_S^\dagger}(\vec{v})/G], \ \ \mathcal{M}_{Q_X^\dagger}(\vec{v}) = [R_{Q_X^\dagger}(\vec{v})/G]$$

are the $\mathbb{C}^*$-rigidified moduli stacks of $Q_S^\dagger$ and $Q_X^\dagger$-representations with dimension vector $(1, \vec{v})$, where 1 is the dimension vector at the vertex $\bullet^\infty$, v_i is the dimension vector at $\bullet^i$.

Let s be the map

$$s \colon R_{Q_S^\dagger}(\vec{v}) \to \mathrm{Hom}(V_1, V_0), \ s(\Xi, A_1, B_1, B_2) = B_1 A_1 B_2 - B_2 A_1 B_1.$$

Then for the derived zero locus $s^{-1}(0) \subset R_{Q_S^\dagger}(\vec{v})$, the derived stack

$$\mathfrak{M}_{A_S}^\dagger(\vec{v}) = [s^{-1}(0)/G] \tag{4.4.3}$$

is the derived moduli stack of pairs $(\mathcal{P}_S \to M)$ in $\mathrm{mod}A_S$, where M has dimension vector $\vec{v}$. Its classical truncation is denoted by $\mathcal{M}_{A_S}^\dagger(\vec{v})$.

By the above description, the (-1)-shifted cotangent derived stack $\Omega_{\mathfrak{M}_{A_S}^\dagger(\vec{v})}[-1]$ is the derived critical locus of

$$w \colon [R_{Q_X^\dagger}(\vec{v})/G] \to \mathbb{C}, \tag{4.4.4}$$

$$w(\Xi, A_1, A_2, B_1, B_2) = \mathrm{tr}(A_2(B_1 A_1 B_2 - B_2 A_1 B_1)).$$

It follows that

$$\mathcal{M}^{\dagger}_{A_X}(\vec{v}) := t_0(\Omega_{\mathfrak{M}^{\dagger}_{A_S}(\vec{v})}[-1]) = \mathrm{Crit}(w) \subset \mathcal{M}_{Q^{\dagger}_X}(\vec{v})$$

is the moduli stack of pairs $(\mathcal{P}_X \to N)$ in $\mathrm{mod}A_X$, where the dimension vector of N is $\vec{v}$. The fiberwise weight two $\mathbb{C}^*$-action on $\mathcal{M}^{\dagger}_{A_X}(\vec{v}) \to \mathcal{M}^{\dagger}_{A_S}(\vec{v})$ is given by

$$t \cdot (\Xi, A_1, A_2, B_1, B_2) = (\Xi, A_1, t^2 A_2, B_1, B_2). \tag{4.4.5}$$

4.4.3 Moduli Stacks of Semistable Representations

Next we discuss King's θ-stability conditions on $Q^{\dagger}_X$-representations [71]. Let us take

$$\theta = (\theta_{\infty}, \theta_0, \theta_1) \in \mathbb{Q}^3.$$

For a dimension vector (v_{∞}, v_0, v_1) of $Q^{\dagger}_X$, we set $\theta(v_{\infty}, v_0, v_1) = \theta_{\infty} v_{\infty} + \theta_0 v_0 + \theta_1 v_1$.

Definition 4.4.1 *([71])* A $Q^{\dagger}_X$-representation E with dimension vector $(1, \vec{v})$ is called θ-(semi)stable if $\theta(1, \vec{v}) = 0$ and for any subrepresentation $0 \neq F \subsetneq E$, we have the inequality $\theta(v(F)) < (\leq)0$, where $v(-)$ is the dimension vector.

Below we fix $\vec{v} = (v_0, v_1)$, and take θ such that $\theta(1, \vec{v}) = 0$ holds. Then θ_{∞} is determined by $\theta_{\infty} = -\theta_0 v_0 - \theta_1 v_1$, so we simply write $\theta = (\theta_0, \theta_1)$.

We denote by

$$\mathcal{M}^{\theta}_{Q^{\dagger}_X}(\vec{v}) = [R^{\theta}_{Q^{\dagger}_X}(\vec{v})/G]$$

the open substack in $\mathcal{M}_{Q^{\dagger}_X}(\vec{v})$ corresponding to θ-semistable $Q^{\dagger}_X$-representations. The stack $\mathcal{M}^{\theta}_{Q^{\dagger}_X}(\vec{v})$ admits the good moduli space

$$\pi_{Q^{\dagger}_X} : \mathcal{M}^{\theta}_{Q^{\dagger}_X}(\vec{v}) \to M^{\theta}_{Q^{\dagger}_X}(\vec{v}) := R^{\theta}_{Q^{\dagger}_X}(\vec{v}) /\!\!/ G \tag{4.4.6}$$

where $M^{\theta}_{Q^{\dagger}_X}(\vec{v})$ parametrizes θ-polystable $Q^{\dagger}_X$-representations with dimension vector $(1, \vec{v})$. Namely a point $p \in M^{\theta}_{Q^{\dagger}_X}(\vec{v})$ corresponds to the direct sum

$$E = F_{\infty} \oplus \bigoplus_{i=1}^{l} U_i \otimes F_i \tag{4.4.7}$$

where $\{F_{\infty}, F_1, \ldots, F_l\}$ are mutually non-isomorphic θ-stable $Q^{\dagger}_X$-representations with $\theta(v(F_i)) = 0$, F_{∞} has dimension vector of the form $(1, *)$, each U_i is a finite dimensional vector space, and E has dimension vector $(1, \vec{v})$.

For a point p as above, we denote by $Q_p^{\dagger}$ the *Ext-quiver* associated with the collection of θ-stable objects $\{F_{\infty}, F_1, \ldots, F_l\}$. Namely the set of vertices of $Q_p^{\dagger}$ is $\{\infty, 1, \ldots, l\}$, and the number of arrows from i to j is the dimension of $\mathrm{Ext}^1(F_i, F_j)$. From the construction of $Q_p^{\dagger}$, note that

$$\sharp(i \to j) = \sharp(j \to i), \ 1 \le i, j \le l, \ \sharp(\infty \to i) - \sharp(i \to \infty) = v_0^{(i)} \ge 0$$
$$(4.4.8)$$

where we have written $v(F_i) = (0, v_0^{(i)}, v_1^{(i)})$ for $1 \le i \le l$. The quiver $Q_p^{\dagger}$ contains the subquiver $Q_p \subset Q_p^{\dagger}$ given by the Ext-quiver of the sub collection $\{F_1, \ldots, F_l\}$.

We set $G_p = \mathrm{Aut}(E)$, which is identified with the stabilizer group of the action of G on $R_{Q_X^{\dagger}}$ at the point corresponding to E. The algebraic group G_p acts on $\mathrm{Ext}^1(E, E)$ by conjugation. Then we set

$$\pi_{Q_p^{\dagger}} : \mathcal{M}_{Q_p^{\dagger}}(\vec{u}) := [\mathrm{Ext}^1(E, E)/G_p] \to M_{Q_p^{\dagger}}(\vec{u}) := \mathrm{Ext}^1(E, E) /\!\!/ G_p.$$

By the decomposition

$$\mathrm{Ext}^1(E, E) = \bigoplus_{i,j} \mathrm{Hom}(U_i, U_j) \otimes \mathrm{Ext}^1(F_i, F_j)$$

where $U_{\infty} = \mathbb{C}$, the stack $\mathcal{M}_{Q_p^{\dagger}}(\vec{u})$ is identified with the stack of $Q_p^{\dagger}$-representations with dimension vector $(1, \vec{u})$, where 1 is the dimension vector at ∞, $\vec{u} = (\dim U_1, \ldots, \dim U_l)$, and $M_{Q_p^{\dagger}}(\vec{u})$ is its good moduli space. We also define the open substacks

$$\mathcal{M}_{Q_p^{\dagger}}^+(\vec{u}) \subset \mathcal{M}_{Q_p^{\dagger}}(\vec{u}), \ \mathcal{M}_{Q_p^{\dagger}}^-(\vec{u}) \subset \mathcal{M}_{Q_p^{\dagger}}(\vec{u})$$

corresponding to $Q_p^{\dagger}$-representations such that the images of the maps $\mathbb{C} \to U_i$ associated with arrows $\infty \to i$ for all $1 \le i \le l$ generate $\oplus_{i=1}^l U_i$ as a $\mathbb{C}[Q_p]$-module (resp. dual of the maps $U_i \to \mathbb{C}$ associated with arrows $i \to \infty$ generate $\oplus_{i=1}^l U_i^{\vee}$ as a $\mathbb{C}[Q_p]$-module). Then we have the following:

Proposition 4.4.2 *There exist analytic open neighborhoods* $p \in T \subset M_{Q_X^{\dagger}}^{\theta}(\vec{v})$, $0 \in U \subset M_{Q_p^{\dagger}}(\vec{u})$, *and commutative isomorphisms*

$$
\begin{array}{ccc}
(\pi_{Q_p^{\dagger}})^{-1}(U) & \xrightarrow{\cong} & (\pi_{Q_X^{\dagger}})^{-1}(T) \\
\downarrow & & \downarrow \\
U & \xrightarrow{\cong} & T.
\end{array}
$$
$$(4.4.9)$$

Moreover the top isomorphism restricts to the isomorphism

$$(\pi_{Q_p^\dagger})^{-1}(U) \cap \mathcal{M}_{Q_p^\dagger}^{\pm}(\bar{u}) \xrightarrow{\cong} (\pi_{Q_X^\dagger})^{-1}(T) \cap \mathcal{M}_{Q_X^\dagger}^{\theta_{\pm}}(\bar{v}). \qquad (4.4.10)$$

Here $\theta_{\pm} = \theta \mp \varepsilon(1, 1)$ for $0 < \varepsilon \ll 1$.

Proof Since $\mathcal{M}_{Q_X^\dagger}^{\theta}(\bar{v})$ is a smooth stack, the isomorphisms (4.4.9) follow from Luna's étale slice theorem. The isomorphism (4.4.10) follows from the similar argument of [137, Theorem 9.11], so we omit details. □

4.4.4　Window Subcategories

We use window theorem for factorizations categories of GIT quotients to approach Conjectures 4.3.3, 4.3.14. We refer to Sect. 6.1 for a review of window theorem.

By King [71], the semistable locus $R_{Q_X^\dagger}^{\theta}(\bar{v})$ is the GIT L_θ-semistable locus for a G-equivariant $\mathbb{Q}$-line bundle L_θ on $R_{Q_X^\dagger}(\bar{v})$, determined by the rational G-character

$$\chi_\theta \colon G = \mathrm{GL}(V_0) \times \mathrm{GL}(V_1) \to \mathbb{C}^*, \ (g_0, g_1) \mapsto \det(g_0)^{-\theta_0} \det(g_1)^{-\theta_1}. \qquad (4.4.11)$$

Let us fix a maximal torus $T \subset G$ and a Weyl-invariant norm $|*|$ on $\mathrm{Hom}_{\mathbb{Z}}(\mathbb{C}^*, T)_{\mathbb{R}}$. There is a Kempf-Ness (KN) stratification (see Sect. 6.1.1)

$$R_{Q_X^\dagger}(\bar{v}) = S_1 \sqcup S_2 \sqcup \cdots \sqcup R_{Q_X^\dagger}^{\theta}(\bar{v}).$$

Let us take $\theta_{\pm} = \theta \mp \varepsilon(1, 1)$ for $0 < \varepsilon \ll 1$. The KN stratifications with respect to $L_{\theta_{\pm}}$ are finer than that with respect to L_θ, so we have the stratifications

$$R_{Q_X^\dagger}^{\theta}(\bar{v}) = S_1^{\pm} \sqcup S_2^{\pm} \sqcup \cdots \sqcup R_{Q_X^\dagger}^{\theta_{\pm}}(\bar{v})$$

with associated one parameter subgroups $\lambda_\alpha^{\pm} \colon \mathbb{C}^* \to T$ and $\lambda_\alpha^{\pm}$-fixed subset $Z_\alpha^{\pm} \subset S_\alpha^{\pm}$. We set

$$\eta_\alpha^{\pm} := \mathrm{wt}_{\lambda_\alpha^{\pm}} \det(N_{S_\alpha^{\pm}/R_{Q_X^\dagger}^{\theta}(\bar{v})}|_{Z_\alpha^{\pm}}).$$

We also set

$$\delta = L_{(1,1)}^{\otimes \varepsilon} \in \mathrm{Pic}_G(R_{Q_X^\dagger}^{\theta}(\bar{v}))_{\mathbb{R}}, \ 0 < \varepsilon \ll 1. \qquad (4.4.12)$$

Then the window subcategories (see Theorem 6.1.2)

$$\mathcal{W}_\delta^{\theta\pm} \subset \mathrm{MF}_{\mathrm{coh}}^{\mathbb{C}^*}(\mathcal{M}_{Q_X^\dagger}^{\theta}(\vec{v}), w)$$

are defined to be the triangulated subcategories of factorizations $\mathcal{P}$ such that for all α we have

$$\mathrm{wt}_{\lambda_\alpha}(\mathcal{P}|_{Z_\alpha}) \subset \mathrm{wt}_{\lambda_\alpha}(\delta|_{Z_\alpha}) + \left[-\frac{\eta_\alpha^\pm}{2}, \frac{\eta_\alpha^\pm}{2} \right). \tag{4.4.13}$$

By Theorem 6.1.2, the compositions

$$\mathcal{W}_\delta^{\theta\pm} \hookrightarrow \mathrm{MF}_{\mathrm{coh}}^{\mathbb{C}^*}(\mathcal{M}_{Q_X^\dagger}^{\theta}(\vec{v}), w) \twoheadrightarrow \mathrm{MF}_{\mathrm{coh}}^{\mathbb{C}^*}(\mathcal{M}_{Q_X^\dagger}^{\theta\pm}(\vec{v}), w) \tag{4.4.14}$$

are equivalences. Here the right arrow is a restriction functor, w is the function (4.4.4), and the right hand sides are the derived categories of $(G \times \mathbb{C}^*)$-equivariant coherent factorizations of w, where the $\mathbb{C}^*$-action is given by (4.4.5).

Proposition 4.4.3 *We have* $\mathcal{W}_\delta^{\theta-} \subset \mathcal{W}_\delta^{\theta+}$. *Hence we have a fully-faithful functor*

$$\mathrm{MF}_{\mathrm{coh}}^{\mathbb{C}^*}(\mathcal{M}_{Q_X^\dagger}^{\theta-}(\vec{v}), w) \hookrightarrow \mathrm{MF}_{\mathrm{coh}}^{\mathbb{C}^*}(\mathcal{M}_{Q_X^\dagger}^{\theta+}(\vec{v}), w).$$

Proof Let $\mathcal{W}_\delta^{\theta\pm, \mathbb{Z}/2}$ be the subcategories

$$\mathcal{W}_\delta^{\theta\pm, \mathbb{Z}/2} \subset \mathrm{MF}_{\mathrm{coh}}^{\mathbb{Z}/2}(\mathcal{M}_{Q_X^\dagger}^{\theta}(\vec{v}), w)$$

defined by the same condition (4.4.13) for factorizations $\mathcal{P}$ without auxiliary $\mathbb{C}^*$-action. Since the condition (4.4.13) holds for $\mathcal{P}$ if and only if the same condition holds after forgetting the $\mathbb{C}^*$-action, it is enough to show the inclusion

$$\mathcal{W}_\delta^{\theta-, \mathbb{Z}/2} \subset \mathcal{W}_\delta^{\theta+, \mathbb{Z}/2}. \tag{4.4.15}$$

Below we follow the same strategy as in [78, Theorem 5.2]. Let us take a point $p \in M_{Q_X^\dagger}(\vec{v})$ corresponding to a polystable object E as in (4.4.7), and take isomorphisms as in Proposition 4.4.2. Let $w_p \colon \pi_{Q_p^\dagger}^{-1}(U) \to \mathbb{C}$ be the pull-back of w under the top isomorphism in Proposition 4.4.2. Since the statement (4.4.15) is a local question on the base of the map (4.4.6), it is enough to show a similar statement for $(\pi_{Q_p^\dagger}^{-1}(U), w_p)$. Let us write $(\pi_{Q_p^\dagger})^{-1}(U) = [R_p/G_p]$ for an analytic open subset $R_p \subset \mathrm{Ext}^1(E, E)$, and denote its intersection with $\mathcal{M}_{Q_p^\dagger}^{\pm}(\vec{u})$ by $[R_p^\pm/G_p]$. We have the KN stratifications

$$R_p = S_{1,p}^\pm \sqcup S_{2,p}^\pm \sqcup \cdots \sqcup R_p^\pm \tag{4.4.16}$$

with respect to the G_p-characters $\chi_{\theta_\pm}|_{G_p}$. Here χ_θ is given by (4.4.11), and we have restricted it to G_p by the inclusion $G_p \subset G$. The G_p-characters $\chi_{\theta_\pm}|_{G_p}$ are written as

$$\chi_{\theta_\pm}|_{G_p} : \prod_{i=1}^{l} \mathrm{GL}(U_i) \to \mathbb{C}^*, \ (h_i) \mapsto \prod_{i=1}^{l} (\det h_i)^{\pm \varepsilon(v_0^{(i)} + v_1^{(i)})}. \tag{4.4.17}$$

Let δ_p be the pull-back of (4.4.12) to a G_p-equivariant $\mathbb{R}$-line bundle on R_p under the top isomorphism of (4.4.9). Then we have the window subcategories

$$\widehat{\mathcal{W}}_{\delta_p}^{\theta_\pm, \mathbb{Z}/2} \subset \mathrm{MF}_{\mathrm{coh}}^{\mathbb{Z}/2}([R_p/G_p], w_p)$$

defined similarly to (4.4.14) with respect to the KN stratifications (4.4.16). By the property of $Q_p^\dagger$ given in (4.4.8) and the description of $\chi_{\theta_\pm}|_{G_p}$ in (4.4.17), we are in the same situation in Proposition 6.1.13, so the inclusion $\widehat{\mathcal{W}}_{\delta_p}^{\theta_-, \mathbb{Z}/2} \subset \widehat{\mathcal{W}}_{\delta_p}^{\theta_+, \mathbb{Z}/2}$ follows from Proposition 6.1.13 (also see Remark 6.1.14). Since this holds for any p, we have (4.4.15). $\qquad\square$

4.4.5   Proof of Conjectures 4.3.3, 4.3.14 Local $(-1, -1)$-Curve

We show the following:

Theorem 4.4.4 *Conjecture 4.3.3 is true for $S = \mathrm{Tot}_{\mathbb{P}^1}(\mathcal{O}_{\mathbb{P}^1}(-1))$, i.e. for $(d, n) \in \mathbb{Z}^2$ with $d \geq 0$, there is a fully-faithful functor*

$$\mathcal{DT}^{\mathbb{C}^*}(P_n(X, d[C])) \hookrightarrow \mathcal{DT}^{\mathbb{C}^*}(I_n(X, d[C])).$$

Proof For $(d, n) \in \mathbb{Z}^2$, we set $\vec{v} = (v_0, v_1) = (n, n - d)$, and

$$\mathcal{M}_{A_X}^{\dagger, \theta}(\vec{v}) = \mathcal{M}_{A_X}^\dagger(\vec{v}) \cap \mathcal{M}_{Q_X^\dagger}^\theta(\vec{v}).$$

For $\theta = (-1, 1)$, it is proved in [94] (see Figure 1 in *loc. cit.*) that the equivalence (4.4.2) induce the isomorphisms

$$\Phi_{X*} : I_n(X, d[C]) \xrightarrow{\cong} \mathcal{M}_{A_X}^{\dagger, \theta_+}(\vec{v}), \ \Phi_{X*} : P_n(X, d[C]) \xrightarrow{\cong} \mathcal{M}_{A_X}^{\dagger, \theta_-}(\vec{v}) \tag{4.4.18}$$

by sending a pair $(\mathcal{O}_X \to F)$ to $(\mathcal{P}_X \to \Phi_X(F))$. Let

$$\mathfrak{M}_S^\dagger(d[C], n)_{\mathrm{qc}} \subset \mathfrak{M}_S^\dagger(d[C], n) \tag{4.4.19}$$

be the derived open substack of pairs $(\mathcal{O}_S \to F)$ on S such that $[F] = (d[C], n)$, and satisfying $\Phi_S(F) \in \mathrm{mod}A_S$. We have the open immersion of derived stacks

$$\Phi_{S*} \colon \mathfrak{M}_S^\dagger(d[C], n)_{\mathrm{qc}} \hookrightarrow \mathfrak{M}_{A_S}^\dagger(\vec{v})$$

defined by sending $(\mathcal{O}_S \to F)$ to $(\mathcal{P}_S \to \Phi_S(F))$. In particular, the derived open substack (4.4.19) is quasi-compact. These morphisms fit into the commutative diagram

$$
\begin{array}{ccc}
I_n(X, d[C]) \overset{\Phi_{X*}}{\hookrightarrow} \mathcal{M}_{A_X}^\dagger(\vec{v}) & \quad & P_n(X, d[C]) \overset{\Phi_{X*}}{\hookrightarrow} \mathcal{M}_{A_X}^\dagger(\vec{v}) \\
\downarrow \qquad\qquad \downarrow & & \downarrow \qquad\qquad \downarrow \\
\mathcal{M}_S^\dagger(d[C], n)_{\mathrm{qc}} \overset{\Phi_{S*}}{\hookrightarrow} \mathcal{M}_{A_S}^\dagger(\vec{v}), & & \mathcal{M}_S^\dagger(d[C], n)_{\mathrm{qc}} \overset{\Phi_{S*}}{\hookrightarrow} \mathcal{M}_{A_S}^\dagger(\vec{v}).
\end{array}
\tag{4.4.20}
$$

Here $\mathcal{M}_S^\dagger(d[C], n)_{\mathrm{qc}} = t_0(\mathfrak{M}_S^\dagger(d[C], n)_{\mathrm{qc}})$. We set

$$\mathcal{Z}^{\theta\text{-us}} := \mathcal{M}_{A_X}^\dagger(\vec{v}) \setminus \mathcal{M}_{A_X}^{\dagger,\theta}(\vec{v}).$$

Then from the isomorphisms (4.4.18) and the diagrams (4.4.20), we obtain the equivalences by Lemma 3.2.9

$$\mathcal{DT}^{\mathbb{C}^*}(I_n(X, d[C])) \overset{\sim}{\to} D_{\mathrm{coh}}^b(\mathfrak{M}_{A_S}(\vec{v}))/\mathcal{C}_{\mathcal{Z}^{\theta+\text{-us}}},$$

$$\mathcal{DT}^{\mathbb{C}^*}(P_n(X, d[C])) \overset{\sim}{\to} D_{\mathrm{coh}}^b(\mathfrak{M}_{A_S}(\vec{v}))/\mathcal{C}_{\mathcal{Z}^{\theta-\text{-us}}}.$$

From the description of the derived moduli stack (4.4.3) and Theorem 2.3.3, we have the equivalence

$$D_{\mathrm{coh}}^b(\mathfrak{M}_{A_S}(\vec{v}))/\mathcal{C}_{\mathcal{Z}^{\theta\pm\text{-us}}} \overset{\sim}{\to} \mathrm{MF}_{\mathrm{coh}}^{\mathbb{C}^*}(\mathcal{M}_{Q_X^\dagger}(\vec{v}) \setminus \mathcal{Z}^{\theta\pm\text{-us}}, w).$$

Since the critical locus of w in $\mathcal{M}_{Q_X^\dagger}^{\theta\pm}(\vec{v})$ is contained in $\mathcal{M}_{Q_X^\dagger}(\vec{v}) \setminus \mathcal{Z}^{\theta\pm\text{-us}}$, the following restriction functors give equivalences (see (2.2.7))

$$\mathrm{MF}_{\mathrm{coh}}^{\mathbb{C}^*}(\mathcal{M}_{Q_X^\dagger}(\vec{v}) \setminus \mathcal{Z}^{\theta\pm\text{-us}}, w) \overset{\sim}{\to} \mathrm{MF}_{\mathrm{coh}}^{\mathbb{C}^*}(\mathcal{M}_{Q_X^\dagger}^{\theta\pm}(\vec{v}), w).$$

Therefore we have the equivalences

$$\mathcal{DT}^{\mathbb{C}^*}(I_n(X, d[C])) \overset{\sim}{\to} \mathrm{MF}_{\mathrm{coh}}^{\mathbb{C}^*}(\mathcal{M}_{Q_X}^{\theta+}(\vec{v}), w),$$

$$\mathcal{DT}^{\mathbb{C}^*}(P_n(X, d[C])) \overset{\sim}{\to} \mathrm{MF}_{\mathrm{coh}}^{\mathbb{C}^*}(\mathcal{M}_{Q_X}^{\theta-}(\vec{v}), w),$$

and the theorem follows from Proposition 4.4.3. $\qquad\qquad\qquad\qquad\square$

The same argument of Theorem 4.4.4 also shows the following:

Theorem 4.4.5 *Conjecture 4.3.14 is true when $S \to \mathbb{C}^2$ is the blow-up at the origin.*

Proof For $t \in \mathbb{Q}_{>0}$ which lies on a wall, we set $\theta = (-t+1, t)$. Then similarly to (4.4.18), we can show that the functor Φ_X induces the isomorphisms

$$\Phi_{X*} \colon P_n^{t\pm}(X, d[C]) \xrightarrow{\cong} \mathcal{M}_{A_X}^{\dagger,\theta\pm}(\vec{v}).$$

The rest of the proof is exactly same as in Theorem 4.4.4.   □

Remark 4.4.6 Let $t = m \in \mathbb{Z}_{>0}$ is a wall. In [140], it is proved that the fully-faithful functor in Conjecture 4.3.14 for the blow-up $S \to \mathbb{C}^2$ is refined to give a semiorthogonal decomposition

$$\mathcal{DT}(P_n^{t+}(X, d[C])) = \left\langle \binom{m}{k}\text{-copies of } \mathcal{DT}(P_{n-km}^{t-}(X, (d-k)[C])) : k \geq 0 \right\rangle.$$

The above semiorthogonal decomposition recovers the wall-crossing formula of numerical DT invariants in [94].

Chapter 5
Categorical Wall-Crossing via Koszul Duality

The main result in this chapter is to prove Conjecture 4.3.14 when β is an irreducible curve class. In this case, we can also describe the semiorthogonal complement of the fully-faithful functor, which recovers the wall-crossing formula of the PT invariants, and also gives a rationality statement of the generating series of PT categories in the Grothendieck group of triangulated categories.

The strategy is to interpret the wall-crossing diagram for an irreducible β in terms of Koszul duality diagram studied by Mirković-Riche [90, 91]. Indeed we will show a general result concerning semiorthogonal decompositions associated with linear Koszul dual pairs in [90, 91], and apply it to prove Conjecture 4.3.14 for an irreducible β. A more general case of the above conjecture will be studied in Chap. 6 via window theorem, and in Chap. 7 via categorified Hall products.

In order to interpret the wall-crossing diagram as a Koszul duality diagram, we need some duality statements for DT categories of D0-D2-D6 bound states, where we use Theorem 4.2.3. In this chapter, we also give a proof of Theorem 4.2.3. Namely we show that the moduli stack of rank one objects in $\mathcal{A}_X$ is isomorphic to the dual obstruction cone over the moduli stack of pairs (F, ξ) for $F \in \mathrm{Coh}_{\leq 1}(S)$ and $\xi : \mathcal{O}_S \to F$. This is not a trivial statement since a priori there is no obvious map from the moduli stack of objects in $\mathcal{A}_X$ to the moduli stack of pairs on S. The main idea is to give an alternative description of the category of $\mathcal{A}_X$ purely in terms of the surface S. More precisely we show that giving a rank one object in $\mathcal{A}_X$ is equivalent to giving a diagram

$$0 \longrightarrow \mathcal{O}_S \longrightarrow \mathcal{U} \longrightarrow F \otimes \omega_S^{-1} \longrightarrow 0$$
$$\mathcal{U} \longrightarrow F$$

Y. Toda, *Categorical Donaldson-Thomas Theory for Local Surfaces*, Lecture Notes in Mathematics 2350, https://doi.org/10.1007/978-3-031-61705-8_5

where $F \in \mathrm{Coh}_{\leq 1}(S)$ and the top sequence is an exact sequence of coherent sheaves on S. The relevant map to the moduli stack of pairs on S is given by sending the above diagram to the dotted arrow.

The organization of this chapter is as follows. In Sect. 5.1, we give some duality statements of DT categories of D0-D2-D6 bound states, and use it to prove Conjecture 4.3.14 for an irreducible β up to the result in Sect. 5.3 on semiorthogonal decomposition under linear Koszul duality. In Sect. 5.2, we give a proof of Theorem 4.2.3. In Sect. 5.3, we prove the existence of semiorthogonal decomposition associated with linear Koszul dual pairs.

5.1 Dualities of DT Categories for D0-D2-D6 Bound States

5.1.1 Moduli Stacks of Dual Pairs

Let S be a smooth projective surface, and $\mathfrak{M}_S(v)$ be the derived moduli stack of one dimensional coherent sheaves on S with numerical class $v \in N_{\leq 1}(S)$, with truncation $\mathcal{M}_S(v)$. In this subsection, we focus on the open substacks

$$\mathfrak{M}_S^{\mathrm{pure}}(v) \subset \mathfrak{M}_S(v), \ \mathcal{M}_S^{\mathrm{pure}}(v) \subset \mathcal{M}_S(v)$$

consisting of pure one dimensional sheaves on S. For a pure one dimensional sheaf F on S, we set

$$F^{\vee} := \mathbb{D}_S(F) \otimes \omega_S[1] = \mathcal{E}xt^1_{\mathcal{O}_S}(F, \omega_S). \tag{5.1.1}$$

Then $F^{\vee}$ is a pure one dimensional sheaf on S with numerical class $v^{\vee}$. Here for $v = (\beta, n) \in N_{\leq 1}(S)$, we write $v^{\vee} = (\beta, -n)$. Hence we have the equivalence of derived stacks

$$\mathbb{D}_S(-) \otimes \omega_S[1] \colon \mathfrak{M}_S^{\mathrm{pure}}(v) \xrightarrow{\sim} \mathfrak{M}_S^{\mathrm{pure}}(v^{\vee}). \tag{5.1.2}$$

Let $\mathfrak{M}_S^{\dagger}(v)$ be the derived moduli stack of pairs as in Sect. 4.2.1. Similarly we denote by

$$\mathfrak{M}_S^{\dagger,\mathrm{pure}}(v) \subset \mathfrak{M}_S^{\dagger}(v), \ \mathcal{M}_S^{\dagger,\mathrm{pure}}(v) \subset \mathcal{M}_S^{\dagger}(v)$$

the open substacks of pairs $(\mathcal{O}_S \to F)$ such that F is a pure one dimensional sheaf. We also define the derived stack $\mathfrak{M}_S^{\sharp,\mathrm{pure}}(v)$ over $\mathfrak{M}_S^{\mathrm{pure}}(v)$ to be

$$\mathfrak{M}_S^{\sharp,\mathrm{pure}}(v) := \mathrm{Spec}_{\mathfrak{M}_S^{\mathrm{pure}}(v)} S(p_{\mathfrak{M}*}\mathfrak{F}[1]).$$

Here $\mathfrak{F}$ is the universal family as in Sect. 4.2.1. Its classical truncation is given by

$$\mathcal{M}_S^{\sharp,\mathrm{pure}}(v) := t_0(\mathfrak{M}_S^{\sharp,\mathrm{pure}}(v)) = \mathrm{Spec}_{\mathcal{M}_S^{\mathrm{pure}}(v)} S(\mathcal{H}^1(\mathbf{R}p_{\mathcal{M}*}\mathcal{F})).$$

Lemma 5.1.1

(i) *The T-valued points of $\mathcal{M}_S^{\sharp,\mathrm{pure}}(v)$ form the groupoid of pairs*

$$(F_T, \xi'), \quad F_T \xrightarrow{\xi'} \omega_S \boxtimes \mathcal{O}_T[1]$$

where F_T is a T-valued point of $\mathcal{M}_S^{\mathrm{pure}}(v)$ and ξ' is a morphism in $D_{\mathrm{coh}}^b(S \times T)$.

(ii) *The fiber of the projection*

$$t_0(\Omega_{\mathfrak{M}_S^{\sharp,\mathrm{pure}}(v)}[-1]) \to \mathcal{M}_S^{\sharp,\mathrm{pure}}(v) \tag{5.1.3}$$

at the pair $(F \xrightarrow{\xi'} \omega_S[1])$ is given by $\mathrm{Hom}(\mathcal{U}, F)$, where $0 \to \mathcal{O}_S \to \mathcal{U} \to F \otimes \omega_S^{-1} \to 0$ is the extension class of ξ'.

Proof

(i) follows from the Serre duality $H^1(F)^\vee = \mathrm{Hom}(F, \omega_S[1])$ for a one dimensional sheaf F. As for (ii), we have the distinguished triangle

$$(\mathbf{R}\mathrm{Hom}(F, F)[1])^\vee \to \mathbb{L}_{\mathfrak{M}_S^{\sharp,\mathrm{pure}}(v)}|_{(F \to \omega_S[1])} \to \mathbf{R}\Gamma(F[1]).$$

By Serre duality, we have

$$(\mathbf{R}\mathrm{Hom}(F, F)[1])^\vee \cong \mathbf{R}\mathrm{Hom}(F, F \otimes \omega_S[1]),$$

so $\mathbb{L}_{\mathfrak{M}_S^{\sharp,\mathrm{pure}}(v)}|_{(F \to \omega_S[1])}$ is obtained by the cone of the morphism $\mathbf{R}\Gamma(F) \to \mathbf{R}\mathrm{Hom}(F, F \otimes \omega_S[1])$ induced by ξ'. Therefore $\mathbb{L}_{\mathfrak{M}_S^{\sharp,\mathrm{pure}}(v)}|_{(F \to \omega_S[1])}$ is quasi-isomorphic to $\mathbf{R}\mathrm{Hom}(\mathcal{U}, F)[1]$. The fiber of (5.1.3) is given by its (-1)-th cohomology, which is $\mathrm{Hom}(\mathcal{U}, F)$. $\qquad\square$

We have the following lemma:

Lemma 5.1.2 *We have the equivalence of derived stacks*

$$\mathbb{D}_S(-) \otimes \omega_S[1] \colon \mathfrak{M}_S^{\dagger,\mathrm{pure}}(v) \xrightarrow{\sim} \mathfrak{M}_S^{\sharp,\mathrm{pure}}(v^\vee) \tag{5.1.4}$$

which on $\mathbb{C}$-valued points given by

$$(\mathcal{O}_S \to F) \mapsto \mathbb{D}_S(\mathcal{O}_S \to F) \otimes \omega_S[1] = (F^\vee \to \omega_S[1]).$$

Moreover we have the commutative diagram

$$
\begin{array}{ccccc}
\mathfrak{M}_S^{\dagger,\mathrm{pure}}(v) & \xrightarrow{\ \rho^\dagger\ } & \mathfrak{M}_S^{\mathrm{pure}}(v) & \xrightarrow{\ i^\dagger\ } & \mathfrak{M}_S^{\dagger,\mathrm{pure}}(v) \\
\mathbb{D}_S(-)\otimes\omega_S[1]\Big\downarrow{\sim} & & \mathbb{D}_S(-)\otimes\omega_S[1]\Big\downarrow{\sim} & & \mathbb{D}_S(-)\otimes\omega_S[1]\Big\downarrow{\sim} \\
\mathfrak{M}_S^{\sharp,\mathrm{pure}}(v^\vee) & \xrightarrow[\ \rho^\sharp\ ]{} & \mathfrak{M}_S^{\mathrm{pure}}(v^\vee) & \xrightarrow[\ i^\sharp\ ]{} & \mathfrak{M}_S^{\sharp,\mathrm{pure}}(v^\vee).
\end{array}
$$

$$(5.1.5)$$

Here $\rho^\dagger$, $\rho^\sharp$ are projections and $i^\dagger$, $i^\sharp$ are the zero sections.

Proof Under the equivalence (5.1.2), the object $(p_{\mathfrak{M}*}\mathfrak{F})^\vee$ on $\mathfrak{M}_S^{\mathrm{pure}}(v)$ corresponds to $p_{\mathfrak{M}*}\mathfrak{F}[1]$ on $\mathfrak{M}_S^{\mathrm{pure}}(v^\vee)$, since for a pure one dimensional sheaf F on S we have

$$
\mathbf{R}\Gamma(F)^\vee = \mathbf{R}\mathrm{Hom}(F, \omega_S[2]) = \mathbf{R}\Gamma(F^\vee)[1]
$$

by the Serre duality. Therefore we have the equivalence (5.1.4). The commutativity of the diagram (5.1.5) is obvious from the constructions.   □

Let $X = \mathrm{Tot}(\omega_S)$ be the local surface. We also denote by

$$
\mathcal{M}_X^{\dagger,\mathrm{pure}}(v) \subset \mathcal{M}_X^\dagger(v)
$$

the open substack of objects in the subcategory

$$
\mathcal{A}_X^{\mathrm{pure}} := \langle \mathcal{O}_{\overline{X}}, \mathrm{Coh}_{\leq 1}^{\mathrm{pure}}(X)[-1]\rangle_{\mathrm{ex}} \subset \mathcal{A}_X \tag{5.1.6}
$$

where $\mathrm{Coh}_{\leq 1}^{\mathrm{pure}}(X)$ is the category of compactly supported pure one dimensional coherent sheaves on X. Note that the map $\pi_*^\dagger$ in Theorem 4.2.3 restricts to the morphism $\pi_*^\dagger \colon \mathcal{M}_X^{\dagger,\mathrm{pure}}(v) \to \mathcal{M}_S^{\dagger,\mathrm{pure}}(v)$ and the isomorphism (4.2.6) restricts to the isomorphism over $\mathcal{M}_S^{\dagger,\mathrm{pure}}(v)$

$$
\eta^\dagger \colon \mathcal{M}_X^{\dagger,\mathrm{pure}}(v) \xrightarrow{\cong} t_0(\Omega_{\mathfrak{M}_S^{\dagger,\mathrm{pure}}(v)}[-1]).
$$

We have a statement similar to Theorem 4.2.3 for the stack $\mathcal{M}_S^{\sharp,\mathrm{pure}}(v)$.

Proposition 5.1.3

(i) *There exists a natural morphism $\pi_*^\sharp \colon \mathcal{M}_X^{\dagger,\mathrm{pure}}(v) \to \mathcal{M}_S^{\sharp,\mathrm{pure}}(v)$ sending a diagram (4.2.4) to the pair $(F \xrightarrow{\xi'} \omega_S[1])$ corresponding to the extension class of the top sequence of (4.2.4).*

(ii) There is an isomorphism over $\mathcal{M}_S^{\sharp,\text{pure}}(v)$

$$\eta^\sharp : \mathcal{M}_X^{\dagger,\text{pure}}(v) \xrightarrow{\cong} t_0(\Omega_{\mathfrak{M}_S^{\sharp,\text{pure}}(v)}[-1])$$

which, over the pair $(F \xrightarrow{\xi'} \omega_S[1])$, *sends a diagram (4.2.4) to the morphism* $\xi : \mathcal{U} \to F$.

Proof The proof for $\mathbb{C}$-valued points is clear from Lemmas 5.1.1 and 5.2.3. The argument for T-valued points is similar to Lemma 5.2.10, so we omit details.   □

Note that we have obtained the commutative diagram

$$
\begin{array}{ccccc}
t_0(\Omega_{\mathfrak{M}_S^{\dagger,\text{pure}}(v)}[-1]) & \xleftarrow[\cong]{\eta^\dagger} & \mathcal{M}_X^{\dagger,\text{pure}}(v) & \xrightarrow[\cong]{\eta^\sharp} & t_0(\Omega_{\mathfrak{M}_S^{\sharp,\text{pure}}(v)}[-1]) \\
\downarrow & \searrow^{\pi_*^\dagger} & \downarrow & \swarrow^{\pi_*^\sharp} & \downarrow \\
\mathcal{M}_S^{\dagger,\text{pure}}(v) & \xrightarrow{\rho_0^\dagger} & \mathcal{M}_S^{\text{pure}}(v) & \xleftarrow{\rho_0^\sharp} & \mathcal{M}_S^{\sharp,\text{pure}}(v).
\end{array}
$$

$$(5.1.7)$$

On the other hand the category $\mathcal{A}_X^{\text{pure}}$ is closed under the derived dual functor $\mathbb{D}_{\overline{X}}$, so we have an isomorphism of stacks

$$\mathbb{D}_{\overline{X}} : \mathcal{M}_X^{\dagger,\text{pure}}(v) \xrightarrow{\cong} \mathcal{M}_X^{\dagger,\text{pure}}(v^\vee).$$

$$(5.1.8)$$

Lemma 5.1.4 *We have the commutative diagram*

$$
\begin{array}{ccc}
\mathcal{M}_X^{\dagger,\text{pure}}(v) & \xrightarrow[\cong]{\mathbb{D}_{\overline{X}}} & \mathcal{M}_X^{\dagger,\text{pure}}(v^\vee) \\
\eta^\dagger \downarrow & & \downarrow \eta^\sharp \\
t_0(\Omega_{\mathfrak{M}_S^{\dagger,\text{pure}}(v)}[-1]) & \xrightarrow{\cong} & t_0(\Omega_{\mathfrak{M}_S^{\sharp,\text{pure}}(v^\vee)}[-1]).
\end{array}
$$

$$(5.1.9)$$

Here the bottom isomorphism is induced by the equivalence (5.1.4).

Proof The commutativity of the diagram (5.1.9) immediately follows from the constructions of $\eta^\dagger$, $\eta^\sharp$ and Lemma 5.2.14.   □

By Proposition 5.1.3, the stack $\mathcal{M}_X^{\dagger,\text{pure}}(v)$ admits two $\mathbb{C}^*$-actions, fiberwise $\mathbb{C}^*$-actions with respect to $\pi_*^\dagger$ and $\pi_*^\sharp$. These two $\mathbb{C}^*$-actions on $\mathcal{M}_X^{\dagger,\text{pure}}(v)$ correspond to scaling actions on the maps $\mathcal{U} \to F$, $\mathcal{U} \to F \otimes \omega_S^{-1}$ in the diagram (4.2.4) respectively. We call a closed substack $\mathcal{Z} \subset \mathcal{M}_X^{\dagger,\text{pure}}(v)$ to be *double conical* if it is closed under both of the above $\mathbb{C}^*$-actions.

Let us take quasi-compact derived open substacks

$$\mathfrak{M}_S(v)_{\mathrm{qc}} \subset \mathfrak{M}_S^{\mathrm{pure}}(v), \ \mathcal{M}_S(v)_{\mathrm{qc}} \subset \mathcal{M}_S^{\mathrm{pure}}(v) \tag{5.1.10}$$

where $\mathcal{M}_S(v)_{\mathrm{qc}} = t_0(\mathfrak{M}_S(v)_{\mathrm{qc}})$. Below we use the subscript qc to indicate the pull-back by the open immersion (5.1.10), e.g. $\mathfrak{M}_S^{\dagger}(v)_{\mathrm{qc}} := \mathfrak{M}_S(v)_{\mathrm{qc}} \times_{\mathfrak{M}_S^{\mathrm{pure}}(v)} \mathfrak{M}_S^{\dagger,\mathrm{pure}}(v)$. By pulling back the diagram (5.1.7) via the open immersion (5.1.10), we obtain the commutative diagram

$$
\begin{array}{ccccc}
t_0(\Omega_{\mathfrak{M}_S^{\dagger}(v)_{\mathrm{qc}}}[-1]) & \overset{\eta^{\dagger}}{\underset{\cong}{\longleftarrow}} & \mathcal{M}_X^{\dagger}(v)_{\mathrm{qc}} & \overset{\eta^{\sharp}}{\underset{\cong}{\longrightarrow}} & t_0(\Omega_{\mathfrak{M}_S^{\sharp}(v)_{\mathrm{qc}}}[-1]) \\
\downarrow & \overset{\pi_*^{\dagger}}{\swarrow} & \downarrow & \overset{\pi_*^{\sharp}}{\searrow} & \downarrow \\
\mathcal{M}_S^{\dagger}(v)_{\mathrm{qc}} & \underset{\rho_0^{\dagger}}{\longrightarrow} & \mathcal{M}_S(v)_{\mathrm{qc}} & \underset{\rho_0^{\sharp}}{\longleftarrow} & \mathcal{M}_S^{\sharp}(v)_{\mathrm{qc}}.
\end{array}
\tag{5.1.11}
$$

Similarly we take a quasi-compact derived open substack $\mathfrak{M}_S(v^{\vee})_{\mathrm{qc}} \subset \mathfrak{M}_S^{\mathrm{pure}}(v^{\vee})$ with truncation $\mathcal{M}_S(v^{\vee})_{\mathrm{qc}}$, such that the equivalence (5.1.2) restricts to the equivalence $\mathfrak{M}_S(v)_{\mathrm{qc}} \overset{\sim}{\to} \mathfrak{M}_S(v^{\vee})_{\mathrm{qc}}$. Then via the isomorphisms $(\mathbb{D}_{\overline{X}}, \mathbb{D}_S(-) \otimes \omega_S)$ in the diagrams (5.1.5) and (5.1.9), we see that the diagram (5.1.11) is isomorphic to the diagram

$$
\begin{array}{ccccc}
t_0(\Omega_{\mathfrak{M}_S^{\sharp}(v^{\vee})_{\mathrm{qc}}}[-1]) & \overset{\eta^{\sharp}}{\underset{\cong}{\longleftarrow}} & \mathcal{M}_X^{\dagger}(v^{\vee})_{\mathrm{qc}} & \overset{\eta^{\dagger}}{\underset{\cong}{\longrightarrow}} & t_0(\Omega_{\mathfrak{M}_S^{\dagger}(v^{\vee})_{\mathrm{qc}}}[-1]) \\
\downarrow & \overset{\pi_*^{\sharp}}{\swarrow} & \downarrow & \overset{\pi_*^{\dagger}}{\searrow} & \downarrow \\
\mathcal{M}_S^{\sharp}(v^{\vee})_{\mathrm{qc}} & \underset{\rho_0^{\sharp}}{\longrightarrow} & \mathcal{M}_S(v^{\vee})_{\mathrm{qc}} & \underset{\rho_0^{\dagger}}{\longleftarrow} & \mathcal{M}_S^{\dagger}(v^{\vee})_{\mathrm{qc}}.
\end{array}
\tag{5.1.12}
$$

We have the following corollary of Propositions 5.1.3 and 5.3.14:

Corollary 5.1.5 *Let* $\mathcal{Z} \subset \mathcal{M}_X^{\dagger}(v)_{\mathrm{qc}}$ *be a double conical closed substack.*

(i) Let $\mathbb{D}_{\overline{X}}(\mathcal{Z}) \subset \mathcal{M}_X^{\dagger}(v^{\vee})_{\mathrm{qc}}$ *be its image under the isomorphism (5.1.8). We have the equivalence*

$$D_{\mathrm{coh}}^b(\mathfrak{M}_S^{\dagger}(v)_{\mathrm{qc}})/\mathcal{C}_{\eta^{\dagger}(\mathcal{Z})} \overset{\sim}{\to} D_{\mathrm{coh}}^b(\mathfrak{M}_S^{\sharp}(v^{\vee})_{\mathrm{qc}})/\mathcal{C}_{\eta^{\sharp}(\mathbb{D}_{\overline{X}}(\mathcal{Z}))}.$$

(ii) We have the equivalence

$$D_{\mathrm{coh}}^b(\mathfrak{M}_S^{\dagger}(v)_{\mathrm{qc}})/\mathcal{C}_{\eta^{\dagger}(\mathcal{Z})} \overset{\sim}{\to} D_{\mathrm{coh}}^b(\mathfrak{M}_S^{\sharp}(v)_{\mathrm{qc}})/\mathcal{C}_{\eta^{\sharp}(\mathcal{Z})}.$$

Proof As for (i), we have the equivalence by Lemma 5.1.2

$$D^b_{\mathrm{coh}}(\mathfrak{M}^\dagger_S(v)_{\mathrm{qc}}) \xrightarrow{\sim} D^b_{\mathrm{coh}}(\mathfrak{M}^\sharp_S(v^\vee)_{\mathrm{qc}}).$$

Then Lemma 5.1.4 implies that the above equivalence restricts to the equivalence between $\mathcal{C}_{\eta^\dagger(\mathcal{Z})}$ and $\mathcal{C}_{\eta^\sharp(\mathbb{D}_{\overline{X}}(\mathcal{Z}))}$. By taking the quotients, we obtain (i).

As for the equivalence in (ii), it is straightforward to check that the following diagram commutes:

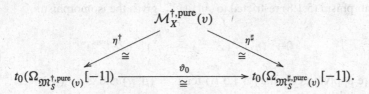

Here ϑ_0 is the isomorphism in Lemma 5.3.13. Therefore the equivalence in (ii) follows from Proposition 5.3.14. □

5.1.2 Wall-Crossing at $t < 0$

Let us return to the situation of Sects. 4.3.2 and 4.3.5. We first give the following lemma:

Lemma 5.1.6 *The moduli stack of $\mu^\dagger_t$-semistable objects $\mathcal{P}^t_n(X, \beta)$ is an open substack in $\mathcal{M}^{\dagger,\mathrm{pure}}_X(\beta, n)$, and the isomorphism (5.1.8) restricts to the isomorphisms*

$$\mathbb{D}_{\overline{X}} \colon \mathcal{P}^t_n(X, \beta) \xrightarrow{\sim} \mathcal{P}^{-t}_{-n}(X, \beta), \quad \mathbb{D}_{\overline{X}} \colon P^t_n(X, \beta) \xrightarrow{\sim} P^{-t}_{-n}(X, \beta).$$

Proof The former statement follows from Lemma 5.2.12 together with the $\mu^\dagger_t$-stability. The latter statement is due to [129, Proposition 5.4 (iii)]. □

Then we have the following duality statement as an application of Corollary 5.1.5:

Theorem 5.1.7 *If $t \in \mathbb{R}$ lies in a chamber, we have the equivalence*

$$\mathcal{DT}^{\mathbb{C}^*}(P^t_n(X, \beta)) \xrightarrow{\sim} \mathcal{DT}^{\mathbb{C}^*}(P^{-t}_{-n}(X, \beta)).$$

In particular we have the equivalence (4.3.7).

Proof Below we use the notation of the diagrams (5.1.11) and (5.1.12). Let us take a quasi-compact open derived substack $\mathfrak{M}_S(\beta, n)_{qc} \subset \mathfrak{M}_S^{pure}(\beta, n)$ such that $\mathcal{M}_S^\dagger(\beta, n)_{qc} = t_0(\mathfrak{M}_S^\dagger(\beta, n)_{qc})$ satisfies the condition (4.3.5). By definition, we have

$$\mathcal{DT}^{\mathbb{C}^*}(P_n^t(X, \beta)) = D_{coh}^b(\mathfrak{M}_S^\dagger(\beta, n)_{qc})/\mathcal{C}_{\mathcal{Z}^{t\text{-us}}}.$$

Note that the unstable locus $(\eta^\dagger)^{-1}(\mathcal{Z}^{t\text{-us}})$ is a double conical closed substack of $\mathcal{M}_X^\dagger(\beta, n)_{qc}$, since the two $\mathbb{C}^*$-actions preserve the $\mu_t^\dagger$-stability. By Lemma 5.1.6 the isomorphism (5.1.8) restricted to $\mathcal{M}_X^\dagger(v)_{qc}$ gives the isomorphism

$$\mathbb{D}_{\overline{X}}\colon (\eta^\dagger)^{-1}(\mathcal{Z}^{t\text{-us}}) \xrightarrow{\cong} (\eta^\dagger)^{-1}(\mathcal{Z}^{-t\text{-us}}).$$

Therefore applying Corollary 5.1.5 (i) for $v = (\beta, n)$ and $\mathcal{Z} = (\eta^\dagger)^{-1}(\mathcal{Z}^{t\text{-us}})$, we have the equivalence

$$D_{coh}^b(\mathfrak{M}_S^\dagger(\beta, n)_{qc})/\mathcal{C}_{\mathcal{Z}^{t\text{-us}}} \xrightarrow{\sim} D_{coh}^b(\mathfrak{M}_S^\sharp(\beta, -n)_{qc})/\mathcal{C}_{\eta^\sharp(\eta^\dagger)^{-1}(\mathcal{Z}^{-t\text{-us}})}.$$

Also applying Corollary 5.1.5 (ii) for $v = (\beta, -n)$ and $\mathcal{Z} = (\eta^\dagger)^{-1}(\mathcal{Z}^{-t\text{-us}})$, we have the equivalence

$$D_{coh}^b(\mathfrak{M}_S^\dagger(\beta, -n)_{qc})/\mathcal{C}_{\mathcal{Z}^{-t\text{-us}}} \xrightarrow{\sim} D_{coh}^b(\mathfrak{M}_S^\sharp(\beta, -n)_{qc})/\mathcal{C}_{\eta^\sharp(\eta^\dagger)^{-1}(\mathcal{Z}^{-t\text{-us}})}.$$

The desired equivalence follows from the above equivalences. □

For $t_0 < 0$, the diagram (4.3.11) is a d-critical anti-flip. Therefore for $t < 0$ wall-crossing, we expect a chain of fully-faithful functors for $0 > s_1 > s_2 > \cdots > s_M$ which lie on chambers

$$\mathcal{DT}^{\mathbb{C}^*}(P_n^{s_1}(X, \beta)) \hookrightarrow \mathcal{DT}^{\mathbb{C}^*}(P_n^{s_2}(X, \beta)) \hookrightarrow \cdots \hookrightarrow \mathcal{DT}^{\mathbb{C}^*}(P_n^{s_M}(X, \beta)).$$

Indeed Theorem 5.1.7 implies that the above chain of fully-faithful functors is equivalent to that of fully-faithful functors (4.3.12) for $(\beta, -n)$.

5.1.3 Wall-Crossing Formula of Categorical PT Theories for Irreducible Curve Classes

We take $(\beta, n) \in N_{\leq 1}(S)$ such that β is an irreducible class and $n \geq 0$. As in Sect. 4.1.3, we use the notation $\mathcal{M}_n(X, \beta)$, $M_n(X, \beta)$ for moduli spaces of one dimensional stable sheaves on X with numerical class (β, n). In the irreducible β case, there is only one wall $t = n/(H \cdot \beta)$ with respect to the $\mu_t^\dagger$-stability, and the diagram (4.3.11) becomes (see [137, Section 9.6])

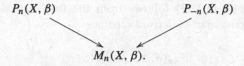

By Toda [137, Theorem 9.22], the above diagram is a simple d-critical flip if $n > 0$, simple d-critical flop if $n = 0$. Below we prove Conjecture 4.3.14 in this case.

As β is irreducible, the derived stack $\mathfrak{M}_S^{\mathrm{pure}}(\beta, n)$ coincides with the derived stack of one dimensional stable sheaves on S with numerical class (β, n). In particular, all of the derived stacks $\mathfrak{M}_S^{\mathrm{pure}}(\beta, n)$, $\mathfrak{M}_S^{\dagger, \mathrm{pure}}(\beta, n)$, $\mathfrak{M}_S^{\sharp}(\beta, n)$ are QCA. In the notation of the diagram (5.1.5), we set

$$\mathfrak{P}_n(S, \beta) := \mathfrak{M}_S^{\dagger, \mathrm{pure}}(\beta, n) \setminus i^{\dagger}(\mathfrak{M}_S^{\mathrm{pure}}(\beta, n)),$$

$$\mathfrak{Q}_n(S, \beta) := \mathfrak{M}_S^{\sharp, \mathrm{pure}}(\beta, n) \setminus i^{\sharp}(\mathfrak{M}_S^{\mathrm{pure}}(\beta, n))$$

which are derived open substacks of $\mathfrak{M}_S^{\dagger, \mathrm{pure}}(\beta, n)$, $\mathfrak{M}_S^{\sharp, \mathrm{pure}}(\beta, n)$ respectively. By the assumption that β is irreducible, the derived stack $\mathfrak{P}_n(S, \beta)$ is the derived moduli scheme of PT stable pairs on S with numerical class (β, n), so its truncation

$$P_n(S, \beta) := t_0(\mathfrak{P}_n(S, \beta))$$

is the moduli space of PT stable pairs on S.

Lemma 5.1.8 *For* $(\beta, n) \in N_{\leq 1}(S)$ *such that* β *is irreducible, we have the equivalences*

$$\mathcal{DT}^{\mathbb{C}^*}(P_n(X, \beta)) \xrightarrow{\sim} D_{\mathrm{coh}}^b(\mathfrak{P}_n(S, \beta)), \qquad (5.1.13)$$

$$\mathcal{DT}^{\mathbb{C}^*}(P_{-n}(X, \beta)) \xrightarrow{\sim} D_{\mathrm{coh}}^b(\mathfrak{Q}_n(S, \beta)).$$

Moreover the following identity holds:

$$\chi(\mathrm{HP}_*(\mathcal{DT}_{\mathrm{dg}}^{\mathbb{C}^*}(P_n(X, \beta)))) = (-1)^{n+\beta^2} P_{n,\beta}. \qquad (5.1.14)$$

Proof In [87, Lemma 5.13] it is proved that, under the assumption that β is irreducible, $P_n(X, \beta)$ is the dual obstruction cone over $P_n(S, \beta)$. It follows that we have

$$P_n(X, \beta) = (\pi_*^{\dagger})^{-1}(P_n(S, \beta))$$

for the map $\pi_*^{\dagger} \colon \mathcal{M}_X^{\dagger}(\beta, n) \to \mathcal{M}_S^{\dagger}(\beta, n)$ in Theorem 4.2.3. Therefore the first equivalence (5.1.13) follows from Lemma 3.2.9 as $\mathfrak{M}_S^{\dagger, \mathrm{pure}}(\beta, n)$ is QCA. The second equivalence of (5.1.13) follows from the first equivalence for $(\beta, -n)$ together with the equivalence of derived schemes

$$\mathfrak{P}_{-n}(S, \beta) \xrightarrow{\sim} \mathfrak{Q}_n(S, \beta)$$

induced by the equivalences in the diagram (5.1.5).

As for the identity (5.1.14), note that the virtual dimension of $P_n(S, \beta)$ is calculated by the Riemann-Roch theorem

$$\text{vdim}\, P_n(S, \beta) = \chi(\mathcal{O}_S \to F, F) = \beta^2 + n$$

for a stable pair $(\mathcal{O}_S \to F)$ on S with $[F] = (\beta, n)$. Then the identity (5.1.14) follows from Proposition 3.3.6, since $\pi_*^\dagger(P_n(X, \beta)) = P_n(S, \beta)$ and $P_n(S, \beta)$ is a projective scheme (see Remark 3.3.7).   □

The following result gives a proof of Conjecture 4.3.14 in the irreducible curve class case:

Theorem 5.1.9 *For $(\beta, n) \in N_{\leq 1}(S)$ such that β is irreducible and $n \geq 0$, there exists a fully-faithful functor*

$$\Phi_P: \mathcal{DT}^{\mathbb{C}^*}(P_{-n}(X, \beta)) \hookrightarrow \mathcal{DT}^{\mathbb{C}^*}(P_n(X, \beta))$$

and a semiorthogonal decomposition

$$\mathcal{DT}^{\mathbb{C}^*}(P_n(X, \beta)) = \langle \Upsilon_{-n+1}, \dots, \Upsilon_0, \text{Im}\, \Phi_P \rangle \qquad (5.1.15)$$

such that each Υ_λ is equivalent to $\mathcal{DT}^{\mathbb{C}^*}(\mathcal{M}_n(X, \beta))_\lambda$.

Proof By Lemma 5.1.8, the result follows from Theorem 5.3.11 by applying it for $\mathfrak{M} = \mathfrak{M}_S^{\text{pure}}(\beta, n)$, $\mathfrak{E} = (p_{\mathfrak{M}*}\mathfrak{F})^\vee$. Here we note that, since the object $\mathfrak{E}$ has weight -1 with respect to the embedding $(\mathbb{C}^*)_{\mathcal{M}} \subset I_{\mathcal{M}}$ given by the scaling action $\mathbb{C}^* \subset \text{Aut}(F)$, we need to apply Theorem 5.3.11 for the composition of the above embedding $(\mathbb{C}^*)_{\mathcal{M}} \subset I_{\mathcal{M}}$ with the isomorphism $(\mathbb{C}^*)_{\mathcal{M}} \xrightarrow{\cong} (\mathbb{C}^*)_{\mathcal{M}}$ sending x to x^{-1}. Then $e = n \geq 0$, and the weights in the semiorthogonal decomposition (5.1.15) have opposite signs from those in Theorem 5.3.11.   □

The above result is regarded as a categorification of the wall-crossing formula

$$P_{n,\beta} - P_{-n,\beta} = (-1)^{n-1} n N_{n,\beta} \qquad (5.1.16)$$

proved in [103], which is a special case of the formula (1.3.6). Indeed we have the following:

Proposition 5.1.10 *In the situation of Theorem 5.1.9, suppose that there is a divisor D on S such that $(D \cdot \beta, n)$ is coprime. Then the semiorthogonal decomposition (5.1.15) implies the formula (5.1.16).*

Proof The assumption together with Proposition 3.2.11 and Lemma 4.1.5 (ii) imply that each $\mathcal{D}_\lambda$ is equivalent to $\mathcal{DT}^{\mathbb{C}^*}(M_n(X, \beta))$. On the other hand, as we discussed in the proof of Lemma 3.3.5, the assignment of dg-categories to mixed complexes of Hochschild complexes takes exact sequences to distinguished triangles (see [69, Theorem 3.1]). Therefore the semiorthogonal decomposition (5.1.15) implies the formula

$$\chi(\mathrm{HP}_*(\mathcal{DT}_{\mathrm{dg}}^{\mathbb{C}^*}(P_n(X, \beta)))) - \chi(\mathrm{HP}_*(\mathcal{DT}_{\mathrm{dg}}^{\mathbb{C}^*}(P_{-n}(X, \beta))))$$

$$= n \cdot \chi(\mathrm{HP}_*(\mathcal{DT}_{\mathrm{dg}}^{\mathbb{C}^*}(M_n(X, \beta)))).$$

The formula (5.1.16) follows from the above identity together with (4.1.4) and (5.1.14). □

Example 5.1.11 Suppose that $H^1(\mathcal{O}_S) = 0$. Then there is unique $L \in \mathrm{Pic}(S)$ such that $c_1(L) = \beta$. Let $|L|$ be the complete linear system, and

$$\pi_\mathcal{C} \colon \mathcal{C} \to |L|$$

the universal curve. We also denote by $g \in \mathbb{Z}$ the arithmetic genus of curves in $|L|$, i.e. $g = 1 + \beta(K_S + \beta)/2$. Then we have an isomorphism (see [103, Appendix])

$$P_n(S, \beta) \cong \mathcal{C}^{[n+g-1]}.$$

Here the right hand side is $\pi_\mathcal{C}$-relative Hilbert scheme of $(n + g - 1)$-points on $\mathcal{C}$. Moreover $P_n(S, \beta)$ has only locally complete intersection singularities, and we have a distinguished triangle in $D^b_{\mathrm{coh}}(P_n(S, \beta))$ (see [87, Section 5.9])

$$\mathcal{V}[1] \to \mathbb{L}_{\mathfrak{P}_n(S,\beta)|P_n(S,\beta)} \to \mathbb{L}_{P_n(S,\beta)}$$

for a locally free sheaf $\mathcal{V}$ on $P_n(S, \beta)$ with rank $h^1(L)$. Therefore the closed immersion $P_n(S, \beta) \hookrightarrow \mathfrak{P}_n(S, \beta)$ is an equivalence if and only if $h^1(L) = 0$.

If $h^1(L) = 0$, then the semiorthogonal decomposition in Theorem 5.1.9 is equivalent to the semiorthogonal decomposition for $n \geq 0$

$$D^b_{\mathrm{coh}}(\mathcal{C}^{[n+g-1]}) = \langle \Upsilon_{-n+1}, \ldots, \Upsilon_0, D^b_{\mathrm{coh}}(\mathcal{C}^{[-n+g-1]}) \rangle. \qquad (5.1.17)$$

Here each $\mathcal{D}_\lambda$ is the bounded derived category of twisted coherent sheaves on the relative compactified Jacobian $\overline{J} \to |L|$. The above semiorthogonal decomposition generalizes [135, Corollary 5.10], where a similar semiorthogonal decomposition is proved under some more additional assumptions. On the other hand if $h^1(L) \neq 0$, the derived scheme $\mathfrak{P}_n(S, \beta)$ has non-trivial derived structures, and the semiorthogonal decomposition in Theorem 5.1.9 is different from (5.1.17).

5.1.4 Application to the Rationality

The wall-crossing formula (5.1.16) was used in [103] to show the rationality of the generating series of PT invariants. Using the semiorthogonal decomposition in Theorem 5.1.9, we also have some rationality statement. First we give the following definition:

Definition 5.1.12 The Grothendieck group of triangulated categories $K(\Delta\text{-}\mathrm{Cat})$ is the abelian group generated by equivalence classes of triangulated categories, with relations $[\mathcal{A}] = [\mathcal{B}] + [\mathcal{C}]$ for semiorthogonal decompositions $\mathcal{A} = \langle \mathcal{B}, \mathcal{C} \rangle$.

Remark 5.1.13 The Grothendieck group of pre-triangulated categories introduced in [21] is a refined version of $K(\Delta\text{-}\mathrm{Cat})$. An advantage of the former is the existence of the product structure. As we will not need the product structure below, we use the simpler version $K(\Delta\text{-}\mathrm{Cat})$.

Let $\mathbb{Q}(q)^{\mathrm{inv}} \subset \mathbb{Q}((q))$ be the subspace of rational functions invariant under $q \leftrightarrow 1/q$. We have the following corollary of Theorem 5.1.9:

Corollary 5.1.14 *Suppose that β is irreducible. Then the generating series*

$$P_\beta^{\mathrm{cat}}(q) := \sum_{n \in \mathbb{Z}} [\mathcal{DT}^{\mathbb{C}^*}(P_n(X, \beta))] q^n \in K(\Delta\text{-}\mathrm{Cat})((q))$$

lies in $K(\Delta\text{-}\mathrm{Cat}) \otimes_{\mathbb{Z}} \mathbb{Q}(q)^{\mathrm{inv}}$.

Proof For a fixed irreducible class β, we set

$$c_{n,\lambda} := [\mathcal{DT}^{\mathbb{C}^*}(\mathcal{M}_n(X, \beta))_{-\lambda}] \in K(\Delta\text{-}\mathrm{Cat}).$$

Then the semiorthogonal decomposition in Theorem 5.1.9 implies the identity in $K(\Delta\text{-}\mathrm{Cat})$

$$[\mathcal{DT}^{\mathbb{C}^*}(P_n(X, \beta))] = [\mathcal{DT}^{\mathbb{C}^*}(P_{-n}(X, \beta))] + \sum_{\lambda=0}^{n-1} c_{n,\lambda} \qquad (5.1.18)$$

for $n \geq 0$. We consider the following generating series

$$N_\beta^{\mathrm{cat}}(q) := \sum_{n \geq 0} \sum_{\lambda=0}^{n-1} c_{n,\lambda} q^n \in K(\Delta\text{-}\mathrm{Cat})[[q]].$$

Then by the relation (5.1.18), we have

$$P_\beta^{\mathrm{cat}}(q) - N_\beta^{\mathrm{cat}}(q) = [\mathcal{DT}^{\mathbb{C}^*}(P_0(X, \beta))] + \sum_{n > 0} [\mathcal{DT}^{\mathbb{C}^*}(P_{-n}(X, \beta))](q^n + q^{-n}).$$

The right hand side lies in $K(\Delta\text{-}\mathrm{Cat})\otimes_{\mathbb{Z}}\mathbb{Q}(q)^{\mathrm{inv}}$, since $P_{-n}(X,\beta)=\emptyset$ for $n\gg 0$. Therefore it is enough to show that $N_{\beta}^{\mathrm{cat}}(q)$ lies in $K(\Delta\text{-}\mathrm{Cat})\otimes_{\mathbb{Z}}\mathbb{Q}(q)^{\mathrm{inv}}$.

Let $d\in\mathbb{Z}_{>0}$ be as in Lemma 4.1.5. Then by Lemma 4.1.5, we have the identities

$$c_{n,\lambda}=c_{n+d,\lambda}=c_{n,\lambda+n}=c_{n,\lambda+d}=c_{-n,-\lambda}. \qquad (5.1.19)$$

Let us set

$$C_n:=\sum_{\lambda=0}^{d-1}c_{n,\lambda},\ \ \overline{C}_n:=\sum_{\lambda=0}^{n-1}c_{n,\lambda}.$$

Then using the relations (5.1.19), a straightforward computation shows that

$$N_{\beta}^{\mathrm{cat}}(q)=C_0\frac{q^d}{(1-q^d)^2}+\sum_{n=1}^{d-1}C_n(q^n+q^{-n})\frac{q^d}{2(1-q^d)^2}+\sum_{n=1}^{d-1}\overline{C}_n\frac{q^n}{2(1-q^d)}\frac{q^{d-n}}{}.$$

Since the right hand side lies in $K(\Delta\text{-}\mathrm{Cat})\otimes_{\mathbb{Z}}\mathbb{Q}(q)^{\mathrm{inv}}$, the assertion holds.   □

5.2 The Category of D0-D2-D6 Bound States

The purpose of this section is to prove Theorem 4.2.3. As we mentioned in Sect. 1.3.2, the key ingredient is to relate rank one objects in the category of D0-D2-D6 bound states on the local surface with the diagram (4.2.4) on the surface. This is similar to Diaconescu's description [35] of D0-D2-D6 bound states on local curves (i.e. non-compact CY three-folds given by total spaces of split rank two vector bundles on smooth projective curves) in terms of ADHM sheaves on curves (though the details of the comparison with $\mathcal{A}_X$ are not available in literatures).

5.2.1 Notation of Local Surfaces

Let S be a smooth projective surface over $\mathbb{C}$, and $X=\mathrm{Tot}_S(\omega_S)$ be the total space of its canonical line bundle. We take its projective compactification $\overline{X}$ and consider the diagram

$$\begin{array}{ccc} X & \xrightarrow{\ j\ } & \overline{X}=\mathbb{P}_S(\omega_S\oplus\mathcal{O}_S)\\ {\scriptstyle\pi}\downarrow & {\scriptstyle\overline{\pi}}\ \ \nearrow & \\ & \nwarrow{\scriptstyle i_0,i_\infty} & \\ S. & & \end{array}$$

Here π, $\overline{\pi}$ are projections, j is an open immersion, i_0 is the zero section of π and i_∞ is the section of $\overline{\pi}$ at the infinity $\overline{X} \setminus X$. We denote by

$$S_0 := i_0(S), \quad S_\infty := i_\infty(S).$$

Below we often identity S_0, S_∞ with S by the morphisms i_0, i_∞ respectively. Note that we have

$$\mathcal{O}_{\overline{X}}(1) = \mathcal{O}_{\overline{X}}(S_\infty), \quad \Omega_{\overline{X}/S} = \mathcal{O}_{\overline{X}}(-S_0 - S_\infty), \quad \overline{\pi}^*\omega_S = \mathcal{O}_{\overline{X}}(S_0 - S_\infty).$$

$$(5.2.1)$$

By taking the restrictions to both of S_0 and S_∞, we have the canonical isomorphism

$$\mathbf{R}\overline{\pi}_*\mathcal{O}_{\overline{X}}(S_\infty) \xrightarrow{\cong} \mathbf{R}\overline{\pi}_*\mathcal{O}_{S_0}(S_\infty) \oplus \mathbf{R}\overline{\pi}_*\mathcal{O}_{S_\infty}(S_\infty) = \mathcal{O}_S \oplus \omega_S^{-1}. \qquad (5.2.2)$$

The above isomorphism will be often used below.

5.2.2   The Category $\mathcal{B}_S$

We define the category $\mathcal{B}_S$ whose objects consist of diagrams

$$
\begin{array}{ccccccccc}
0 & \longrightarrow & \mathcal{V} & \xrightarrow{\alpha} & \mathcal{U} & \xrightarrow{\psi} & F \otimes \omega_S^{-1} & \longrightarrow & 0 \\
& & & & \downarrow{\phi} & & & & \\
& & & & F & & & &
\end{array}
$$

$$(5.2.3)$$

where $\mathcal{V} \in \langle \mathcal{O}_S \rangle_{\mathrm{ex}}$, $F \in \mathrm{Coh}_{\leq 1}(S)$ and the top sequence is an exact sequence of coherent sheaves on S. The morphisms between two diagrams (5.2.3) are given by commutative diagrams

$$
\begin{array}{ccccccccc}
0 & \longrightarrow & \mathcal{V} & \xrightarrow{\alpha} & \mathcal{U} & \xrightarrow{\psi} & F \otimes \omega_S^{-1} & \longrightarrow & 0 \\
& & \downarrow{a_\mathcal{V}} & & \downarrow{a_\mathcal{U}} & & \downarrow{a_F \otimes \mathrm{id}} & & \\
0 & \longrightarrow & \mathcal{V}' & \xrightarrow{\alpha'} & \mathcal{U}' & \xrightarrow{\psi'} & F' \otimes \omega_S^{-1} & \longrightarrow & 0,
\end{array}
\qquad
\begin{array}{ccc}
\mathcal{U} & \xrightarrow{\phi} & F \\
\downarrow{a_\mathcal{U}} & & \downarrow{a_F} \\
\mathcal{U}' & \xrightarrow{\phi'} & F',
\end{array}
$$

Lemma 5.2.1 *The category $\mathcal{B}_S$ is an abelian category.*

Proof Because $\mathrm{Hom}(F, \mathcal{O}_S) = 0$ for $F \in \mathrm{Coh}_{\leq 1}(S)$, the snake lemma easily implies that the termwise kernels and cokernels give diagrams in $\mathcal{B}_S$. □

For a diagram (5.2.3), its rank is defined to be the rank of the sheaf $\mathcal{V}$. We denote by

$$\mathcal{B}_S^{\leq 1} \subset \mathcal{B}_S$$

the subcategory of $\mathcal{B}_S$ consisting of diagrams (5.2.3) with rank less than or equal to one. For a rank one diagram, we have another equivalent way to give it.

Lemma 5.2.2 *Giving a rank one diagram (5.2.3) is equivalent to giving a pair $(\mathcal{O}_S \xrightarrow{\xi} F)$ for $F \in \mathrm{Coh}_{\leq 1}(S)$ together with a morphism $\vartheta \colon F \otimes \omega_S^{-1} \to I^\bullet[1]$ in $D_{\mathrm{coh}}^b(S)$. Here $I^\bullet = (\mathcal{O}_S \xrightarrow{\xi} F)$ is the two term complex such that $\mathcal{O}_S$ is located in degree zero.*

Proof First suppose that we are given a digram (5.2.3) with $\mathcal{V} = \mathcal{O}_S$. Then we have the associated pair

$$(\xi \colon \mathcal{O}_S \xrightarrow{\alpha} \mathcal{U} \xrightarrow{\phi} F).$$

The top sequence in (5.2.3) gives a quasi-isomorphism $(\mathcal{O}_S \xrightarrow{\alpha} \mathcal{U})[1] \xrightarrow{\sim} F \otimes \omega_S^{-1}$ and we have the morphism of complexes

$$
\begin{array}{ccc}
\mathcal{O}_S & \xrightarrow{\alpha} & \mathcal{U} \\
{\scriptstyle\mathrm{id}}\downarrow & & \downarrow{\scriptstyle\phi} \\
\mathcal{O}_S & \xrightarrow{\xi} & F.
\end{array}
$$

The above diagram gives a morphism $\vartheta \colon F \otimes \omega_S^{-1} \to I^\bullet[1]$ in $D_{\mathrm{coh}}^b(S)$.

Conversely, suppose that we are given a pair $(\xi \colon \mathcal{O}_S \to F)$ together with a morphism $\vartheta \colon F \otimes \omega_S^{-1} \to I^\bullet[1]$. Then we have the commutative diagram

$$
\begin{array}{ccccccc}
\mathcal{O}_S & \xrightarrow{\alpha} & \mathcal{U} & \xrightarrow{\psi} & F \otimes \omega_S^{-1} & \longrightarrow & \mathcal{O}_S[1] \\
{\scriptstyle\mathrm{id}}\downarrow & & \downarrow{\scriptstyle\phi} & & \downarrow{\scriptstyle\vartheta} & & \downarrow{\scriptstyle\mathrm{id}} \\
I^\bullet \xrightarrow{\beta} \mathcal{O}_S & \xrightarrow{\xi} & F & \longrightarrow & I^\bullet[1] & \longrightarrow & \mathcal{O}_S[1].
\end{array}
\qquad (5.2.4)
$$

Here horizontal sequences are distinguished triangles. Therefore there exists a morphism $\phi \colon \mathcal{U} \to F$ which makes the above diagram commutative. We need to show that ϕ is uniquely determined by the above commutativity. Suppose that there exists another $\phi' \colon \mathcal{U} \to F$ which makes the above diagram commutative. Below we show that $\phi = \phi'$. By the commutativity of (5.2.4), $\phi - \phi'$ is written as

$$\phi - \phi' = \xi \circ \gamma, \quad \mathcal{U} \xrightarrow{\gamma} \mathcal{O}_S \xrightarrow{\xi} F$$

for some morphism $\gamma: \mathcal{U} \to \mathcal{O}_S$. We have $\phi = \phi'$ if $\xi = 0$, so we may assume that $\xi \neq 0$. The commutativity of (5.2.4) implies that

$$\xi \circ \gamma \circ \alpha = (\phi - \phi') \circ \alpha = 0, \ \mathcal{O}_S \to F.$$

Therefore $\gamma \circ \alpha: \mathcal{O}_S \to \mathcal{O}_S$ is written as

$$\gamma \circ \alpha = \beta \circ w, \ \mathcal{O}_S \overset{w}{\to} I^\bullet \overset{\beta}{\to} \mathcal{O}_S$$

for some morphism $w: \mathcal{O}_S \to I^\bullet$. However from the distinguished triangle $F[-1] \to I^\bullet \to \mathcal{O}_S$ we have the exact sequence

$$0 = \mathrm{Hom}(\mathcal{O}_S, F[-1]) \to \mathrm{Hom}(\mathcal{O}_S, I^\bullet) \to \mathbb{C} \to \mathrm{Hom}(\mathcal{O}_S, F)$$

where the right arrow takes 1 to ξ, which is injective by the assumption $\xi \neq 0$. Therefore $\mathrm{Hom}(\mathcal{O}_S, I^\bullet) = 0$, hence $w = 0$. It follows that $\gamma \circ \alpha = 0$, hence γ is written as

$$\gamma = \iota \circ \psi, \ \mathcal{U} \overset{\psi}{\to} F \otimes \omega_S^{-1} \overset{\iota}{\to} \mathcal{O}_S$$

for some morphism ι. But ι must be a zero map as F is a torsion sheaf. Therefore $\gamma = 0$ and $\phi = \phi'$ follows.   □

We also have the following description of rank one objects in $\mathcal{B}_S$, which is obvious.

Lemma 5.2.3 *Giving a rank one diagram (5.2.3) is equivalent to giving a pair* $(\xi: F \otimes \omega_S^{-1} \to \mathcal{O}_S[1])$ *together with a morphism* $\phi: \mathcal{U} \to F$, *where* $0 \to \mathcal{O}_S \to \mathcal{U} \to F \otimes \omega_S^{-1} \to 0$ *is the extension determined by* ξ.

5.2.3 The Functor from $\mathcal{A}_X$ to $\mathcal{B}_S$

We have the following natural diagram in $D^b_{\mathrm{coh}}(\overline{X})$

$$
\begin{array}{ccc}
\mathcal{O}_{S_\infty} \longrightarrow \mathcal{O}_{\overline{X}}(-S_0 - S_\infty)[1] \longrightarrow \mathcal{O}_{\overline{X}}(-S_0)[1] \\
\downarrow \\
\mathcal{O}_{\overline{X}}(-S_\infty)[1]
\end{array}
$$

$$(5.2.5)$$

where the top arrow is a distinguished triangle. For $E \in D^b_{coh}(\overline{X})$, by taking the tensor product with (5.2.5) and push-forward to S, we obtain the following diagram in $D^b_{coh}(S)$

$$\mathbf{R}\overline{\pi}_*(E|_{S_\infty}) \longrightarrow \mathbf{R}\overline{\pi}_*(E(-S_0 - S_\infty)[1]) \longrightarrow \mathbf{R}\overline{\pi}_*(E(-S_0)[1])$$

$$\mathbf{R}\overline{\pi}_*(E(-S_\infty)[1]) \tag{5.2.6}$$

Let $\mathcal{A}_X$ be the abelian subcategory of $D^b_{coh}(\overline{X})$ defined in Definition 4.2.2. We have the following lemma:

Lemma 5.2.4 *If $E \in \mathcal{A}_X$, then the diagram (5.2.6) determines an object in $\mathcal{B}_S$.*

Proof We check that

$$\mathbf{R}\overline{\pi}_*(E|_{S_\infty}) \in \langle \mathcal{O}_S \rangle_{ex}, \quad \mathbf{R}\overline{\pi}_*(E(-S_\infty)[1]) \in \mathrm{Coh}_{\leq 1}(S).$$

The former one follows from the definition of $\mathcal{A}_X$ and the latter one follows from $\mathbf{R}\overline{\pi}_*(\mathcal{O}_{\overline{X}}(-S_\infty)) = 0$. Moreover noting that $\overline{\pi}^*\omega_S = \mathcal{O}_{\overline{X}}(S_0 - S_\infty)$, we have

$$\mathbf{R}\overline{\pi}_*(E(-S_0)[1]) = \mathbf{R}\overline{\pi}_*(E(-S_\infty)[1]) \otimes \omega_S^{-1}.$$

Therefore the lemma follows.   □

Let $\mathcal{A}_X^{\leq 1} \subset \mathcal{A}_X$ be the subcategory consisting of objects E with $\mathrm{rank}(E) \leq 1$. By Lemma 5.2.4, we have the functors

$$\Phi \colon \mathcal{A}_X \to \mathcal{B}_S, \quad \Phi \colon \mathcal{A}_X^{\leq 1} \to \mathcal{B}_S^{\leq 1}$$

sending $E \in \mathcal{A}_X$ to the diagram (5.2.6).

5.2.4   The Functor from $\mathcal{B}_S$ to $D^b_{coh}(\overline{X})$

As for the other direction of functors, we define

$$\Psi \colon \mathcal{B}_S \to D^b_{coh}(\overline{X}) \tag{5.2.7}$$

by sending a diagram (5.2.3) to the two term complex

$$\left(\overline{\pi}^*\mathcal{U} \xrightarrow{\eta} \overline{\pi}^*F \otimes \mathcal{O}_{\overline{X}}(S_\infty) \right). \tag{5.2.8}$$

Here $\overline{\pi}^*\mathcal{U}$ is located in degree zero and η is determined by the adjoint of the following map

$$(\phi, \psi)\colon \mathcal{U} \to \mathbf{R}\overline{\pi}_*(\overline{\pi}^* F \otimes \mathcal{O}_{\overline{X}}(S_\infty)) \xrightarrow{\cong} F \oplus (F \otimes \omega_S^{-1})$$

where ϕ, ψ are maps in the diagram (5.2.3), and the second arrow is given by the projection formula together with the isomorphism (5.2.2).

Lemma 5.2.5 *For $E \in \mathcal{A}_X$, we have a functorial isomorphism $\Psi \circ \Phi(E) \xrightarrow{\cong} E$.*

Proof The object $\Psi \circ \Phi(E)$ is given by the two term complex concentrated in degree $[-1, 0]$

$$\left(\overline{\pi}^*\mathbf{R}\overline{\pi}_* E(-S_0 - S_\infty) \xrightarrow{\eta} \overline{\pi}^*(\mathbf{R}\overline{\pi}_* E(-S_\infty))(S_\infty)\right). \qquad (5.2.9)$$

Here η is adjoint to the map

$$\mathbf{R}\overline{\pi}_* E(-S_0 - S_\infty) \to \mathbf{R}\overline{\pi}_* E(-S_\infty) \otimes (\mathcal{O}_S \oplus \omega_S^{-1})$$

$$\xrightarrow{\cong} \mathbf{R}\overline{\pi}_* E(-S_0) \oplus \mathbf{R}\overline{\pi}_* E(-S_\infty)$$

induced by the sum of the natural inclusions $\mathcal{O}_{\overline{X}}(-S_0 - S_\infty) \to \mathcal{O}_{\overline{X}}(-S_0) \oplus \mathcal{O}_{\overline{X}}(-S_\infty)$. On the other hand, we have the Koszul resolution of the diagonal $\Delta \subset \overline{X} \times_S \overline{X}$

$$0 \to \Omega_{\overline{X}/S}(1) \boxtimes \mathcal{O}_{\overline{X}}(-1) \to \mathcal{O}_{\overline{X} \times_S \overline{X}} \to \mathcal{O}_\Delta \to 0.$$

Note that by (5.2.1), we have

$$\Omega_{\overline{X}/S}(1) \boxtimes \mathcal{O}_{\overline{X}}(-1) = \mathcal{O}_{\overline{X}}(-S_0) \boxtimes \mathcal{O}_{\overline{X}}(-S_\infty).$$

By pulling back $E(-S_\infty)$ to $\overline{X} \times_S \overline{X}$ by the first projection, tensoring with the above exact sequence and pushing forward by the second projection, we obtain the distinguished triangle

$$\overline{\pi}^*\mathbf{R}\overline{\pi}_* E(-S_0 - S_\infty) \to \overline{\pi}^*(\mathbf{R}\overline{\pi}_* E(-S_\infty))(S_\infty) \to E.$$

It is straightforward to check that the first arrow is given by η in (5.2.9). Therefore the lemma holds.   □

Lemma 5.2.6 *The functor Ψ in (5.2.7) restricts to the functor*

$$\Psi\colon \mathcal{B}_S^{\leq 1} \to \mathcal{A}_{\overline{X}}^{\leq 1}.$$

Proof For a diagram (5.2.3), let E be the two term complex (5.2.8)

$$E = \left(\pi^* \mathcal{U} \xrightarrow{\eta} \pi^* F \otimes \mathcal{O}_{\overline{X}}(S_\infty) \right).$$

Below we show that if the rank of the diagram (5.2.3) is less than or equal to one, then we have $E \in \mathcal{A}_{\overline{X}}^{\leq 1}$. Note that E is concentrated in degrees $[0, 1]$. By restricting E to S_∞, we have the distinguished triangle

$$\mathbf{L} i_\infty^* E \to \mathcal{U} \xrightarrow{\psi} F \otimes \omega_S^{-1}.$$

The above sequence is isomorphic to the top sequence in (5.2.3), therefore we have

$$\mathbf{L} i_\infty^* E = \mathcal{V} \in \langle \mathcal{O}_S \rangle_{\mathrm{ex}}. \tag{5.2.10}$$

In particular, $\mathcal{H}^1(E) = \mathrm{Cok}(\eta)$ is zero on S_∞, so it is supported away from S_∞. On the other hand, we have the surjection $\pi^* F(S_\infty) \twoheadrightarrow \mathrm{Cok}(\eta)$, so we have

$$\mathrm{Supp}(\mathcal{H}^1(E)) \subset \overline{\pi}^{-1}(\mathrm{Supp}(F)).$$

Therefore if $\mathcal{H}^1(E)$ has two dimensional support, it is of the form $\overline{\pi}^{-1}(Z)$ for a one dimensional support $Z \subset \mathrm{Supp}(F)$. It contradicts to that $\mathcal{H}^1(E)$ is supported away from S_∞, so $\mathcal{H}^1(E)$ is at most one dimensional and we have $\mathcal{H}^1(E) \in \mathrm{Coh}_{\leq 1}(X)$.

Next we show that $\mathcal{H}^0(E) = \mathrm{Ker}(\eta)$ is a torsion free sheaf on $\overline{X}$. By (5.2.10), the sheaf $\mathcal{H}^0(E)$ is a locally free sheaf in a neighborhood of S_∞. Suppose that $\mathcal{H}^0(E)$ has a torsion subsheaf. Then it is a subsheaf of the torsion part of $\pi^* \mathcal{U}$, therefore its support is of the form $\overline{\pi}^{-1}(Z)$ for $Z \subset S$ with $\dim Z \leq 1$. It contradicts to that $\mathcal{H}^0(E)$ is locally free near S_∞, hence $\mathcal{H}^0(E)$ is torsion free.

Now suppose that the diagram (5.2.3) has rank one, so E is of rank one. As $\mathcal{H}^0(E)$ is torsion free, we have the exact sequence of sheaves

$$0 \to \mathcal{H}^0(E) \to \mathcal{H}^0(E)^{\vee\vee} \to Q \to 0$$

for $Q \in \mathrm{Coh}_{\leq 1}(\overline{X})$. As $\mathcal{H}^0(E)$ is a line bundle in a neighborhood of S_∞, the sheaf Q is supported away from S_∞, i.e. $Q \in \mathrm{Coh}_{\leq 1}(X)$. We also have an isomorphism $\mathcal{H}^0(E)^{\vee\vee} \cong \mathcal{O}_{\overline{X}}$, as it is a rank one reflexive sheaf with trivial determinant. Now the object E is filtered by objects $Q[-1], \mathcal{O}_{\overline{X}}, \mathcal{H}^1(E)[-1]$, so it is an object in $\mathcal{A}_{\overline{X}}^{\leq 1}$. The rank zero case is easier and we omit details. $\qquad\square$

Theorem 5.2.7 *The functor* $\Phi: \mathcal{A}_{\overline{X}}^{\leq 1} \to \mathcal{B}_S^{\leq 1}$ *is an equivalence of categories whose quasi-inverse is given by* $\Psi: \mathcal{B}_S^{\leq 1} \to \mathcal{A}_{\overline{X}}^{\leq 1}$.

Proof By Lemma 5.2.5, it is enough to show that there exists an isomorphism of functors $\Phi \circ \Psi \cong \mathrm{id}$. For a diagram (5.2.3), let E be the object given by (5.2.8). Then we have the distinguished triangle

$$\overline{\pi}^* F \otimes \mathcal{O}_{\overline{X}}(S_\infty)[-1] \to E \to \overline{\pi}^* \mathcal{U}.$$

Using the above triangle, we see that there exist natural isomorphisms

$$\mathbf{R}\overline{\pi}_*(E(-S_0 - S_\infty)[1]) \xrightarrow{\cong} \mathcal{U} \otimes \mathbf{R}\overline{\pi}_* \mathcal{O}_{\overline{X}}(-S_0 - S_\infty)[1] \xrightarrow{\cong} \mathcal{U}.$$

Similarly we have isomorphisms

$$\mathcal{V} \xrightarrow{\cong} \mathbf{R}\overline{\pi}_*(E|_{S_\infty}), \quad F \xrightarrow{\cong} \mathbf{R}\overline{\pi}_*(E(-S_\infty)[1]), \quad F \otimes \omega_S^{-1} \xrightarrow{\cong} \mathbf{R}\overline{\pi}_*(E(-S_0)[1]).$$

Then it is straightforward to check that the diagram (5.2.6) is isomorphic to the original diagram (5.2.3). This shows that $\Phi \circ \Psi \cong \mathrm{id}$. □

Finally in this subsection, we give a characterization of rank one objects in $\mathcal{A}_X$ given by pairs $(\mathcal{O}_X \to F)$ in terms of objects in $\mathcal{B}_S$:

Lemma 5.2.8 *For a pair $(\mathcal{O}_X \to F)$ with $F \in \mathrm{Coh}_{\leq 1}(X)$, it determines an object $(\mathcal{O}_{\overline{X}} \to F)$ in $\mathcal{A}_{\overline{X}}^{\leq 1}$. Under the equivalence in Theorem 5.2.7, a rank one diagram (5.2.3) corresponds to an object in $\mathcal{A}_X$ of the form $(\mathcal{O}_{\overline{X}} \to F)$ if and only if the top sequence of (5.2.3) splits.*

Proof For a pair $(\mathcal{O}_X \to F)$ with $F \in \mathrm{Coh}_{\leq 1}(X)$, we have the exact sequence in $\mathcal{A}_X$

$$0 \to F[-1] \to (\mathcal{O}_{\overline{X}} \to F) \to \mathcal{O}_{\overline{X}} \to 0.$$

The first statement follows from the above sequence. By the above exact sequence, it is clear that a rank one object $E \in \mathcal{A}_X$ is of the form $(\mathcal{O}_{\overline{X}} \to F)$ if and only if it admits a surjection $E \twoheadrightarrow \mathcal{O}_{\overline{X}}$ in $\mathcal{A}_X$. Therefore via the equivalence in Theorem 5.2.7, such an object corresponds to a diagram (5.2.3) with $\mathcal{V} = \mathcal{O}_S$, which admits a surjection to the following diagram

$$
\begin{array}{ccccccc}
0 & \longrightarrow & \mathcal{O}_S & \xrightarrow{\ \mathrm{id}\ } & \mathcal{O}_S & \longrightarrow & 0 & \longrightarrow & 0 \\
& & & & \downarrow & & & & \\
& & & & 0. & & & &
\end{array}
$$

The latter condition is equivalent to that the top sequence of the diagram (5.2.3) with $\mathcal{V} = \mathcal{O}_S$ splits. □

5.2.5 Moduli Stacks of Rank One Objects in $\mathcal{A}_X$

Let $\mathcal{M}_X^{\dagger}$ be the moduli stack of rank one objects in $\mathcal{A}_X$, defined in Sect. 4.2.2. Also let $\mathcal{M}_S^{\dagger}$ be the moduli stack of pairs $(\mathcal{O}_S \to F)$ considered in Sect. 4.2.1, and denote by $\mathcal{M}_{S,\leq 1}^{\dagger} \subset \mathcal{M}_S^{\dagger}$ the open and closed substack of pairs $(\mathcal{O}_S \to F)$ such that $F \in \mathrm{Coh}_{\leq 1}(S)$. A family version of the arguments in the previous section immediately yields the following corollary of Theorem 5.2.7:

Corollary 5.2.9 *The stack $\mathcal{M}_X^{\dagger}$ is isomorphic to the stack whose T-valued points for an affine $\mathbb{C}$-scheme T form the groupoid of diagrams*

$$
\begin{array}{ccccccccc}
0 & \longrightarrow & \mathcal{O}_{S \times T} & \longrightarrow & \mathcal{U}_T & \overset{\psi}{\longrightarrow} & F_T \boxtimes \omega_S^{-1} & \longrightarrow & 0 \\
& & & \overset{\xi}{\searrow} & & \downarrow{\phi} & & & \\
& & & & F_T. & & & &
\end{array}
\tag{5.2.11}
$$

Here the top arrow is an exact sequence of sheaves on $S \times T$ and $F_T \in \mathrm{Coh}(S \times T)$ is T-flat such that $F|_{S \times \{x\}} \in \mathrm{Coh}_{\leq 1}(S)$ for any closed point $x \in T$. The isomorphisms are given by

$$
\begin{array}{ccccccccc}
0 & \longrightarrow & \mathcal{O}_{S \times T} & \longrightarrow & \mathcal{U}_T & \overset{\psi}{\longrightarrow} & F_T \boxtimes \omega_S^{-1} & \longrightarrow & 0 \\
& & \mathrm{id}\downarrow & & a_{\mathcal{U}}\downarrow{\cong} & & a_F \otimes \mathrm{id}\downarrow{\cong} & & \\
0 & \longrightarrow & \mathcal{O}_{S \times T} & \longrightarrow & \mathcal{U}_T' & \overset{\psi'}{\longrightarrow} & F_T' \boxtimes \omega_S^{-1} & \longrightarrow & 0,
\end{array}
\qquad
\begin{array}{ccc}
\mathcal{U}_T & \overset{\phi}{\longrightarrow} & F_T \\
a_{\mathcal{U}}\downarrow{\cong} & \cdot & a_F\downarrow{\cong} \\
\mathcal{U}_T' & \overset{\phi'}{\longrightarrow} & F_T'.
\end{array}
$$

In particular by sending a diagram (5.2.11) to the pair $\xi: \mathcal{O}_{S \times T} \to F_T$, we obtain the natural morphism

$$
\pi_*^{\dagger}: \mathcal{M}_X^{\dagger} \to \mathcal{M}_{S,\leq 1}^{\dagger}.
$$

Let $\mathcal{E}^{\dagger \bullet}$ be the perfect obstruction theory on $\mathcal{M}_{S,\leq 1}^{\dagger}$ given in (4.2.2). We consider the associated dual obstruction cone

$$
\mathrm{Obs}^*(\mathcal{E}^{\dagger \bullet}) \to \mathcal{M}_{S,\leq 1}^{\dagger}.
$$

Lemma 5.2.10 *For an affine $\mathbb{C}$-scheme T, the T-valued points of $\mathrm{Obs}^*(\mathcal{E}^{\dagger \bullet})$ form the groupoid of data*

$$
\left\{ (\mathcal{O}_{S \times T} \overset{\xi}{\to} F_T), \ \xi': F_T \boxtimes \omega_S^{-1} \to I_T^{\bullet}[1] \right\}
$$

where $(\mathcal{O}_{S \times T} \xrightarrow{\xi} F_T)$ gives a T-valued point of $\mathcal{M}_{S, \leq 1}^{\dagger}$, $I_T^{\bullet} = (\mathcal{O}_{S \times T} \xrightarrow{\xi} F_T)$ is a two term complex with $\mathcal{O}_{S \times T}$ located in degree zero, and ξ' is a morphism in $D_{\mathrm{coh}}^b(S \times T)$.

Proof By its definition, the T-valued points of $\mathrm{Obs}^*(\mathcal{E}^{\dagger \bullet})$ form the groupoid of data

$$\left\{ f : T \to \mathcal{M}_{S, \leq 1}^{\dagger}, \ f^* \mathcal{H}^1((\mathcal{E}^{\dagger \bullet})^{\vee}) \to \mathcal{O}_T \right\}.$$

Since $(\mathcal{E}^{\dagger \bullet})^{\vee}$ is perfect of cohomological amplitude $[-1, 1]$, we have $f^* \mathcal{H}^1((\mathcal{E}^{\dagger \bullet})^{\vee})$ $= \mathcal{H}^1(\mathbf{L}f^*(\mathcal{E}^{\dagger \bullet})^{\vee})$. Also we have

$$\mathbf{L}f^* \mathcal{E}^{\dagger \bullet} = \left(\mathbf{R}p_{T*} \mathbf{R}\mathcal{H}om_{S \times T}(I_T^{\bullet}, F_T) \right)^{\vee}$$

$$\cong \mathbf{R}p_{T*} \mathbf{R}\mathcal{H}om_{S \times T}(F_T \boxtimes \omega_S^{-1}, I_T^{\bullet}[2]).$$

Here $p_T : S \times T \to T$ is the projection and the last isomorphism follows from the Grothendieck duality. Using above, we see that

$$\mathrm{Hom}(f^* \mathcal{H}^1((\mathcal{E}^{\dagger \bullet})^{\vee}), \mathcal{O}_T) = \mathrm{Hom}(\mathcal{H}^1(\mathbf{L}f^*(\mathcal{E}^{\dagger \bullet})^{\vee}), \mathcal{O}_T)$$

$$= \mathrm{Hom}(\mathbf{L}f^*(\mathcal{E}^{\dagger \bullet})^{\vee}, \mathcal{O}_T[-1])$$

$$= \mathrm{Hom}(\mathcal{O}_T[1], \mathbf{L}f^* \mathcal{E}^{\dagger \bullet})$$

$$= \mathrm{Hom}(F_T \boxtimes \omega_S^{-1}, I_T^{\bullet}[1]).$$

Therefore the lemma holds.   □

Now we finish the proof of Theorem 4.2.3:

Proposition 5.2.11 *We have an isomorphism over* $\mathcal{M}_{S, \leq 1}^{\dagger}$

$$\mathcal{M}_X^{\dagger} \xrightarrow{\cong} \mathrm{Obs}^*(\mathcal{E}^{\dagger \bullet}).$$

Proof For each affine $\mathbb{C}$-scheme T, a T-valued point of $\mathcal{M}_X^{\dagger}$ is identified with a diagram (5.2.11) by Lemma 5.2.9. A T-flat version of the argument of Lemma 5.2.2 shows that giving a diagram (5.2.11) is equivalent to giving a T-valued point $(\mathcal{O}_{S \times T} \xrightarrow{\xi} F_T)$ of $\mathcal{M}_{S, \leq 1}^{\dagger}$ together with a morphism in $D_{\mathrm{coh}}^b(S \times T)$

$$\xi' : F_T \boxtimes \omega_S^{-1} \to I_T^{\bullet}[1], \ I_T^{\bullet} = (\mathcal{O}_{S \times T} \xrightarrow{\xi} F_T)$$

where $\mathcal{O}_{S \times T}$ located in degree zero. By Lemma 5.2.10, this is equivalent to giving a T-valued point of $\mathrm{Obs}^*(\mathcal{E}^{\dagger \bullet})$.   □

5.2.6 Comparison of Dualities

Let $\mathcal{A}_X^{\text{pure}} \subset \mathcal{A}_X$ be defined in (5.1.6), and $\mathcal{B}_S^{\text{pure}} \subset \mathcal{B}_S$ the subcategory consisting of diagrams (5.2.3) such that F is a pure one dimensional sheaf. We set

$$\mathcal{A}_X^{\text{pure},\leq 1} := \mathcal{A}_{\overline{X}}^{\leq 1} \cap \mathcal{A}_X^{\text{pure}}, \quad \mathcal{B}_S^{\text{pure},\leq 1} := \mathcal{B}_S^{\leq 1} \cap \mathcal{B}_S^{\text{pure}}.$$

We will give two lemmas on the above subcategories.

Lemma 5.2.12 *An object $E \in \mathcal{A}_{\overline{X}}^{\leq 1}$ is an object in $\mathcal{A}_X^{\text{pure},\leq 1}$ if and only if $\text{Hom}(Q[-1], E) = 0$ for any zero dimensional sheaf Q on X.*

Proof The only if direction is obvious. As for the if direction, the statement is obvious for rank zero objects. Let us take a rank one object $E \in \mathcal{A}_X$ satisfying $\text{Hom}(Q[-1], E) = 0$ for any zero dimensional sheaf Q on X, and take the maximal zero dimensional subsheaf $T \subset \mathcal{H}^1(E)$. Then $F = \mathcal{H}^1(E)/T$ is a pure one dimensional sheaf, and we have the surjection $E \twoheadrightarrow F[-1]$ in $\mathcal{A}_X$. Let E' be the kernel of the above surjection. Then $\mathcal{H}^0(E')$ is zero dimensional, hence Lemma 4.3.2 implies that E' is isomorphic to a two term complex $(\mathcal{O}_{\overline{X}} \to F')$ for a one dimensional sheaf F'. If F' is not pure, then E' contains a subobject of the form $Q[-1]$ for a zero dimensional sheaf Q, which contradicts to the assumption. Therefore F' is pure, and $E \in \mathcal{A}_X^{\text{pure},\leq 1}$ holds. □

Lemma 5.2.13 *The equivalence Φ in Theorem 5.2.7 restricts to the equivalence $\Phi: \mathcal{A}_X^{\text{pure},\leq 1} \xrightarrow{\sim} \mathcal{B}_S^{\text{pure},\leq 1}$.*

Proof The construction of Φ obviously implies that it takes $\mathcal{A}_X^{\text{pure},\leq 1}$ to $\mathcal{B}_S^{\text{pure},\leq 1}$. Note that we have

$$\text{Hom}(\Phi(Q[-1]), \mathcal{B}_S^{\text{pure},\leq 1}) = \text{Hom}(Q[-1], \Psi(\mathcal{B}_S^{\text{pure},\leq 1})) = 0$$

for any zero dimensional sheaf Q on X. Therefore by Lemma 5.2.12, the functor Ψ in Lemma 5.2.6 sends $\mathcal{B}_S^{\text{pure},\leq 1}$ to $\mathcal{A}_X^{\text{pure},\leq 1}$. □

The category $\mathcal{A}_X^{\text{pure},\leq 1}$ is closed under the derived dual $\mathbb{D}_{\overline{X}}$. Below we describe the corresponding duality on $\mathcal{B}_S^{\text{pure},\leq 1}$ under the equivalence in Lemma 5.2.13. For a diagram (5.2.3) such that F is a pure one dimensional sheaf, let us apply the derived dual functor $\mathbb{D}_S(-)$ on it. Then we obtain the diagram

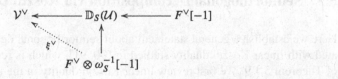

where $F^\vee$ is given as in (5.1.1). By taking the cones, we obtain the following diagram (unique up to isomorphisms)

$$0 \longrightarrow \mathcal{V}^\vee \longrightarrow \mathrm{Cone}(\xi^\vee) \longrightarrow F^\vee \otimes \omega_S^{-1} \longrightarrow 0$$

$$\downarrow$$

$$F^\vee. \tag{5.2.12}$$

Here the top sequence corresponds to the extension class $\xi^\vee$. By sending a diagram (5.2.3) to the diagram (5.2.12), we obtain the functor

$$\mathbb{D}_{\mathcal{B}} \colon \mathcal{B}_S^{\mathrm{pure}} \to (\mathcal{B}_S^{\mathrm{pure}})^{\mathrm{op}}.$$

The above functor obviously preserves the subcategory $\mathcal{B}_S^{\mathrm{pure}, \leq 1}$. It is easy to see that $\mathbb{D}_{\mathcal{B}} \circ \mathbb{D}_{\mathcal{B}} = \mathrm{id}$, so in particular $\mathbb{D}_{\mathcal{B}}$ is an equivalence.

Lemma 5.2.14 *The following diagram is commutative:*

$$\begin{array}{ccc} \mathcal{A}_X^{\mathrm{pure}, \leq 1} & \xrightarrow{\mathbb{D}_{\overline{X}}} & (\mathcal{A}_X^{\mathrm{pure}, \leq 1})^{\mathrm{op}} \\ \Phi \downarrow & & \downarrow \Phi \\ \mathcal{B}_S^{\mathrm{pure}, \leq 1} & \xrightarrow{\mathbb{D}_{\mathcal{B}}} & (\mathcal{B}_S^{\mathrm{pure}, \leq 1})^{\mathrm{op}}. \end{array}$$

Proof We only check the commutativity for rank one object $E \in \mathcal{A}_X^{\mathrm{pure}, \leq 1}$. The diagram $\mathbb{D}_{\mathcal{B}} \circ \Phi(E)$ is by the construction

$$\mathcal{O}_S \longrightarrow \mathbb{D}_S \circ \mathbf{R}\overline{\pi}_*(E) \longrightarrow \mathbb{D}_S \circ \mathbf{R}\overline{\pi}_*(E(-S_\infty)[1])$$

$$\downarrow$$

$$\mathbb{D}_S \circ \mathbf{R}\overline{\pi}_*(E(-S_0)[1]).$$

By the Grothendieck duality $\mathbf{R}\overline{\pi}_*(\mathbb{D}_{\overline{X}}(-) \otimes \Omega_{\overline{X}/S}[1]) \cong \mathbb{D}_S \circ \mathbf{R}\overline{\pi}_*(-)$, the above diagram is nothing but $\Phi \circ \mathbb{D}_{\overline{X}}(E)$. $\square$

5.3 Semiorthogonal Decomposition via Koszul Duality

Here we establish a general statement about semiorthogonal decomposition associated with linear Koszul duality studied in [90, 91], which is required for the proof of Theorem 5.1.9. We first review linear Koszul duality in the local case, globalize it, and then show the existence of certain semiorthogonal decomposition under the linear Koszul duality. We also prove a certain comparison result of singular supports under linear Koszul duality, which is essential for the proof of Corollary 5.1.5.

5.3.1 Linear Koszul Duality: Local Case

Here we review the linear Koszul duality proved in [90, 91], and prove its variants. Suppose that Y is a smooth affine $\mathbb{C}$-scheme, and take a two term complex of vector bundles on it

$$\mathcal{E} = (\mathcal{E}^{-1} \xrightarrow{\phi} \mathcal{E}^0). \tag{5.3.1}$$

Here $\mathcal{E}^i$ is located in degree i. We take a trivial $\mathbb{C}^*$-action on Y and weight two $\mathbb{C}^*$-actions on fibers of $\mathcal{E}^{-1}$ and $\mathcal{E}^0$. Let $\mathcal{A}^\dagger$, $\mathcal{A}^\natural$ and $\mathcal{A}^\sharp$ be the sheaves of dg-algebras on Y defined by

$$\mathcal{A}^\dagger := S_{\mathcal{O}_Y}(\mathcal{E}), \ \mathcal{A}^\natural := S_{\mathcal{O}_Y}(\mathcal{E}^\vee[-1]), \ \mathcal{A}^\sharp := S_{\mathcal{O}_Y}(\mathcal{E}^\vee[1]).$$

For $\star \in \{\dagger, \natural, \sharp\}$, we denote by $D_{\mathrm{fg}}^{\mathbb{C}^*}(\mathcal{A}^\star\text{-mod})$ the $\mathbb{C}^*$-equivariant derived category of finitely generated dg-modules over $\mathcal{A}^\star$.

Theorem 5.3.1 ([90, 91]) *There exists an equivalence*

$$\overline{\Phi}_Y^{\mathrm{op}} : D_{\mathrm{fg}}^{\mathbb{C}^*}(\mathcal{A}^\dagger\text{-mod})^{\mathrm{op}} \xrightarrow{\sim} D_{\mathrm{fg}}^{\mathbb{C}^*}(\mathcal{A}^\sharp\text{-mod}) \tag{5.3.2}$$

sending $\mathcal{A}^\dagger$ to $\mathcal{O}_Y$ and $\mathcal{O}_Y$ to $\mathcal{A}^\sharp$.

Below we recall the construction of the equivalence (5.3.2) and its quasi-inverse. For a $\mathbb{C}^*$-equivariant dg-module M over $\mathcal{A}^\dagger$, we write its $\mathbb{C}^*$-weight j part by M_j which is a direct summand of M as $\mathcal{O}_Y$-module. Then its $\mathcal{O}_Y$-dual $\mathbb{D}_Y(M) := \oplus_j \mathbb{D}_Y(M_j)$ is naturally equipped with a $\mathbb{C}^*$-equivariant dg-module structure over $\mathcal{A}^\dagger$, where $\mathbb{D}_Y(M_j)$ is of $\mathbb{C}^*$-weight $-j$. We set

$$\mathcal{K} := \mathcal{A}^\natural \otimes_{\mathcal{O}_Y} \mathcal{A}^\dagger, \ \mathcal{K}' := \mathbb{D}_Y(\mathcal{A}^\dagger) \otimes_{\mathcal{O}_Y} \mathcal{A}^\natural$$

with differentials given by

$$d_{\mathcal{K}} = d_{\mathcal{A}^\natural \otimes_{\mathcal{O}_Y} \mathcal{A}^\dagger} + \eta, \ d_{\mathcal{K}'} = d_{\mathbb{D}_Y(\mathcal{A}^\dagger) \otimes_{\mathcal{O}_Y} \mathcal{A}^\natural} + \eta'$$

where $\eta \in \mathcal{E}^\vee \otimes \mathcal{E}, \eta' \in \mathcal{E} \otimes \mathcal{E}^\vee$ are degree one elements of $\mathcal{A}^\natural \otimes_{\mathcal{O}_Y} \mathcal{A}^\dagger$, $\mathcal{A}^\dagger \otimes_{\mathcal{O}_Y} \mathcal{A}^\natural$, corresponding to the identity map id: $\mathcal{E} \to \mathcal{E}$. Then both of $\mathcal{K}, \mathcal{K}'$ are dg-modules over $\mathcal{A}^\natural \otimes_{\mathcal{O}_Y} \mathcal{A}^\dagger$, $\mathcal{A}^\dagger \otimes_{\mathcal{O}_Y} \mathcal{A}^\natural$, respectively. For a $\mathbb{C}^*$-equivariant dg-module M over $\mathcal{A}^\dagger$ and a $\mathbb{C}^*$-equivariant dg-module N over $\mathcal{A}^\natural$, we set

$$\Phi'_Y(M) = \mathcal{K} \otimes_{\mathcal{A}^\dagger} M, \ \Psi'_Y(N) = \mathcal{K}' \otimes_{\mathcal{A}^\natural} N.$$

Then $\Phi'_Y(M)$ is a dg-module over $\mathcal{A}^\natural$, and $\Psi'_Y(N)$ is a dg-module over $\mathcal{A}^\dagger$.

Remark 5.3.2 The functor Φ'_Y is described in another way as follows. Let $\overline{\mathcal{K}} = S(\mathcal{E}[1]) \otimes_{\mathcal{O}_Y} \mathcal{A}^\dagger$ be the Koszul resolution of $\mathcal{O}_Y$ by free $\mathcal{A}^\dagger$-modules. Then we have

$$\Phi'_Y(M) = \mathrm{Hom}_{\mathcal{A}^\dagger}(\overline{\mathcal{K}}, M) = \mathbf{R}\mathrm{Hom}_{\mathcal{A}^\dagger}(\mathcal{O}_Y, M),$$

i.e. Φ'_Y is a derived Morita-type functor with respect to $\mathcal{O}_Y$.

Finally there is a regrading equivalence ξ

$$\xi \colon D_{\mathrm{fg}}^{\mathbb{C}^*}(\mathcal{A}^\natural\text{-mod}) \overset{\sim}{\to} D_{\mathrm{fg}}^{\mathbb{C}^*}(\mathcal{A}^\sharp\text{-mod}) \tag{5.3.3}$$

sending a $\mathbb{C}^*$-equivariant $\mathcal{A}^\natural$-module M to the $\mathbb{C}^*$-equivariant $\mathcal{A}^\sharp$-module $\xi(M)$ given by $\xi(M)^i_j = M^{i-j}_j$, where i is the cohomological grading and j is the $\mathbb{C}^*$-weight. The equivalence $\overline{\Phi}_Y^{\mathrm{op}}$ and its quasi-inverse $\overline{\Psi}_Y^{\mathrm{op}}$ are given by

$$\overline{\Phi}_Y^{\mathrm{op}} = \xi \circ \Phi'_Y \circ \mathbb{D}_Y, \ \overline{\Psi}_Y^{\mathrm{op}} := \mathbb{D}_Y \circ \Psi'_Y \circ \xi^{-1}.$$

We will give some variants of Theorem 5.3.1. We set the following affine derived schemes over Y:

$$Y^\dagger := \mathrm{Spec}\,\mathcal{A}^\dagger, \ Y^\sharp := \mathrm{Spec}\,\mathcal{A}^\sharp.$$

Instead of the weight two action of $\mathbb{C}^*$ on $\mathcal{E}$, we take the weight one action on $\mathcal{E}$, and take the quotient stacks $[Y^\dagger/\mathbb{C}^*]$, $[Y^\sharp/\mathbb{C}^*]$.

Proposition 5.3.3 *There is an equivalence*

$$\Phi_Y^{\mathrm{op}} \colon D_{\mathrm{coh}}^b([Y^\dagger/\mathbb{C}^*])^{\mathrm{op}} \overset{\sim}{\to} D_{\mathrm{coh}}^b([Y^\sharp/\mathbb{C}^*]). \tag{5.3.4}$$

Proof We denote by $[Y^\dagger /\!\!/ \mathbb{C}^*]$, $[Y^\sharp /\!\!/ \mathbb{C}^*]$ the stack quotients by the weight two $\mathbb{C}^*$-actions on $\mathcal{E}^\bullet$. The result of Theorem 5.3.1 means the equivalence

$$\overline{\Phi}_Y^{\mathrm{op}} \colon D_{\mathrm{coh}}^b([Y^\dagger /\!\!/ \mathbb{C}^*])^{\mathrm{op}} \overset{\sim}{\to} D_{\mathrm{coh}}^b([Y^\sharp /\!\!/ \mathbb{C}^*]). \tag{5.3.5}$$

Now the natural map $\mathbb{C}^* \to \mathbb{C}^*$ given by $t \mapsto t^2$ induces the map

$$\rho \colon [Y^\star /\!\!/ \mathbb{C}^*] \to [Y^\star/\mathbb{C}^*], \ \star \in \{\dagger, \sharp\}$$

which is a μ_2-gerbe. Then similarly to (3.2.10) we have the decomposition

$$D^b_{\mathrm{coh}}([Y^\star /\!\!/ \mathbb{C}^*]) = D^b_{\mathrm{coh}}([Y^\star /\!\!/ \mathbb{C}^*])_{\lambda=0} \oplus D^b_{\mathrm{coh}}([Y^\star /\!\!/ \mathbb{C}^*])_{\lambda=1} \qquad (5.3.6)$$

such that $D^b_{\mathrm{coh}}([Y^\star /\!\!/ \mathbb{C}^*])_{\lambda=0}$ is equivalent to $D^b_{\mathrm{coh}}([Y^\star /\mathbb{C}^*])$ via ρ^*. The equivalence (5.3.5) preserves the decomposition (5.3.6), as it preserves the parity of the $\mathbb{C}^*$-weights. Therefore we obtain the equivalence (5.3.4). Explicitly, Φ_Y^{op} and its quasi-inverse Ψ_Y^{op} are given by

$$\Phi_Y^{\mathrm{op}} = \xi' \circ \Phi_Y' \circ \mathbb{D}_Y, \quad \Psi_Y^{\mathrm{op}} = \mathbb{D}_Y \circ \Psi_Y' \circ \xi'. \qquad (5.3.7)$$

Here the regrading equivalence (5.3.3) is replaced by ξ' determined by $\xi'(M)^i_j = M^{i-2j}_j$. $\qquad\qquad\square$

Let us take an affine derived scheme $\mathfrak{U} = \mathrm{Spec}\,\mathcal{R}(V \to Y, s)$ for a vector bundle $V \to Y$ with a section s as in (2.1.1). We take $\mathfrak{E} = \mathcal{E} \otimes_{\mathcal{O}_Y} \mathcal{O}_{\mathfrak{U}}$ where $\mathcal{E}$ is a two term complex of vector bundles on Y as in (5.3.1), which is equipped with a weight one $\mathbb{C}^*$-action. We set

$$\mathfrak{U}^\dagger := \mathrm{Spec}\,S_{\mathcal{O}_{\mathfrak{U}}}(\mathfrak{E}) = \mathrm{Spec}\,\mathcal{A}^\dagger_{\mathfrak{U}}, \quad \mathfrak{U}^\sharp := \mathrm{Spec}\,S_{\mathcal{O}_{\mathfrak{U}}}(\mathfrak{E}^\vee[1]) = \mathrm{Spec}\,\mathcal{A}^\sharp_{\mathfrak{U}}.$$

Here for a $\mathcal{O}_Y$-module $(-)$, we have written $(-)_{\mathfrak{U}} = (-) \overset{\mathbf{L}}{\otimes}_{\mathcal{O}_Y} \mathcal{O}_{\mathfrak{U}}$, and we use the same notation below. We have the following Cartesian diagrams for $\star \in \{\dagger, \sharp\}$

$$
\begin{array}{ccccc}
\mathfrak{U}^\star & \xrightarrow{\rho^\star_{\mathfrak{U}}} & \mathfrak{U} & \xrightarrow{i^\star_{\mathfrak{U}}} & \mathfrak{U}^\star \\
{\scriptstyle j^\star}\downarrow & \square & \downarrow{\scriptstyle j} & \square & \downarrow{\scriptstyle j^\star} \\
Y^\star & \xrightarrow{\rho^\star_Y} & Y & \xrightarrow{i^\star_Y} & Y^\star.
\end{array}
\qquad (5.3.8)
$$

Here j is the closed immersion of affine derived schemes, $\rho^\star_{\mathfrak{U}}$, $\rho^\star_Y$ are projections and $i^\star_{\mathfrak{U}}$, $i^\star_Y$ are their zero sections. We have the following slight generalization of Proposition 5.3.3:

Lemma 5.3.4 *There is an equivalence*

$$\Phi^{\mathrm{op}}_{\mathfrak{U}} : D^b_{\mathrm{coh}}([\mathfrak{U}^\dagger/\mathbb{C}^*])^{\mathrm{op}} \xrightarrow{\sim} D^b_{\mathrm{coh}}([\mathfrak{U}^\sharp/\mathbb{C}^*]) \qquad (5.3.9)$$

which fits into the commutative diagram

$$\begin{array}{ccc} D_{\mathrm{coh}}^b([\mathfrak{U}^\dagger/\mathbb{C}^*])^{\mathrm{op}} & \xrightarrow{\Phi_{\mathfrak{U}}^{\mathrm{op}}} & D_{\mathrm{coh}}^b([\mathfrak{U}^\sharp/\mathbb{C}^*]) \\ {\scriptstyle j_*^\dagger} \Big\uparrow & & \Big\downarrow {\scriptstyle j_*^\sharp} \\ D_{\mathrm{coh}}^b([Y^\dagger/\mathbb{C}^*])^{\mathrm{op}} & \xrightarrow[\widetilde{\Phi}_Y^{\mathrm{op}}]{} & D_{\mathrm{coh}}^b([Y^\sharp/\mathbb{C}^*]). \end{array}$$

$$(5.3.10)$$

Here $\widetilde{\Phi}_Y^{\mathrm{op}} := \Phi_Y^{\mathrm{op}} \circ \otimes_{\mathcal{O}_Y} \det V^\vee [-\operatorname{rank} V].$

Proof Since the equivalence in Proposition 5.3.3 is defined over Y, the equivalence $\Phi_{\mathfrak{U}}$ is obtained by pulling back the equivalence in Proposition 5.3.3 by the closed immersion of derived schemes $j \colon \mathfrak{U} \to Y$ (cf. [80, Theorem 6.4]).

More precisely, the proof is as follows. For a dg-module M over $\mathcal{A}_{\mathfrak{U}}^\dagger$ and a dg-module N over $\mathcal{A}_{\mathfrak{U}}^\sharp$, we set

$$\Phi_{\mathfrak{U}}'(M) = \mathcal{K}_{\mathfrak{U}} \otimes_{\mathcal{A}_{\mathfrak{U}}^\dagger} M, \ \ \Psi_{\mathfrak{U}}'(N) = \mathcal{K}_{\mathfrak{U}}' \otimes_{\mathcal{A}_{\mathfrak{U}}^\sharp} N.$$

Then similarly to (5.3.7), the functors $\Phi_{\mathfrak{U}}^{\mathrm{op}}$ and $\Psi_{\mathfrak{U}}^{\mathrm{op}}$ are defined to be

$$\Phi_{\mathfrak{U}}^{\mathrm{op}} := \xi' \circ \Phi_{\mathfrak{U}}' \circ \mathbb{D}_{\mathfrak{U}}, \ \ \Psi_{\mathfrak{U}}^{\mathrm{op}} := \mathbb{D}_{\mathfrak{U}} \circ \Psi_{\mathfrak{U}}' \circ \xi'^{-1}. \qquad (5.3.11)$$

Here for a dg-module M over $\mathcal{A}_{\mathfrak{U}}^\dagger$, its $\mathcal{O}_{\mathfrak{U}}$-dual $\mathbb{D}_{\mathfrak{U}}(M)$ is naturally regarded as a dg-module over $\mathcal{A}_{\mathfrak{U}}^\dagger$. Note that $\Phi_{\mathfrak{U}}^{\mathrm{op}}(M)$, $\Psi_{\mathfrak{U}}^{\mathrm{op}}(N)$ are dg-modules over $\mathcal{A}_{\mathfrak{U}}^\sharp$, $\mathcal{A}_{\mathfrak{U}}^\dagger$ respectively.

We first show that $\Phi_{\mathfrak{U}}^{\mathrm{op}}$ determines a functor (5.3.9), i.e. it preserves the coherence and boundedness. Let us take an object $M \in D_{\mathrm{coh}}^b([\mathfrak{U}^\dagger/\mathbb{C}^*])$. By the constructions of Φ_Y^{op} and $\Phi_{\mathfrak{U}}^{\mathrm{op}}$, using the projection formula together with the Grothendieck duality on $D_{\mathrm{coh}}^b(\mathfrak{U})$

$$j_* \circ \mathbb{D}_{\mathfrak{U}}(-) \cong \mathbb{D}_Y \circ (j_*(-) \otimes \det V^\vee [-\operatorname{rank} V]),$$

we see that there is a quasi-isomorphism of dg-modules over $\mathcal{A}^\sharp$

$$j_*^\sharp \circ \Phi_{\mathfrak{U}}^{\mathrm{op}}(M) \xrightarrow{\sim} \Phi_Y^{\mathrm{op}} \circ (j_*^\dagger(M) \otimes \det V^\vee [-\operatorname{rank} V]).$$

Therefore by Proposition 5.3.3, we have $j_*^\sharp \circ \Phi_{\mathfrak{U}}^{\mathrm{op}}(M) \in D_{\mathrm{coh}}^b([Y^\sharp/\mathbb{C}^*])$. It follows that we have $\Phi_{\mathfrak{U}}^{\mathrm{op}}(M) \in D_{\mathrm{coh}}^b([\mathfrak{U}^\sharp/\mathbb{C}^*])$, hence we have the functor (5.3.9) and the commutative diagram (5.3.10).

Similarly $\Psi_{\mathfrak{U}}^{\mathrm{op}}$ determines the functor

$$\Psi_{\mathfrak{U}}^{\mathrm{op}} \colon D_{\mathrm{coh}}^b([\mathfrak{U}^\sharp/\mathbb{C}^*]) \to D_{\mathrm{coh}}^b([\mathfrak{U}^\dagger/\mathbb{C}^*])^{\mathrm{op}}.$$

In order to show that $\Psi_{\mathfrak{U}}^{\mathrm{op}} \circ \Phi_{\mathfrak{U}}^{\mathrm{op}}$ and $\Phi_{\mathfrak{U}}^{\mathrm{op}} \circ \Psi_{\mathfrak{U}}^{\mathrm{op}}$ are identity functors, it is enough to show that

$$\mathcal{K}'_{\mathfrak{U}} \otimes_{\mathcal{A}_{\mathfrak{U}}^{\natural}} \mathcal{K}_{\mathfrak{U}} \cong \mathcal{A}_{\mathfrak{U}}^{\dagger}, \ \mathcal{K}_{\mathfrak{U}} \otimes_{\mathcal{A}_{\mathfrak{U}}^{\dagger}} \mathcal{K}_{\mathfrak{U}} \cong \mathcal{A}_{\mathfrak{U}}^{\natural}$$

as dg-modules over $\mathcal{A}_{\mathfrak{U}}^{\dagger} \otimes_{\mathcal{O}_{\mathfrak{U}}} \mathcal{A}_{\mathfrak{U}}^{\dagger}$, $\mathcal{A}_{\mathfrak{U}}^{\natural} \otimes_{\mathcal{O}_{\mathfrak{U}}} \mathcal{A}_{\mathfrak{U}}^{\natural}$, respectively. These isomorphisms follow by pulling back the isomorphisms $\mathcal{K}' \otimes_{\mathcal{A}^{\natural}} \mathcal{K} \cong \mathcal{A}^{\dagger}$, $\mathcal{K} \otimes_{\mathcal{A}^{\dagger}} \mathcal{K}' \cong \mathcal{A}^{\natural}$ proved in [91, Proposition 1.3.1] by the map $j \colon \mathfrak{U} \to Y$ respectively. $\qquad \square$

5.3.2 Linear Koszul Duality: Global Case

Let $\mathfrak{M}$ be a quasi-smooth and QCA derived stack with truncation $\mathcal{M} = t_0(\mathfrak{M})$. Here we assume that it admits an embedding $(\mathbb{C}^*)_{\mathfrak{M}} \hookrightarrow I_{\mathfrak{M}}$ as in Sect. 3.2.4 so that we have the $\mathbb{C}^*$-rigidification $\mathfrak{M} \to \mathfrak{M}^{\mathbb{C}^*\text{-rig}}$. Let us take a perfect object

$$\mathfrak{E} \in \mathrm{Perf}(\mathfrak{M})$$

which is of cohomological amplitude $[-1, 0]$, and of $\mathbb{C}^*$-weight one with respect to the inertia action. We define

$$\mathfrak{M}^{\dagger} = \mathrm{Spec}_{\mathfrak{M}}(S(\mathfrak{E})), \ \mathfrak{M}^{\natural} = \mathrm{Spec}_{\mathfrak{M}}(S(\mathfrak{E}^{\vee}[1])).$$

Below we will use the following diagram of derived stacks over $\mathfrak{M}^{\mathbb{C}^*\text{-rig}}$

$$
\begin{array}{ccccc}
\mathfrak{M}^{\dagger} & \underset{\rho^{\dagger}}{\overset{i^{\dagger}}{\rightleftarrows}} & \mathfrak{M} & \underset{\rho^{\natural}}{\overset{i^{\natural}}{\leftrightarrows}} & \mathfrak{M}^{\natural} \\
 & \searrow & \downarrow & \swarrow & \\
 & & \mathfrak{M}^{\mathbb{C}^*\text{-rig}}. & &
\end{array}
\tag{5.3.12}
$$

Here $\rho^{\dagger}$, $\rho^{\natural}$ are projections and $i^{\dagger}$, $i^{\natural}$ are their zero sections. The following is a global version of Proposition 5.3.3:

Proposition 5.3.5 *There is an equivalence*

$$\Phi_{\mathfrak{M}}^{\mathrm{op}} \colon D_{\mathrm{coh}}^{b}(\mathfrak{M}^{\dagger})^{\mathrm{op}} \overset{\sim}{\to} D_{\mathrm{coh}}^{b}(\mathfrak{M}^{\natural}). \tag{5.3.13}$$

Proof Let $\mathfrak{U} = \mathrm{Spec}\, \mathcal{R}(V \to Y, s)$ be an affine derived scheme as in (2.1.1), and take a smooth morphism $\alpha \colon \mathfrak{U} \to \mathfrak{M}^{\mathbb{C}^*\text{-rig}}$ such that $\mathfrak{M} \times_{\mathfrak{M}^{\mathbb{C}^*\text{-rig}}} \mathfrak{U}$ is a trivial $\mathbb{C}^*$-

gerbe over $\mathfrak{U}$, i.e. it is equivalent to $[\mathfrak{U}/\mathbb{C}^*]$ where $\mathbb{C}^*$ acts on $\mathfrak{U}$ trivially. Then the pull-back of the diagram (5.3.12) by $\alpha \colon \mathfrak{U} \to \mathfrak{M}^{\mathbb{C}^*\text{-rig}}$ is equivalent to the diagram

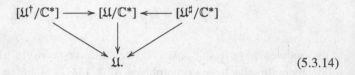

$$(5.3.14)$$

By shrinking $\mathfrak{U}$ if necessary, we may assume that the object $\alpha^*\mathfrak{E}$ on $[\mathfrak{U}/\mathbb{C}^*]$ is quasi-isomorphic to $\mathcal{E} \otimes_{\mathcal{O}_Y} \mathcal{O}_{\mathfrak{U}}$ for a two term complex of vector bundles (5.3.1) on Y with $\mathbb{C}^*$-weight one. Then we have the equivalence $\Phi_{\mathfrak{U}}^{\mathrm{op}}$ in Lemma 5.3.4, which naturally lifts to a weak equivalence of dg-categories by the construction (5.3.11)

$$\Phi_{\mathfrak{U},\mathrm{dg}}^{\mathrm{op}} \colon L_{\mathrm{coh}}([\mathfrak{U}^\dagger/\mathbb{C}^*])^{\mathrm{op}} \xrightarrow{\sim} L_{\mathrm{coh}}([\mathfrak{U}^\sharp/\mathbb{C}^*]). \tag{5.3.15}$$

Note that for $\star \in \{\dagger, \sharp\}$, we can write

$$L_{\mathrm{coh}}(\mathfrak{M}^\star) = \lim_{\mathfrak{U} \to \mathfrak{M}^{\mathbb{C}^*\text{-rig}}} L_{\mathrm{coh}}([\mathfrak{U}^\star/\mathbb{C}^*]).$$

Here the limit is taken for all smooth morphisms $\alpha \colon \mathfrak{U} \to \mathfrak{M}^{\mathbb{C}^*\text{-rig}}$. We will see that the dg-functor $\Phi_{\mathfrak{U},\mathrm{dg}}^{\mathrm{op}}$ in (5.3.15) is naturally globalized to a dg-functor

$$\Phi_{\mathfrak{M},\mathrm{dg}}^{\mathrm{op}} = \xi' \circ \Phi_{\mathfrak{M}}' \circ \mathbb{D}_{\mathfrak{M}} \colon L_{\mathrm{coh}}(\mathfrak{M}^\dagger)^{\mathrm{op}} \to L_{\mathrm{coh}}(\mathfrak{M}^\sharp). \tag{5.3.16}$$

Below we explain each term of (5.3.16) in the global setting. First for a $S(\mathfrak{E})$-module $\mathcal{F}$ on $\mathfrak{M}$, its $\mathcal{O}_{\mathfrak{M}}$-dual is given by

$$\mathbb{D}_{\mathfrak{M}}(\mathcal{F}) = \bigoplus_{\lambda \in \mathbb{Z}} \mathbb{D}_{\mathfrak{M}}(\mathcal{F}_\lambda)$$

where $\mathcal{F}_\lambda$ is the weight λ part of $\mathcal{F}$. Then $\mathbb{D}_{\mathfrak{M}}(\mathcal{F})$ is naturally equipped with a $S(\mathfrak{E})$-module structure. Next, we set $\overline{\mathcal{K}}_{\mathfrak{M}}$ to be the totalization of the Koszul complex in the dg-category $L_{\mathrm{qcoh}}(\mathfrak{M})$

$$\cdots \to \bigwedge^2 \mathfrak{E} \otimes_{\mathcal{O}_{\mathfrak{M}}} S(\mathfrak{E}) \to \mathfrak{E} \otimes_{\mathcal{O}_{\mathfrak{M}}} S(\mathfrak{E}) \to S(\mathfrak{E}).$$

The above object naturally defines an object in $L_{\mathrm{coh}}(\mathfrak{M}^\dagger)$ which is equivalent to $i_*^\dagger \mathcal{O}_{\mathfrak{M}}$. Then for a $S(\mathfrak{E})$-module $\mathcal{F}$ on $\mathfrak{M}$, we set

$$\Phi_{\mathfrak{M}}'(\mathcal{F}) = \mathcal{H}om_{S(\mathfrak{E})}(\overline{\mathcal{K}}_{\mathfrak{M}}, \mathcal{F})$$

which is equipped with a natural $S(\mathcal{E}^\vee[-1])$-module structure. Note that the above object represents $\mathbf{R}\mathcal{H}om_{\mathfrak{M}^\dagger}(i_*^\dagger \mathcal{O}_{\mathfrak{M}}, \mathcal{F})$ (see Remark 5.3.2). Finally for a $S(\mathcal{E}^\vee[-1])$-module on $\mathfrak{M}$, the regrading functor ξ' is given by

$$\xi'(\mathcal{F}) = \bigoplus_\lambda \mathcal{F}_\lambda[-2\lambda],$$

which is equipped with a natural $S(\mathcal{E}^\vee[1])$-module structure. The composition $\xi' \circ \Phi'_{\mathfrak{M}} \circ \mathbb{D}_{\mathfrak{M}}$ preserves objects with bounded coherent cohomologies, since this is true locally on $\mathfrak{M}^{\mathbb{C}^*\text{-rig}}$ by Lemma 5.3.4. Therefore we obtain the functor (5.3.16).

Similarly we can globalize $\Psi_{\mathfrak{U}}^{\mathrm{op}}$ in (5.3.11) to a dg-functor $\Psi_{\mathfrak{M},\mathrm{dg}}^{\mathrm{op}}$ from $L_{\mathrm{coh}}(\mathfrak{M}^\sharp)$ to $L_{\mathrm{coh}}(\mathfrak{M}^\dagger)^{\mathrm{op}}$. By taking the induced functors on homotopy categories, we obtain the functors

$$\Phi_{\mathfrak{M}}^{\mathrm{op}} \colon D_{\mathrm{coh}}^b(\mathfrak{M}^\dagger)^{\mathrm{op}} \to D_{\mathrm{coh}}^b(\mathfrak{M}^\sharp), \quad \Psi_{\mathfrak{M}}^{\mathrm{op}} \colon D_{\mathrm{coh}}^b(\mathfrak{M}^\sharp) \to D_{\mathrm{coh}}^b(\mathfrak{M}^\dagger)^{\mathrm{op}}. \qquad (5.3.17)$$

They are adjoints each other, and quasi-inverses each other as this is true locally on $\mathfrak{M}^{\mathbb{C}^*\text{-rig}}$ by Lemma 5.3.4. In particular, the functors (5.3.17) are equivalences.   □

Since $\mathfrak{M}^\dagger$ is quasi-smooth, the derived dual $\mathbb{D}_{\mathfrak{M}^\dagger}$ is a contravariant autoequivalence of $D_{\mathrm{coh}}^b(\mathfrak{M}^\dagger)$. Therefore we also have a covariant equivalence,

$$\Phi_{\mathfrak{M}} := \Phi_{\mathfrak{M}}^{\mathrm{op}} \circ \mathbb{D}_{\mathfrak{M}^\dagger} \colon D_{\mathrm{coh}}^b(\mathfrak{M}^\dagger) \xrightarrow{\sim} D_{\mathrm{coh}}^b(\mathfrak{M}^\sharp) \qquad (5.3.18)$$

whose quasi-inverse is $\Psi_{\mathfrak{M}} := \mathbb{D}_{\mathfrak{M}^\dagger} \circ \Psi_{\mathfrak{M}}^{\mathrm{op}}$.

5.3.3   Semiorthogonal Decomposition

In the diagram (5.3.12), we set

$$\mathfrak{M}_\circ^\dagger := \mathfrak{M}^\dagger \setminus i^\dagger(\mathfrak{M}), \quad \mathfrak{M}_\circ^\sharp := \mathfrak{M}^\sharp \setminus i^\sharp(\mathfrak{M})$$

which are open substacks of $\mathfrak{M}^\dagger$, $\mathfrak{M}^\sharp$ respectively. We have the following proposition, which is a consequence of [48].

Proposition 5.3.6 *For each $\lambda \in \mathbb{Z}$, the functors*

$$i_*^\dagger \colon D_{\mathrm{coh}}^b(\mathfrak{M})_\lambda \to D_{\mathrm{coh}}^b(\mathfrak{M}^\dagger), \quad i_*^\sharp \colon D_{\mathrm{coh}}^b(\mathfrak{M})_\lambda \to D_{\mathrm{coh}}^b(\mathfrak{M}^\sharp) \qquad (5.3.19)$$

are fully-faithful. Moreover we have semiorthogonal decomposition

$$D_{\mathrm{coh}}^b(\mathfrak{M}^\dagger) = \langle \ldots, i_*^\dagger D_{\mathrm{coh}}^b(\mathfrak{M})_{-1}, \mathcal{D}^\dagger, i_*^\dagger D_{\mathrm{coh}}^b(\mathfrak{M})_0, i_*^\dagger D_{\mathrm{coh}}^b(\mathfrak{M})_1, \ldots \rangle, \tag{5.3.20}$$

$$D_{\mathrm{coh}}^b(\mathfrak{M}^\sharp) = \langle \ldots, i_*^\sharp D_{\mathrm{coh}}^b(\mathfrak{M})_1, \mathcal{D}^\sharp, i_*^\sharp D_{\mathrm{coh}}^b(\mathfrak{M})_0, i_*^\sharp D_{\mathrm{coh}}^b(\mathfrak{M})_{-1}, \ldots \rangle$$

such that the restriction functors give equivalences

$$\mathcal{D}^\dagger \xrightarrow{\sim} D_{\mathrm{coh}}^b(\mathfrak{M}_\circ^\dagger), \quad \mathcal{D}^\sharp \xrightarrow{\sim} D_{\mathrm{coh}}^b(\mathfrak{M}_\circ^\sharp).$$

Proof The proposition is a consequence of [48, Theorem 3.2] by applying it for the Θ-stratifications $\mathfrak{M}^\dagger = \mathfrak{M}_\circ^\dagger \cup i^\dagger(\mathfrak{M})$, $\mathfrak{M}^\sharp = \mathfrak{M}_\circ^\dagger \cup i^\sharp(\mathfrak{M})$. Here we note that the weights in the above semiorthogonal components have opposite signs for $\mathfrak{M}^\dagger$ and $\mathfrak{M}^\sharp$, because the $\mathbb{C}^*$-weights on $\mathfrak{E}$ and $\mathfrak{E}^\vee[1]$ have the opposite sign. □

Proposition 5.3.7 *The functors*

$$\rho^{\dagger*}: D_{\mathrm{coh}}^b(\mathfrak{M})_\lambda \to D_{\mathrm{coh}}^b(\mathfrak{M}^\dagger), \quad \rho^{\sharp*}: D_{\mathrm{coh}}^b(\mathfrak{M})_\lambda \to D_{\mathrm{coh}}^b(\mathfrak{M}^\sharp) \tag{5.3.21}$$

are fully-faithful. Moreover for the equivalence $\Phi_{\mathfrak{M}}^{\mathrm{op}}$ in (5.3.13), we have

$$\Phi_{\mathfrak{M}}^{\mathrm{op}}(\rho^{\dagger*} D_{\mathrm{coh}}^b(\mathfrak{M})_\lambda) = i_*^\sharp D_{\mathrm{coh}}^b(\mathfrak{M})_{-\lambda}, \quad \Phi_{\mathfrak{M}}^{\mathrm{op}}(i_*^\dagger D_{\mathrm{coh}}^b(\mathfrak{M})_\lambda) = \rho^{\sharp*} D_{\mathrm{coh}}^b(\mathfrak{M})_{-\lambda}.$$

Proof The right adjoint functor of $\rho^{\dagger*}$ in (5.3.21) is given by $(\rho_*^\dagger)_\lambda$, where $(-)_\lambda$ means taking the weight λ part. There is a natural transform

$$(-) \to (\rho_*^\dagger)_\lambda \circ \rho^{\dagger*}(-) = \big((-) \otimes_{\mathcal{O}_{\mathfrak{M}}} S(\mathfrak{E})\big)_\lambda$$

which is obviously an isomorphism as $\mathfrak{E}$ is of $\mathbb{C}^*$-weight one. Therefore $\rho^{\dagger*}$ is fully-faithful, and the proof for $\rho^{\sharp*}$ is similar.

As for the second statement, the statement is local on $\mathfrak{M}^{\mathbb{C}^*\text{-rig}}$, so we can assume that $\mathfrak{M} = [\mathfrak{U}/\mathbb{C}^*]$ where $\mathfrak{U}$ is given as in (2.1.1) with trivial $\mathbb{C}^*$-action, and $\mathfrak{E} = \mathcal{E} \otimes_{\mathcal{O}_Y} \mathcal{O}_{\mathfrak{U}}$ for a two term complex of vector bundles $\mathcal{E}$ as in (5.3.1). We first consider the case that $\mathfrak{U} = Y$. For $m \in \mathbb{Z}$ and $M \in D_{\mathrm{qcoh}}([Y/\mathbb{C}^*])$, we write $M(m) = M \otimes_{\mathbb{C}} \mathbb{C}(m)$ where $\mathbb{C}(m)$ is the one dimensional $\mathbb{C}^*$-representation with weight m. In the notation of the proof of Proposition 5.3.3, we have

$$\Phi_Y^{\mathrm{op}}(\mathcal{O}_Y(\lambda)) = \xi' \circ \Phi' \circ \mathbb{D}_Y(\mathcal{O}_Y(\lambda)) = \mathcal{O}_{Y^\sharp}(-\lambda).$$

Similarly we have $\Phi_Y^{\mathrm{op}}(\mathcal{O}_{Y^\dagger}(\lambda)) = \mathcal{O}_Y(-\lambda)$. Since $D_{\mathrm{coh}}^b(Y)$ is locally generated by $\mathcal{O}_Y$, for $\mathcal{F} \in D_{\mathrm{coh}}^b([Y/\mathbb{C}^*])_\lambda$ we have

$$\Phi_Y^{\mathrm{op}}(\rho_Y^{\dagger*}\mathcal{F}) \in i_{Y*}^\sharp D_{\mathrm{coh}}^b([Y/\mathbb{C}^*])_{-\lambda}, \quad \Phi_Y^{\mathrm{op}}(i_{Y*}^\dagger \mathcal{F}) \in \rho_Y^{\sharp*} D_{\mathrm{coh}}^b([Y/\mathbb{C}^*])_{-\lambda}. \tag{5.3.22}$$

In general, let us take an object $\mathcal{F} \in D^b_{\mathrm{coh}}([\mathfrak{U}/\mathbb{C}^*])_\lambda$. By the commutative diagrams (5.3.8) and (5.3.10) together with (5.3.22), we have

$$j^\sharp_* \Phi^{\mathrm{op}}_\mathfrak{U}(\rho^{\dagger*}_\mathfrak{U} \mathcal{F}) \cong \widetilde{\Phi}^{\mathrm{op}}_Y(j^\dagger_* \rho^{\dagger*}_\mathfrak{U} \mathcal{F}) \cong \widetilde{\Phi}^{\mathrm{op}}_Y(\rho^{\dagger*}_Y j_* \mathcal{F}) \in i^\sharp_{Y*} D^b_{\mathrm{coh}}([Y/\mathbb{C}^*])_{-\lambda},$$

$$j^\sharp_* \Phi^{\mathrm{op}}_\mathfrak{U}(i^\dagger_{\mathfrak{U}*} \mathcal{F}) \cong \widetilde{\Phi}^{\mathrm{op}}_Y(j^\dagger_* i^\dagger_{\mathfrak{U}*} \mathcal{F}) \cong \widetilde{\Phi}^{\mathrm{op}}_Y(i^\dagger_{Y*} j_* \mathcal{F}) \in \rho^{\sharp*}_Y D^b_{\mathrm{coh}}([Y/\mathbb{C}^*])_{-\lambda}.$$

Then by Lemma 5.3.8 below, we have the inclusions

$$\Phi^{\mathrm{op}}_\mathfrak{U}(\rho^{\dagger*}_\mathfrak{U} D^b_{\mathrm{coh}}([\mathfrak{U}/\mathbb{C}^*])_\lambda) \subset i^\sharp_{\mathfrak{U}*} D^b_{\mathrm{coh}}([\mathfrak{U}/\mathbb{C}^*])_{-\lambda},$$

$$\Phi^{\mathrm{op}}_\mathfrak{U}(i^\dagger_{\mathfrak{U}*} D^b_{\mathrm{coh}}([\mathfrak{U}/\mathbb{C}^*])_\lambda) \subset \rho^{\sharp*}_\mathfrak{U} D^b_{\mathrm{coh}}([\mathfrak{U}/\mathbb{C}^*])_{-\lambda}$$

respectively. By applying the same argument for $\Psi^{\mathrm{op}}_\mathfrak{M}$ given in (5.3.17), we see that the above inclusions are identities. Therefore the proposition holds. □

Here we have used the following lemma.

Lemma 5.3.8 *In the proof of Proposition 5.3.7, we have the following:*

(i) *An object $M \in D^b_{\mathrm{coh}}([\mathfrak{U}^\sharp/\mathbb{C}^*])$ is an object in $i^\sharp_{\mathfrak{U}*} D^b_{\mathrm{coh}}([\mathfrak{U}/\mathbb{C}^*])_\lambda$ if and only if $j^\sharp_* M$ is an object in $i^\sharp_{Y*} D^b_{\mathrm{coh}}([Y/\mathbb{C}^*])_\lambda$.*

(ii) *An object $M \in D^b_{\mathrm{coh}}([\mathfrak{U}^\sharp/\mathbb{C}^*])$ is an object in $\rho^{\sharp*}_\mathfrak{U} D^b_{\mathrm{coh}}([\mathfrak{U}/\mathbb{C}^*])_\lambda$ if and only if $j^\sharp_* M$ is an object in $\rho^{\sharp*}_Y D^b_{\mathrm{coh}}([Y/\mathbb{C}^*])_\lambda$.*

Proof As for (i), since the functors (5.3.19) are fully-faithful, the subcategory $i^\sharp_* D^b_{\mathrm{coh}}(\mathfrak{M})_\lambda$ in $D^b_{\mathrm{coh}}(\mathfrak{M}^\sharp)$ is identified with the subcategory of objects $\mathcal{F} \in D^b_{\mathrm{coh}}(\mathfrak{M}^\sharp)$ such that each $\mathcal{H}^i(\mathcal{F})$ is an $\mathcal{O}_\mathcal{M}$-module with $\mathbb{C}^*$-weight λ. Then (i) follows from this fact.

As for (ii), suppose that M satisfies the latter condition. Then the natural map

$$j^\sharp_* \rho^{\sharp*}_\mathfrak{U}(\rho^\sharp_{\mathfrak{U}*})_\lambda M \cong \rho^{\sharp*}_Y(\rho^\sharp_{Y*})_\lambda j^\sharp_* M \to j^\sharp_* M$$

is an isomorphism. Here the first isomorphism follows from base change with respect to the diagram (5.3.8). Therefore the natural map $\rho^{\sharp*}_\mathfrak{U}(\rho^\sharp_{\mathfrak{U}*})_\lambda M \to M$ is also an isomorphism, since the functor $j^\sharp_* : D^b_{\mathrm{coh}}([\mathfrak{U}^\sharp/\mathbb{C}^*]) \to D^b_{\mathrm{coh}}([Y^\sharp/\mathbb{C}^*])$ is conservative. It follows that M is an object in $\rho^{\sharp*}_\mathfrak{U} D^b_{\mathrm{coh}}([\mathfrak{U}/\mathbb{C}^*])_\lambda$. □

Using Proposition 5.3.7, we show the following lemma.

Lemma 5.3.9 *We have the semiorthogonal decomposition*

$$D^b_{\mathrm{coh}}(\mathfrak{M}^\dagger) = \langle \ldots, \rho^{\dagger*} D^b_{\mathrm{coh}}(\mathfrak{M})_1, \rho^{\dagger*} D^b_{\mathrm{coh}}(\mathfrak{M})_0, \mathcal{T}^\dagger, \rho^{\dagger*} D^b_{\mathrm{coh}}(\mathfrak{M})_{-1}, \ldots \rangle$$

$$(5.3.23)$$

together with an equivalence $\mathcal{T}^\dagger \xrightarrow{\sim} D^b_{\mathrm{coh}}(\mathfrak{M}^\sharp_0)$.

Proof We apply the functor $\Psi_{\mathfrak{M}}^{\mathrm{op}}$ in (5.3.17) to the semiorthogonal decomposition of $D_{\mathrm{coh}}^b(\mathfrak{M}^\sharp)$ in Proposition 5.3.6. By Proposition 5.3.7, we obtain the semiorthogonal decomposition (5.3.23) for $\mathcal{T}^\dagger = \Psi_{\mathfrak{M}}^{\mathrm{op}}(\mathcal{D}^\sharp)$. Then $\mathcal{T}^\dagger$ is equivalent to $D_{\mathrm{coh}}^b(\mathfrak{M}_\circ^\sharp)^{\mathrm{op}}$, and applying $\mathbb{D}_{\mathfrak{M}_\circ^\sharp}$ we obtain the equivalence $\mathcal{T}^\dagger \xrightarrow{\sim} D_{\mathrm{coh}}^b(\mathfrak{M}_\circ^\sharp)$.   □

We set $e \in \mathbb{Z}$ to be

$$e := \mathrm{rank}(\mathfrak{E}|_{\mathcal{M}}) = \mathrm{wt}_{\mathbb{C}^*}(\det \mathfrak{E}|_{\mathcal{M}}).$$

Lemma 5.3.10 *For each $\lambda \in \mathbb{Z}$, we have*

$$\mathbb{D}_{\mathfrak{M}^\dagger}(i_*^\dagger D_{\mathrm{coh}}^b(\mathfrak{M})_\lambda) = i_*^\dagger D_{\mathrm{coh}}^b(\mathfrak{M})_{-\lambda-e}.$$

Proof For $\mathcal{F} \in D_{\mathrm{coh}}^b(\mathfrak{M})_\lambda$, we have

$$\mathbf{R}\mathcal{H}om_{\mathfrak{M}^\dagger}(i_*^\dagger \mathcal{F}, \mathcal{O}_{\mathfrak{M}^\dagger}) \cong i_*^\dagger \mathbf{R}\mathcal{H}om_{\mathfrak{M}}(\mathcal{F}, i^{\dagger!}\mathcal{O}_{\mathfrak{M}^\dagger}) \cong i_*^\dagger \mathbb{D}_{\mathfrak{M}}(\mathcal{F}) \otimes \det \mathfrak{E}^\vee[-e].$$

The lemma holds since $\mathbb{D}_{\mathfrak{M}}(\mathcal{F})$ is of $\mathbb{C}^*$-weight $-\lambda$ and $\det \mathfrak{E}^\vee$ is of $\mathbb{C}^*$-weight $-e$.   □

The following is the main result in this section.

Theorem 5.3.11 *Suppose that $e \geq 0$. Then we have the semiorthogonal decomposition of the form*

$$D_{\mathrm{coh}}^b(\mathfrak{M}_0^\dagger) = \langle \rho^{\dagger*} D_{\mathrm{coh}}^b(\mathfrak{M})_{e-1}, \ldots, \rho^{\dagger*} D_{\mathrm{coh}}^b(\mathfrak{M})_0, D_{\mathrm{coh}}^b(\mathfrak{M}_\circ^\sharp) \rangle.$$

Proof By Proposition 5.3.6 and Lemma 5.3.9, it is enough to show the identity of subcategories in $D_{\mathrm{coh}}^b(\mathfrak{M}^\dagger)$

$$\mathcal{D}^\dagger = \langle \rho^{\dagger*} D_{\mathrm{coh}}^b(\mathfrak{M})_{e-1}, \ldots, \rho^{\dagger*} D_{\mathrm{coh}}^b(\mathfrak{M})_0, \mathcal{T}^\dagger \rangle. \tag{5.3.24}$$

We prove the above identity following the argument of [100, Theorem 2.5], by replacing $\mathcal{S}_\lambda$, $\mathcal{P}_\lambda$ in *loc. cit.* by $i_*^\dagger D_{\mathrm{coh}}^b(\mathfrak{M})_\lambda$, $\rho^{\dagger*} D_{\mathrm{coh}}^b(\mathfrak{M})_\lambda$ respectively. For each $\lambda \in \mathbb{Z}$, we set $\mathcal{S}_\lambda$, $\mathcal{P}_\lambda$ as above, and define $\mathcal{S}_{\leq\lambda}$, $\mathcal{P}_{\leq\lambda}$ to be the triangulated subcategories of $D_{\mathrm{coh}}^b(\mathfrak{M}^\dagger)$ generated by $\mathcal{S}_{\lambda'}$, $\mathcal{P}_{\lambda'}$ for $\lambda' \leq \lambda$ respectively. The subcategories $\mathcal{S}_{>\lambda}$, $\mathcal{P}_{>\lambda}$ are defined similarly. We also define $D_{\mathrm{coh}}^b(\mathfrak{M}^\dagger)_{\geq\lambda}$ to be the subcategory of $D_{\mathrm{coh}}^b(\mathfrak{M}^\dagger)$ consisting of objects M such that $\rho_*^\dagger M \in D_{\mathrm{qcoh}}(\mathfrak{M})_{\geq\lambda}$. Then by Lemma 5.3.12 below, we have the semiorthogonal decompositions

$$D_{\mathrm{coh}}^b(\mathfrak{M}^\dagger) = \langle \mathcal{S}_{<0}, D_{\mathrm{coh}}^b(\mathfrak{M}^\dagger)_{\geq 0} \rangle = \langle D_{\mathrm{coh}}^b(\mathfrak{M}^\dagger)_{\geq 0}, \mathcal{P}_{<0} \rangle. \tag{5.3.25}$$

By comparing with semiorthogonal decomposition in (5.3.20) and (5.3.23), we have

$$D_{\mathrm{coh}}^b(\mathfrak{M}^\dagger)_{\geq 0} = \langle \mathcal{D}^\dagger, \mathcal{S}_{\geq 0} \rangle = \langle \mathcal{P}_{\geq 0}, \mathcal{T}^\dagger \rangle. \tag{5.3.26}$$

Therefore we have the semiorthogonal decomposition

$$D_{\mathrm{coh}}^b(\mathfrak{M}^\dagger) = \langle \mathcal{S}_{<0}, \mathcal{P}_{\geq 0}, \mathcal{T}^\dagger \rangle. \tag{5.3.27}$$

On the other hand, we apply $\mathbb{D}_{\mathfrak{M}^\dagger} \circ \otimes_{\mathbb{C}} \mathbb{C}(-e+1)$ to the semiorthogonal decompositions (5.3.20) and (5.3.23). By noting that $\mathbb{D}_{\mathfrak{M}^\dagger}(\mathcal{S}_{\geq \lambda}) = \mathcal{S}_{\leq -\lambda-e}$ by Lemma 5.3.10, we have

$$D_{\mathrm{coh}}^b(\mathfrak{M}^\dagger) = \langle \mathcal{S}_{<0}, \mathbb{D}_{\mathfrak{M}^\dagger}(\mathcal{D}^\dagger(-e+1)), \mathcal{S}_{\geq 0} \rangle = \langle \mathcal{P}_{\geq e}, \mathbb{D}_{\mathfrak{M}^\dagger}(\mathcal{T}^\dagger(-e+1)), \mathcal{P}_{<e} \rangle.$$

By comparing with (5.3.20), we have $\mathcal{D}^\dagger = \mathbb{D}_{\mathfrak{M}^\dagger}(\mathcal{D}^\dagger(-e+1))$. Similarly applying $\mathbb{D}_{\mathfrak{M}^\dagger} \circ \otimes_{\mathbb{C}} \mathbb{C}(-e+1)$ to the semiorthogonal decomposition (5.3.26), we have

$$\mathbb{D}_{\mathfrak{M}^\dagger}(D_{\mathrm{coh}}^b(\mathfrak{M})_{\geq -e+1}) = \langle \mathcal{S}_{<0}, \mathbb{D}_{\mathfrak{M}^\dagger}(\mathcal{D}^\dagger(-e+1)) \rangle = \langle \mathbb{D}_{\mathfrak{M}^\dagger}(\mathcal{T}^\dagger(-e+1)), \mathcal{P}_{<e} \rangle.$$

It follows that we have

$$D_{\mathrm{coh}}^b(\mathfrak{M}^\dagger) = \langle \mathcal{P}_{\geq e}, \mathcal{S}_{<0}, \mathcal{D}^\dagger \rangle = \langle \mathcal{S}_{<0}, \mathcal{P}_{\geq e}, \mathcal{D}^\dagger \rangle.$$

Here the second identity follows from $\mathrm{Hom}(\mathcal{P}_{\geq e}, \mathcal{S}_{<0}) = 0$ since $e \geq 0$. By comparing with (5.3.27), we obtain the identity (5.3.24). $\qquad\square$

We have used the following lemma, which is a generalization of [100, Lemma 2.3].

Lemma 5.3.12 *We have semiorthogonal decompositions (5.3.25).*

Proof We first show the left semiorthogonal decomposition in (5.3.25). There is no non-zero morphism from $D_{\mathrm{coh}}^b(\mathfrak{M}^\dagger)_{\geq 0}$ to $\mathcal{S}_{<0}$ since locally on $\mathfrak{M}$ any object in $D_{\mathrm{coh}}^b(\mathfrak{M}^\dagger)_{\geq 0}$ admits a bounded above resolution of free $S(\mathfrak{E})$-modules with non-negative $\mathbb{C}^*$-weights. For any object $A \in D_{\mathrm{coh}}^b(\mathfrak{M}^\dagger)$, we have the split distinguished triangle

$$(\rho_*^\dagger A)_{\geq 0} \to \rho_*^\dagger A \to (\rho_*^\dagger A)_{<0} \tag{5.3.28}$$

in $D_{\mathrm{qcoh}}(\mathfrak{M})$ such that $(\rho_*^\dagger A)_{\geq 0}$ has non-negative $\mathbb{C}^*$-weights and $(\rho_*^\dagger A)_{<0}$ has negative $\mathbb{C}^*$-weights. Since $S(\mathfrak{E})$ has non-negative $\mathbb{C}^*$-weights, the $S(\mathfrak{E})$-module structure on $\rho_*^\dagger A$ induces a $S(\mathfrak{E})$-module structures on $(\rho_*^\dagger A)_{\geq 0}$. Therefore the distinguished triangle (5.3.28) lifts to a distinguished triangle $A_{\geq 0} \to A \to A_{<0}$ in $D_{\mathrm{coh}}^b(\mathfrak{M}^\dagger)$ such that $A_{\geq 0} \in D_{\mathrm{coh}}^b(\mathfrak{M}^\dagger)_{\geq 0}$, $A_{<0} \in \mathcal{S}_{<0}$.

We next show the right semiorthogonal decomposition in (5.3.25). It is obvious that there is no non-zero morphism from $\mathcal{P}_{<0}$ to $D_{\mathrm{coh}}^b(\mathfrak{M}^\dagger)_{\geq 0}$ from the definition of $D_{\mathrm{coh}}^b(\mathfrak{M}^\dagger)_{\geq 0}$. For an object $A \in D_{\mathrm{coh}}^b(\mathfrak{M}^\dagger)$, let λ be the minimum $\mathbb{C}^*$-weight of

$\rho_*^{\dagger} A$ and assume that $\lambda < 0$. We have the morphism $\rho^{\dagger *}(\rho_*^{\dagger} A)_\lambda \to A$ in $D_{\mathrm{coh}}^b(\mathfrak{M}^{\dagger})$, and let A' be its cone. Then we have the distinguished triangle in $D_{\mathrm{qcoh}}(\mathfrak{M})$

$$(\rho_*^{\dagger} A)_\lambda \otimes S(\mathfrak{E}) \to \rho_*^{\dagger} A \to \rho_*^{\dagger} A'.$$

From the above distinguished triangle, the minimum $\mathbb{C}^*$-weight of $\rho_*^{\dagger} A'$ is strictly bigger than λ. Therefore by repeating this argument, we find a distinguished triangle $A'_{<0} \to A \to A'_{\geq 0}$ such that $A'_{<0} \in \mathcal{P}_{<0}$ and $A_{\geq 0} \in D_{\mathrm{coh}}^b(\mathfrak{M}^{\dagger})_{\geq 0}$.

$\square$

5.3.4 Singular Supports Under Linear Koszul Duality

Let $\mathfrak{M}^{\dagger}$, $\mathfrak{M}^{\sharp}$ be the derived stacks as in the diagram (5.3.12). Since $\mathbb{L}_{\mathfrak{M}^{\dagger}/\mathfrak{M}} = \rho^{\dagger *}\mathfrak{E}$ and $\mathbb{L}_{\mathfrak{M}^{\sharp}/\mathfrak{M}} = \rho^{\sharp *}\mathfrak{E}^{\vee}[1]$, we have natural equivalences

$$\Omega_{\mathfrak{M}^{\dagger}/\mathfrak{M}}[-1] \xrightarrow{\sim} \mathfrak{M}^{\dagger} \times_{\mathfrak{M}} \mathfrak{M}^{\sharp} \xleftarrow{\sim} \Omega_{\mathfrak{M}^{\sharp}/\mathfrak{M}}[-1]. \qquad (5.3.29)$$

We show that the above natural equivalence lifts to the absolute (-1)-shifted cotangent stacks on classical truncations:

Lemma 5.3.13 *There is a natural isomorphism* $\vartheta_0 \colon t_0(\Omega_{\mathfrak{M}^{\dagger}}[-1]) \xrightarrow{\cong} t_0(\Omega_{\mathfrak{M}^{\sharp}}[-1])$ *which fits into the commutative diagram*

$$(5.3.30)$$

Here the vertical arrows are natural maps induced by $\mathbb{L}_{\mathfrak{M}^{\dagger}} \to \mathbb{L}_{\mathfrak{M}^{\dagger}/\mathfrak{M}}$, $\mathbb{L}_{\mathfrak{M}^{\sharp}} \to \mathbb{L}_{\mathfrak{M}^{\sharp}/\mathfrak{M}}$, *and the bottom arrows are induced by* (5.3.29).

Proof We first consider the situation of the diagram (5.3.14), i.e. $\mathfrak{M} = [\mathfrak{U}/\mathbb{C}^*]$ where $\mathfrak{U} = \operatorname{Spec} \mathcal{R}(V \to Y, s)$ is given as in (2.1.1) with trivial $\mathbb{C}^*$-action, and $\mathfrak{E} = \mathcal{E} \otimes_{\mathcal{O}_Y} \mathcal{O}_{\mathfrak{U}}$ for a two term complex of vector bundles $\mathcal{E}$ as in (5.3.1). We set $\mathcal{E}_0 := (\mathcal{E}^0)^{\vee}$ and $\mathcal{E}_1 := (\mathcal{E}^1)^{\vee}$. Note that the structure sheaves of $\mathfrak{U}^{\dagger}$ and $\mathfrak{U}^{\sharp}$ are Koszul complexes

$$\mathcal{O}_{\mathfrak{U}^{\dagger}} = S_{\mathcal{O}_{\mathfrak{U}}}(\mathfrak{E}) = \mathcal{R}(\mathcal{E}_0 \times_Y \mathcal{E}_1 \times_Y V \to \mathcal{E}_0, (\phi, s)),$$

$$\mathcal{O}_{\mathfrak{U}^{\sharp}} = S_{\mathcal{O}_{\mathfrak{U}}}(\mathfrak{E}^{\vee}[1]) = \mathcal{R}(\mathcal{E}^{-1} \times_Y \mathcal{E}^0 \times_Y V \to \mathcal{E}^{-1}, (\phi^{\vee}, s))$$

where differentials are induced by

$$\mathcal{E}^{-1} \oplus V^\vee \overset{(\phi,s)}{\to} \mathcal{E}^0 \oplus \mathcal{O}_Y \subset \mathcal{O}_{\mathcal{E}_0}, \quad \mathcal{E}_0 \oplus V^\vee \overset{(\phi^\vee,s)}{\to} \mathcal{E}_1 \oplus \mathcal{O}_Y \subset \mathcal{O}_{\mathcal{E}_{-1}}$$

respectively. By the above descriptions, both of (-1)-shifted cotangent derived schemes $\Omega_{\mathfrak{U}^\dagger}[-1]$, $\Omega_{\mathfrak{U}^\sharp}[-1]$ are given by the derived critical locus of the function

$$w \colon \mathcal{E}_0 \times_Y \mathcal{E}^{-1} \times_Y V^\vee \to \mathbb{C}, \quad w(x, e, e', v) = \langle \phi|_x(e'), e \rangle + \langle s(x), v \rangle \quad (5.3.31)$$

for $x \in Y$, $e \in \mathcal{E}_0|_x$, $e' \in \mathcal{E}^{-1}|_x$, $v \in V^\vee|_x$. Therefore the isomorphism ϑ_0 is given by the composition of isomorphisms

$$t_0(\Omega_{\mathfrak{U}^\dagger}[-1]) \overset{\cong}{\to} \mathrm{Crit}(w) \overset{\cong}{\leftarrow} t_0(\Omega_{\mathfrak{U}^\sharp}[-1]). \quad (5.3.32)$$

Also both of $\Omega_{\mathfrak{U}^\dagger/\mathfrak{U}}[-1]$ and $\Omega_{\mathfrak{U}^\sharp/\mathfrak{U}}[-1]$ are the derived zero locus of

$$(\phi^\vee, \phi, s) \colon \mathcal{E}_0 \times_Y \mathcal{E}^{-1} \to \mathcal{E}_1 \times_Y \mathcal{E}^0 \times_Y V$$

and the left and right maps in (5.3.30) are induced by the projection $\mathcal{E}_0 \times_Y \mathcal{E}^{-1} \times_Y V \to \mathcal{E}_0 \times_Y \mathcal{E}^{-1}$. Therefore the diagram (5.3.30) commutes.

In order to globalize the above isomorphism, it is enough to show that the isomorphism ϑ_0 is independent of a presentation of $\mathfrak{U}$ and $\mathcal{E}$. Here we check the latter independence. The check for the former independence is similarly discussed. Let

$$\mathcal{E} = (\mathcal{E}^{-1} \to \mathcal{E}^0) \to \mathcal{E}' = (\mathcal{E}'^{-1} \to \mathcal{E}'^0) \quad (5.3.33)$$

be a quasi-isomorphism of two term complexes of vector bundles on Y. We set

$$\mathcal{E}' = \mathcal{E}' \otimes_{\mathcal{O}_Y} \mathcal{O}_{\mathfrak{U}}, \quad \mathfrak{U}'^\dagger = \mathrm{Spec}\, S_{\mathcal{O}_{\mathfrak{U}}}(\mathcal{E}'), \quad \mathfrak{U}'^\sharp = \mathrm{Spec}\, S_{\mathcal{O}_{\mathfrak{U}}}(\mathcal{E}'^\vee[1]).$$

We also have the function w' on $\mathcal{E}_0' \times_Y \mathcal{E}'^{-1} \times_Y V^\vee$ as in (5.3.31), and isomorphisms

$$t_0(\Omega_{\mathfrak{U}'^\dagger}[-1]) \overset{\cong}{\to} \mathrm{Crit}(w') \overset{\cong}{\leftarrow} t_0(\Omega_{\mathfrak{U}'^\sharp}[-1]) \quad (5.3.34)$$

as in (5.3.32). On the other hand, the quasi-isomorphism (5.3.33) induces equivalences $\mathfrak{U}'^\dagger \overset{\sim}{\to} \mathfrak{U}^\dagger$ and $\mathfrak{U}^\sharp \overset{\sim}{\to} \mathfrak{U}'^\sharp$, which induce isomorphisms

$$t_0(\Omega_{\mathfrak{U}'^\dagger}[-1]) \overset{\cong}{\to} t_0(\Omega_{\mathfrak{U}^\dagger}[-1]), \quad t_0(\Omega_{\mathfrak{U}'^\sharp}[-1]) \overset{\cong}{\to} t_0(\Omega_{\mathfrak{U}^\sharp}[-1]). \quad (5.3.35)$$

Via the isomorphisms in (5.3.32) and (5.3.34), the above isomorphisms give two isomorphisms $\mathrm{Crit}(w') \overset{\cong}{\to} \mathrm{Crit}(w)$, and we need to check that they are indeed the same isomorphism. Namely we claim that there is an isomorphism $\gamma \colon \mathrm{Crit}(w') \xrightarrow{\cong} \mathrm{Crit}(w)$ such that the following diagram commutes

$$
\begin{array}{ccc}
t_0(\Omega_{\mathfrak{U}'^{\dagger}}[-1]) \xrightarrow{\cong} \mathrm{Crit}(w') \xleftarrow{\cong} t_0(\Omega_{\mathfrak{U}'^{\sharp}}[-1]) \\
\cong \downarrow \qquad\qquad \gamma \downarrow \cong \qquad\qquad \downarrow \cong \\
t_0(\Omega_{\mathfrak{U}^{\dagger}}[-1]) \xrightarrow{\cong} \mathrm{Crit}(w) \xleftarrow{\cong} t_0(\Omega_{\mathfrak{U}^{\sharp}}[-1]).
\end{array} \tag{5.3.36}
$$

Here the horizontal arrows are given by (5.3.32) and (5.3.34), and the left and right vertical arrows are given by (5.3.35).

The quasi-isomorphism $\mathcal{E} \to \mathcal{E}'$ induces the diagram

$$
\mathcal{E}'_0 \times_Y \mathcal{E}'^{-1} \times_Y V^{\vee} \xleftarrow{h'} \mathcal{E}'_0 \times_Y \mathcal{E}^{-1} \times_Y V^{\vee} \xrightarrow{h} \mathcal{E}_0 \times_Y \mathcal{E}^{-1} \times_Y V^{\vee}.
$$

By Lemmas 2.4.2 and 3.1.2, we have the isomorphism

$$
\gamma \colon \mathrm{Crit}(w') \xleftarrow[h']{\cong} (h')^{-1}(\mathrm{Crit}(w')) \cap h^{-1}(\mathrm{Crit}(w)) \xrightarrow[h]{\cong} \mathrm{Crit}(w),
$$

which fits into the commutative diagram (5.3.36). Therefore we have the independence of ϑ_0 for a presentation of $\mathcal{E}$.   □

We have the isomorphism by Lemma 5.3.13

$$
\vartheta_0 \colon t_0(\Omega_{\mathfrak{M}^{\dagger}}[-1]) \xrightarrow{\cong} t_0(\Omega_{\mathfrak{M}^{\sharp}}[-1]).
$$

In particular they admit two fiberwise $\mathbb{C}^*$-actions with respect to projections to $t_0(\mathfrak{M}^{\dagger})$, $t_0(\mathfrak{M}^{\sharp})$ respectively. We have the following proposition:

Proposition 5.3.14 *Let $\mathcal{Z} \subset t_0(\Omega_{\mathfrak{M}^{\dagger}}[-1])$ be a double conical closed substack. Under the equivalence $\Phi_{\mathfrak{M}}$ in (5.3.18), we have $\Phi_{\mathfrak{M}}(\mathcal{C}_{\mathcal{Z}}) = \mathcal{C}_{\vartheta_0(\mathcal{Z})}$. Therefore we have the equivalence*

$$
\Phi_{\mathfrak{M}} \colon D^b_{\mathrm{coh}}(\mathfrak{M}^{\dagger})/\mathcal{C}_{\mathcal{Z}} \xrightarrow{\sim} D^b_{\mathrm{coh}}(\mathfrak{M}^{\sharp})/\mathcal{C}_{\vartheta_0(\mathcal{Z})}.
$$

Proof The statement is local on $\mathfrak{M}^{\mathbb{C}^*\text{-rig}}$, so we can assume the situation of the diagram (5.3.14). Below, we use notation of the proof of Lemma 5.3.13. By the local computation in *loc. cit.*, singular supports for objects in $D^b_{\mathrm{coh}}(\mathfrak{U}^{\star})$ with $\star \in \{\dagger, \sharp\}$ are closed subsets in $\mathcal{E}_0 \times_Y \mathcal{E}^{-1} \times_Y V^{\vee}$. By Lemma 2.1.10, they are determined by natural maps to relative Hochschild cohomologies

$$
\mathcal{E}^0 \oplus \mathcal{E}_1 \oplus V \to \mathrm{HH}^*(\mathfrak{U}^{\dagger}/\mathcal{E}_0), \quad \mathcal{E}^0 \oplus \mathcal{E}_1 \oplus V \to \mathrm{HH}^*(\mathfrak{U}^{\sharp}/\mathcal{E}^{-1}) \tag{5.3.37}
$$

respectively.

Let $\Phi_{\mathfrak{U}}$ be the equivalence

$$\Phi_{\mathfrak{U}} = \Phi_{\mathfrak{U}}^{\mathrm{op}} \circ \mathbb{D}_{\mathfrak{U}^\dagger} \colon D^b_{\mathrm{coh}}([\mathfrak{U}^\dagger/\mathbb{C}^*]) \xrightarrow{\sim} D^b_{\mathrm{coh}}([\mathfrak{U}^\sharp/\mathbb{C}^*]).$$

From the construction of the equivalence $\Phi_{\mathfrak{U}}$, it commutes with the $\mathbb{C}^*$-weight shift up to cohomological shift (cf. [91, Theorem 1.7.1]): we have $\Phi_{\mathfrak{U}}((-)[m](j)) = \Phi_{\mathfrak{U}}(-)[m+2j](j)$. So the equivalence $\Phi_{\mathfrak{U}}$ identifies the natural transforms

$$\mathrm{Nat}_{D^b_{\mathrm{coh}}([\mathfrak{U}^\dagger/\mathbb{C}^*])}(\mathrm{id}, \mathrm{id}[m](j)) \xrightarrow{\sim} \mathrm{Nat}_{D^b_{\mathrm{coh}}([\mathfrak{U}^\sharp/\mathbb{C}^*])}(\mathrm{id}, \mathrm{id}[m+2j](j)). \tag{5.3.38}$$

On the other hand for $\star \in \{\dagger, \sharp\}$, we have

$$\mathrm{HH}^*(\mathfrak{U}^\star/Y) := \mathrm{Hom}^*_{\mathfrak{U}^\star \times_Y \mathfrak{U}^\star}(\Delta_* \mathcal{O}_{\mathfrak{U}^\star}, \Delta_* \mathcal{O}_{\mathfrak{U}^\star})$$

$$= \bigoplus_{(m,j) \in \mathbb{Z}^2} \mathrm{Hom}_{[\mathfrak{U}^\star/\mathbb{C}^*] \times_{[Y/\mathbb{C}^*]} [\mathfrak{U}^\star/\mathbb{C}^*]}(\Delta_* \mathcal{O}_{\mathfrak{U}^\star}, \Delta_* \mathcal{O}_{\mathfrak{U}^\star}[m](j)).$$

Therefore as in the proof of Proposition 2.3.9, one can lift the direct sum of (5.3.38) for all (m, j) to the isomorphism of vector spaces

$$\Phi_{\mathfrak{U}}^{\mathrm{HH}} \colon \mathrm{HH}^*(\mathfrak{U}^\dagger/Y) \xrightarrow{\cong} \mathrm{HH}^*(\mathfrak{U}^\sharp/Y)$$

compatible with maps to natural transforms in (5.3.38). Note that $\Phi_{\mathfrak{U}}^{\mathrm{HH}}$ does not preserve the cohomological grading $*$.

By ignoring cohomological grading, it is enough to show that the following diagram is commutative:

$$
\begin{array}{ccccc}
\mathcal{E}^0 \oplus \mathcal{E}_1 \oplus V & \longrightarrow & \mathrm{HH}^*(\mathfrak{U}^\dagger/\mathcal{E}_0) & \longrightarrow & \mathrm{HH}^*(\mathfrak{U}^\dagger/Y) \\
\| & & & & \downarrow{\scriptstyle \Phi_{\mathfrak{U}}^{\mathrm{HH}}} \\
\mathcal{E}^0 \oplus \mathcal{E}_1 \oplus V & \longrightarrow & \mathrm{HH}^*(\mathfrak{U}^\sharp/\mathcal{E}^{-1}) & \longrightarrow & \mathrm{HH}^*(\mathfrak{U}^\sharp/Y).
\end{array}
\tag{5.3.39}
$$

Here the left horizontal arrows are the maps (5.3.37), and the right horizontal arrows are natural maps via the projections $\mathcal{E}_0 \to Y$, $\mathcal{E}^{-1} \to Y$. The relative Hochschild cohomologies $\mathrm{HH}^*(\mathfrak{U}^\star/Y)$ and the map $\Phi_{\mathfrak{U}}^{\mathrm{HH}}$ can be computed by the similar argument as in Proposition 2.3.9. Namely using the automorphism $(x, y) \mapsto (x+y, x-y)/2$ on $V^{\oplus 2}$ and $\mathcal{E}^{\oplus 2}$, we have equivalences

$$\mathfrak{U}^\dagger \times_Y \mathfrak{U}^\dagger \xrightarrow{\sim} Y^\dagger \times_Y Y^\dagger \times_Y \mathfrak{U}^\flat \times_Y \mathfrak{U},$$

$$\mathfrak{U}^\sharp \times_Y \mathfrak{U}^\sharp \xrightarrow{\sim} Y^\sharp \times_Y Y^\sharp \times_Y \mathfrak{U}^\flat \times_Y \mathfrak{U}$$

where $\mathcal{O}_{\mathfrak{U}^\flat} = S(V^\vee[1])$ with zero differential. Under the above equivalences, the objects $\Delta_* \mathcal{O}_{\mathfrak{U}^\dagger}, \Delta_* \mathcal{O}_{\mathfrak{U}^\sharp}$ correspond to $\mathcal{O}_{Y^\dagger} \boxtimes \mathcal{O}_Y \boxtimes \mathcal{O}_Y \boxtimes \mathcal{O}_{\mathfrak{U}}, \mathcal{O}_Y \boxtimes \mathcal{O}_{Y^\sharp} \boxtimes \mathcal{O}_Y \boxtimes \mathcal{O}_{\mathfrak{U}}$

respectively. Therefore $\mathrm{HH}^*(\mathfrak{U}^\dagger/Y)$, $\mathrm{HH}^*(\mathfrak{U}^\sharp/Y)$ are computed by cohomologies of the complexes

$$\mathbf{R}\mathrm{Hom}_{Y^\dagger}(\mathcal{O}_{Y^\dagger}, \mathcal{O}_{Y^\dagger}) \overset{\mathbf{L}}{\otimes}_{\mathcal{O}_Y} \mathbf{R}\mathrm{Hom}_{Y^\dagger}(\mathcal{O}_Y, \mathcal{O}_Y) \tag{5.3.40}$$

$$\overset{\mathbf{L}}{\otimes}_{\mathcal{O}_Y} \mathrm{RHom}_{\mathfrak{U}^\flat}(\mathcal{O}_Y, \mathcal{O}_Y) \overset{\mathbf{L}}{\otimes}_{\mathcal{O}_Y} \mathcal{O}_{\mathfrak{U}},$$

$$\mathrm{RHom}_{Y^\sharp}(\mathcal{O}_Y, \mathcal{O}_Y) \overset{\mathbf{L}}{\otimes}_{\mathcal{O}_Y} \mathrm{RHom}_{Y^\sharp}(\mathcal{O}_Y^\sharp, \mathcal{O}_Y^\sharp)$$

$$\overset{\mathbf{L}}{\otimes}_{\mathcal{O}_Y} \mathrm{RHom}_{\mathfrak{U}^\flat}(\mathcal{O}_Y, \mathcal{O}_Y) \overset{\mathbf{L}}{\otimes}_{\mathcal{O}_Y} \mathcal{O}_{\mathfrak{U}}$$

respectively. By taking Koszul resolutions

$$S(\mathcal{E}[1]) \otimes_{\mathcal{O}_Y} \mathcal{O}_{Y^\dagger} \overset{\sim}{\to} \mathcal{O}_Y, \; S(\mathcal{E}^\vee[2]) \otimes_{\mathcal{O}_Y} \mathcal{O}_{Y^\sharp} \overset{\sim}{\to} \mathcal{O}_Y, \; S(V^\vee[2]) \otimes_{\mathcal{O}_Y} \mathcal{O}_{\mathfrak{U}^\flat} \overset{\sim}{\to} \mathcal{O}_Y$$

the complexes (5.3.40) are quasi-isomorphic to

$$S(\mathcal{E}) \otimes_{\mathcal{O}_Y} S(\mathcal{E}^\vee[-1]) \otimes_{\mathcal{O}_Y} S(V[-2]) \otimes_{\mathcal{O}_Y} \mathcal{O}_{\mathfrak{U}}, \tag{5.3.41}$$

$$S(\mathcal{E}[-2]) \otimes_{\mathcal{O}_Y} S(\mathcal{E}^\vee[1]) \otimes_{\mathcal{O}_Y} S(V[-2]) \otimes_{\mathcal{O}_Y} \mathcal{O}_{\mathfrak{U}}$$

respectively. Through the above identifications, the map $\Phi_{\mathfrak{U}}^{\mathrm{HH}}$ is obtained by naturally identifying the complexes in (5.3.41) ignoring cohomological gradings.

On the other hand from the constructions of top horizontal arrows of (5.3.39) in Sect. 2.1.2, it is straightforward to see that the compositions of left and right horizontal arrows in (5.3.39) are induced by maps to (5.3.41) given by

$$\mathcal{E}^0 \oplus \mathcal{E}_1 \oplus V \ni (x, y, z) \mapsto x \otimes y \otimes z \otimes 1.$$

This applies to both of the top and bottom horizontal arrows in (5.3.39). Therefore we conclude that the diagram (5.3.39) commutes.   □

Chapter 6
Window Theorem for DT Categories

In this chapter, we prove window theorem for DT categories for (-1)-shifted cotangent stacks, and apply it to prove some conjectures proposed in the previous chapters. For a quasi-smooth and QCA derived stack $\mathfrak{M}$ with $\mathcal{M} = t_0(\mathfrak{M})$ and $l \in$ $\mathrm{Pic}(\mathcal{M})_{\mathbb{R}}$, its pull-back to the (-1)-shifted cotangent determines the l-semistable locus

$$\mathcal{N}^{l\text{-ss}} \subset \mathcal{N} := t_0(\Omega_{\mathfrak{M}}[-1])$$

which is an open substack of $\mathcal{N}$. We expect the existence of a triangulated subcategory $\mathcal{W}(\mathfrak{M}) \subset D^b_{\mathrm{coh}}(\mathfrak{M})$ such that the composition

$$\mathcal{W}(\mathfrak{M}) \hookrightarrow D^b_{\mathrm{coh}}(\mathfrak{M}) \twoheadrightarrow \mathcal{DT}^{\mathbb{C}^*}(\mathcal{N}^{l\text{-ss}}) \tag{6.0.1}$$

is an equivalence. We will develop such a theory in the case that $\mathcal{M}$ admits a good moduli space $\mathcal{M} \to M$. More precisely we show that, given a symmetric structure $\mathbb{S}$ on $\mathfrak{M}$ there exists a subcategory $\mathcal{W}(\mathfrak{M})$ as above such that the composition (6.0.1) is fully-faithful, and an equivalence if l is also compatible with $\mathbb{S}$. Here we refer to Sect. 6.2.1 for the notions of symmetric structures and compatibility with them.

An idea for the construction is as follows. Since we have the good moduli space $\mathcal{M} \to M$, the category $D^b_{\mathrm{coh}}(\mathfrak{M})$ is obtained as a limit of $D^b_{\mathrm{coh}}(\mathfrak{M}_U)$ for all étale maps $U \to M$ as in Theorem 3.1.4, where $\mathfrak{M}_U$ is obtained in Proposition 3.1.5 so it is of the form $[\mathfrak{U}/G]$ for a reductive algebraic group G and a G-equivariant tuple (Y, V, s). Then we have the equivalence by Theorem 2.3.3

$$\Phi : D^b_{\mathrm{coh}}([\mathfrak{U}/G]) \xrightarrow{\sim} \mathrm{MF}^{\mathbb{C}^*}_{\mathrm{coh}}([V^\vee/G], w). \tag{6.0.2}$$

Y. Toda, *Categorical Donaldson-Thomas Theory for Local Surfaces*, Lecture Notes in Mathematics 2350, https://doi.org/10.1007/978-3-031-61705-8_6

We then construct some subcategory (called *intrinsic window subcategory*) $\mathcal{W}(\mathfrak{M}_U)$ in $D^b_{\mathrm{coh}}([\mathfrak{U}/G])$ such that the equivalence (6.0.2) restricts to the fully-faithful functor

$$\Phi\colon \mathcal{W}(\mathfrak{M}_U) \hookrightarrow \mathcal{W}^l([V^\vee/G], w),$$

which is an equivalence if l is compatible with $\mathbb{S}$. Here the right hand side is the window subcategory for derived categories of factorizations constructed in [11, 45]. We then show that intrinsic window subcategory is independent of a presentation of $\mathfrak{M}_U$ as $[\mathfrak{U}/G]$, so that they glue to define the subcategory

$$\mathcal{W}(\mathfrak{M}) := \lim_{U \to M} \mathcal{W}(\mathfrak{M}_U) \subset D^b_{\mathrm{coh}}(\mathfrak{M}).$$

The above subcategory gives a desired window subcategory. The proof of the presentation independence of $\mathcal{W}(\mathfrak{M}_U)$ and its comparison with window subcategory under Koszul duality are technically hard parts of this chapter, which will be discussed in Sect. 6.2.

As applications of window theorem for DT categories, we prove conjectures in the previous sections under some setting. The first one is a proof of Conjecture 4.1.2, when strictly semisimple sheaves on X push-forward to semistable sheaves on S at the wall. The next one is a proof of Conjecture 4.3.3 when the curve class is reduced. The latter can be also proved using categorified Hall products in Chap. 7.

The organization of this chapter is as follows. In Sect. 6.1, we review the original window theorem for derived categories of factorizations with respect to an action of a reductive algebraic group on a smooth affine scheme, and prove some of its variants. In Sect. 6.2, we introduce intrinsic window subcategories and prove the window theorem for DT categories. In Sect. 6.4, we apply the window theorem for Conjecture 4.1.2. In Sect. 6.5, we apply the window theorem for Conjecture 4.3.3.

6.1 Window Theorem for GIT Quotient

6.1.1 Kempf-Ness Stratification

Let G be a reductive algebraic group, with maximal torus $T \subset G$. We always denote by M the character lattice of T and N the cocharacter lattice of T, i.e.

$$M = \mathrm{Hom}_\mathbb{Z}(T, \mathbb{C}^*), \quad N = \mathrm{Hom}_\mathbb{Z}(\mathbb{C}^*, T).$$

The subspace $M^W_\mathbb{R} \subset M_\mathbb{R}$ is defined to be the Weyl-invariant subspace, which is identified with $\mathrm{Pic}(BG)_\mathbb{R}$. Note that M, N are finitely generated free abelian groups with a perfect pairing

$$\langle -, - \rangle \colon M \times N \to \mathbb{Z}.$$

Below we follow the convention of [45, Section 2.1] for Kempf-Ness stratification associated with GIT quotients. Let Y be a smooth affine variety with a G-action. For an element $l \in \mathrm{Pic}([Y/G])_{\mathbb{R}}$, we have the open subset of l-semistable points

$$Y^{l\text{-ss}} \subset Y. \tag{6.1.1}$$

By the Hilbert-Mumford criterion, $Y^{l\text{-ss}}$ is characterized by the set of points $y \in Y$ such that for any one parameter subgroup $\lambda \colon \mathbb{C}^* \to G$ such that the limit $z = \lim_{t \to 0} \lambda(t)(y)$ exists in Y, we have $\mathrm{wt}(l|_z) \geq 0$.

We will often take $l \in M_{\mathbb{R}}^W = \mathrm{Pic}(BG)_{\mathbb{R}}$ and regard it as an element of $\mathrm{Pic}([Y/G])_{\mathbb{R}}$ by the pull-back of the natural morphism

$$a \colon [Y/G] \to BG.$$

In this case, the condition $\mathrm{wt}(l|_z) \geq 0$ is equivalent to $\langle l, \lambda \rangle \geq 0$. In some cases we may assume that l is pulled back from $\mathrm{Pic}(BG)_{\mathbb{R}}$ by the following lemma:

Lemma 6.1.1 *Let $x \in Y$ be fixed by G and set $y = p(x)$ for the quotient morphism $p \colon Y \to Y /\!\!/ G$. Let $\mu \colon BG \to [Y/G]$ be the map sending a point to x and identity on stabilizer groups. Then for any $l \in \mathrm{Pic}([Y/G])$ there exists a Zariski open subset $y \in U \subset Y /\!\!/ G$ such that $l|_{[p^{-1}(U)/G]}$ is isomorphic to $a^*\mu^*(l)|_{[p^{-1}(U)/G]}$.*

Proof Let us set $\mathcal{L} := l \otimes a^*\mu^*(l)^{-1} \in \mathrm{Pic}([Y/G])$. Then $\mu^*\mathcal{L}$ is a trivial line bundle on BG, or in other word $H^0(\mathcal{L}|_x)^G = \mathbb{C}$. Since Y is affine and G is reductive, the functor $H^0(-)^G$ on $\mathrm{Coh}([Y/G])$ is an exact functor. Therefore by applying the above functor to the surjection $\mathcal{L} \twoheadrightarrow \mathcal{L}|_x$, we obtain a surjection $H^0(\mathcal{L})^G \twoheadrightarrow H^0(\mathcal{L}|_x)^G$. In particular there is $s \in H^0(\mathcal{L})^G$ which is non-zero on x. It gives a G-equivariant map $s \colon \mathcal{O}_Y \to \mathcal{L}$, and let $Z \subset Y$ be the zero locus of s. Then Z is a G-invariant closed subset of Y which does not contain x. However the map $[Y/G] \to Y /\!\!/ G$ is a good moduli space for $[Y/G]$, so in particular it is universally closed (see [4, Theorem 4.16]). Therefore there exists a Zariski open subset $y \in U \subset Y /\!\!/ G$ such that $Z \cap p^{-1}(U) = \emptyset$, which implies that s is an isomorphism on $p^{-1}(U)$. $\qquad\square$

By fixing a Weyl-invariant norm $|*|$ on $N_{\mathbb{R}}$, we have the associated Kempf-Ness (KN) stratification

$$Y = S_1 \sqcup S_2 \sqcup \cdots \sqcup S_N \sqcup Y^{l\text{-ss}}. \tag{6.1.2}$$

Here for each α there exists a one parameter subgroup $\lambda_\alpha \colon \mathbb{C}^* \to T \subset G$, a connected component Z_α of the λ_α-fixed part of $Y \setminus \cup_{\alpha' < \alpha} S_{\alpha'}$ such that

$$S_\alpha = G \cdot Y_\alpha, \quad Y_\alpha := \{y \in Y : \lim_{t \to 0} \lambda_\alpha(t)(y) \in Z_\alpha\}.$$

We call Z_α the *center* and Y_α the *attracting locus*. Moreover by setting the slope to be

$$\mu_\alpha := -\frac{\mathrm{wt}(l|_{Z_\alpha})}{|\lambda_\alpha|} \in \mathbb{R} \tag{6.1.3}$$

we have the inequalities $\mu_1 > \mu_2 > \cdots > 0$. By taking the quotient stacks of the stratification (6.1.2), we have the stratification of the quotient stack $\mathcal{Y} = [Y/G]$

$$\mathcal{Y} = \mathcal{S}_1 \sqcup \mathcal{S}_2 \sqcup \cdots \sqcup \mathcal{S}_N \sqcup \mathcal{Y}^{l\text{-ss}}.$$

Using Hilbert-Mumford criterion, the notion of semistability can be generalized to an arbitrary Artin stack $\mathcal{Y}$ and $l \in \mathrm{Pic}(\mathcal{Y})_{\mathbb{R}}$ (see [49, Definition 1.13]). Namely a point $p \in \mathcal{Y}$ is *l-semistable* if for any map $f : [\mathbb{A}^1/\mathbb{C}^*] \to \mathcal{Y}$ with $f(1) \sim p$, we have $\mathrm{wt}(f(0)^*(l)) \geq 0$. Here $\mathbb{C}^*$ acts on $\mathbb{A}^1$ by weight one. The set of l-semistable points is denoted by

$$\mathcal{Y}^{l\text{-ss}} \subset \mathcal{Y}.$$

When $\mathcal{Y} = [Y/G]$, then $\mathcal{Y}^{l\text{-ss}} = [Y^{l\text{-ss}}/G]$ where $Y^{l\text{-ss}}$ is the GIT semistable locus (6.1.1).

6.1.2 Semiorthogonal Decomposition via KN Stratification

In the setting of Sect. 6.1.1, suppose furthermore that Y is a smooth affine variety. Given $l \in \mathrm{Pic}([Y/G])_{\mathbb{R}}$, we have a KN-stratification (6.1.2) with one parameter subgroup $\lambda_\alpha : \mathbb{C}^* \to G$. We have the following subgroups in G

$$P_\alpha := \{g \in G : \text{there exists } \lim_{t \to 0} \lambda_\alpha(t)g\lambda_\alpha^{-1}(t) \in G\},$$

$$G_\alpha := \{g \in G : \lambda_\alpha(t)g\lambda_\alpha^{-1}(t) = g\}.$$

We have the following diagram (see [45, Definition 2.2])

$$\begin{array}{ccc}
[Y_\alpha/P_\alpha] \xrightarrow{\cong} [S_\alpha/G] \xhookrightarrow{q_\alpha} [(Y \setminus \cup_{\alpha' < \alpha} S_{\alpha'})/G] \\
\downarrow{\scriptstyle p_\alpha} \qquad\qquad \searrow^{i_\alpha} \qquad\qquad \downarrow{\scriptstyle w} \\
[Z_\alpha/G_\alpha] \xrightarrow{\quad w|_{Z_\alpha}\quad} \mathbb{C}.
\end{array} \tag{6.1.4}$$

Here the left arrow is induced by morphisms $Y_\alpha \to Z_\alpha$ and $P_\alpha \to G_\alpha$ given by taking $t \to 0$ limit of the action of $\lambda_\alpha(t)$ for $t \in \mathbb{C}^*$, and i_α, q_α are induced by the embedding $Z_\alpha \hookrightarrow Y$, $S_\alpha \hookrightarrow Y$ respectively. Let $\eta_\alpha \in \mathbb{Z}$ be defined by

$$\eta_\alpha := \mathrm{wt}_{\lambda_\alpha}(\det(N^\vee_{S_\alpha/Y}|_{Z_\alpha})). \tag{6.1.5}$$

As in Sect. 2.2.1, let $\mathbb{C}^*$ acts on Y which commutes with the G-action, and $w\colon Y \to \mathbb{C}^*$ is a G-invariant function with $\mathbb{C}^*$-weight two. We will use the following version of window theorem.

Theorem 6.1.2 ([11, 45]) *For each α, we take $k_\alpha \in \mathbb{R}$.*

(i) For each $j \in \mathbb{Z}$, the composition

$$q_{\alpha*}p_\alpha^*\colon \mathrm{MF}^{\mathbb{C}^*}_{\mathrm{coh}}([Z_\alpha/G_\alpha], w|_{Z_\alpha})_{\lambda_\alpha\text{-wt}=j} \to \mathrm{MF}^{\mathbb{C}^*}_{\mathrm{coh}}([S_\alpha/G], w|_{S_\alpha})$$

$$\to \mathrm{MF}^{\mathbb{C}^*}_{\mathrm{coh}}([(Y \setminus \cup_{\alpha'<\alpha} S_{\alpha'})/G], w)$$

is fully-faithful, whose essential image is denoted by $\Upsilon^{l,\alpha}_j$.

(ii) There exist semiorthogonal decomposition

$$\mathrm{MF}^{\mathbb{C}^*}_{\mathrm{coh}}([(Y \setminus \cup_{\alpha'<\alpha} S_{\alpha'})/G], w)$$

$$= \langle \ldots, \Upsilon^{l,\alpha}_{\lceil k_\alpha \rceil - 2}, \Upsilon^{l,\alpha}_{\lceil k_\alpha \rceil - 1}, \mathcal{W}^{l,\alpha}_{k_\alpha}, \Upsilon^{l,\alpha}_{\lceil k_\alpha \rceil}, \Upsilon^{l,\alpha}_{\lceil k_\alpha \rceil + 1}, \ldots \rangle.$$

Here $\mathcal{W}^{l,\alpha}_{k_\alpha}$ consists of factorizations $(\mathcal{P}, d_{\mathcal{P}})$ as in (2.2.2) satisfying that

$$i^*_\alpha(\mathcal{P}, d_{\mathcal{P}}) \in \bigoplus_{j \in [k_\alpha, k_\alpha + \eta_\alpha)} \mathrm{MF}^{\mathbb{C}^*}_{\mathrm{coh}}([Z_\alpha/G_\alpha], w|_{Z_\alpha})_{\lambda_\alpha\text{-wt}=j}. \tag{6.1.6}$$

(iii) The composition functor

$$\mathcal{W}^{l,\alpha}_{k_\alpha} \hookrightarrow \mathrm{MF}^{\mathbb{C}^*}_{\mathrm{coh}}([(Y \setminus \cup_{\alpha'<\alpha} S_{\alpha'})/G], w) \twoheadrightarrow \mathrm{MF}^{\mathbb{C}^*}_{\mathrm{coh}}([(Y \setminus \cup_{\alpha'\leq\alpha} S_{\alpha'})/G], w)$$

is an equivalence

As a consequence of Theorem 6.1.2, we have the following window theorem. Let

$$\mathcal{W}^l_{k_\bullet}([Y/G], w) \subset \mathrm{MF}^{\mathbb{C}^*}_{\mathrm{coh}}([Y/G], w)$$

be the subcategory of objects $(\mathcal{P}, d_{\mathcal{P}})$ satisfying the condition (6.1.6) for all α. Then the composition functor

$$\mathcal{W}^l_{k_\bullet}([Y/G], w) \hookrightarrow \mathrm{MF}^{\mathbb{C}^*}_{\mathrm{coh}}([Y/G], w) \twoheadrightarrow \mathrm{MF}^{\mathbb{C}^*}_{\mathrm{coh}}([Y^{l\text{-ss}}/G], w)$$

is an equivalence. The above window subcategory depends on a choice of $k_\bullet$. For $\delta \in \mathrm{Pic}([Y/G])_\mathbb{R}$, we often use the following special choice

$$\mathcal{W}_\delta^l([Y/G], w) := \mathcal{W}_{k_\bullet}^l([Y/G], w), \ k_\alpha = -\frac{1}{2}\eta_\alpha + \mathrm{wt}_{\lambda_\alpha}(\delta). \tag{6.1.7}$$

We will apply Theorem 6.1.2 for a KN stratification of $\mathrm{Crit}(w)$

$$\mathrm{Crit}(w) = S_1' \sqcup S_2' \sqcup \cdots \sqcup S_N' \sqcup \mathrm{Crit}(w)^{l\text{-ss}}$$

in the following way. After discarding KN strata $S_\alpha \subset Y$ with $\mathrm{Crit}(w) \cap S_\alpha = \emptyset$, the above filtration is obtained by restricting a KN filtration (6.1.2) for Y to $\mathrm{Crit}(w)$. Let $\lambda_\alpha : \mathbb{C}^* \to G$ be a one parameter subgroup for S_α' with center $Z_\alpha' \subset S_\alpha'$. We define $\overline{Z}_\alpha \subset Y$ to be the connected component of the λ_α-fixed part of Y which contains Z_α', and $\overline{Y}_\alpha \subset Y$ is the set of point $y \in Y$ with $\lim_{t\to 0}\lambda_\alpha(t)y \in \overline{Z}_\alpha$. Similarly to (6.1.4), we have the diagram

$$
\begin{array}{ccc}
[\overline{Y}_\alpha/P_\alpha] & \xrightarrow{\ \overline{q}_\alpha\ } & [Y/G] \\
{\scriptstyle \overline{p}_\alpha}\downarrow & \ \nearrow{\scriptstyle \overline{i}_\alpha} & \\
[\overline{Z}_\alpha/G_\alpha]. &
\end{array}
\tag{6.1.8}
$$

By noting the equivalence (2.2.6) together with $\mathrm{Crit}(w) \cap \overline{Z}_\alpha = \mathrm{Crit}(w|_{\overline{Z}_\alpha})$, the result of Theorem 6.1.2 implies the semiorthogonal decomposition

$$\mathrm{MF}_{\mathrm{coh}}^{\mathbb{C}^*}([(Y \setminus \cup_{\alpha' < \alpha} S_{\alpha'}')/G], w)$$

$$= \langle \ldots, \Upsilon_{\lceil k_\alpha\rceil-2}^{l,\alpha}, \Upsilon_{\lceil k_\alpha\rceil-1}^{l,\alpha}, \mathcal{W}_{k_\alpha}^{l,\alpha}, \Upsilon_{\lceil k_\alpha\rceil}^{l,\alpha}, \Upsilon_{\lceil k_\alpha\rceil+1}^{l,\alpha}, \ldots \rangle$$

with equivalences

$$\overline{q}_{\alpha*}\overline{p}_\alpha^* : \mathrm{MF}_{\mathrm{coh}}^{\mathbb{C}^*}([(\overline{Z}_\alpha \setminus \cup_{\alpha' < \alpha} S_{\alpha'}')/G_\alpha], w|_{\overline{Z}_\alpha})_{\lambda_\alpha\text{-wt}=j} \xrightarrow{\sim} \Upsilon_j^{l,\alpha}. \tag{6.1.9}$$

The subcategory $\mathcal{W}_{k_\alpha}^{l,\alpha}$ consists of factorizations $(\mathcal{P}, d_\mathcal{P})$ such that

$$\overline{i}_\alpha^*(\mathcal{P}, d_\mathcal{P})|_{[(\overline{Z}_\alpha\setminus\cup_{\alpha'<\alpha}S_{\alpha'}')/G_\alpha]}$$

$$\in \bigoplus_{j\in[k_\alpha, k_\alpha+\overline{\eta}_\alpha)} \mathrm{MF}_{\mathrm{coh}}^{\mathbb{C}^*}([(\overline{Z}_\alpha \setminus \cup_{\alpha'<\alpha}S_{\alpha'}')/G_\alpha], w|_{\overline{Z}_\alpha})_{\lambda_\alpha\text{-wt}=j}$$

where $\overline{\eta}_\alpha = \mathrm{wt}_{\lambda_\alpha}\det(\mathbb{L}_{\overline{q}_\alpha})^\vee|_{\overline{Z}_\alpha}$, and the composition functor

$$\mathcal{W}_{k_\alpha}^{l,\alpha} \hookrightarrow \mathrm{MF}_{\mathrm{coh}}^{\mathbb{C}^*}([(Y \setminus \cup_{\alpha'<\alpha}S_{\alpha'}')/G], w) \twoheadrightarrow \mathrm{MF}_{\mathrm{coh}}^{\mathbb{C}^*}([(Y \setminus \cup_{\alpha'\le\alpha}S_{\alpha'}')/G], w)$$

is an equivalence.

6.1.3 Magic Window Subcategories

Suppose that Y is an affine space, i.e. $Y = \mathbb{A}^n$ for some n, and it is a G-representation. Let us take a decomposition into a direct sum of G-representations

$$Y = \mathbb{S} \oplus \mathbb{U} \tag{6.1.10}$$

such that $\mathbb{S}$ is a symmetric, i.e. $\mathbb{S} \cong \mathbb{S}^\vee$ as G-representations. We call such a decomposition $Y = \mathbb{S} \oplus \mathbb{U}$ as a *symmetric structure* of Y, and refer to it as $\mathbb{S}$.

For each one parameter subgroup $\lambda \colon \mathbb{C}^* \to T$ and an element $\mathbb{L} \in K_0(\mathrm{Rep}(T)) = \mathbb{Z}[M]$, we define $\mathbb{L}^{\lambda > 0}$ to be the projection of this class onto the subspace spanned by weights which pair positively with λ. We define $\nabla_{\mathbb{S}} \subset M_{\mathbb{R}}$ to be

$$\nabla_{\mathbb{S}} := \left\{ \chi \in M_{\mathbb{R}} : \langle \chi, \lambda \rangle \in \left[-\frac{1}{2} \left\langle \mathbb{L}^{\lambda > 0}_{[\mathbb{S}/G]}|_0, \lambda \right\rangle, \frac{1}{2} \left\langle \mathbb{L}^{\lambda > 0}_{[\mathbb{S}/G]}|_0, \lambda \right\rangle \right] \right.$$

$$\text{for all } \lambda \colon \mathbb{C}^* \to T \right\}.$$

Here $\mathbb{L}_{[\mathbb{S}/G]}|_0$ is the cotangent complex of the quotient stack $[\mathbb{S}/G]$ restricted to the origin, whose K-theory class is

$$\mathbb{L}_{[\mathbb{S}/G]}|_0 = [\mathbb{S}^\vee] - [\mathfrak{g}^\vee] \in K_0(\mathrm{Rep}(T)).$$

Here $\mathfrak{g}$ is the Lie algebra of G, and regarded as a T-representation by the adjoint representation. If $\lambda = \lambda_\alpha$ for the KN stratification (6.1.2), then we have the (in)equalities

$$\left\langle \mathbb{L}^{\lambda > 0}_{[\mathbb{S}/G]}|_0, \lambda \right\rangle \leq \left\langle \mathbb{L}^{\lambda > 0}_{[Y/G]}|_0, \lambda \right\rangle = \eta_\alpha. \tag{6.1.11}$$

Here η_α is defined in (6.1.5), and the second identity is due to [45, Equation (4)].

We introduce some conditions for elements in $M_{\mathbb{R}}^W$. Let $\overline{\Sigma}_{\mathbb{S}} \subset M_{\mathbb{R}}$ be the convex hull of the T-characters of $\bigwedge^*(\mathbb{S})$.

Definition 6.1.3 For $l, \delta \in M_{\mathbb{R}}^W$, we say

(i) l is $\mathbb{S}$-*generic* if it is contained in the linear span of $\overline{\Sigma}_{\mathbb{S}}$ but is not parallel to any face of $\overline{\Sigma}_{\mathbb{S}}$.

(ii) δ is l-*generic* if $\langle \delta, \lambda \rangle \notin \mathbb{Q}$ for any one parameter subgroup $\lambda \colon \mathbb{C}^* \to T$ such that $\langle l, \lambda \rangle \neq 0$.

(iii) l is compatible with the symmetric structure $\mathbb{S}$ in (6.1.10) if we have

$$Y^{l\text{-ss}} = \mathbb{S}^{l\text{-ss}} \oplus \mathbb{U}.$$

The genericity condition for (i) is introduced in [50, Section 2]. As for (ii), for example if $l \in M_{\mathbb{Q}}^W$ then $\delta = \varepsilon \cdot l$ is l-generic for any $\varepsilon \in \mathbb{R} \setminus \mathbb{Q}$. Note that if δ is l-generic, then $\langle \delta, \lambda_\alpha \rangle \notin \mathbb{Q}$ for any one parameter subgroup λ_α which appears in a KN stratification (6.1.2). Later we will use the following lemma:

Lemma 6.1.4 *For $l_1, l_2 \in M_{\mathbb{R}}^W$, there is an uncountable dense subset $U \subset \mathbb{R}^2$ such that for $(\alpha_1, \alpha_2) \in U$, $\alpha_1 l_1 + \alpha_2 l_2$ is l_1-generic and l_2-generic.*

Proof For a one parameter subgroup $\lambda \colon \mathbb{C}^* \to T$ such that either $\langle l_1, \lambda \rangle \neq 0$ or $\langle l_2, \lambda \rangle \neq 0$, the set of $(\alpha_1, \alpha_2) \in \mathbb{R}^2$ such that $\alpha_1 \langle l_1, \lambda \rangle + \alpha_2 \langle l_2, \lambda \rangle \notin \mathbb{Q}$ is a complement of countable number of lines in $\mathbb{R}^2$, so it is uncountable and dense. As such one parameter subgroups λ are also countable many, we obtain the lemma.   □

The magic window subcategory is introduced in [50] in order to give a stability independent description of window subcategories. Here we define its variant as follows:

Definition 6.1.5 For $\delta \in M_{\mathbb{R}}^W$, the subcategory (called *magic window over* $\mathbb{S}$)

$$\mathcal{W}_\delta^{\mathrm{mag}/\mathbb{S}}([Y/G], w) \subset \mathrm{MF}_{\mathrm{coh}}^{\mathbb{C}^*}([Y/G], w) \quad.$$

is defined to be split generated by factorizations $(\mathcal{P}, d_{\mathcal{P}})$ as in (2.2.2), such that $\mathcal{P}$ is isomorphic to $W \otimes_{\mathbb{C}} \mathcal{O}_Y$ for a $(G \times \mathbb{C}^*)$-representation W whose $(T \times \{1\})$-weights are contained in $\delta + \nabla_{\mathbb{S}}$.

The following result is proved in the proof of [78, Proposition 2.6], which is itself based on magic window theorem in [50, 122]. Its formal fiber version will be proved in Proposition 6.1.7 by the same argument.

Proposition 6.1.6 ([78, Proposition 2.6]) *Let us take $l, \delta \in M_{\mathbb{R}}^W$ such that δ is l-generic. Then we have the inclusion*

$$\mathcal{W}_\delta^{\mathrm{mag}/\mathbb{S}}([Y/G], w) \subset \mathcal{W}_\delta^l([Y/G], w)$$

which is identity if l is $\mathbb{S}$-generic and compatible with $\mathbb{S}$.

6.1.4   Window Subcategories for Formal Completions

Let Y be a smooth affine scheme with an action of a reductive algebraic group G, and $V \to Y$ a G-equivariant vector bundle. In this book, we often work with the formal fiber of the map

$$[V/G] \to [Y/G] \to Y /\!\!/ G.$$

We will use the following notation. For $y \in Y /\!\!/ G$, we denote by $\widehat{Y}_y /\!\!/ G$ be the formal completion of $Y /\!\!/ G$ at y,

$$\widehat{Y}_y /\!\!/ G := \operatorname{Spec} \widehat{\mathcal{O}}_{Y /\!\!/ G, y}.$$

Then by setting

$$\widehat{Y}_y := Y \times_{Y /\!\!/ G} \widehat{Y}_y /\!\!/ G, \ \ \widehat{V}_y := V \times_{Y /\!\!/ G} \widehat{Y}_y /\!\!/ G,$$

we have the Cartesian squares

$$
\begin{array}{ccccc}
[\widehat{V}_y / G] & \longrightarrow & [\widehat{Y}_y / G] & \longrightarrow & \widehat{Y}_y /\!\!/ G \\
\downarrow & \square & \downarrow & \square & \downarrow \\
[V / G] & \longrightarrow & [Y / G] & \longrightarrow & Y /\!\!/ G.
\end{array}
\tag{6.1.12}
$$

The KN stratification of V with respect to the above G-action on V and $l \in M_{\mathbb{R}}^W$ is restricted to a stratification on $\widehat{V}_y$

$$\widehat{V}_y = \widehat{S}_1 \sqcup \widehat{S}_2 \sqcup \cdots \sqcup \widehat{S}_N \sqcup \widehat{V}_y^{l\text{-ss}}. \tag{6.1.13}$$

Suppose furthermore that Y is a finite dimensional G-representation, and $V \to Y$ a G-equivariant vector bundle. Then the total space of V is a direct sum of G-representations $V|_0 \oplus Y$ and the projection $V \to Y$ is identified with the second projection. Below we take a symmetric structure $\mathbb{S}$ of V,

$$V|_0 \oplus Y = \mathbb{S} \oplus \mathbb{U} \tag{6.1.14}$$

and prove a version of Theorem 6.1.6 for the formal fiber $[\widehat{V}_y / G]$ at $y = 0 \in Y /\!\!/ G$. Let $\mathbb{C}^*$ acts on the fibers of $V \to Y$ and $\widehat{w}_0 : [\widehat{V}_0 / G] \to \mathbb{C}$ a function of $\mathbb{C}^*$-weight two. Then for each $l, \delta \in M_{\mathbb{R}}^W$, the window subcategory

$$\mathcal{W}_\delta^l([\widehat{V}_0 / G], \widehat{w}_0) \subset \mathrm{MF}_{\mathrm{coh}}^{\mathbb{C}^*}([\widehat{V}_0 / G], \widehat{w}_0) \tag{6.1.15}$$

is defined in the same way as (6.1.7), using the induced stratification (6.1.13). The magic window subcategory

$$\mathcal{W}_\delta^{\mathrm{mag}/\mathbb{S}}([\widehat{V}_0 / G], \widehat{w}_0) \subset \mathrm{MF}_{\mathrm{coh}}^{\mathbb{C}^*}([\widehat{V}_0 / G], \widehat{w}_0)$$

is also defined to be split generated by $(\mathcal{P}, d_{\mathcal{P}})$ where $\mathcal{P}$ is isomorphic to $W \otimes_{\mathbb{C}} \mathcal{O}_{\widehat{V}_0}$ for a $(G \times \mathbb{C}^*)$-representation W whose $(T \times \{1\})$-weights are contained in $\delta + \nabla_{\mathbb{S}}$. We will use the following variant of Theorem 6.1.6:

Proposition 6.1.7 *Let us take* $l, \delta \in M_{\mathbb{R}}^W$ *such that* δ *is* l-*generic. Then we have the inclusion*

$$\mathcal{W}_\delta^{\mathrm{mag}/\mathbb{S}}([\widehat{V}_0/G], \widehat{w}_0) \subset \mathcal{W}_\delta^l([\widehat{V}_0/G], \widehat{w}_0)$$

which is identity if l *is* $\mathbb{S}$-*generic and compatible with* $\mathbb{S}$.

Proof The proof is almost same as Proposition 6.1.6, given in [78, Proposition 2.6]. Let $\lambda_\alpha \colon \mathbb{C}^* \to T$ be a one parameter subgroup which appears in a KN stratification of V with respect to the G-action on it, and set η_α as in (6.1.5) for the above KN stratification. By the assumption that δ is l-generic, we have $\langle \delta, \lambda_\alpha \rangle \pm \eta_\alpha/2 \notin \mathbb{Z}$. Together with the (in)equality in (6.1.11)

$$\left\langle \mathbb{L}_{[\mathbb{S}/G]}^{\lambda_\alpha > 0} |_0, \lambda_\alpha \right\rangle \leq \left\langle \mathbb{L}_{[V/G]}^{\lambda_\alpha > 0} |_0, \lambda_\alpha \right\rangle = \eta_\alpha$$

we have the inclusion

$$\mathcal{W}_\delta^{\mathrm{mag}}([\widehat{V}_0/G], \widehat{w}_0) \subset \mathcal{W}_\delta^l([\widehat{V}_0/G], \widehat{w}_0) \tag{6.1.16}$$

by the definition of these window subcategories. The composition

$$\mathcal{W}_\delta^l([\widehat{V}_0/G], \widehat{w}_0) \hookrightarrow \mathrm{MF}_{\mathrm{coh}}^{\mathbb{C}^*}([\widehat{V}_0/G], \widehat{w}_0) \twoheadrightarrow \mathrm{MF}_{\mathrm{coh}}^{\mathbb{C}^*}([\widehat{V}_0^{l\text{-ss}}/G], \widehat{w}_0) \tag{6.1.17}$$

is an equivalence by Theorem 6.1.2. Therefore the composition

$$\mathcal{W}_\delta^{\mathrm{mag}/\mathbb{S}}([\widehat{V}_0/G], \widehat{w}_0) \hookrightarrow \mathrm{MF}_{\mathrm{coh}}^{\mathbb{C}^*}([\widehat{V}_0/G], \widehat{w}_0) \twoheadrightarrow \mathrm{MF}_{\mathrm{coh}}^{\mathbb{C}^*}([\widehat{V}_0^{l\text{-ss}}/G], \widehat{w}_0) \tag{6.1.18}$$

is fully-faithful.

Now suppose that l is $\mathbb{S}$-generic and compatible with $\mathbb{S}$, where $\mathbb{S}$ is the symmetric structure in (6.1.14). Let $\Delta_\mathbb{S}$ be the set of isomorphism classes of finite dimensional $(G \times \mathbb{C}^*)$-representations W whose $(T \times \{1\})$-weights are contained in $\delta + \nabla_\mathbb{S}$. Using the $\mathbb{S}$-genericity of l, it is proved in [50, Proposition 3.11] that any vector bundle $W' \otimes \mathcal{O}_\mathbb{S}$ on $\mathbb{S}$ for a finite dimensional $(G \times \mathbb{C}^*)$-representation W' admits a $(G \times \mathbb{C}^*)$-equivariant resolution by vector bundles $W \otimes \mathcal{O}_\mathbb{S}$ for $W \in \Delta_\mathbb{S}$, modulo sheaves on $\mathbb{S}$ supported on l-unstable locus. By pulling it back to V via the projection $V \to \mathbb{S}$ and using the assumption that l is compatible with $\mathbb{S}$, any vector bundle $W' \otimes \mathcal{O}_V$ for a finite dimensional $(G \times \mathbb{C}^*)$-representation W' admits a $(G \times \mathbb{C}^*)$-equivariant resolution by vector bundles $W \otimes \mathcal{O}_V$ for $W \in \Delta_\mathbb{S}$, modulo sheaves on V supported on l-unstable locus. By restricting it to $\widehat{V}_0$, the same also applies to $W' \otimes \mathcal{O}_{\widehat{V}_0}$. Therefore $D_{\mathrm{coh}}^b([\widehat{V}_0^{l\text{-ss}}/(G \times \mathbb{C}^*)])$ is split generated by vector bundles $W \otimes \mathcal{O}_{\widehat{V}_0^{l\text{-ss}}}$ for $W \in \Delta_\mathbb{S}$. It follows that for any object $(\mathcal{P}, d_\mathcal{P})$ in $\mathrm{MF}_{\mathrm{coh}}^{\mathbb{C}^*}([\widehat{V}_0^{l\text{-ss}}/G], \widehat{w}_0)$, the underlying $(G \times \mathbb{C}^*)$-equivariant sheaf $\mathcal{P}$ is a direct summand of a bounded complex

$\mathcal{W} = W^{\bullet} \otimes \mathcal{O}_{\widehat{V}_0^{l\text{-ss}}}$ for $W^i \in \Delta_{\mathbb{S}}$ in the derived category $D_{\mathrm{coh}}^b([\widehat{V}_0^{l\text{-ss}}/(G \times \mathbb{C}^*)])$. Note that we have

$$\mathrm{Ext}_{[\widehat{V}_0^{l\text{-ss}}/(G \times \mathbb{C}^*)]}^{>0}(W_1 \otimes \mathcal{O}_{\widehat{V}_0^{l\text{-ss}}}, W_2 \otimes \mathcal{O}_{\widehat{V}_0^{l\text{-ss}}})$$

$$\xrightarrow{\cong} \mathrm{Ext}_{[\widehat{V}_0/(G \times \mathbb{C}^*)]}^{>0}(W_1 \otimes \mathcal{O}_{\widehat{V}_0}, W_2 \otimes \mathcal{O}_{\widehat{V}_0}) = 0$$

for $W_1, W_2 \in \Delta_{\mathbb{S}}$. Here the first isomorphism follows since the composition (6.1.17) without super-potential case is also fully-faithful, and the second identity follows since $\widehat{V}_0$ is affine and $G \times \mathbb{C}^*$ is reductive. Therefore using the resolution property of factorizations (see [10, Theorem 3.9], [2, Lemma 4.10]), the differential $d_{\mathcal{P}}$ can be lifted to a differential $d_{\mathcal{W}}$ on the totalization of $\mathcal{W}$ so that $(\mathcal{W}, d_{\mathcal{W}}) \in \mathrm{MF}_{\mathrm{coh}}^{\mathbb{C}^*}([\widehat{V}_0^{l\text{-ss}}/G], \widehat{w}_0)$ and $(\mathcal{P}, d_{\mathcal{P}})$ is a direct summand of $(\mathcal{W}, d_{\mathcal{W}})$. Therefore the composition (6.1.18) is essentially surjective, so the inclusion (6.1.16) is identity.

$\square$

The following is an obvious corollary of Proposition 6.1.7:

Corollary 6.1.8 *In the setting of Proposition 6.1.7, let $l, l', \delta \in M_{\mathbb{R}}^W$ satisfy that δ is l-generic, l'-generic, and l is $\mathbb{S}$-generic and compatible with $\mathbb{S}$. Then we have the inclusion*

$$\mathcal{W}_{\delta}^{l}([\widehat{V}_0/G], \widehat{w}_0) \subset \mathcal{W}_{\delta}^{l'}([\widehat{V}_0/G], \widehat{w}_0)$$

which is identity if l' is also $\mathbb{S}$-generic and compatible with $\mathbb{S}$.

6.1.5 KN Stratifications for Some Representations of Quivers

We will use the following example of KN stratifications for moduli spaces of representations of quivers. Let Q be a quiver

$$Q = (V(Q), E(Q), s, t), \quad s, t \colon E(Q) \to V(Q)$$

where $V(Q)$ is the set of vertices, $E(Q)$ is the set of edges and s, t are the maps which correspond to the source and target of the edge. Its dual quiver $Q^{\vee}$ is defined by

$$Q^{\vee} = (V(Q), E(Q), s^{\vee}, t^{\vee}), \quad s^{\vee} = t, \quad t^{\vee} = s.$$

For a quiver Q, we denote by $E_{i,j}$ the set of $e \in E(Q)$ such that $s(e) = i$ and $t(e) = j$. A quiver Q is called *symmetric* if $\sharp E_{i,j} = \sharp E_{j,i}$ for all $i, j \in V(Q)$.

Let $\mathrm{Rep}(Q)$ be the category of finite dimensional Q-representations. We have the map taking the dimension vectors

$$\mathbf{dim}\colon K(\mathrm{Rep}(Q)) \to \Gamma_Q := \bigoplus_{i \in V(Q)} \mathbb{Z} \cdot \mathbf{e}_i.$$

For $\vec{v} = (v_i)_{i \in V(Q)} \in \Gamma_Q$, we set $|\vec{v}| := \sum_i v_i$.

Let us fix $0 \in V(Q)$ and take the full subquiver $Q_0 \subset Q$ whose vertex set $V(Q_0)$ is $V(Q) \setminus \{0\}$. For each $i \in V(Q_0)$, let V_i be a finite dimensional vector space with dimension v_i. For $\vec{v} = (v_i)_{i \in V(Q_0)}$, we set

$$R^{\dagger}(Q_0, \vec{v}) := \bigoplus_{e \in E(Q_0)} \mathrm{Hom}(V_{s(e)}, V_{t(e)}) \oplus \bigoplus_{0 \to i} V_i \oplus \bigoplus_{i \to 0} V_i^{\vee} \oplus \mathbb{C}^{\sharp(0 \to 0)}.$$

The algebraic group $G = \prod_{i \in V(Q_0)} \mathrm{GL}(V_i)$ acts on $R^{\dagger}(Q_0, \vec{v})$ by the conjugation. The quotient stack

$$\mathcal{M}^{\dagger}(Q_0, \vec{v}) = \left[R^{\dagger}(Q_0, \vec{v})/G \right]$$

is the $\mathbb{C}^*$-rigidified moduli stack of Q-representations of dimension vector $(1, \vec{v})$, where 1 is the dimension vector at 0.

Let us take $\theta = (\theta_i)_{i \in V(Q_0)} \in \mathbb{Z}^{V(Q_0)}$ and set

$$\chi_{\theta}\colon G \to \mathbb{C}^*, \quad (g_i) \mapsto \prod_{i \in V(Q_0)} \det(g_i)^{\theta_i}.$$

We also set $\chi_0 := \chi_{\theta_i = 1}$, i.e. $\chi_0((g_i)_{i \in V(Q_0)}) = \prod_{i \in V(Q_0)} \det(g_i)$. Let us fix basis of V_i as $\{e_i^{(j)} : 1 \le j \le v_i\}$ and take the maximal torus

$$T = \prod_{i \in V(Q_0)} \prod_{j=1}^{v_i} \mathbb{C}^* \cdot e_i^{(j)\vee} \otimes e_i^{(j)} \subset G.$$

For a one parameter subgroup $\lambda\colon \mathbb{C}^* \to T$ given by $t \mapsto (t^{\lambda_i^{(j)}})$, we set

$$|\lambda| := \sqrt{\sum_{i \in V(Q_0)} \sum_{j=1}^{v_i} (\lambda_i^{(j)})^2} \tag{6.1.19}$$

which determines a Weyl-invariant norm on $N_{\mathbb{R}}$.

We describe the KN stratifications of $R^\dagger(Q_0, \vec{v})$ with respect to $\chi_0^\pm$ and the norm (6.1.19). Let us take a point

$$x = ((A_e)_{e \in E(Q_0)}, (b_i)_{i \in V(Q_0)}, (c_i)_{i \in V(Q_0)}, u) \in R^\dagger(Q_0, \vec{v}), \qquad (6.1.20)$$

where each elements are

$$A_e \in \mathrm{Hom}(V_{s(e)}, V_{t(e)}), \ b_i \in V_i, \ c_i \in V_i^\vee, \ u \in \mathbb{C}^{\sharp(0 \to 0)}.$$

Then $(A_e)_{e \in E(Q_0)}$ determine the $\mathbb{C}[Q_0]$-module structure on $\oplus_{i \in V(Q_0)} V_i$, and $(A_e^\vee)_{e \in E(Q_0)}$ determine the $\mathbb{C}[Q_0^\vee]$-module structure on $\oplus_{i \in V(Q_0)} V_i^\vee$. We define the following sub vector spaces

$$M_x^+ := \mathbb{C}[Q_0]\langle b_i : i \in V(Q_0)\rangle \subset \oplus_{i \in V(Q_0)} V_i,$$
$$M_x^- := \mathbb{C}[Q_0^\vee]\langle c_i : i \in V(Q_0)\rangle \subset \oplus_{i \in V(Q_0)} V_i^\vee.$$

Namely M_x^+ is the subspace generated by b_i as $\mathbb{C}[Q_0]$-module, and M_x^- is the subspace generated by c_i as $\mathbb{C}[Q_0^\vee]$-module. For $\vec{w} = (w_i)_{i \in V(Q_0)}$ with $w_i \in \mathbb{Z}_{\geq 0}$, we set

$$S_{\vec{w}}^\pm := \{x \in R^\dagger(Q_0, \vec{v}) : \mathbf{dim}(M_x^\pm) = \vec{w}\}, \ S_k^\pm := \coprod_{|\vec{w}|=k} S_{\vec{w}}^\pm.$$

Lemma 6.1.9 *The stratifications*

$$R^\dagger(Q_0, \vec{v}) = S_0^\pm \sqcup S_1^\pm \sqcup \cdots \sqcup S_{|\vec{v}|-1}^\pm \sqcup S_{|\vec{v}|}^\pm$$

are KN stratifications of $R^\dagger(Q_0, \vec{v})$ with respect to $\chi_0^{\pm 1}$ and the norm (6.1.19), where $S_{|\vec{v}|}^\pm$ are semistable loci $R^\dagger(Q_0, \vec{v})^{\chi_0^{\pm 1}\text{-ss}}$. The corresponding one parameter subgroup $\lambda_{\vec{w}}^\pm$ for $S_{\vec{w}}^\pm$ with $|\vec{w}| < |\vec{v}|$ can be taken as

$$\lambda_{\vec{w}}^\pm \colon \mathbb{C}^* \to T, \ t \mapsto (t^{\lambda_i^{(j)}}), \ \lambda_i^{(j)} = \begin{cases} 0, & 1 \leq j \leq w_i, \\ \mp 1, & w_i < j \leq v_i. \end{cases}$$

Proof We only prove the case of χ_0. For a one parameter subgroup $\lambda \colon t \mapsto t^{(\lambda_i^{(j)})}$, the slope in (6.1.3) is given by

$$\mu_\lambda := -\frac{\sum_{i,j} \lambda_i^{(j)}}{\sqrt{\sum_{i,j} (\lambda_i^{(j)})^2}}.$$

Suppose by induction that stratas $S_0^+ \sqcup \cdots \sqcup S_{k-1}^+$ are chosen and let $\lambda \colon t \mapsto t^{(\lambda_i^{(j)})}$ be a one parameter subgroup for next strata. For $i \in Q_0$, we set w_i to be the number of $1 \le j \le v_i$ such that $\lambda_i^{(j)} = 0$. Then for $\vec{w} = (w_i)_{i \in V(Q_0)}$, if $|\vec{w}| < k$ then the λ-fixed locus is contained in $S_0^+ \sqcup \cdots \sqcup S_{k-1}^+$. Therefore $|\vec{w}| \ge k$. Among such λ, the slope μ_λ takes the maximal value $\sqrt{|\vec{v}| - k}$ only if $|\vec{w}| = k$ and $\lambda_i^{(j)} = c$ for some constant $c < 0$ if $\lambda_i^{(j)} \ne 0$. We have the decomposition $V_i = W_i \oplus W_i^\perp$, where

$$W_i = \bigoplus_{\lambda_i^{(j)} = 0} \mathbb{C} \cdot e_i^{(j)}, \ W_i^\perp = \bigoplus_{\lambda_i^{(j)} \ne 0} \mathbb{C} \cdot e_i^{(j)}.$$

Then the λ-fixed locus consists of x as in (6.1.20) such that $b_i \in W_i$, $c_i \in W_i^\vee$ and $A_{ij} \in \mathrm{Hom}(W_i, W_j) \oplus \mathrm{Hom}(W_i^\perp, W_j^\perp)$. The attracting locus is then $b_i \in W_i$ and $A_{ij}(W_i) \subset W_i$. By applying the G-action, we obtain the strata $S_{\vec{w}}^+$. $\square$

We also have the following lemma:

Lemma 6.1.10 For $\theta = (\theta_i)_{i \in V(Q_0)} \in \mathbb{Z}^{V(Q_0)}$ with $\theta_i > 0$ for all i, we have

$$R^\dagger(Q_0, \vec{v})^{\chi_\theta^\pm \text{-ss}} = R^\dagger(Q_0, \vec{v})^{\chi_0^\pm \text{-ss}}.$$

Proof From the correspondence between GIT stability and θ-stability proved in [71, Theorem 4.1], the χ_θ-semistable locus in $R^\dagger(Q_0, \vec{v})$ corresponds to Z_θ-semistable representations, where $(\mathrm{Rep}(Q), Z_\theta)$ are Bridgeland stability conditions [24] with central charges given by

$$Z_\theta \colon \Gamma_Q \to \mathbb{C}, \ \mathbf{e}_i \mapsto \theta_i + \sqrt{-1} \ (i \ne 0), \mathbf{e}_0 \mapsto \theta_0 + \sqrt{-1}.$$

Here θ_0 is determined by $\theta_0 + \sum_{i \in V(Q_0)} \theta_i v_i = 0$. Namely a Q-representation R is Z_θ-semistable if and only if for any subrepresentation $R' \subset R$, we have $\arg Z_\theta(R') \le \arg Z_\theta(R)$ in $(0, \pi]$. Then by the condition $\theta_i > 0$, a Q-representation R of dimension vector $(1, \vec{v})$ is Z_θ-semistable if and only if there is no non-trivial surjection $R \twoheadrightarrow R''$ where R'' has dimension vector $(0, *)$. Indeed if there is such a surjection, then $\arg Z_\theta(R'') < \arg Z_\theta(R)$ which destabilizes R. Conversely if there is no such a surjection, then any subobject $R' \subset R$ has dimension vector $(0, *)$ so $\arg Z_\theta(R') < \arg Z_\theta(R)$ holds. In other words, a Q-representation of dimension vector $(1, \vec{v})$ is Z_θ-semistable if and only if it corresponds to a point x as in (6.1.20) so that $M_x^+ = \oplus_{i \in V(Q_0)} V_i$. This is independent of a choice of $\theta_i > 0$, so the stability for θ is equivalent to the stability for $\theta_i = 1$. The case of $\chi_\theta^{-1} = \chi_{-\theta}$-stability is similarly discussed. $\square$

Remark 6.1.11 Although the lemma implies that semistable locus does not depend on $\theta_i > 0$, the KN stratification depends on θ. The KN stratifications in Lemma 6.1.9 are nothing but Harder-Narasimhan filtrations with respect to Z_θ-stability for $\theta_i = \pm 1$.

Suppose that we have the following condition

$$\sharp E_{i,j} - \sharp E_{j,i} \begin{cases} = 0, & i, j \in V(Q_0), \\ \geq 0, & i = 0, j \in V(Q_0). \end{cases} \tag{6.1.21}$$

In particular, Q_0 is symmetric. We fix subsets $E'_{0,i} \subset E_{0,i}$ for $i \in V(Q_0)$ such that $\sharp E'_{0,i} = \sharp E_{i,0}$. Then we have the symmetric structure $R^\dagger(Q_0, \vec{v}) = \mathbb{S} \oplus \mathbb{U}$, where $\mathbb{S}$ and $\mathbb{U}$ are given by

$$\mathbb{S} = \bigoplus_{e \in E(Q_0)} \mathrm{Hom}(V_{s(e)}, V_{t(e)}) \oplus \bigoplus_{(0 \to i) \in E'_{0,i}} V_i \oplus \bigoplus_{(i \to 0) \in E_{i,0}} V_i^\vee \oplus \mathbb{C}^{\sharp(0 \to 0)},$$

$$\tag{6.1.22}$$

$$\mathbb{U} = \bigoplus_{(0 \to i) \in E_{0,i} \setminus E'_{0,i}} V_i.$$

Lemma 6.1.12 *For* $\theta = (\theta_i)_{i \in V(Q_0)} \in \mathbb{Z}^{V(Q_0)}$ *with* $\theta_i > 0$, *the* G-*characters* $\chi_\theta^\pm$ *are* $\mathbb{S}$-*generic, and* χ_θ^- *is compatible with* $\mathbb{S}$.

Proof Let $Q' \subset Q$ be a symmetric subquiver such that $V(Q') = V(Q)$, and the edges of Q' are the same as Q except that the set of edges from 0 to i is $E'_{0,i}$. Then Lemma 6.1.15 below shows that χ_θ is $\mathbb{S}$-generic, where we apply it for the quiver Q' with dimension vector $(1, \vec{v})$ and $\theta_0 = -\sum_{i \in V(Q_0)} \theta_i v_i$. Therefore χ_θ^{-1} is also $\mathbb{S}$-generic.

By Lemmas 6.1.9 and 6.1.10, the χ_θ^{-1}-stability on $R^\dagger(Q_0, \vec{v})$ imposes constraints only on $V_i^\vee$ and $\mathrm{End}(V_i)$-factors, and does not impose any constraint on V_i-factors. The same also applies to the χ_θ^{-1}-stability on $\mathbb{S}$. Since $\mathbb{S}$ is obtained by extracting some of V_i-factors from $R^\dagger(Q_0, \vec{v})$, the condition

$$R^\dagger(Q_0, \vec{v})^{\chi_\theta^{-1}\text{-ss}} = \mathbb{S}^{\chi_\theta^{-1}\text{-ss}} \oplus \mathbb{U}$$

is satisfied. Therefore χ_θ^{-1} is compatible with $\mathbb{S}$. $\square$

Let $\mathbb{C}^*$ acts on $R^\dagger(Q_0, \vec{v})$ which commutes with the G-action, and $w: R^\dagger(Q_0, \vec{v}) \to \mathbb{C}$ be a G-invariant function with $\mathbb{C}^*$-weight two. As an application of Proposition 6.1.6, we have the following:

Proposition 6.1.13 *For* $\theta = (\theta_i)_{i \in V(Q_0)} \in \mathbb{Z}^{V(Q_0)}$ *with* $\theta_i > 0$ *and* $\delta \in M_\mathbb{R}^W$ *which is* χ_θ-*generic (e.g.* $\delta = \varepsilon \cdot \chi_\theta$ *with* $\varepsilon \in \mathbb{R} \setminus \mathbb{Q}$*), we have the inclusion in* $\mathrm{MF}_{\mathrm{coh}}^{\mathbb{C}^*}(\mathcal{M}^\dagger(Q_0, \vec{v}), w)$

$$\mathcal{W}_\delta^{\chi_\theta^{-1}}(\mathcal{M}^\dagger(Q_0, \vec{v}), w) \subset \mathcal{W}_\delta^{\chi_\theta}(\mathcal{M}^\dagger(Q_0, \vec{v}), w). \tag{6.1.23}$$

In particular, we have a fully-faithful functor

$$\mathrm{MF}_{\mathrm{coh}}^{\mathbb{C}^*}(\mathcal{M}^\dagger(Q_0, \vec{v})^{\chi_\theta^{-1}\text{-ss}}, w) \hookrightarrow \mathrm{MF}_{\mathrm{coh}}^{\mathbb{C}^*}(\mathcal{M}^\dagger(Q_0, \vec{v})^{\chi_\theta\text{-ss}}, w).$$

Proof By Lemma 6.1.12, we can apply Proposition 6.1.6 (also see Corollary 6.1.8) and obtain the inclusion (6.1.23). The last assertion follows from Theorem 6.1.2.
□

Remark 6.1.14 The $\mathbb{C}^*$-action on $R^\dagger(Q_0^\dagger, \vec{v})$ is not essential in the above arguments, so we also have $\mathbb{Z}/2$-periodic version of Proposition 6.1.13

$$\mathcal{W}_\delta^{\chi_\theta^{-1}, \mathbb{Z}/2}(\mathcal{M}^\dagger(Q_0, \vec{v}), w) \subset \mathcal{W}_\delta^{\chi_\theta, \mathbb{Z}/2}(\mathcal{M}^\dagger(Q_0, \vec{v}), w).$$

Here both sides are window subcategories in $\mathrm{MF}_{\mathrm{coh}}^{\mathbb{Z}/2}(\mathcal{M}^\dagger(Q_0, \vec{v}), w)$ defined by the same condition as (6.1.6) replacing $\mathbb{C}^*$ with $\mathbb{Z}/2$. Moreover as in Proposition 6.1.7, we also have the formal fiber version

$$\mathcal{W}_\delta^{\chi_\theta^{-1}, \mathbb{Z}/2}(\widehat{\mathcal{M}}^\dagger(Q_0, \vec{v})_y, w) \subset \mathcal{W}_\delta^{\chi_\theta, \mathbb{Z}/2}(\widehat{\mathcal{M}}^\dagger(Q_0, \vec{v})_y, w).$$

Here both sides are window subcategories in $\mathrm{MF}_{\mathrm{coh}}^{\mathbb{Z}/2}(\widehat{\mathcal{M}}^\dagger(Q_0, \vec{v})_y, w)$ for $y \in R^\dagger(Q_0, \vec{v}) /\!\!/ G$ defined by the same condition (6.1.6).

We have used the following lemma.

Lemma 6.1.15 *Let Q be a symmetric quiver whose vertex set, edge set, are denoted by $V(Q)$, $E(Q)$, respectively. Let $\vec{v} = (v_i)_{i \in V(Q)}$ be a dimension vector of Q, and $\{\theta_i\}_{i \in V(Q)}$ with $\theta_i \in \mathbb{Z}$ satisfy that*

$$\sum_{i \in V(Q)} \theta_i \cdot v_i = 0, \quad \sum_{i \in V(Q)} \theta_i \cdot v_i' \neq 0 \tag{6.1.24}$$

for any $\vec{v}' = (v_i')_{i \in V(Q)}$ such that $0 < \vec{v}' < \vec{v}$. Here $0 < \vec{v}'$ means $v_i' \geq 0$ for any $i \in V(Q)$ and $\vec{v}' \neq 0$, and $\vec{v}' < \vec{v}$ means $\vec{v} - \vec{v}' > 0$. Let V_i for $i \in V(Q)$ be vector spaces with dimension v_i. Then for $G = \prod_{i \in V(Q)} \mathrm{GL}(V_i)$ and $G' = G/\mathbb{C}^*$, where $\mathbb{C}^* \subset G$ is the diagonal torus, the G'-character

$$\chi_\theta : G' \to \mathbb{C}^*, \quad (g_i)_{i \in V(Q)} \mapsto \prod_{i \in V(Q)} \det(g_i)^{\theta_i}$$

is $\mathbb{S}$-generic with respect to the symmetric G'-representation

$$\mathbb{S} = \bigoplus_{e \in E(Q)} \mathrm{Hom}(V_{s(e)}, V_{t(e)}). \tag{6.1.25}$$

Here G' acts on (6.1.25) by conjugation.

Proof The proof is a slight modification of [78, Lemma 3.3]. The maximal torus of G' is given by $T' = T/\mathbb{C}^*$ where $T = \prod_{i\in V(Q)} T_i$ for the maximal torus $T_i \subset GL(V_i)$, and the character lattice M' of M is given by the kernel of $M \to \mathbb{Z}$ dual to the diagonal embedding $\mathbb{C}^* \to T$. Let $\{e_i^{(j)} : 1 \leq j \leq v_i\}$ be a basis of V_i. By fixing $i_0 \in V(Q)$ and $1 \leq k_0 \leq v_{i_0}$, we can write $M_{\mathbb{R}}'$ as

$$M_{\mathbb{R}}' = \bigoplus_{i\in V(Q)} \bigoplus_{1\leq k\leq v_i} \mathbb{R}(e_i^{(k)} - e_{i_0}^{(k_0)}).$$

Let $\gamma_1, \ldots, \gamma_d \in M'$ be the T'-weights of the G'-representation (6.1.25). Then any non-zero T'-character γ_j is of the form $e_i^{(k)} - e_{i'}^{(k')}$. By Halpern-Leistner and Sam [50, Proposition 2.1], the genericity of χ_θ is equivalent to that for any proper subspace $H \subsetneq M_{\mathbb{R}}'$, there is a one parameter subgroup $\lambda: \mathbb{C}^* \to T'$ such that $\langle \gamma_j, \lambda \rangle = 0$ for any $\gamma_j \in H$ and $\langle \chi_\theta, \lambda \rangle \neq 0$. Given H as above, we set $\lambda: \mathbb{C}^* \to T'$ be to be

$$\lambda(t) = (t^{\lambda_i^{(k)}})_{i\in V(Q),1\leq k\leq v_i}, \quad \lambda_i^{(k)} = \begin{cases} 0, & \text{if } e_i^{(k)} - e_{i_0}^{(k_0)} \in H, \\ 1, & \text{if } e_i^{(k)} - e_{i_0}^{(k_0)} \notin H. \end{cases}$$

Then $\langle \gamma_j, \lambda \rangle = 0$ for any $\gamma_j \in H$. As $\chi_\theta = \sum_{i,k} \theta_i \cdot (e_i^{(k)} - e_{i_0}^{(k_0)})$, we have

$$\langle \chi_\theta, \lambda \rangle = \sum_{i\in V(Q)} \theta_i \cdot \sharp\{1 \leq k \leq v_i : e_i^{(k)} - e_{i_0}^{(k_0)} \notin H\} \neq 0.$$

Here the latter inequality follows from (6.1.24). Therefore χ_θ is $\mathbb{S}$-generic. $\square$

6.2 Intrinsic Window Subcategories

6.2.1 Symmetric Structures of Derived Stacks

Let $\mathfrak{M}$ be a quasi-smooth and QCA derived stack such that $\mathcal{M} = t_0(\mathfrak{M})$ admits a good moduli space $\pi_{\mathcal{M}}: \mathcal{M} \to M$. Recall that for any closed point $x \in \mathcal{M}$, its automorphism group $\mathrm{Aut}(x)$ is a reductive algebraic group by [4, Proposition 12.14]. We introduce the notion of symmetric structures for derived stacks.

Definition 6.2.1 A symmetric structure $\mathbb{S}$ of $\mathfrak{M}$ is a choice of symmetric structures of $\mathcal{H}^0(\mathbb{T}_{\mathfrak{M}}|_x) \oplus \mathcal{H}^1(\mathbb{T}_{\mathfrak{M}}|_x)^\vee$ at each closed point $x \in \mathcal{M}$, i.e. a direct sum of $\mathrm{Aut}(x)$-representations

$$\mathcal{H}^0(\mathbb{T}_{\mathfrak{M}}|_x) \oplus \mathcal{H}^1(\mathbb{T}_{\mathfrak{M}}|_x)^\vee = \mathbb{S}_x \oplus \mathbb{U}_x \qquad (6.2.1)$$

such that $\mathbb{S}_x$ is a symmetric $\mathrm{Aut}(x)$-representation.

A derived stack $\mathfrak{M}$ is called symmetric if $\mathcal{H}^0(\mathbb{T}_{\mathfrak{M}}|_x) \oplus \mathcal{H}^1(\mathbb{T}_{\mathfrak{M}}|_x)^\vee$ is a symmetric $\mathrm{Aut}(x)$-representation for any closed point $x \in \mathcal{M}$. In this case, we have a symmetric structure (6.2.1) such that $\mathbb{U}_x = 0$, which we call a maximal symmetric structure.

Recall that we defined G-equivariant tuple (Y, V, s) in Definition 2.3.1. In this chapter, we always assume that G is reductive. We will use the following lemma on symmetric structures.

Lemma 6.2.2 *Let (Y, V, s) be a G-equivariant tuple for a reductive G, and $\mathfrak{M} = [\mathfrak{U}/G]$ the associated derived stack as in Definition 2.3.1. Suppose that a symmetric structure of $\mathfrak{M}$ is given as in (6.2.1). Let $x \in [\mathcal{U}/G]$ be a closed point, and $G_x \subset G$ the stabilizer subgroup of x. Then there is a symmetric structure of the G_x-representation $T_x Y \oplus V|_x^\vee$ of the form*

$$T_x Y \oplus V|_x^\vee \cong (\mathbb{S}_x \oplus P_x \oplus P_x^\vee \oplus \mathfrak{g}/\mathfrak{g}_x) \oplus \mathbb{U}_x$$

for some G_x-representation P_x. Here $\mathfrak{g}$, $\mathfrak{g}_x$ are the Lie algebras of G, G_x, and $\mathbb{S}_x \oplus P \oplus P^\vee \oplus \mathfrak{g}/\mathfrak{g}_x$ is the symmetric part.

Proof The tangent complex of $\mathfrak{M}$ at x is given by

$$\mathbb{T}_{\mathfrak{M}}|_x = (\mathfrak{g} \to T_x Y \overset{ds|_x}{\to} V|_x).$$

Here the kernel of the first map is $\mathfrak{g}_x$. Since G_x is reductive, we have an isomorphism of complexes of G_x-representations

$$(T_x Y \overset{ds|_x}{\to} V|_x) \cong (\mathrm{Ker}\, ds|_x \overset{0}{\to} \mathrm{Cok}\, ds|_x) \oplus (P_x \overset{\mathrm{id}}{\to} P_x)$$

for some G_x-representation P_x. We also have a splitting $\mathrm{Ker}\, ds|_x \cong \mathcal{H}^0(\mathbb{T}_{\mathfrak{M}}|_x) \oplus \mathfrak{g}/\mathfrak{g}_x$ as G_x-representations. Then we have

$$T_x Y \oplus V|_x^\vee \cong \mathcal{H}^0(\mathbb{T}_{\mathfrak{M}}|_x) \oplus \mathcal{H}^1(\mathbb{T}_{\mathfrak{M}}|_x)^\vee \oplus P_x \oplus P_x^\vee \oplus \mathfrak{g}/\mathfrak{g}_x.$$

Therefore the lemma holds.   □

We next introduce the notion of 'formal neighborhood theorem' for quasi-smooth derived stacks with good moduli spaces.

Definition 6.2.3 We say that $\mathfrak{M}$ satisfies formal neighborhood theorem if for any closed point $x \in \mathcal{M}$ with $y = \pi_{\mathcal{M}}(x) \in M$, there exists an $\mathrm{Aut}(x)$-equivariant morphism

$$\kappa : \widehat{\mathcal{H}^0(\mathbb{T}_{\mathfrak{M}}|_x)}_0 \to \mathcal{H}^1(\mathbb{T}_{\mathfrak{M}}|_x) \tag{6.2.2}$$

with $\kappa(0) = 0$ such that, by setting $\widehat{\mathcal{N}}_0 \hookrightarrow \widehat{\mathcal{H}^0(\mathbb{T}_{\mathfrak{M}}|_x)}_0$ to be the classical zero locus of κ, we have commutative isomorphisms

$$
\begin{array}{ccc}
[\widehat{\mathcal{N}}_0/\operatorname{Aut}(x)] & \xrightarrow{\;\cong\;} & \mathcal{M} \times_M \operatorname{Spec} \widehat{\mathcal{O}}_{M,y} \\
\downarrow & & \downarrow \\
\widehat{\mathcal{N}}_0 /\!/ \operatorname{Aut}(x) & \xrightarrow{\;\cong\;} & \operatorname{Spec} \widehat{\mathcal{O}}_{M,y}.
\end{array}
\qquad (6.2.3)
$$

Here the top isomorphism sends 0 to x, and identity on stabilizer groups at these points.

Remark 6.2.4 Similarly to (6.1.12), the scheme $\widehat{\mathcal{H}^0(\mathbb{T}_{\mathfrak{M}}|_x)}_0$ is defined to be the formal fiber of the morphism

$$
\mathcal{H}^0(\mathbb{T}_{\mathfrak{M}}|_x) \to \mathcal{H}^0(\mathbb{T}_{\mathfrak{M}}|_x) /\!/ \operatorname{Aut}(x)
$$

at the origin. This should not be confused with the formal completion of $\mathcal{H}^0(\mathbb{T}_{\mathfrak{M}}|_x)$ at the origin.

The following lemma is proved along with the similar argument of Proposition 8.3.2, so we omit its proof.

Lemma 6.2.5 *Suppose that $\mathfrak{M}$ satisfies formal neighborhood theorem. Then the diagram (6.2.3) can be extended to a Cartesian diagram*

$$
\begin{array}{ccc}
[\widehat{\mathcal{N}}_0/\operatorname{Aut}(x)] & \longrightarrow & \mathcal{M} \\
\big\uparrow & \square & \big\uparrow \\
[\widehat{\mathfrak{N}}_0/\operatorname{Aut}(x)] & \longrightarrow & \mathfrak{M}.
\end{array}
$$

Here the vertical arrows are closed immersions given by taking the classical truncations and $\widehat{\mathfrak{N}}_0$ is the derived zero locus of (6.2.2).

We also introduce some conditions for $\mathbb{R}$-line bundles on the classical stack $\mathcal{M} = t_0(\mathfrak{M})$. For a closed point $x \in \mathcal{M}$, we denote by $\mu_x \colon B\operatorname{Aut}(x) \to \mathcal{M}$ the map sending a point to x and identity on the automorphism groups. For $l \in \operatorname{Pic}(\mathcal{M})_{\mathbb{R}}$, we set $l_x := \mu_x^* l \in \operatorname{Pic}(B\operatorname{Aut}(x))_{\mathbb{R}}$. Recall the conditions for the characters of reductive algebraic groups in Definition 6.1.3.

Definition 6.2.6 For $l, \delta \in \operatorname{Pic}(\mathcal{M})_{\mathbb{R}}$, we say

(i) l is $\mathbb{S}$-generic for a symmetric structure $\mathbb{S}$ given in (6.2.1) if for any closed point $x \in \mathcal{M}$, the $\mathbb{R}$-line bundle l_x on $B\operatorname{Aut}(x)$ is $\mathbb{S}_x$-generic.
(ii) δ is l-generic if δ_x is l_x-generic for any closed point $x \in \mathcal{M}$.
(iii) l is compatible with the symmetric structure $\mathbb{S}$ in (6.2.1) if l_x is compatible with the symmetric structure (6.2.1) for any closed point $x \in \mathcal{M}$.

For an $\mathbb{R}$-line bundle $l \in \mathrm{Pic}(\mathcal{M})_{\mathbb{R}}$, we have the open substack of l-semistable locus

$$\mathcal{N}^{l\text{-ss}} \subset \mathcal{N} = t_0(\Omega_{\mathfrak{M}}[-1]).$$

Here we have regarded l as an $\mathbb{R}$-line bundle on $\mathcal{N}$ by the pull-back for the projection $\mathcal{N} \to \mathcal{M}$. We now state the main result in this chapter:

Theorem 6.2.7 *Let* $\mathfrak{M}$ *be a quasi-smooth and QCA derived stack with a good moduli space* $M = t_0(\mathfrak{M}) \to M$. *Suppose that a symmetric structure* $\mathbb{S}$ *of* $\mathfrak{M}$ *is given as (6.2.1), and* $\mathfrak{M}$ *satisfies formal neighborhood theorem. Let us take* $l, \delta \in \mathrm{Pic}(\mathcal{M})_{\mathbb{R}}$ *such that* δ *is* l-generic. *Then there exists a triangulated subcategory* $\mathcal{W}^{\mathrm{int}/\mathbb{S}}_{\delta}(\mathfrak{M}) \subset D^b_{\mathrm{coh}}(\mathfrak{M})$ *such that the composition*

$$\Theta_l : \mathcal{W}^{\mathrm{int}/\mathbb{S}}_{\delta}(\mathfrak{M}) \hookrightarrow D^b_{\mathrm{coh}}(\mathfrak{M}) \to \mathcal{DT}^{\mathbb{C}^*}(\mathcal{N}^{l\text{-ss}})$$

is fully-faithful, which is an equivalence if l *is* $\mathbb{S}$-generic *and compatible with* $\mathbb{S}$.

Proof The proof will be given in Sect. 6.3.4 (see Theorem 6.3.13).   □

We have the following corollary of the above theorem.

Corollary 6.2.8 *Under the assumption of Theorem 6.2.7, let us take* $l_1, l_2 \in \mathrm{Pic}(\mathcal{M})_{\mathbb{R}}$ *such that* l_1 *is compatible with the symmetric structure* $\mathbb{S}$. *Then there exists a fully-faithful functor*

$$\Theta_{l_1, l_2} : \mathcal{DT}^{\mathbb{C}^*}(\mathcal{N}^{l_1\text{-ss}}) \hookrightarrow \mathcal{DT}^{\mathbb{C}^*}(\mathcal{N}^{l_2\text{-ss}})$$

which is an equivalence if l_2 *is also compatible with* $\mathbb{S}$. *In particular* Θ_{l_1, l_2} *is an equivalence if* $\mathfrak{M}$ *is symmetric and* l_i *are generic with respect to the maximal symmetric structure.*

Proof By Lemma 6.1.4, for each closed point $x \in \mathcal{M}$ there is an uncountable many $(\varepsilon_1, \varepsilon_2) \in \mathbb{R}^2$ such that for $\delta = \varepsilon_1 l_1 + \varepsilon_2 l_2$, δ_x is $l_{1,x}$-generic and $l_{2,x}$-generic. Since the set of isomorphism classes of reductive algebraic groups $\mathrm{Aut}(x)$ together with their representations $\mathcal{H}^0(T_{\mathfrak{M}}|_x) \oplus \mathcal{H}^1(T_{\mathfrak{M}}|_x)^{\vee}$ and their symmetric structures is at most countable many, we can take $(\varepsilon_1, \varepsilon_2)$ to be independent of x. Then the desired fully-faithful functor Θ_{l_1, l_2} is given by $\Theta_{l_2} \circ \Theta_{l_1}^{-1}$ for the above choice of δ, which is an equivalence if l_2 is also compatible with $\mathbb{S}$. In the case that $\mathfrak{M}$ is symmetric, then each l_i is compatible with the maximal symmetric structure, so Θ_{l_1, l_2} is an equivalence.   □

Remark 6.2.9 In the situation of Corollary 6.2.8, let $\mathfrak{M}' \subset \mathfrak{M}$ be an open immersion with $\mathcal{M}' = t_0(\mathfrak{M}')$, $\mathcal{N}' = t_0(\Omega_{\mathfrak{M}'}[-1])$. We set $\mathcal{N}'^i := \mathcal{N}^{l_i\text{-ss}} \times_{\mathcal{M}} \mathcal{M}'$. If $\mathcal{N}'^1 = \mathcal{N}'^2$, then we have the commutative diagram

$$\begin{array}{ccc} \mathcal{DT}^{\mathbb{C}^*}(\mathcal{N}^{l_1\text{-ss}}) & \xrightarrow{\Theta_{l_1,l_2}} & \mathcal{DT}^{\mathbb{C}^*}(\mathcal{N}^{l_2\text{-ss}}) \\ \downarrow & & \downarrow \\ \mathcal{DT}^{\mathbb{C}^*}(\mathcal{N}'^1) & \xrightarrow{\;\sim\;} & \mathcal{DT}^{\mathbb{C}^*}(\mathcal{N}'^2). \end{array} \qquad (6.2.4)$$

Here the vertical arrows are restriction functors, and the bottom arrow is a natural equivalence given by $\mathcal{N}'^1 = \mathcal{N}'^2$. The commutative diagram (6.2.4) follows since the compositions

$$\mathcal{W}_\delta^{\text{int}/\mathbb{S}}(\mathfrak{M}) \hookrightarrow D^b_{\text{coh}}(\mathfrak{M}) \to \mathcal{DT}^{\mathbb{C}^*}(\mathcal{N}^{l_i\text{-ss}}) \to \mathcal{DT}^{\mathbb{C}^*}(\mathcal{N}'^i)$$

for $i = 1, 2$ are identified if $\mathcal{N}'^1 = \mathcal{N}'^2$.

6.2.2 Intrinsic Window Subcategories

In this subsection, we define the intrinsic window subcategory of $D^b_{\text{coh}}([\mathfrak{U}/G])$ for a derived stack $[\mathfrak{U}/G]$ as in Definition 2.3.1 in terms of weight conditions for objects in $D^b_{\text{coh}}([\mathfrak{U}/G])$ under the push-forward to $[Y/G]$.

We first prepare some notation. Let $\mathfrak{M}$ be a quasi-smooth derived stack such that $\mathcal{M} = t_0(\mathfrak{M})$ admits a good moduli space $\mathcal{M} \to M$. Assume that $\mathfrak{M}$ satisfies the formal neighborhood theorem (see Definition 6.2.3) and fix its symmetric structure (6.2.1). For each map $\lambda \colon B\mathbb{C}^* \to \mathcal{M}$, it induces the map of good moduli spaces $\operatorname{Spec} \mathbb{C} \to M$ whose image is denoted by $y(\lambda) \in M$. We define $x(\lambda) \in \mathcal{M}$ to be the unique closed point of the fiber of $\pi_{\mathcal{M}} \colon \mathcal{M} \to M$ at $y(\lambda)$. Then by the diagram (6.2.3), λ factors through the map

$$\lambda \colon B\mathbb{C}^* \to [\widehat{\mathcal{N}}_0/\operatorname{Aut}(x(\lambda))].$$

Note that any $\operatorname{Aut}(x(\lambda))$-representation induces a vector bundle on $[\widehat{\mathcal{N}}_0/\operatorname{Aut}(x(\lambda))]$, so in particular $\mathbb{U}_{x(\lambda)}$ in the decomposition (6.2.1) is regarded as a vector bundle on $[\widehat{\mathcal{N}}_0/\operatorname{Aut}(x(\lambda))]$. Under the above preparation, we introduce the following definition:

Definition 6.2.10 We define $\mu_\lambda^\pm \in \mathbb{Z}$ to be

$$\mu_\lambda^+ := \operatorname{wt} \det((\lambda^*\mathbb{U}_{x(\lambda)})^{\text{wt}>0}), \quad \mu_\lambda^- := \operatorname{wt} \det((\lambda^*\mathbb{U}_{x(\lambda)})^{\text{wt}<0}).$$

Remark 6.2.11 The integers $\mu_\lambda^\pm$ are independent of choices of isomorphisms in (6.2.3) by Lemma 6.3.6.

Let (Y, V, s) be a G-equivariant tuple for a reductive G, and consider the associated derived stack $[\mathfrak{U}/G]$. Note that we have the following diagram

$$
\begin{array}{ccc}
\mathcal{V} & =\!=\!= & [V/G] \\
& & \Big\downarrow \Big)_s \\
[\mathfrak{U}/G] \overset{j}{\hookrightarrow} \mathcal{Y} & =\!=\!= & [Y/G]
\end{array}
\tag{6.2.5}
$$

such that $[\mathfrak{U}/G]$ is the derived zero locus of s. Below we also fix a symmetric structure $\mathbb{S}$ of $[\mathfrak{U}/G]$, i.e. decompositions

$$
\mathcal{H}^0(\mathbb{T}_{[\mathfrak{U}/G]}|_x) \oplus \mathcal{H}^1(\mathbb{T}_{[\mathfrak{U}/G]}|_x)^\vee = \mathbb{S}_x \oplus \mathbb{U}_x
\tag{6.2.6}
$$

of $\mathrm{Aut}(x)$-representations for each closed point $x \in [\mathcal{U}/G]$ such that $\mathbb{S}_x$ is symmetric. The intrinsic window subcategory with respect to the above symmetric structure is given as follows.

Definition 6.2.12 For an element $\delta \in \mathrm{Pic}([\mathcal{U}/G])_{\mathbb{R}}$, we define

$$
\mathcal{W}_\delta^{\mathrm{int}/\mathbb{S}}([\mathfrak{U}/G]) \subset D_{\mathrm{coh}}^b([\mathfrak{U}/G])
\tag{6.2.7}
$$

to be the triangulated subcategory consisting of $\mathcal{E} \in D_{\mathrm{coh}}^b([\mathfrak{U}/G])$ such that for any morphism $\lambda \colon B\mathbb{C}^* \to [\mathcal{U}/G]$ the set of weights $\mathrm{wt}(\lambda^* j_* \mathcal{E})$ is contained in

$$
\mathrm{wt}(\lambda^*\delta) + \left[\frac{1}{2}\mathrm{wt}\det((\lambda^*\mathbb{L}_\mathcal{V}|_\mathcal{Y})^{\mathrm{wt}<0}) - \frac{1}{2}\mu_\lambda^-, \ \frac{1}{2}\mathrm{wt}\det((\lambda^*\mathbb{L}_\mathcal{V}|_\mathcal{Y})^{\mathrm{wt}>0}) - \frac{1}{2}\mu_\lambda^+ \right].
\tag{6.2.8}
$$

Here $\mathbb{L}_\mathcal{V}|_\mathcal{Y}$ is the cotangent complex on $\mathcal{V}$ restricted to the zero section of $\mathcal{V} \to \mathcal{Y}$, and we have used the same notation λ for the composition $B\mathbb{C}^* \overset{\lambda}{\to} [\mathcal{U}/G] \overset{j}{\hookrightarrow} \mathcal{Y}$.

In what follows, for an equivalence of derived stacks $\mathbf{f} \colon [\mathfrak{U}/G] \overset{\sim}{\to} [\mathfrak{U}'/G']$, we always take the symmetric structure $\mathbb{S}'$ of $[\mathfrak{U}'/G']$ induced by the symmetric structure (6.2.6) and isomorphisms $\mathbf{f}^*\mathcal{H}^i(\mathbb{T}_{[\mathfrak{U}'/G']}) \overset{\cong}{\to} \mathcal{H}^i(\mathbb{T}_{[\mathfrak{U}/G]})$. We have the following lemma.

Lemma 6.2.13 *Suppose that we have a commutative diagram*

$$
\begin{array}{ccccccc}
[\mathcal{U}/G] & \hookrightarrow & [\mathfrak{U}/G] & \overset{j}{\hookrightarrow} & [Y/G] & \overset{s}{\underset{\longleftarrow}{\longrightarrow}} & [V/G] \\
f \downarrow \cong & & \mathbf{f} \downarrow \sim & & f \downarrow \cong & & g \downarrow \cong \\
[\mathcal{U}'/G'] & \hookrightarrow & [\mathfrak{U}'/G'] & \overset{j'}{\hookrightarrow} & [Y'/G'] & \underset{\underset{s'}{\longrightarrow}}{\longleftarrow} & [V'/G'].
\end{array}
$$

Here (Y', V', s') is a G'-equivariant tuple for a reductive G', and $[\mathfrak{U}'/G']$ is the associated derived stack in Definition 2.3.1. The right vertical arrow is an isomorphism of vector bundles, the equivalence $\mathbf{f}$ is induced by the right commutative isomorphisms, and the left vertical arrow is the induced isomorphism on classical truncations.

For $\delta' \in \mathrm{Pic}([\mathcal{U}'/G'])_{\mathbb{R}}$, we take $\delta = f^*\delta' \in \mathrm{Pic}([\mathcal{U}/G])_{\mathbb{R}}$. Then we have the equivalences

$$\mathbf{f}_*\colon \mathcal{W}_\delta^{\mathrm{int}/\mathbb{S}}([\mathfrak{U}/G]) \xrightarrow{\sim} \mathcal{W}_{\delta'}^{\mathrm{int}/\mathbb{S}'}([\mathfrak{U}'/G']), \quad \mathbf{f}^*\colon \mathcal{W}_{\delta'}^{\mathrm{int}/\mathbb{S}'}([\mathfrak{U}'/G']) \xrightarrow{\sim} \mathcal{W}_\delta^{\mathrm{int}/\mathbb{S}}([\mathfrak{U}/G]).$$

Proof The lemma is obvious since the category $\mathcal{W}_\delta^{\mathrm{int}/\mathbb{S}}([\mathfrak{U}/G])$ is defined in terms of intrinsic properties of isomorphism classes of the diagram of stacks (6.2.5). $\quad\square$

Below we show that the category $\mathcal{W}_\delta^{\mathrm{int}/\mathbb{S}}([\mathfrak{U}/G])$ is also independent of possibly non-isomorphic presentation (6.2.5) for a fixed G. Let (Y', V', s') be another G-equivariant tuple, and suppose that we have a morphism of G-equivariant tuples (see Definition 2.4.1)

$$
s\left(\begin{array}{ccc} V & \xrightarrow{g} & V' \\ \uparrow\downarrow & & \downarrow\uparrow \\ Y & \xrightarrow{f} & Y', \end{array}\right)s'
\qquad
s\left(\begin{array}{ccc} \mathcal{V} & \xrightarrow{g} & \mathcal{V}' \\ \downarrow\uparrow & & \downarrow\uparrow \\ \mathcal{Y} & \xrightarrow{f} & \mathcal{Y}'. \end{array}\right)s'
\tag{6.2.9}
$$

Here the right diagram is obtained from the left one by taking the quotients by G, where $\mathcal{Y} = [Y/G]$, $\mathcal{Y}' = [\mathcal{Y}'/G]$, $\mathcal{V} = [V/G]$ and $\mathcal{V}' = [V'/G]$. We assume that the induced morphism of derived stacks is an equivalence

$$\mathbf{f}\colon [\mathfrak{U}/G] \xrightarrow{\sim} [\mathfrak{U}'/G] \tag{6.2.10}$$

where both sides are the associated derived stacks in Definition 2.3.1. As in the diagram (2.4.3), the diagram (6.2.9) also induces the following G-equivariant commutative diagram

$$
\begin{array}{ccccc}
 & & \mathbb{C} & & \\
 & {}^{w}\nearrow & {\uparrow}{\scriptstyle \overline{w}} & \nwarrow{}^{w'} & \\
V^\vee & \xleftarrow{\;g\;} & f^*V'^\vee & \xrightarrow{\;f\;} & V'^\vee \\
{\scriptstyle p}\downarrow & & {\scriptstyle p''}\downarrow \quad \square & & \downarrow{\scriptstyle p'} \\
Y & =\!\!=\!\!= & Y & \xrightarrow{\;f\;} & Y'.
\end{array}
\tag{6.2.11}
$$

Here w' is defined as in (2.1.3) from (Y', s'). We prepare two lemmas.

Lemma 6.2.14 *Suppose that we have a morphism of G-equivariant tuples (6.2.9) which induces an equivalence of derived stacks (6.2.10). Moreover suppose that the morphism f in the diagram (6.2.9) is a closed immersion. Then for an object $\mathcal{E} \in D_{\mathrm{coh}}^b(\mathcal{Y})$ and $\gamma \in \mathbb{R}$, we have the condition*

$$\mathrm{wt}(\lambda^*\mathcal{E}) \subset \gamma + \left[\frac{1}{2}\mathrm{wt}\det((\lambda^*\mathbb{L}_{\mathcal{V}}|_{\mathcal{Y}})^{\mathrm{wt}<0}), \frac{1}{2}\mathrm{wt}\det((\lambda^*\mathbb{L}_{\mathcal{V}}|_{\mathcal{Y}})^{\mathrm{wt}>0})\right]$$

for any $\lambda: B\mathbb{C}^* \to \mathcal{Y}$ if and only if $f_*\mathcal{E} \in D_{\mathrm{coh}}^b(\mathcal{Y}')$ satisfies the condition

$$\mathrm{wt}(\lambda'^* f_*\mathcal{E}) \subset \gamma + \left[\frac{1}{2}\mathrm{wt}\det((\lambda'^*\mathbb{L}_{\mathcal{V}'}|_{\mathcal{Y}'})^{\mathrm{wt}<0}), \frac{1}{2}\mathrm{wt}\det((\lambda'^*\mathbb{L}_{\mathcal{V}'}|_{\mathcal{Y}'})^{\mathrm{wt}>0})\right]$$

for any $\lambda: B\mathbb{C}^* \to \mathcal{Y}$, where $\lambda' = \lambda \circ f: B\mathbb{C}^* \to \mathcal{Y}'$.

Proof Since f is a closed immersion, for $E \in \mathrm{Coh}(\mathcal{Y})$ we have

$$\mathcal{H}^{-k}(f^*f_*E) = \mathcal{T}or_k^{\mathcal{O}_{\mathcal{Y}'}}(f_*E, f_*\mathcal{O}_{\mathcal{Y}}) \cong E \otimes \bigwedge^k N_{\mathcal{Y}/\mathcal{Y}'}^\vee.$$

By the above isomorphism, for any $\mathcal{E} \in D_{\mathrm{coh}}^b(\mathcal{Y})$ the object $f^*f_*\mathcal{E} \in D_{\mathrm{coh}}^b(\mathcal{Y})$ fits into a finite sequence of distinguished triangles

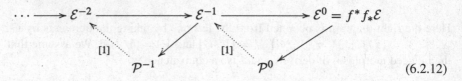

$$(6.2.12)$$

such that $\mathcal{P}^{-k} \cong \mathcal{E} \otimes \bigwedge^k N_{\mathcal{Y}/\mathcal{Y}'}^\vee[k]$. For an object $\mathcal{F} \in D_{\mathrm{coh}}^b(B\mathbb{C}^*)$, we denote by $\mathrm{wt}^{\mathrm{max}}(\mathcal{F}) \in \mathbb{Z}$ (resp. $\mathrm{wt}^{\mathrm{min}}(\mathcal{F}) \in \mathbb{Z}$) the maximal (resp. minimal) $\mathbb{C}^*$-weight of $\mathcal{H}^\bullet(\mathcal{F})$. By the distinguished triangles (6.2.12), for any map $\lambda: B\mathbb{C}^* \to \mathcal{Y}$ we have

$$\mathrm{wt}^{\mathrm{max}}(\lambda'^* f_*\mathcal{E}) = \mathrm{wt}^{\mathrm{max}}(\lambda^*\mathcal{E}) + \mathrm{wt}\det((\lambda^* N_{\mathcal{Y}/\mathcal{Y}'}^\vee)^{\mathrm{wt}>0}). \qquad (6.2.13)$$

Moreover since the diagram (6.2.9) induces an equivalence of derived zero loci (6.2.10), by comparing the cotangent complexes of $[\mathfrak{U}/G]$ and $[\mathfrak{U}'/G]$ we have the identity in $K(B\mathbb{C}^*)$

$$\lambda^*\mathbb{L}_{\mathcal{Y}} - \lambda^*\mathcal{V}^\vee = \lambda'^*\mathbb{L}_{\mathcal{Y}'} - \lambda'^*\mathcal{V}'^\vee.$$

Therefore we have the identities in $K(B\mathbb{C}^*)$

$$\lambda'^* L_{\mathcal{V}'}|_{\mathcal{Y}'} = \lambda'^* L_{\mathcal{Y}'} + \lambda'^* \mathcal{V}'^\vee$$

$$= \lambda^* L_{\mathcal{Y}} + \lambda'^* \mathcal{V}'^\vee - \lambda^* \mathcal{V}^\vee + \lambda'^* \mathcal{V}'^\vee$$

$$= \lambda^* (L_{\mathcal{Y}} + \mathcal{V}^\vee) + 2(\lambda^* \mathcal{V}'^\vee - \lambda^* \mathcal{V}^\vee)$$

$$= \lambda^* L_{\mathcal{V}}|_{\mathcal{Y}} + 2\lambda^* N^\vee_{\mathcal{Y}/\mathcal{Y}'}.$$

From (6.2.13), it follows that

$$\mathrm{wt}^{\max}(\lambda'^* f_* \mathcal{E}) = \mathrm{wt}^{\max}(\lambda^* \mathcal{E}) + \frac{1}{2}\mathrm{wt}\det((\lambda'^* L_{\mathcal{V}'}|_{\mathcal{Y}'})^{\mathrm{wt}>0})$$

$$- \frac{1}{2}\mathrm{wt}\det((\lambda^* L_{\mathcal{V}}|_{\mathcal{Y}})^{\mathrm{wt}>0}).$$

Similarly we have

$$\mathrm{wt}^{\min}(\lambda'^* f_* \mathcal{E}) = \mathrm{wt}^{\min}(\lambda^* \mathcal{E}) + \frac{1}{2}\mathrm{wt}\det((\lambda'^* L_{\mathcal{V}'}|_{\mathcal{Y}'})^{\mathrm{wt}<0})$$

$$- \frac{1}{2}\mathrm{wt}\det((\lambda^* L_{\mathcal{V}}|_{\mathcal{Y}})^{\mathrm{wt}<0}).$$

The lemma follows from the above two identities. $\square$

Later we will reduce some statements to the case that f is a closed immersion using the following lemma:

Lemma 6.2.15 *Suppose that we have a morphism of G-equivariant tuples (6.2.9) which induces an equivalence of derived stacks (6.2.10). Then there exists another G-equivariant tuple (Y'', V'', s'') and morphisms of G-equivariant tuples*

$$
\begin{array}{ccc}
V & \xrightarrow{g''} & V'' \\
s \Big\uparrow & & \Big\uparrow s'' \\
Y & \xrightarrow{f''} & Y'',
\end{array}
\qquad
\begin{array}{ccc}
V' & \xrightarrow{g'} & V'' \\
s' \Big\uparrow & & \Big\uparrow s'' \\
Y' & \xrightarrow{f'} & Y'',
\end{array}
\qquad (6.2.14)
$$

satisfying the followings:

(i) *The diagrams (6.2.14) induce equivalences of derived stacks*

$$\mathbf{f}'' : [\mathfrak{U}/G] \overset{\sim}{\to} [\mathfrak{U}''/G], \ \mathbf{f}' : [\mathfrak{U}'/G] \overset{\sim}{\to} [\mathfrak{U}''/G]$$

which commute with the equivalence (6.2.10) in the ∞-category of derived stacks, i.e. $\mathbf{f}' \circ \mathbf{f} \sim \mathbf{f}''$. Here $[\mathfrak{U}''/G]$ is the derived stack associated with G-equivariant tuple (Y'', V'', s'').

(ii) *The morphisms f', f'' are closed immersions.*

Proof The proof will be given in Sect. 8.3.4.   □

The following proposition shows that the intrinsic window subcategories are independent of presentations as derived critical loci for a fixed G.

Proposition 6.2.16 *Suppose that we have a commutative diagram*

$$
\begin{array}{ccccccc}
BG & \longleftarrow & [\mathfrak{U}/G] & \overset{j}{\longhookrightarrow} & [Y/G] & \overset{s}{\underset{}{\longleftarrow}} & [V/G] \\
{\scriptstyle h}\downarrow{\scriptstyle \cong} & & {\scriptstyle f}\downarrow{\scriptstyle \sim} & & & & \\
BG & \longleftarrow & [\mathfrak{U}'/G] & \overset{j'}{\longhookrightarrow} & [Y'/G] & \underset{s'}{\longleftarrow} & [V'/G]
\end{array}
$$

(6.2.15)

where $\mathbf{f}$ *is an equivalence of derived stacks and* $[\mathfrak{U}/G]$, $[\mathfrak{U}'/G]$ *are derived zero loci of* s, s'*, and the left horizontal arrows are given by canonical G-torsors* $\mathfrak{U} \to [\mathfrak{U}/G]$, $\mathfrak{U}' \to [\mathfrak{U}'/G]$*. Then by setting* $\delta = \mathbf{f}^*\delta'$ *for* $\delta' \in \mathrm{Pic}([\mathcal{U}'/G])_{\mathbb{R}}$*, we have the equivalences*

$$
\mathbf{f}_*: \mathcal{W}_\delta^{\mathrm{int}/\mathbb{S}}([\mathfrak{U}/G]) \overset{\sim}{\to} \mathcal{W}_{\delta'}^{\mathrm{int}/\mathbb{S}'}([\mathfrak{U}'/G]), \ \mathbf{f}^*: \mathcal{W}_{\delta'}^{\mathrm{int}/\mathbb{S}'}([\mathfrak{U}'/G]) \overset{\sim}{\to} \mathcal{W}_\delta^{\mathrm{int}/\mathbb{S}}([\mathfrak{U}/G]).
$$

(6.2.16)

Proof We first remark that since $\mathbf{f}$ is an equivalence, the functors

$$
\mathbf{f}_*: D^b_{\mathrm{coh}}([\mathfrak{U}/G]) \to D^b_{\mathrm{coh}}([\mathfrak{U}'/G]), \ \mathbf{f}^*: D^b_{\mathrm{coh}}([\mathfrak{U}'/G]) \to D^b_{\mathrm{coh}}([\mathfrak{U}/G])
$$

are equivalences which are quasi-inverse each other. So we have the left equivalence in (6.2.16) if and only if we have the right equivalence in (6.2.16). Moreover let

$$
\mathbf{f}: [\mathfrak{U}/G] \overset{\mathbf{f}'}{\to} [\mathfrak{U}''/G] \overset{\mathbf{f}''}{\to} [\mathfrak{U}'/G]
$$

be a factorization of $\mathbf{f}$, i.e. $\mathbf{f} \sim \mathbf{f}'' \circ \mathbf{f}'$, such that $\mathbf{f}'$, $\mathbf{f}''$ are equivalences of derived stacks. Then if two of three pairs $(\mathbf{f}_*, \mathbf{f}^*)$, $(\mathbf{f}'_*, \mathbf{f}'^*)$, $(\mathbf{f}''_*, \mathbf{f}''^*)$ satisfy the equivalences (6.2.16), then the rest of them also satisfies (6.2.16).

By taking the pull-back of the bottom horizontal diagram in (6.2.15) via $h: BG \to BG$, we have the commutative diagram

$$
\begin{array}{ccccccc}
BG & \longleftarrow & [\mathfrak{U}/G] & \overset{j}{\longhookrightarrow} & [Y/G] & \overset{s}{\underset{}{\longleftarrow}} & [V/G] \\
{\scriptstyle \mathrm{id}}\downarrow{\scriptstyle =} & & {\scriptstyle \mathbf{f}\circ h^{-1}}\downarrow{\scriptstyle \sim} & & & & \\
BG & \longleftarrow & [\mathfrak{U}'/G] & \overset{j'}{\longhookrightarrow} & [Y'/G] & \longleftarrow & [V'/G] \\
{\scriptstyle h}\downarrow{\scriptstyle \cong} & & {\scriptstyle h}\downarrow{\scriptstyle \sim} & & {\scriptstyle h}\downarrow{\scriptstyle \cong}\ {\scriptstyle s'} & & {\scriptstyle h}\downarrow{\scriptstyle \cong} \\
BG & \longleftarrow & [\mathfrak{U}'/G] & \overset{j'}{\longhookrightarrow} & [Y'/G] & \underset{s'}{\longleftarrow} & [V'/G].
\end{array}
$$

Here the G-actions on $\mathfrak{U}'$, Y' and V' in the middle horizontal diagram are twisted by $h \in \mathrm{Aut}(G)$. By the above remark together with Lemma 6.2.13, we may assume that $h = \mathrm{id}$.

By Lemma 6.2.17 below and the first remark in the proof of this proposition, we can assume that $\mathbf{f}$ fits into a commutative diagram

$$
\begin{array}{ccc}
[\mathfrak{U}/G] \overset{j}{\hookrightarrow} [Y/G] & \overset{s}{\longleftarrow} & [V/G] \\
\mathbf{f} \downarrow \sim \qquad f \downarrow & & \downarrow \\
[\mathfrak{U}'/G] \overset{j'}{\hookrightarrow} [Y'/G] & \underset{s'}{\longleftrightarrow} & [V'/G].
\end{array}
$$

Here the right diagram is induced by a morphism of G-equivariant tuples. Then using Lemma 6.2.15 and the above mentioned remark, we may also assume that f is a closed immersion. Then the left equivalence in (6.2.16) follows from Lemma 6.2.14, noting the presentation independence of $\mu_\lambda^{\pm}$. Therefore the proposition holds.   □

We have used the following lemma:

Lemma 6.2.17 *Suppose that $h = \mathrm{id}$ in the diagram (6.2.15). Then there exists a G-equivariant tuple $(\widetilde{Y}, \widetilde{V}, \widetilde{s})$ and morphisms of G-equivariant tuples*

$$
s \left(
\begin{array}{ccc}
V & \longleftarrow & \widetilde{V} \longrightarrow V' \\
\uparrow & \widetilde{s} \uparrow & \downarrow \\
\downarrow & \downarrow & \downarrow \\
Y & \underset{\widetilde{f}}{\longleftarrow} & \widetilde{Y} \underset{\widetilde{f}'}{\longrightarrow} Y'
\end{array}
\right) s'
\tag{6.2.17}
$$

such that, by setting $\widetilde{\mathfrak{U}}$ to be the derived zero locus of $\widetilde{s}$, the above diagram induces equivalences $\widetilde{\mathbf{f}} \colon [\widetilde{\mathfrak{U}}/G] \overset{\sim}{\to} [\mathfrak{U}/G]$, $\widetilde{\mathbf{f}}' \colon [\widetilde{\mathfrak{U}}/G] \overset{\sim}{\to} [\mathfrak{U}'/G]$ which commute with $\mathbf{f}$, i.e. $\mathbf{f} \circ \widetilde{\mathbf{f}} \sim \widetilde{\mathbf{f}}'$.

Proof The proof will be given in Sect. 8.3.5.   □

6.3   Window Theorem for DT Categories

In this section, we compare intrinsic window subcategories with the original window subcategories on derived categories of factorizations under Koszul duality in Theorem 2.3.3, and use it to prove window theorem for DT categories.

6.3.1 Window Subcategories Under Koszul Duality (Linear Case)

Let (Y, V, s) be a G-equivariant tuple for a reductive G, and $[\mathfrak{U}/G]$ is the associated derived stack. In this subsection, we assume that $Y = \mathbb{A}^n$ is a G-representation and $s(0) = 0$. Let $\mathbb{S}$ be a symmetric structure of $[\mathfrak{U}/G]$ as in (6.2.6). So for each closed point $x \in [\mathcal{U}/G]$, we have

$$\mathcal{H}^0(\mathbb{T}_{[\mathfrak{U}/G]}|_x) \oplus \mathcal{H}^1(\mathbb{T}_{[\mathfrak{U}/G]}|_x)^\vee = \mathbb{S}_x \oplus \mathbb{U}_x.$$

Let $G_x \subset G$ be the stabilizer subgroup of x which is also reductive. By writing the total space of $V \to Y$ as a direct sum of G-representations $V|_0 \oplus Y$, by Lemma 6.2.2 we have the induced symmetric structure on the total space of $V^\vee \to Y$ as G_x-representations

$$V|_0^\vee \oplus Y \cong (\mathbb{S}_x \oplus P_x \oplus P_x^\vee \oplus \mathfrak{g}/\mathfrak{g}_x) \oplus \mathbb{U}_x \qquad (6.3.1)$$

for some G_x-representation P_x. We denote by $\widetilde{\mathbb{S}}_x = \mathbb{S}_x \oplus P_x \oplus P_x^\vee \oplus \mathfrak{g}/\mathfrak{g}_x$ its symmetric part.

As in the diagram (6.2.5), we set $\mathcal{Y} = [Y/G]$ and $\mathcal{V} = [V/G]$. For a one parameter subgroup $\lambda\colon \mathbb{C}^* \to T_x := T \cap G_x$, we use the same notation $\lambda\colon B\mathbb{C}^* \to \mathcal{Y}$ for the corresponding map sending a point to x, and the map on stabilizer groups is given by $\lambda\colon \mathbb{C}^* \to T_x$. It factors through $\lambda\colon \mathbb{C}^* \to [\mathcal{U}/G]$, and we have $\mu_\lambda^\pm \in \mathbb{Z}$ as in Definition 6.2.10. We have the following lemma:

Lemma 6.3.1 *For a one parameter subgroup $\lambda\colon \mathbb{C}^* \to T_x$ and the corresponding map $\lambda\colon B\mathbb{C}^* \to \mathcal{Y}$, we have the identities*

$$\operatorname{wt}\det((\lambda^*\mathbb{L}_\mathcal{V}|_\mathcal{Y})^{\operatorname{wt}>0}) - \mu_\lambda^+ = \left\langle \mathbb{L}_{[\widetilde{\mathbb{S}}_x/G]}^{\lambda>0}, \lambda \right\rangle + \operatorname{wt}\lambda^* K_\mathcal{Y}, \qquad (6.3.2)$$

$$\operatorname{wt}\det((\lambda^*\mathbb{L}_\mathcal{V}|_\mathcal{Y})^{\operatorname{wt}<0}) - \mu_\lambda^- = -\left\langle \mathbb{L}_{[\widetilde{\mathbb{S}}_x/G]}^{\lambda>0}, \lambda \right\rangle + \operatorname{wt}\lambda^* K_\mathcal{Y}. \qquad (6.3.3)$$

Here (though $\widetilde{\mathbb{S}}_x$ is not G-representation) we denoted $\mathbb{L}_{[\widetilde{\mathbb{S}}_x/G]}^{\lambda>0} := (\widetilde{\mathbb{S}}_x^\vee)^{\lambda>0} - (\mathfrak{g}^\vee)^{\lambda>0}$ as an element of $K(BT_x)$, and $K_\mathcal{Y} := \det(\mathbb{L}_\mathcal{Y}) \in \operatorname{Pic}(\mathcal{Y})$.

Proof Since $\widetilde{\mathbb{S}}_x$ is symmetric, we have the identity in $K(BG_x)$

$$V|_0^\vee + Y - \mathbb{U}_x = V|_0 + Y^\vee - \mathbb{U}_x^\vee.$$

Therefore for any $\lambda \colon \mathbb{C}^* \to T_x$ we have $\langle Y, \lambda \rangle = \langle V|_0 + \mathbb{U}_x, \lambda \rangle$. Using the above identity, we have the identities

$$
\begin{aligned}
\mathrm{wt}\det((\lambda^* \mathbb{L}_{\mathcal{V}}|_{\mathcal{Y}})^{\mathrm{wt}>0}) &= \langle (V|_0^{\vee})^{\lambda>0} + (Y^{\vee})^{\lambda>0} - (\mathfrak{g}^{\vee})^{\lambda>0}, \lambda \rangle \\
&= \langle V|_0^{\lambda>0} + (Y^{\vee})^{\lambda>0} - (\mathfrak{g}^{\vee})^{\lambda>0}, \lambda \rangle - \langle V|_0, \lambda \rangle \\
&= \langle (\widetilde{\mathbb{S}}_x^{\vee})^{\lambda>0} - (\mathfrak{g}^{\vee})^{\lambda>0}, \lambda \rangle + \langle (\mathbb{U}_x^{\vee})^{\lambda>0}, \lambda \rangle \\
&\quad - \langle Y, \lambda \rangle + \langle \mathbb{U}_x, \lambda \rangle \\
&= \left\langle \mathbb{L}_{[\widetilde{\mathbb{S}}_x/G]}^{\lambda>0}, \lambda \right\rangle + \mathrm{wt}\lambda^* K_{\mathcal{Y}} + \mu_{\lambda}^+.
\end{aligned}
$$

Therefore the identity (6.3.2) holds. The identity (6.3.3) also holds by the same computation.   □

Recall that we have the Koszul duality equivalence

$$
\Phi \colon D^b_{\mathrm{coh}}([\mathfrak{U}/G]) \xrightarrow{\sim} \mathrm{MF}^{\mathbb{C}^*}_{\mathrm{coh}}([V^{\vee}/G], w) \tag{6.3.4}
$$

from Theorem 2.3.3. The symmetric structure (6.3.1) for $x = 0$ gives a symmetric structure $V^{\vee} = \widetilde{\mathbb{S}}_0 \oplus \mathbb{U}_0$ of G-representations, and we have the magic window subcategory over $\widetilde{\mathbb{S}}_0$ in the RHS of (6.3.4) as in Definition 6.1.5. We also have another window subcategory in the RHS of (6.3.4) defined as in (6.1.7). So for $\delta, l \in M^W_{\mathbb{R}}$, we have subcategories (see Proposition 6.1.6)

$$
\mathcal{W}^{\mathrm{mag}/\widetilde{\mathbb{S}}_0}_{\delta}([V^{\vee}/G], w) \subset \mathcal{W}^l_{\delta}([V^{\vee}/G], w) \subset \mathrm{MF}^{\mathbb{C}^*}_{\mathrm{coh}}([V^{\vee}/G], w).
$$

On the other hand, we have subcategories

$$
\mathcal{W}^{\mathrm{int}/\mathbb{S}}_{\delta}([\mathfrak{U}/G]) \subset \mathcal{W}^{\mathrm{int}/\mathbb{S}_0}_{\delta}([\mathfrak{U}/G]) \subset D^b_{\mathrm{coh}}([\mathfrak{U}/G]).
$$

Here the intermediate one is defined by the condition (6.2.8) for all $\lambda \colon B\mathbb{C}^* \to [\mathcal{U}/G]$ which is mapped to 0 by the composition $[\mathcal{U}/G] \to \mathcal{U} /\!\!/ G$. The following proposition gives a comparison of these window subcategories under the equivalence (6.3.4).

Proposition 6.3.2 *We take $l, \delta \in M^W_{\mathbb{R}}$ such that δ is l-generic. Then under the equivalence Φ in (6.3.4), we have*

$$
\Phi(\mathcal{W}^{\mathrm{int}/\mathbb{S}}_{\delta}([\mathfrak{U}/G])) \subset \mathcal{W}^l_{\delta+K_{\mathcal{Y}}/2}([V^{\vee}/G], w), \tag{6.3.5}
$$

$$
\mathcal{W}^{\mathrm{mag}/\widetilde{\mathbb{S}}_0}_{\delta+K_{\mathcal{Y}}/2}([V^{\vee}/G], w) \subset \Phi(\mathcal{W}^{\mathrm{int}/\mathbb{S}_0}_{\delta}([\mathfrak{U}/G])). \tag{6.3.6}
$$

Proof We first show the inclusion (6.3.5). Let $\lambda = \lambda_\alpha \colon \mathbb{C}^* \to T$ be a one parameter subgroup which appears in a KN stratification (6.1.2) for the G-action on $V^{\vee}$ with respect to $l \in M_{\mathbb{R}}^W$, and $(V^\vee)^\lambda \to Y^\lambda$ the restriction of the projection $V \to Y$ to the λ-fixed loci. Let η_α be defined as in (6.1.5) for the G-action on $V^\vee$. By the l-genericity of δ, we have $\langle \delta + K_Y/2, \lambda \rangle \pm \eta_\alpha/2 \notin \mathbb{Z}$. Therefore for an object $\mathcal{E} \in \mathcal{W}_\delta^{\mathrm{int}/\mathbb{S}}([\mathfrak{U}/G])$, it is enough to show that

$$\Phi(\mathcal{E})|_{(V^\vee)^\lambda} \in \bigoplus_{j \in I} \mathrm{MF}_{\mathrm{coh}}^{\mathbb{C}^*}([(V^\vee)^\lambda/G^\lambda], w|_{(V^\vee)^\lambda})_{\lambda\text{-wt}=j}. \tag{6.3.7}$$

Here $G^\lambda \subset G$ is the center of λ, and $I \subset \mathbb{R}$ is the interval

$$I = \left\langle \delta + \frac{K_Y}{2}, \lambda \right\rangle + \left[-\frac{1}{2}\eta_\alpha, \frac{1}{2}\eta_\alpha \right].$$

By restricting the left hand side of (6.3.7) to the zero section $Y^\lambda \hookrightarrow (V^\vee)^\lambda$ and noting that $\omega|_{Y^\lambda} = 0$, we obtain the object

$$\Phi(\mathcal{E})|_{Y^\lambda} \in \mathrm{MF}_{\mathrm{coh}}^{\mathbb{C}^*}([Y^\lambda/G^\lambda], 0).$$

By Lemma 2.4.5, under the tautological equivalence

$$D_{\mathrm{coh}}^b([Y^\lambda/G^\lambda]) \xrightarrow{\sim} \mathrm{MF}_{\mathrm{coh}}^{\mathbb{C}^*}([Y^\lambda/G^\lambda], 0)$$

we have $\Phi(\mathcal{E})|_{Y^\lambda} \cong (j_*\mathcal{E})|_{Y^\lambda}$. Then by Lemma 6.3.1 and noting that (see (6.1.11))

$$\left\langle \mathbb{L}_{[\mathbb{S}_x/G]}^{\lambda>0}|0, \lambda \right\rangle \leq \left\langle \mathbb{L}_{[V^\vee/G]}^{\lambda>0}|0, \lambda \right\rangle = \eta_\alpha$$

the condition (6.2.8) implies that

$$\Phi(\mathcal{E})|_{Y^\lambda} \in \bigoplus_{j \in I} \mathrm{MF}_{\mathrm{coh}}^{\mathbb{C}^*}([Y^\lambda/G^\lambda], 0)_{\lambda\text{-wt}=j}. \tag{6.3.8}$$

By comparing (6.3.7) with (6.3.8), it is enough to show that the pull-back by the zero section

$$0_{Y^\lambda}^* \colon \mathrm{MF}_{\mathrm{coh}}^{\mathbb{C}^*}([(V^\vee)^\lambda/G^\lambda], w|_{(V^\vee)^\lambda}) \to \mathrm{MF}_{\mathrm{coh}}^{\mathbb{C}^*}([Y^\lambda/G^\lambda], 0)$$

is conservative. This also follows from Lemma 2.4.5, since the push-forward j_* in the diagram (2.4.11) is conservative as j is a closed immersion.

We next show the inclusion (6.3.6). Let us take an object

$$\mathcal{P} \in \mathcal{W}_{\delta + K_Y/2}^{\mathrm{mag}/\widetilde{\mathbb{S}}_0}([V^\vee/G], w).$$

By Lemma 2.4.5 and the definition of the magic window subcategory, the object $j_* \Phi^{-1}(\mathcal{P}) \in D^b_{\mathrm{coh}}([Y/G])$ is split generated by $W \otimes \mathcal{O}_Y$ for G-representations W whose T-weights are contained in $\delta + K_Y/2 + \nabla_{\widehat{\mathbb{S}}_0}$. It follows that, by Lemma 6.3.1, for any map $\lambda \colon B\mathbb{C}^* \to [\mathcal{U}/G]$ which is mapped to 0 by $[\mathcal{U}/G] \to \mathcal{U}/\!\!/ G$ the object $\lambda^* j_* \Phi^{-1}(\mathcal{P})$ satisfies the weight condition (6.2.8). Therefore we have $\Phi^{-1}(\mathcal{P}) \in \mathcal{W}^{\mathrm{int}/\mathbb{S}_0}_{\delta}([\mathcal{U}/G])$ from its definition. $\qquad\square$

6.3.2 Window Subcategories Under Koszul Duality (Formal Fiber Case)

We also have the formal fiber version of the results in the previous subsections. Let Y be a smooth affine scheme with a G-action, where G is a reductive algebraic group. We take the formal completion of $Y/\!\!/ G$ at y, that is $\widehat{Y}_y/\!\!/ G := \operatorname{Spec} \widehat{\mathcal{O}}_{Y/\!\!/ G, y}$ as in Sect. 6.1.4. Let $V \to Y$ be a G-equivariant vector bundle, and

$$[\widehat{V}_y/G] \to [\widehat{Y}_y/G]$$

the formal fibers at y as in the diagram (6.1.12). Let $\widehat{s}_y$ be a section of the above vector bundle and $[\widehat{\mathcal{U}}_y/G]$ the derived zero locus of $\widehat{s}_y$. Similarly to (2.3.1), we have the commutative diagram

$$(6.3.9)$$

The proof of Theorem 2.3.3 applies to this setting, so we have the equivalence

$$\widehat{\Phi}_y \colon D^b_{\mathrm{coh}}([\widehat{\mathcal{U}}_y/G]) \xrightarrow{\sim} \mathrm{MF}^{\mathbb{C}^*}_{\mathrm{coh}}([\widehat{V}^\vee_y/G], \widehat{w}_y). \tag{6.3.10}$$

Let x be the unique closed point in $[\widehat{\mathcal{U}}_y/G]$. Then for a symmetric structure $\mathbb{S}$ of $[\mathcal{U}/G]$ as in (6.2.6), its restriction to x determines the symmetric structure $\mathbb{S}_x$ of $[\widehat{\mathcal{U}}_y/G]$. For $\delta \in \operatorname{Pic}([\mathcal{U}/G])_{\mathbb{R}}$, the intrinsic window subcategory

$$\mathcal{W}^{\mathrm{int}/\mathbb{S}_x}_{\delta}([\widehat{\mathcal{U}}_y/G]) \subset D^b_{\mathrm{coh}}([\widehat{\mathcal{U}}_y/G]) \tag{6.3.11}$$

is defined similarly to (6.2.7), using the closed immersion $[\widehat{\mathfrak{U}}_y/G] \hookrightarrow [\widehat{Y}_y/G]$ and the symmetric structure (6.2.6) at x. We have the following lemma which relates window subcategories and those on formal fibers.

Lemma 6.3.3 *Let s be a section of $[V/G] \to [Y/G]$, and $[\mathfrak{U}/G] \hookrightarrow [Y/G]$ its derived zero locus. For each $y \in Y /\!\!/ G$, let $\widehat{s}_y$ be the section of $[\widehat{V}_y/G] \to [\widehat{Y}_y/G]$ induced from s. Then for an object $\mathcal{E} \in D^b_{coh}([\mathfrak{U}/G])$, we have*

$$\mathcal{E} \in \mathcal{W}^{\mathrm{int}/\mathbb{S}}_\delta([\mathfrak{U}/G]), \quad (resp.\ \Phi(\mathcal{E}) \in \mathcal{W}^l_{\delta+K_y/2}([V^\vee/G], w))$$

if and only if for any $y \in \mathcal{U} /\!\!/ G$ we have

$$\widehat{\mathcal{E}}_y \in \mathcal{W}^{\mathrm{int}/\mathbb{S}_x}_\delta([\widehat{\mathfrak{U}}_y/G]), \quad (resp.\ \widehat{\Phi}_y(\widehat{\mathcal{E}}_y) \in \mathcal{W}^l_{\delta+K_y/2}([\widehat{V}^\vee_y/G], \widehat{w}_y)).$$

Here $\widehat{\mathcal{E}}_y$ is the pull-back of $\mathcal{E}$ to $[\widehat{\mathfrak{U}}_y/G]$, and the pull-backs of l, δ, K_y to the formal fibers are also denoted as l, δ, K_y.

Proof By the construction of the equivalence in Theorem 2.3.3, we have the commutative diagram

$$
\begin{array}{ccc}
D^b_{coh}([\mathfrak{U}/G]) & \xrightarrow{\Phi} & \mathrm{MF}^{\mathbb{C}^*}_{coh}([V^\vee/G], w) \\
\downarrow & & \downarrow \\
D^b_{coh}([\widehat{\mathfrak{U}}_y/G]) & \xrightarrow{\widehat{\Phi}_y} & \mathrm{MF}^{\mathbb{C}^*}_{coh}([\widehat{V}^\vee_y/G], \widehat{w}_y).
\end{array}
$$

Here the vertical arrows are pull-back functors. The lemma follows from the above commutative diagram, since the defining conditions of the relevant window subcategories are local on $\mathcal{U} /\!\!/ G$.   □

Suppose that $Y = \mathbb{A}^n$ is a G-representation and take the formal fibers at $0 \in Y /\!\!/ G$. Let $\widetilde{\mathbb{S}}_0$ be the symmetric structure as in (6.3.1) for $x = 0$. In this case, we have the following formal fiber version of Proposition 6.3.2.

Proposition 6.3.4 *We take $l, \delta \in M^W_\mathbb{R}$ such that δ is l-generic. Then under the equivalence $\widehat{\Phi}_0$ in (6.3.10), we have*

$$\mathcal{W}^{\mathrm{mag}/\widetilde{\mathbb{S}}_0}_{\delta+K_y/2}([\widehat{V}^\vee_0/G], \widehat{w}_0) \subset \widehat{\Phi}_0(\mathcal{W}^{\mathrm{int}/\mathbb{S}_0}_\delta([\widehat{\mathfrak{U}}_0/G])) \subset \mathcal{W}^l_{\delta+K_y/2}([\widehat{V}^\vee_0/G], \widehat{w}_0).$$

In particular if furthermore l is $\widetilde{\mathbb{S}}_0$-generic and compatible with $\widetilde{\mathbb{S}}_0$, then

$$\mathcal{W}^{\mathrm{mag}/\widetilde{\mathbb{S}}_0}_{\delta+K_y/2}([\widehat{V}^\vee_0/G], \widehat{w}_0) = \widehat{\Phi}_0(\mathcal{W}^{\mathrm{int}/\mathbb{S}_0}_\delta([\widehat{\mathfrak{U}}_0/G])) = \mathcal{W}^l_{\delta+K_y/2}([\widehat{V}^\vee_0/G], \widehat{w}_0).$$

Proof The argument of Proposition 6.3.2 applies verbatim. Here we have a chain of inclusions since $\mathcal{W}^{\mathrm{int}/\mathbb{S}_0}_{\delta}([\widehat{\mathfrak{U}}_0/G]) = \mathcal{W}^{\mathrm{int}/\mathbb{S}}_{\delta}([\widehat{\mathfrak{U}}_0/G])$ for the formal fiber case. Then the second statement follows from Proposition 6.1.7. $\quad\square$

We have the following formal fiber version of Proposition 6.2.16:

Lemma 6.3.5 *Let Y, Y' be smooth affine schemes of finite presentation over $\mathbb{C}$ with G-actions, and $V \to Y$, $V' \to Y'$ be G-equivariant vector bundles. Suppose that we have the following diagram for $y \in Y /\!\!/ G$, $y' \in Y' /\!\!/ G$*

$$
\begin{array}{ccccc}
[\widehat{\mathfrak{U}}_y/G] & \xrightarrow{\ j\ } & [\widehat{Y}_y/G] & \xleftarrow{\ \widehat{s}_y\ } & [\widehat{V}_y/G] \\
\widehat{f}\downarrow{\scriptstyle \sim} & & & & \\
[\widehat{\mathfrak{U}}'_{y'}/G] & \xrightarrow{\ j'\ } & [\widehat{Y}'_{y'}/G] & \xleftarrow[\ \widehat{s}'_{y'}\ ]{} & [\widehat{V}'_{y'}/G]
\end{array}
\tag{6.3.12}
$$

where $\widehat{f}$ is an equivalence of derived stacks and $[\widehat{\mathfrak{U}}_y/G]$, $[\widehat{\mathfrak{U}}'_{y'}/G]$ are derived zero loci of the sections $\widehat{s}_y$, $\widehat{s}'_{y'}$, respectively. We assume that $\widehat{s}_y(x) = 0$, $\widehat{s}'_{y'}(x') = 0$, $G = \mathrm{Aut}(x) = \mathrm{Aut}(x')$, where x, x' are unique closed points of $[\widehat{Y}_y/G]$, $[\widehat{Y}'_{y'}/G]$ respectively. Then for $\delta' \in \mathrm{Pic}([\widehat{\mathfrak{U}}'_{y'}/G])_{\mathbb{R}}$ and $\delta = \widehat{f}^*\delta'$, we have equivalences

$$
\widehat{f}_* : \mathcal{W}^{\mathrm{int}/\mathbb{S}_x}_{\delta}([\widehat{\mathfrak{U}}_y/G]) \xrightarrow{\sim} \mathcal{W}^{\mathrm{int}/\mathbb{S}'_{x'}}_{\delta'}([\widehat{\mathfrak{U}}'_{y'}/G]),
$$

$$
\widehat{f}^* : \mathcal{W}^{\mathrm{int}/\mathbb{S}'_{x'}}_{\delta'}([\widehat{\mathfrak{U}}'_{y'}/G]) \xrightarrow{\sim} \mathcal{W}^{\mathrm{int}/\mathbb{S}_x}_{\delta}([\widehat{\mathfrak{U}}_y/G]).
$$

Proof The proof of Proposition 6.2.16 applies verbatim, using Lemma 8.3.3 and Lemma 8.3.4, instead of Lemma 6.2.17 and Lemma 6.2.15. One subtle difference is that, in the setting of lemma, we have $\widehat{f}(x) = x'$ and this implies that $\widehat{f}$ fits into a left commutative diagram in (6.2.15). Namely, let $\widehat{f}(x) \colon BG \to BG$ be the induced morphism at the closed points. Then the diagram

$$
\begin{array}{ccc}
[\widehat{\mathfrak{U}}_y/G] & \longrightarrow & BG \\
\widehat{f}\downarrow & & \downarrow\widehat{f}(x) \\
[\widehat{\mathfrak{U}}'_{y'}/G] & \longrightarrow & BG
\end{array}
\tag{6.3.13}
$$

commutes. Here the horizontal arrows are given by canonical G-torsors $\widehat{\mathfrak{U}}_y \to [\widehat{\mathfrak{U}}_y/G]$, $\widehat{\mathfrak{U}}'_{y'} \to [\widehat{\mathfrak{U}}'_{y'}/G]$. Then the commutative diagram (6.3.13) follows from Lemma 6.3.6 below. $\quad\square$

We have used the following lemma:

Lemma 6.3.6 *For morphisms $f, f' : [\widehat{\mathfrak{U}}_y/G] \to BG$, suppose that $f \circ \mu \cong f' \circ \mu$ as morphisms $BG \to BG$, where $\mu : BG \to [\widehat{\mathfrak{U}}_y/G]$ sends the point to x and identity on the stabilizer groups. Then we have $f \sim f'$.*

Proof The proof will be given in Sect. 8.3.7.   □

6.3.3 Window Subcategories Under Koszul Duality (Affine Case)

Let G be a reductive algebraic group, and (Y, V, s) a G-equivariant tuple as in Definition 2.3.1. We consider the associated derived stack $[\mathfrak{U}/G]$ with a symmetric structure $\mathbb{S}$ as in (6.2.6). In Proposition 6.3.2, we compared window subcategories under Koszul duality when Y is a G-representation. By applying the results for the formal fibers and using étale slice theorem, we prove a similar comparison result for an affine Y. We first prove the following proposition:

Proposition 6.3.7 *Let us take $l, \delta \in \text{Pic}([\mathcal{U}/G])_{\mathbb{R}}$ such that δ is l-generic, and they are extended to $\mathbb{R}$-line bundles on $[Y/G]$ which we use the same notation l, δ. Then the equivalence Φ in Theorem 2.3.3 restricts to the fully-faithful functor*

$$\Phi : \mathcal{W}_{\delta}^{\text{int}/\mathbb{S}}([\mathfrak{U}/G]) \hookrightarrow \mathcal{W}_{\delta+K_{\mathcal{Y}}/2}^{l}([V^{\vee}/G], w). \tag{6.3.14}$$

Here we have regarded $\mathbb{R}$-line bundles on Y as $\mathbb{R}$-line bundles on $V^{\vee}$ by the pullback of the projection $V^{\vee} \to Y$.

Proof Let us take a closed point $y \in \mathcal{U}/\!\!/G$, and a unique closed point $x \in [\mathcal{U}/G]$ in the fiber of $[\mathcal{U}/G] \to \mathcal{U}/\!\!/G$ at y. We denote by $G_x \subset G$ the stabilizer subgroup of x, which is also reductive. Below we take a representative of x in $\mathcal{U}$ and write $x \in \mathcal{U}$. By Luna's étale slice theorem for the G-action on Y (see [5, 85]), there is a G_x-invariant locally closed subscheme $x \in Z \subset Y$ and Cartesian diagrams

$$
\begin{array}{ccccc}
[(V, x)/G] & \longleftarrow & [(V|_Z, x)/G_x] & \longrightarrow & [(T_x(V|_Z), 0)/G_x] \\
\downarrow & \square & \quad\downarrow \pi_Z & \square & \downarrow \\
[(Y, x)/G] & \longleftarrow & [(Z, x)/G_x] & \longrightarrow & [(T_x Z, 0)/G_x] \\
\quad\downarrow \pi_Y & \square & \quad\downarrow \pi_Z & \square & \quad\downarrow \pi_T \\
(Y/\!\!/G, y) & \longleftarrow & (Z/\!\!/G_x, y) & \longrightarrow & (T_x Z/\!\!/G_x, 0).
\end{array}
$$

Here each horizontal arrows are étale morphisms, and $T_x Z$, $T_x(V|_Z)$ are the Zariski tangent spaces of Z, $V|_Z$ at x, where we regard x as a point of $V|_Z$ by the zero section of $V|_Z \to Z$. Note that Z is smooth since Y is smooth and $[Z/G_x] \to [Y/G]$ is étale. Also note that $T_x(V|_Z) = V|_x \oplus T_x Z$ as G_x-representations.

By taking the formal fibers of left arrows at $y \in Y /\!\!/ G$ and the right arrows at $0 \in T_x Z /\!\!/ G_x$, we obtain the commutative diagram (see the diagram (6.3.9) for the notation)

$$
\begin{array}{ccc}
[\widehat{V}_y / G] & \xrightarrow{\cong} & [(V|_x \times \widehat{(T_x Z)_0}) / G_x] \\
{\scriptstyle \widehat{s}_y} \Big\uparrow \Big\downarrow & & \Big\downarrow {\scriptstyle \widehat{t}_0} \\
[\widehat{Y}_y / G] & \xrightarrow{\cong} & [\widehat{(T_x Z)_0} / G_x].
\end{array}
\tag{6.3.15}
$$

Here $\widehat{s}_y$ is induced by the section $s \colon Y \to V$, and $\widehat{t}_0$ is defined by the commutative diagram (6.3.15). In particular, $[\widehat{\mathfrak{U}}_y / G]$ is equivalent to the derived zero locus of $\widehat{t}_0$. By Lemma 6.3.3, in order to show that Φ restricts to the functor (6.3.14), it is enough to show that the functor $\widehat{\Phi}_y$ in (6.3.10) restricts to the functor

$$
\widehat{\Phi}_y \colon \mathcal{W}_\delta^{\mathrm{int}/\mathbb{S}_x}([\widehat{\mathfrak{U}}_y / G]) \to \mathcal{W}_{\delta + K_y / 2}^l([\widehat{V}_y^\vee / G], \widehat{w}_y).
\tag{6.3.16}
$$

Here both sides are the window subcategories (6.3.11) and (6.1.15). By the commutative diagram (6.3.15) and Lemma 6.2.13, we can reduce the above claim to the corresponding claim in the right hand side of (6.3.15), which is the case of the formal fiber for a linear representation. Namely let $[\widehat{\mathfrak{T}}_0 / G_x]$ be the derived zero locus of $\widehat{t}_0$, and consider the Koszul duality equivalence in Theorem 2.3.3

$$
\widehat{\Phi}_T \colon D_{\mathrm{coh}}^b([\widehat{\mathfrak{T}}_0 / G_x]) \xrightarrow{\sim} \mathrm{MF}_{\mathrm{coh}}^{\mathbb{C}^*}([(V^\vee|_x \times \widehat{(T_x Z)_0}) / G_x], \widehat{w}_0).
$$

Here $\widehat{w}_0$ is defined from $\widehat{t}_0$ as in (2.1.3). Then Proposition 6.3.4 implies that $\widehat{\Phi}_T$ restricts to the functor

$$
\widehat{\Phi}_T \colon \mathcal{W}_\delta^{\mathrm{int}/\mathbb{S}_x}([\widehat{\mathfrak{T}}_0 / G_x]) \to \mathcal{W}_{\delta + K_y / 2}^l([(V^\vee|_x \times \widehat{(T_x Z)_0}) / G_x], \widehat{w}_0).
\tag{6.3.17}
$$

Here we have regarded $\delta, l, \mathbb{S}_x, K_y$ as objects in the right hand side of (6.3.15) by the isomorphisms in (6.3.15). By Lemma 6.2.13, we conclude that $\widehat{\Phi}_y$ restricts to the functor (6.3.16) as desired. Therefore the functor Φ in Theorem 2.3.3 restricts to the functor (6.3.14), which is fully-faithful.

$$\square$$

In the next lemma, we show that if the Koszul duality equivalence restricts to the equivalence of window subcategories for some presentation as a derived zero locus, then the same property also holds for other presentations.

Lemma 6.3.8 *In the situation of Proposition 6.2.16, assume that $h = \mathrm{id}$ in the diagram (6.2.15). Let Φ, Φ' be equivalences in Theorem 2.3.3 applied for $[\mathfrak{U}/G]$, $[\mathfrak{U}'/G]$. We take $l, \delta \in M_{\mathbb{R}}^W$ such that δ is l-generic. Then Φ restricts to the equivalence*

$$
\Phi \colon \mathcal{W}_\delta^{\mathrm{int}/\mathbb{S}}([\mathfrak{U}/G]) \xrightarrow{\sim} \mathcal{W}_{\delta + K_y / 2}^l([V^\vee / G], w)
$$

if and only if Φ' restricts to the equivalence

$$\Phi': \mathcal{W}_\delta^{\mathrm{int}/\mathbb{S}'}([\mathfrak{U}'/G]) \xrightarrow{\sim} \mathcal{W}_{\delta+K_{y'}/2}^l([V'^\vee/G], w').$$

Proof As in the proof of Proposition 6.2.16, we may assume that **f** is induced by a G-equivariant diagram (6.2.9) such that f is a closed immersion. Then in the notation of the diagram (6.2.11), we have the commutative diagram by Lemma 2.4.4

$$
\begin{array}{ccc}
D_{\mathrm{coh}}^b([\mathfrak{U}/G]) & \xrightarrow{\;\Phi\;} & \mathrm{MF}_{\mathrm{coh}}^{\mathbb{C}^*}([V^\vee/G], w) \\
{\scriptstyle \mathbf{f}_*}\downarrow & & \downarrow{\scriptstyle f_*g^*} \\
D_{\mathrm{coh}}^b([\mathfrak{U}'/G]) & \xrightarrow{\;\Phi'\;} & \mathrm{MF}_{\mathrm{coh}}^{\mathbb{C}^*}([V'^\vee/G], w').
\end{array}
\tag{6.3.18}
$$

Together with Proposition 6.3.7, we have the commutative diagram

(6.3.19)

We show that there is an equivalence Θ in the dotted arrow which makes the above diagram commutative. We set

$$Z^{l\text{-us}} = \mathrm{Crit}(w) \setminus \mathrm{Crit}(w)^{l\text{-ss}}, \quad Z'^{l\text{-us}} = \mathrm{Crit}(w') \setminus \mathrm{Crit}(w')^{l\text{-ss}}.$$

Since we have $\mathrm{Crit}(w)^{l\text{-ss}} = (V^\vee)^{l\text{-ss}} \cap \mathrm{Crit}(w)$, we have the open immersion

$$(V^\vee)^{l\text{-ss}} \hookrightarrow V^\vee \setminus Z^{l\text{-us}} \tag{6.3.20}$$

such that we have

$$\mathrm{Crit}(w) \cap (V^\vee)^{l\text{-ss}} = \mathrm{Crit}(w) \cap (V^\vee \setminus Z^{l\text{-us}}) = \mathrm{Crit}(w)^{l\text{-ss}}.$$

Since a derived factorization category depends only on an open neighborhood of the critical locus (see (2.2.7)), the restriction along the open immersion (6.3.20) gives an equivalence

$$\mathrm{MF}_{\mathrm{coh}}^{\mathbb{C}^*}([(V^\vee \setminus Z^{l\text{-us}})/G], w) \xrightarrow{\sim} \mathrm{MF}_{\mathrm{coh}}^{\mathbb{C}^*}([(V^\vee)^{l\text{-ss}}/G], w). \tag{6.3.21}$$

On the other hand, the equivalence of derived stacks $\mathbf{f}$ in the diagram (6.2.15) induces the isomorphism

$$[\mathrm{Crit}(w)/G] \xrightarrow{\cong} [\mathrm{Crit}(w')/G]$$

which sends a conical closed substack $\mathcal{Z}^{l\text{-us}} := [Z^{l\text{-us}}/G]$ to $\mathcal{Z}'^{l\text{-us}} := [Z'^{l\text{-us}}/G]$. Therefore the equivalence $\mathbf{f}_* \colon D^b_{\mathrm{coh}}([\mathfrak{U}/G]) \xrightarrow{\sim} D^b_{\mathrm{coh}}([\mathfrak{U}'/G])$ restricts to the equivalence

$$\mathbf{f}_* \colon \mathcal{C}_{\mathcal{Z}^{l\text{-us}}} \xrightarrow{\sim} \mathcal{C}_{\mathcal{Z}'^{l\text{-us}}}.$$

By Proposition 2.3.9 and the commutative diagram (6.3.18), the equivalence $f_* g^*$ in the diagram (6.3.19) restricts to the equivalence

$$f_* g^* \colon \mathrm{MF}^{\mathbb{C}^*}_{\mathrm{coh}}([V^\vee/G], w)_{\mathcal{Z}^{l\text{-us}}} \xrightarrow{\sim} \mathrm{MF}^{\mathbb{C}^*}_{\mathrm{coh}}([V'^\vee/G], w')_{\mathcal{Z}'^{l\text{-us}}}.$$

By taking the Verdier quotients as in (2.2.5) and using (6.3.21), we obtain the desired equivalence Θ.

Note that the functors res, res$'$ in the diagram (6.3.19) are equivalences by Theorem 6.1.2. Using the equivalence Θ, we have the commutative diagram

$$
\begin{array}{ccc}
\mathcal{W}^{\mathrm{int}/\mathbb{S}}_{\delta}([\mathfrak{U}/G]) & \overset{\Phi}{\hookrightarrow} & \mathcal{W}^{l}_{\delta+K_{\mathcal{Y}}/2}([V^\vee/G], w) \\
{\scriptstyle \mathbf{f}_*} \downarrow {\scriptstyle \sim} & & {\scriptstyle \sim} \downarrow {\scriptstyle \mathrm{res}'^{-1}\circ\Theta\circ\mathrm{res}} \\
\mathcal{W}^{\mathrm{int}/\mathbb{S}'}_{\delta}([\mathfrak{U}'/G]) & \overset{\Phi'}{\hookrightarrow} & \mathcal{W}^{l}_{\delta+K_{\mathcal{Y}'}/2}([V'^\vee/G], w').
\end{array}
$$

The lemma follows from the above commutative diagram. $\qquad\square$

We also have the following formal fiber version of Lemma 6.3.8 (see (6.3.9) for the notation of formal fibers):

Lemma 6.3.9 *In the setting of Lemma 6.3.5, assume that $\widehat{\mathbf{f}}(x) = \mathrm{id}$ in the diagram (6.3.13). We take $l, \delta \in M^W_{\mathbb{R}}$ such that δ is l-generic. Then the functor $\widehat{\Phi}_y$ in (6.3.10) restricts to the equivalence*

$$\widehat{\Phi}_y \colon \mathcal{W}^{\mathrm{int}/\mathbb{S}_x}_{\delta}([\widehat{\mathfrak{U}}_y/G]) \xrightarrow{\sim} \mathcal{W}^{l}_{\delta+K_{\mathcal{Y}}/2}([\widehat{V}^\vee_y/G], \widehat{w}_y)$$

if and only if $\widehat{\Phi}'_{y'}$ restricts to the equivalence

$$\widehat{\Phi}'_{y'} \colon \mathcal{W}^{\mathrm{int}/\mathbb{S}'_{x'}}_{\delta}([\widehat{\mathfrak{U}}'_{y'}/G]) \xrightarrow{\sim} \mathcal{W}^{l}_{\delta+K_{\mathcal{Y}'}/2}([\widehat{V}'^\vee_{y'}/G], \widehat{w}'_{y'}).$$

Proof The proof of Proposition 6.3.7 shows that the functor $\widehat{\Phi}_y$ restricts to the fully-faithful functor

$$\widehat{\Phi}_y \colon \mathcal{W}_\delta^{\mathrm{int}/\mathbb{S}_x}([\widehat{\mathfrak{U}}_y/G]) \hookrightarrow \mathcal{W}_{\delta+K_y/2}^l([\widehat{V}_y^\vee/G], \widehat{w}_y),$$

and the same also applies to $\widehat{\Phi}_{y'}$. Therefore the argument of Lemma 6.3.8 applies verbatim, using Lemma 6.3.5 instead of Proposition 6.2.16, and also using Lemmas 8.3.3 and 8.3.4, instead of Lemmas 6.2.17 and 6.2.15 for the reduction to the case of closed immersion.   □

Using the above argument for formal fibers, we prove the following proposition for affine case:

Proposition 6.3.10 *In the setting of Proposition 6.3.7, suppose that* $[\mathfrak{U}/G]$ *satisfies formal neighborhood theorem, l is $\mathbb{S}$-generic and compatible with $\mathbb{S}$. Then the functor (6.3.14) is an equivalence.*

Proof We use the notation in the proof of Proposition 6.3.7. By Lemma 6.3.3, it is enough to prove that the functor $\widehat{\Phi}_y$ in (6.3.16) is an equivalence

$$\widehat{\Phi}_y \colon \mathcal{W}_\delta^{\mathrm{int}/\mathbb{S}_x}([\widehat{\mathfrak{U}}_y/G]) \xrightarrow{\sim} \mathcal{W}_{\delta+K_y/2}^l([\widehat{V}_y^\vee/G], \widehat{w}_y). \tag{6.3.22}$$

We set

$$Y' := \mathcal{H}^0(\mathbb{T}_{[\mathfrak{U}/G]}|_x), \ V' := \mathcal{H}^1(\mathbb{T}_{[\mathfrak{U}/G]}|_x) \oplus \mathcal{H}^0(\mathbb{T}_{[\mathfrak{U}/G]}|_x).$$

Note that Y' is a G_x-representation, and V' is regarded as a G_x-equivariant vector bundle on Y' by the second projection $V' \to Y'$. As we assume that $[\mathfrak{U}/G]$ satisfies the formal neighborhood theorem, Lemma 6.2.5 implies the following: there exists a diagram of formal fibers at $0 \in Y' /\!\!/ G_x$

$$
\begin{array}{c}
[\widehat{V}_0'/G_x] \\
\downarrow {\scriptstyle \widehat{s}_0'} \\
[\widehat{\mathfrak{U}}_0'/G_x] \lhook\joinrel\longrightarrow [\widehat{Y}_0'/G_x]
\end{array}
\tag{6.3.23}
$$

where $[\widehat{\mathfrak{U}}_0'/G_x]$ is the derived zero locus of the section $\widehat{s}_0'$, such that there is an equivalence

$$[\widehat{\mathfrak{U}}_y/G] \xrightarrow{\sim} [\widehat{\mathfrak{U}}_0'/G_x] \tag{6.3.24}$$

where $x \in \mathcal{U}$ corresponds to $0 \in Y'$. By Lemma 6.1.1, the $\mathbb{R}$-line bundles $l|_{\widehat{\mathcal{U}}_y}, \delta|_{\widehat{\mathcal{U}}_y} \in \mathrm{Pic}([\widehat{\mathcal{U}}_y/G])_\mathbb{R}$ correspond to $l_x, \delta_x \in \mathrm{Pic}(BG_x)_\mathbb{R}$ under the equivalence (6.3.24). Note that by the genericity assumption on l, δ, the element l_x is $\mathbb{S}_x$-generic and δ_x is l_x-generic.

Let $\widehat{\Phi}'_0$ be the Koszul duality equivalence in Theorem 2.3.3 applied for the diagram (6.3.23)

$$\widehat{\Phi}'_0: D^b_{\mathrm{coh}}([\widehat{\mathfrak{U}}'_0/G_x]) \xrightarrow{\sim} \mathrm{MF}^{\mathbb{C}^*}_{\mathrm{coh}}([\widehat{V}_0^{'\vee}/G_x], \widehat{w}'_0).$$

Here $\widehat{w}'_0$ is defined from $\widehat{s}'_0$ in the diagram (6.3.23) as in (2.1.3). By Proposition 6.3.4, the genericity condition on l_x and δ_x, and the assumption that l_x is compatible with $\mathbb{S}_x$, the equivalence $\widehat{\Phi}'_0$ restricts to the equivalence

$$\widehat{\Phi}'_0: \mathcal{W}^{\mathrm{int}/\mathbb{S}_x}_{\delta_x}([\widehat{\mathfrak{U}}'_0/G_x]) \xrightarrow{\sim} \mathcal{W}^{l_x}_{\delta_x+K_{y'}/2}([\widehat{V}_0^{'\vee}/G_x], \widehat{w}'_0). \tag{6.3.25}$$

On the other hand, by the equivalence (6.3.24) and the diagram (6.3.15), we have an equivalence

$$[\widehat{\mathfrak{U}}'_0/G_x] \sim [\widehat{\mathfrak{T}}_0/G_x],$$

such that the induced map $BG_x \to BG_x$ on closed points is the identity. By Lemma 6.3.9 and the equivalence (6.3.25), we conclude that the functor (6.3.17) is an equivalence. The equivalence (6.3.22) now follows from the diagram (6.3.15) and Lemma 6.2.13. □

We have the following corollary of the above proposition:

Corollary 6.3.11 *In the situation of Proposition 6.3.7, let*

$$\mathcal{Z}^{l\text{-us}} \subset [\mathrm{Crit}(w)/G]$$

be the conical closed substack of l-unstable points. Then the composition

$$\mathcal{W}^{\mathrm{int}/\mathbb{S}}_\delta([\mathfrak{U}/G]) \hookrightarrow D^b_{\mathrm{coh}}([\mathfrak{U}/G]) \twoheadrightarrow D^b_{\mathrm{coh}}([\mathfrak{U}/G])/\mathcal{C}_{\mathcal{Z}^{l\text{-us}}}$$

is fully-faithful. Moreover it is an equivalence under the situation of Proposition 6.3.10.

Proof By Proposition 6.3.7, we have the commutative diagram

$$\begin{array}{ccc}
\mathcal{W}^{\mathrm{int}/\mathbb{S}}_\delta([\mathfrak{U}/G]) \hookrightarrow & D^b_{\mathrm{coh}}([\mathfrak{U}/G]) & \twoheadrightarrow D^b_{\mathrm{coh}}([\mathfrak{U}/G])/\mathcal{C}_{\mathcal{Z}^{l\text{-us}}} \\
\Phi\downarrow & \Phi\downarrow\sim & \Phi\downarrow\sim \\
\mathcal{W}^l_{\delta+K_y/2}([V^\vee/G], w) \hookrightarrow & \mathrm{MF}^{\mathbb{C}^*}_{\mathrm{coh}}([V^\vee/G], w]) & \twoheadrightarrow \mathrm{MF}^{\mathbb{C}^*}_{\mathrm{coh}}([V^\vee/G] \setminus \mathcal{Z}^{l\text{-us}}, w).
\end{array}$$

Since have an equivalence (6.3.21), the bottom composition is an equivalence by Theorem 6.1.2. Therefore the corollary follows from Proposition 6.3.7 and Proposition 6.3.10.   □

6.3.4   Proof of Window Theorem for DT Categories

Finally, we give a proof of Theorem 6.2.7 by taking the limits of the results in the previous subsections. Let $\mathfrak{M}$ be a quasi-smooth derived stack such that $\mathcal{M} = t_0(\mathfrak{M})$ admits a good moduli space $\mathcal{M} \to M$ and satisfies the formal neighborhood theorem. Let $\mathbb{S}$ be a symmetric structure of $\mathfrak{M}$ as in (6.2.1). Note that for an étale morphism $\iota_{\mathfrak{M}} \colon \mathfrak{M}_U \to \mathfrak{M}$ in the diagram (3.1.12), we have the induced symmetric structure $\mathbb{S}_U$ since $\iota_{\mathfrak{M}}$ induces the quasi-isomorphisms of tangent complexes at each closed points. Using Proposition 3.1.5, the definition of intrinsic window subcategory in $D^b_{\mathrm{coh}}(\mathfrak{M})$ is defined as a globalization of Definition 4.4.14:

Definition 6.3.12   For $\delta \in \mathrm{Pic}(\mathcal{M})_{\mathbb{R}}$, we define the triangulated subcategory

$$\mathcal{W}^{\mathrm{int}/\mathbb{S}}_{\delta}(\mathfrak{M}) \subset D^b_{\mathrm{coh}}(\mathfrak{M})$$

to be consisting of objects $\mathcal{E} \in D^b_{\mathrm{coh}}(\mathfrak{M})$ such that for any étale morphism $\iota \colon U \to M$ from an affine scheme U which fits into a diagram (3.1.12), we have $\iota^*_{\mathfrak{M}}\mathcal{E} \in \mathcal{W}^{\mathrm{int}/\mathbb{S}_U}_{\iota^*_{\mathfrak{M}}\delta}(\mathfrak{M}_U)$.

In the following, we show that the above intrinsic window subcategory gives a desired subcategory in Theorem 6.2.7.

Theorem 6.3.13   *Let us take $l, \delta \in \mathrm{Pic}(\mathcal{M})_{\mathbb{R}}$ such that δ is l-generic. Then the composition*

$$\Theta_l \colon \mathcal{W}^{\mathrm{int}/\mathbb{S}}_{\delta}(\mathfrak{M}) \hookrightarrow D^b_{\mathrm{coh}}(\mathfrak{M}) \to \mathcal{DT}^{\mathbb{C}^*}(\mathcal{N}^{l\text{-ss}}) \tag{6.3.26}$$

is fully-faithful, which is an equivalence if l is $\mathbb{S}$-generic and compatible with $\mathbb{S}$.

Proof   Let $\mathcal{D}_{\text{ét}}/M$ be the category of étale morphisms $\iota \colon U \to M$ as in Sect. 3.1.4, and $\mathcal{D}'_{\text{ét}}/M \subset \mathcal{D}_{\text{ét}}/M$ the subcategory satisfying the condition in Theorem 3.1.4. For each $(\iota \colon U \to M) \in \mathcal{D}'_{\text{ét}}/M$, we have the induced étale morphism $\iota_{\mathfrak{M}} \colon \mathfrak{M}_U \to \mathfrak{M}$ in the diagram (3.1.12). Moreover for any morphism $\rho \colon U' \to U$ in $\mathcal{D}'_{\text{ét}}/M$, we have the induced étale morphism $\rho_{\mathfrak{M}} \colon \mathfrak{M}_{U'} \to \mathfrak{M}_U$ as in the diagram (3.1.13). Since we have the étale cover $\coprod_{(U \xrightarrow{\iota} M) \in \mathcal{D}'_{\text{ét}}/M} \mathfrak{M}_U \xrightarrow{\iota_{\mathfrak{M}}} \mathfrak{M}$ of $\mathfrak{M}$, we have an equivalence

$$D^b_{\mathrm{coh}}(\mathfrak{M}) \xrightarrow{\sim} \lim_{(U \xrightarrow{\iota} M) \in \mathcal{D}'_{\text{ét}}/M} D^b_{\mathrm{coh}}(\mathfrak{M}_U). \tag{6.3.27}$$

By Lemma 6.3.14 below, for a morphism $\rho \colon U' \to U$ the pull-back $\rho^*_{\mathfrak{M}}$ restricts to the functor

$$\rho^*_{\mathfrak{M}} \colon \mathcal{W}^{\mathrm{int}/\mathbb{S}_U}_{\iota^*_{\mathfrak{M}}\delta}(\mathfrak{M}_U) \to \mathcal{W}^{\mathrm{int}/\mathbb{S}_{U'}}_{\iota'^*_{\mathfrak{M}}\delta}(\mathfrak{M}_{U'}).$$

Therefore from (6.3.27) and the definition of $\mathcal{W}^{\mathrm{int}/\mathbb{S}}_{\delta}(\mathfrak{M})$, the equivalence (6.3.27) restricts to the equivalence

$$\mathcal{W}^{\mathrm{int}/\mathbb{S}}_{\delta}(\mathfrak{M}) \xrightarrow{\sim} \lim_{(U \xrightarrow{\iota} M) \in \mathcal{D}'_{\mathrm{\acute{e}t}}/M} \mathcal{W}^{\mathrm{int}/\mathbb{S}_U}_{\iota^*_{\mathfrak{M}}\delta}(\mathfrak{M}_U). \qquad (6.3.28)$$

On the other hand the assumption on $\mathfrak{M}$ together with the genericity of l, δ imply that, for each $(U \xrightarrow{\iota} M) \in \mathcal{D}'_{\mathrm{\acute{e}t}}/M$, the derived stack $\mathfrak{M}_U$ together with $\iota^*_{\mathfrak{M}}l$, $\iota^*_{\mathfrak{M}}\delta$ satisfy the assumption of Proposition 6.3.7. Therefore by Corollary 6.3.11, the composition

$$\mathcal{W}^{\mathrm{int}/\mathbb{S}_U}_{\iota^*_{\mathfrak{M}}\delta}(\mathfrak{M}_U) \hookrightarrow D^b_{\mathrm{coh}}(\mathfrak{M}_U) \twoheadrightarrow D^b_{\mathrm{coh}}(\mathfrak{M}_U)/\mathcal{C}_{\iota^*_{\mathfrak{M}}z_{l\text{-us}}} \qquad (6.3.29)$$

is fully-faithful, and an equivalence if l is compatible with $\mathbb{S}$. By taking the limit for $(U \xrightarrow{\iota} M) \in \mathcal{D}'_{\mathrm{\acute{e}t}}/M$ and using the equivalences (6.3.27) and (6.3.28), the composition

$$\mathcal{W}^{\mathrm{int}/\mathbb{S}}_{\delta}(\mathfrak{M}) \hookrightarrow D^b_{\mathrm{coh}}(\mathfrak{M}) \to \lim_{(U \xrightarrow{\iota} M) \in \mathcal{D}'_{\mathrm{\acute{e}t}}/M} \left(D^b_{\mathrm{coh}}(\mathfrak{M}_U)/\mathcal{C}_{\iota^*_{\mathfrak{M}}z_{l\text{-us}}} \right)$$

is full-faithful, and an equivalence if l is compatible with $\mathbb{S}$. Now the above composition functor factors as

$$\mathcal{W}^{\mathrm{int}/\mathbb{S}}_{\delta}(\mathfrak{M}) \to \mathcal{DT}^{\mathbb{C}^*}(\mathcal{N}^{l\text{-ss}}) \to \lim_{(U \xrightarrow{\iota} M) \in \mathcal{D}'_{\mathrm{\acute{e}t}}/M} \left(D^b_{\mathrm{coh}}(\mathfrak{M}_U)/\mathcal{C}_{\iota^*_{\mathfrak{M}}z_{l\text{-us}}} \right).$$

Here the first arrow is the composition functor (6.3.26). The right arrow is fully-faithful by Lemma 8.2.3, and the above composition is fully-faithful, so the first arrow is also fully-faithful. Moreover if l is compatible with $\mathbb{S}$, then the first arrow is also an equivalence since each arrow is fully-faithful and the composition is an equivalence. □

We have used the following lemma:

Lemma 6.3.14 *Let* $[\mathfrak{U}/G]$, $[\mathfrak{U}'/G']$ *be derived stacks as in Definition 2.3.1 for reductive* G, G'. *Suppose that we have a commutative diagram*

$$
\begin{array}{ccccc}
\mathcal{U}'/\!\!/G' & \longleftarrow & [\mathcal{U}'/G'] & \lhook\joinrel\longrightarrow & [\mathfrak{U}'/G'] \\
\downarrow f & \square & \downarrow f & \square & \downarrow \mathbf{f} \\
\mathcal{U}/\!\!/G & \longleftarrow & [\mathcal{U}/G] & \lhook\joinrel\longrightarrow & [\mathfrak{U}/G]
\end{array}
\tag{6.3.30}
$$

where each square is a Cartesian and the vertical arrows are étale. Then for $\delta \in \mathrm{Pic}([\mathcal{U}/G])_{\mathbb{R}}$ *and* $\delta' = f^*\delta \in \mathrm{Pic}([\mathcal{U}'/G'])_{\mathbb{R}}$, *the functor* $\mathbf{f}^*\colon D^b_{\mathrm{coh}}([\mathfrak{U}/G]) \to D^b_{\mathrm{coh}}([\mathfrak{U}'/G'])$ *restricts to the functor*

$$
\mathbf{f}^*\colon \mathcal{W}^{\mathrm{int}/\mathbb{S}}_{\delta}([\mathfrak{U}/G]) \to \mathcal{W}^{\mathrm{int}/\mathbb{S}'}_{\delta'}([\mathfrak{U}'/G']).
$$

Here $\mathbb{S}'$ *is induced from* $\mathbb{S}$ *by the étale morphism* f.

Proof For a closed point $y' \in \mathcal{U}'/\!\!/G'$ and $y = f(y') \in \mathcal{U}/\!\!/G$, the diagram (6.3.30) induces an equivalence

$$
\widehat{\mathbf{f}}_y\colon [\widehat{\mathfrak{U}}'_{y'}/G'] \xrightarrow{\sim} [\widehat{\mathfrak{U}}_y/G].
$$

Here we have used the notation in (6.3.9). Let $x \in \mathcal{U}$, $x' \in \mathcal{U}'$ be closed points in the closed orbits of the fibers of $\mathcal{U} \to \mathcal{U}/\!\!/G, \mathcal{U}' \to \mathcal{U}'/\!\!/G'$ at y, y', respectively. Let $x \in Z \subset Y$, $x' \in Z' \subset Y'$ be étale slices as in the proof of Proposition 6.3.7. Then by the diagram (6.3.15), we have the diagram

$$
\begin{array}{ccccc}
& & \overset{\widehat{\imath}_0'}{\text{---------}\!\!\nearrow} & & \\
[\widehat{\mathfrak{U}}'_{y'}/G'] & \overset{j'}{\lhook\joinrel\longrightarrow} & [(\widehat{T_{x'}Z'})_0/G'_{x'}] & \longleftarrow & [V'|_{x'} \oplus (\widehat{T_{x'}Z'})_0/G'_{x'}] \\
\widehat{\mathbf{f}}_y \downarrow \sim & & & & \\
[\widehat{\mathfrak{U}}_y/G] & \overset{j}{\lhook\joinrel\longrightarrow} & [(\widehat{T_xZ})_0/G_x] & \longleftarrow & [(V|_x \oplus (\widehat{T_xZ})_0)/G_x] \\
& & \underset{\widehat{\imath}_0}{\text{---------}\!\!\nearrow} & &
\end{array}
$$

such that $[\widehat{\mathfrak{U}}'_{y'}/G']$, $[\widehat{\mathfrak{U}}_y/G]$ are equivalent to derived zero loci of $\widehat{\imath}_0'$, $\widehat{\imath}_0$ respectively.

Let us take an object $\mathcal{E} \in \mathcal{W}^{\mathrm{int}/\mathbb{S}}_{\delta}([\mathfrak{U}/G])$. Then we have $\widehat{\mathcal{E}}_y \in \mathcal{W}^{\mathrm{int}/\mathbb{S}_x}_{\delta}([\widehat{\mathfrak{U}}_y/G])$ by Lemma 6.3.3. Since f induces the isomorphism $G'_{x'} \xrightarrow{\cong} G_x$, by Lemmas 6.2.13 and 6.3.5 we conclude that

$$
(\widehat{\mathbf{f}^*\mathcal{E}})_{y'} = \widehat{\mathbf{f}}^*_y(\widehat{\mathcal{E}}_y) \in \mathcal{W}^{\mathrm{int}/\mathbb{S}'_{x'}}_{\delta'}([\widehat{\mathfrak{U}}'_{y'}/G']).
$$

Since this holds for any $y' \in \mathcal{U}' /\!\!/ G'$, we have $\mathbf{f}^* \mathcal{E} \in \mathcal{W}_{\delta'}^{\mathrm{int}/\mathbb{S}'}([\mathfrak{U}'/G'])$ by Lemma 6.3.3 and the diagram (6.3.15). □

6.4 Applications of Window Theorem

In this section, we use Corollary 6.2.8 to give an application to Conjecture 4.1.2.

6.4.1 Derived Moduli Stacks of One Dimensional Semistable Sheaves on Surfaces

Let S be a smooth projective surface. As in Sect. 3.4, we consider the derived stack $\mathfrak{M}_S$ of coherent sheaves on S. We take a stability condition $\sigma = B + iH \in A(S)_{\mathbb{C}}$ and an element $v \in N_{\leq 1}(S)$. As in (3.4.12), we have the derived open substack

$$\mathfrak{M}_S^\sigma(v) \subset \mathfrak{M}_S$$

consisting of one dimensional σ-semistable sheaves with numerical class v.

Lemma 6.4.1 *Both of derived stacks $\mathfrak{M}_S^\sigma(v)$ and $\mathfrak{M}_S^\sigma(v)^{\mathbb{C}^*\text{-rig}}$ are symmetric and satisfy formal neighborhood theorem.*

Proof A closed point $x \in \mathcal{M}_S^\sigma(v)$ corresponds to a σ-polystable sheaf F on S, which is of the form

$$F = \bigoplus_{i=1}^m V_i \otimes F_i \tag{6.4.1}$$

where each F_i is a σ-stable sheaf on S, V_i is a finite dimensional vector space such that F_i is not isomorphic to F_j for $i \neq j$, and $\mu_\sigma(F_i) = \mu_\sigma(F_j)$ for all i, j, where μ_σ is defined in (4.1.1). By the description of the cotangent complex (3.4.3), we have

$$\mathcal{H}^0(\mathbb{T}_{\mathfrak{M}_S^\sigma(v)}|_x) \oplus \mathcal{H}^1(\mathbb{T}_{\mathfrak{M}_S^\sigma(v)}|_x)^\vee = \mathrm{Ext}_S^1(F, F) \oplus \mathrm{Ext}_S^2(F, F)^\vee$$

$$= \bigoplus_{a,b} \mathrm{Hom}(V_a, V_b) \otimes (\mathrm{Ext}_S^1(F_a, F_b)$$

$$\oplus \mathrm{Ext}_S^2(F_b, F_a)^\vee). \tag{6.4.2}$$

The automorphism group of $\mathcal{M}_S^\sigma(v)$ at x is given by

$$\mathrm{Aut}(x) = \mathrm{Aut}(F) = \prod_{i=1}^{m} \mathrm{GL}(V_i)$$

and its acts on (6.4.2) by the conjugation. The dual representation of (6.4.2) is given by

$$\bigoplus_{a,b} \mathrm{Hom}(V_a, V_b) \otimes (\mathrm{Ext}_S^1(F_b, F_a)^\vee \oplus \mathrm{Ext}_S^2(F_a, F_b)).$$

Therefore in order to show that (6.4.2) is a symmetric representation of $\mathrm{Aut}(x)$, we need to show that

$$\mathrm{ext}_S^1(F_a, F_b) + \mathrm{ext}_S^2(F_b, F_a) = \mathrm{ext}_S^1(F_b, F_a) + \mathrm{ext}_S^2(F_a, F_b). \qquad (6.4.3)$$

By the Riemann-Roch theorem and the stability for F_a, by writing $[F_a] = (\beta_a, n_a)$ we have

$$\mathrm{ext}_S^1(F_a, F_b) - \mathrm{ext}_S^2(F_a, F_b) = \delta_{ab} + \beta_a \cdot \beta_b$$

which is symmetric in a and b. Therefore (6.4.3) holds, and $\mathfrak{M}_S^\sigma(v)$ is symmetric. Similarly (6.4.2) is a symmetric $\mathrm{Aut}(x)/\mathbb{C}^*$-representation, so $\mathfrak{M}_S^\sigma(v)^{\mathbb{C}^*\text{-rig}}$ is symmetric.

The fact that the derived stack $\mathfrak{M}_S^\sigma(v)$ satisfies the formal neighborhood theorem follows from Theorem 8.4.2. Here note that Theorem 8.4.2 is formulated for Gieseker stability (i.e. $\sigma = iH$ for an ample divisor H), but the same argument applies for one dimensional μ_σ-semistable sheaves by the existence of the good moduli space for $\mathcal{M}_S^\sigma(v)$ (see [132, Lemma 7.4]). Then the derived stack $\mathfrak{M}_S^\sigma(v)^{\mathbb{C}^*\text{-rig}}$ also satisfies the formal neighborhood theorem by taking the $\mathbb{C}^*$-rigidifications of top isomorphism in the diagram (8.4.9).   □

Remark 6.4.2 Let $X = \mathrm{Tot}_S(\omega_S)$ and $i: S \hookrightarrow X$ the zero section. The $\mathrm{Aut}(x)$-representation (6.4.2) is isomorphic to the conjugate $\mathrm{Aut}(x)$-action on

$$\mathrm{Ext}_X^1(i_*F, i_*F) = \bigoplus_{e \in E(Q_{E_\bullet})} \mathrm{Hom}(V_{s(e)}, V_{t(e)})$$

where $Q_{E_\bullet}$ is the Ext-quiver associated with the collection $E_\bullet = (i_*F_1, \ldots, i_*F_k)$ (see Sect. 8.4.2). This is because of the isomorphisms

$$\mathrm{Ext}_X^1(i_*F_a, i_*F_b) \cong \mathrm{Ext}_S^1(i^*i_*F_a, F_b)$$

$$\cong \mathrm{Ext}_S^1(F_a, F_b) \oplus \mathrm{Hom}(F_a, F_b \otimes \omega_S)$$

$$\cong \mathrm{Ext}_S^1(F_a, F_b) \oplus \mathrm{Ext}_S^2(F_b, F_a)^\vee.$$

6.4.2 Line Bundles on Moduli Stacks

We now define some line bundle on $\mathcal{M}_S(v)$ associated with an integral class $\sigma \in A(S)_{\mathbb{C}}$.

Definition 6.4.3 For an integral class $\sigma = B + iH \in A(S)_{\mathbb{C}}$ such that H is an effective class, we define $l(\sigma) \in \mathrm{Pic}(\mathcal{M}_S(v))$ by

$$l(\sigma) = (\det \mathbf{R}p_{\mathcal{M}*}(\mathcal{F} \boxtimes \mathcal{O}_S(-B)))^{-\beta \cdot H} \otimes (\det \mathbf{R}p_{\mathcal{M}*}(\mathcal{F} \boxtimes \mathcal{O}_H))^{n - B \cdot \beta}.$$

Here $\mathcal{F}$ is a universal sheaf (3.4.2). Its pull-back to $\mathcal{M}_X(v)$, and also its restriction to an open substack of $\mathcal{M}_X(v)$ are also denoted by $l(\sigma)$.

The line bundle in Definition 6.4.3 descends to the line bundle in the $\mathbb{C}^*$-rigidification:

Lemma 6.4.4 *The line bundles $l(\sigma)$ on $\mathcal{M}_S(v)$, $\mathcal{M}_X(v)$ descend to line bundles on $\mathcal{M}_S^{\mathbb{C}^*\text{-rig}}(v)$, $\mathcal{M}_X^{\mathbb{C}^*\text{-rig}}(v)$.*

Proof At each point $[F] \in \mathcal{M}_S(v)$, the inertial $\mathbb{C}^*$-weight of $l(\sigma)|_{[F]}$ is

$$- (\beta \cdot H)\chi(F \otimes \mathcal{O}_S(-B)) + (n - B \cdot \beta)\chi(F \otimes \mathcal{O}_H)$$
$$= -(\beta \cdot H)(n - B \cdot \beta) + (n - B \cdot \beta)(\beta \cdot H) = 0.$$

Therefore $l(\sigma)$ descends to $\mathcal{M}_S^{\mathbb{C}^*\text{-rig}}(v)$. The case for $\mathcal{M}_X(v)$ follows from the same argument. $\qquad\square$

Suppose that $\sigma \in A(S)_{\mathbb{C}}$ lies on a wall and take $\sigma_\pm = B_\pm + iH_\pm \in A(S)_{\mathbb{C}}$ which lie on its adjacent chambers. Note that we have open immersions

$$\mathcal{M}_X^{\sigma_\pm}(v) \subset \mathcal{M}_X^{\sigma}(v).$$

Since each chamber contains dense rational points, by taking small deformations of $\sigma_\pm$ and rescaling we may assume that $B_\pm$, $H_\pm$ are integral and $H_\pm$ are effective without changing $\mathcal{M}_X^{\sigma_\pm}(v)$. Then by Definition 6.4.3, we have the line bundles $l(\sigma_\pm)$ on $\mathcal{M}_S^{\sigma}(v)$, $\mathcal{M}_X^{\sigma}(v)$. By Lemma 6.4.4, they descend to line bundles on $\mathcal{M}_S^{\sigma}(v)^{\mathbb{C}^*\text{-rig}}$, $\mathcal{M}_X^{\sigma}(v)^{\mathbb{C}^*\text{-rig}}$, which we also denote by $l(\sigma_\pm)$.

The following lemma shows that the open substacks $\mathcal{M}_X^{\sigma_\pm}(v) \subset \mathcal{M}_X^{\sigma}(v)$ coincide with $l(\sigma_\pm)$-semistable loci.

Lemma 6.4.5 *We have the identity of open substacks in $\mathcal{M}_X^{\sigma}(v)$,*

$$\mathcal{M}_X^{\sigma}(v)^{l(\sigma_\pm)\text{-ss}} = \mathcal{M}_X^{\sigma_\pm}(v). \tag{6.4.4}$$

Proof It is enough to prove the identity (6.4.4) on each fiber of the good moduli space morphism $\pi_{\mathcal{M}_X} : \mathcal{M}_X^{\sigma}(v) \to M_X^{\sigma}(v)$. Let $y \in M_X^{\sigma}(v)$ corresponds to a σ-polystable sheaf on X of the form

$$E = \bigoplus_{i=1}^{m} V_i \otimes E_i. \tag{6.4.5}$$

Here each V_i is a finite dimensional vector space and $\{E_1, \ldots, E_m\}$ are mutually non-isomorphic σ-stable sheaves such that $\mu_{\sigma}(E_i) = \mu_{\sigma}(E_j)$ for all i, j. Let $Q_{E_{\bullet}}$ be the Ext-quiver associated with the collection $(E_1, \ldots, E_m)$ (see Sect. 8.4.2). Then the fiber of $\pi_{\mathcal{M}_X}$ at y is the closed substack of the nilpotent $Q_{E_{\bullet}}$-representations with dimension vector $(\dim V_i)_{1 \le i \le m}$ (see Sect. 8.4.2)

$$\left[\{ \oplus_{(i \to j) \in Q_{E_{\bullet}}} \operatorname{Hom}(V_i, V_j) \}^{\mathrm{nil}} / G \right]. \tag{6.4.6}$$

Here $G = \prod_{i=1}^{m} \operatorname{GL}(V_i)$ and the subscript 'nil' means nilpotent $Q_{E_{\bullet}}$-representations. Let us write $[\pi_* E_i] = (\beta_i, n_i)$ in $N_{\le 1}(S)$. We define the following group homomorphisms $W_{\pm}$

$$W_{\pm} : K(Q_{E_{\bullet}}) \xrightarrow{\mathbf{dim}} \bigoplus_{i=1}^{m} \mathbb{Z} \cdot \mathbf{e}_i \to \mathbb{C}.$$

Here the first arrow is taking the dimension vector, and the second arrow is given by

$$\mathbf{e}_i \mapsto -n_i + B_{\pm} \cdot \beta_i + (H_{\pm} \cdot \beta_i)\sqrt{-1} \in \mathbb{C}.$$

Then $W_{\pm}$ determine Bridgeland stability conditions [24] on the abelian category of finite dimensional $Q_{E_{\bullet}}$-representations: a finite dimensional $Q_{E_{\bullet}}$-representation R is $W_{\pm}$-(semi)stable if for any non-zero subrepresentation $R' \subsetneq R$, we have

$$\arg W_{\pm}(R') < (\le) \arg W_{\pm}(R)$$

in $(0, \pi]$. By Toda [132, Lemma 7.8], the intersection $\pi_{\mathcal{M}_X}^{-1}(p) \cap \mathcal{M}_X^{\sigma\pm}(v)$ corresponds to $W_{\pm}$-semistable $Q_{E_{\bullet}}$-representations inside the stack (6.4.6). An easy calculation shows that a $Q_{E_{\bullet}}$-representation R of dimension vector $(\dim V_i)_{1 \le i \le m}$ is $W_{\pm}$-(semi)stable if and only if for any non-zero subrepresentation $R' \subsetneq R$, we have

$$\theta_{\pm}(R') := \sum_{i=1}^{m} \theta_{\pm,i} \cdot r_i > (\ge) 0 = \theta_{\pm}(R).$$

Here $(r_i)_{1 \leq i \leq m}$ is the dimension vector of R' and $\theta_{\pm, i} \in \mathbb{Z}$ is given by

$$\theta_{\pm, i} = (B_\pm \cdot \beta_i - n_i) \cdot (H_\pm \cdot \beta) + (n - B_\pm \cdot \beta) \cdot (H_\pm \cdot \beta_i). \qquad (6.4.7)$$

By the relation of θ-stability and GIT stability proved by King [71, Theorem 4.1], the $\theta_\pm$-(semi)stable loci in (6.4.6) correspond to GIT (semi)stable loci with respect to the characters

$$G \to \mathbb{C}^*, \ (g_i)_{1 \leq i \leq m} \mapsto \prod_{i=1}^{m} \det(g_i)^{\theta_{\pm, i}}. \qquad (6.4.8)$$

On the other hand, let $x \in \mathcal{M}_X^\sigma(v)$ be the closed point corresponding to the polystable sheaf (6.4.5). Then we have $G = \mathrm{Aut}(x)$. In the notation of Definition 6.2.6, the pull-backs $l(\sigma_\pm)_x \in \mathrm{Pic}(BG)$ are described as

$$l(\sigma_\pm)_x$$

$$= \det \mathbf{R}\Gamma \left(\bigoplus_{i=1}^{m} V_i \otimes \pi_* E_i \otimes \mathcal{O}_S(-B_\pm) \right)^{-H_\pm \cdot \beta}$$

$$\otimes \det \cdot \mathbf{R}\Gamma \left(\bigoplus_{i=1}^{m} V_i \otimes \pi_* E_i \otimes \mathcal{O}_{H_\pm} \right)^{n - B_\pm \cdot \beta}$$

$$= \bigotimes_{i=1}^{m} (\det V_i)^{-(H_\pm \cdot \beta) \cdot \chi(\pi_* E_i \otimes \mathcal{O}_S(-B_\pm))} \otimes \bigotimes_{i=1}^{m} (\det V_i)^{(H_\pm \cdot \beta_i) \cdot (n - B_\pm \cdot \beta)}$$

$$= \bigotimes_{i=1}^{m} \det(V_i)^{\theta_{+, i}}.$$

Therefore $l(\sigma_\pm)_x$ are induced by the G-characters (6.4.8). Together with using Lemma 6.1.1, the line bundles $l(\sigma_\pm)$ on $\mathcal{M}_X^\sigma(v)$ are induced by the G-characters (6.4.8) on the fiber $\pi_{\mathcal{M}_X}^{-1}(p)$, so the identity (6.4.4) holds on $\pi_{\mathcal{M}_X}^{-1}(p)$. □

Since the derived stacks $\mathfrak{M}_S^\sigma(v)$, $\mathfrak{M}_S^\sigma(v)^{\mathbb{C}^*\text{-rig}}$ are symmetric by Lemma 6.4.1, we take their maximal symmetric structures $\mathbb{S}$ as in Definition 6.2.1. The following lemma shows that the line bundles $l(\sigma_\pm)$ satisfy the genericity condition in Definition 6.2.6.

Lemma 6.4.6 *The line bundles* $l(\sigma_\pm) \in \mathrm{Pic}(\mathcal{M}_S^\sigma(v)^{\mathbb{C}^*\text{-rig}})$ *are* $\mathbb{S}$-*generic.*

Proof Let us take a closed point $x \in \mathcal{M}_S^\sigma(v)^{\mathbb{C}^*\text{-rig}}$ corresponding to a polystable sheaf (6.4.1). Then $G' := \mathrm{Aut}(x) = G/\mathbb{C}^*$ where $\mathbb{C}^* \subset G = \prod_{i=1}^{m} \mathrm{GL}(V_i)$ is the diagonal torus. From the proof of Lemma 6.4.5, the element $l(\sigma_\pm)_x \in \mathrm{Pic}(B\,\mathrm{Aut}(x))$ corresponds to a G'-character of the form (6.4.8), where $\theta_\pm$ are given as in (6.4.7) for $[F_i] = (\beta_i, n_i)$. Here note that the G-character (6.4.8) descends to the G'-character since it restricts to the trivial character on the diagonal torus $\mathbb{C}^* \subset G$ (see Lemma 6.4.4). By the assumption that $\sigma_\pm$ do not lie on walls, we have $\theta_\pm(\vec{v}') \neq 0$ for any $0 < \vec{v}' < \vec{v}$ where $\vec{v} = (\dim V_i)_{1 \le i \le m}$. Then the lemma follows from Remark 6.4.2 and Lemma 6.1.15.   □

6.4.3   Equivalences of DT Categories for One Dimensional Stable Sheaves

Applying Corollary 6.2.8, we have the following result which gives an evidence of Conjecture 4.1.2.

Theorem 6.4.7 *Let $\sigma \in A(S)_\mathbb{C}$ lies on a wall with respect to $v \in N_{\le 1}(S)$ and $\sigma_\pm \in A(S)_\mathbb{C}$ lie on its adjacent chambers. Moreover assume that*

$$\mathcal{M}_X^\sigma(v) \subset \pi_*^{-1}(\mathcal{M}_S^\sigma(v)). \tag{6.4.9}$$

Then there exists an equivalence

$$\mathcal{DT}^{\mathbb{C}^*}(M_X^{\sigma_+}(v)) \xrightarrow{\sim} \mathcal{DT}^{\mathbb{C}^*}(M_X^{\sigma_-}(v)).$$

Proof The condition (6.4.9) implies that $\mathcal{M}_X^\sigma(v) = \pi_*^{-1}(\mathcal{M}_S^\sigma(v))$ by Lemma 3.4.7. Therefore by Lemma 6.4.5, we have

$$M_X^{\sigma_\pm}(v) = t_0(\Omega_{\mathfrak{M}_S^\sigma(v)^{\mathbb{C}^*-\mathrm{rig}}}[-1])^{l(\sigma_\pm)-\mathrm{ss}}.$$

Therefore the theorem is a consequence of Lemmas 6.4.1, 6.4.6, and Corollary 6.2.8.
                                                                                  □

If we impose some further assumption, the result of Theorem 6.4.7 is described in terms of derived moduli spaces of stable sheaves on S.

Corollary 6.4.8 *Under the assumption of Theorem 6.4.7, suppose furthermore that the following condition also holds:*

$$\mathcal{M}_X^{\sigma_\pm}(v) \subset \pi_*^{-1}(\mathcal{M}_S^{\sigma_\pm}(v)). \tag{6.4.10}$$

Then we have an equivalence

$$\Theta_{\sigma_+,\sigma_-} : D_{\mathrm{coh}}^b(\mathfrak{M}_S^{\sigma_+}(v)^{\mathbb{C}^*-\mathrm{rig}}) \xrightarrow{\sim} D_{\mathrm{coh}}^b(\mathfrak{M}_S^{\sigma_-}(v)^{\mathbb{C}^*-\mathrm{rig}}) \tag{6.4.11}$$

such that we have the commutative diagram

$$
\begin{array}{ccc}
D^b_{\mathrm{coh}}(\mathfrak{M}^{\sigma+}_S(v)^{\mathbb{C}^*\text{-rig}}) & \xrightarrow[\sim]{\Theta_{\sigma+,\sigma-}} & D^b_{\mathrm{coh}}(\mathfrak{M}^{\sigma-}_S(v)^{\mathbb{C}^*\text{-rig}}) \\
\downarrow & & \downarrow \\
D^b_{\mathrm{coh}}(\mathfrak{M}^{\sigma\text{-st}}_S(v)^{\mathbb{C}^*\text{-rig}}) & \xrightarrow[\sim]{id} & D^b_{\mathrm{coh}}(\mathfrak{M}^{\sigma\text{-st}}_S(v)^{\mathbb{C}^*\text{-rig}}).
\end{array}
\tag{6.4.12}
$$

Here the vertical arrows are restriction functors.

Proof If the inclusion (6.4.10) holds, then it is identity by Lemma 3.4.7, and we have equivalences

$$
\mathcal{DT}^{\mathbb{C}^*}(M^{\sigma\pm}_X(v)) \xrightarrow{\sim} D^b_{\mathrm{coh}}(\mathfrak{M}^{\sigma\pm}_S(v)^{\mathbb{C}^*\text{-rig}})
$$

by Lemma 3.4.8. Therefore by Theorem 6.4.7, we have the equivalence (6.4.11). The commutative diagram (6.4.12) follows from Remark 6.2.9. $\qquad\square$

6.4.4 Examples and Applications

In this subsection, we give several examples where the conditions (6.4.9) and (6.4.10) hold so that we can apply Corollary 6.4.8. We first consider the case of $v = (\beta, n) \in N_{\leq 1}(S)$ such that β is a reduced class. Note that v is primitive for any n.

Lemma 6.4.9 *For $v = (\beta, n) \in N_{\leq 1}(S)$, suppose that β is reduced. Then the condition (6.4.9) is satisfied for any $\sigma \in A(S)_{\mathbb{C}}$.*

Proof Note that giving a compactly supported coherent sheaf on X is equivalent to giving a pair (F, θ), where $F \in \mathrm{Coh}(S)$ and $\theta \in \mathrm{Hom}(F, F \otimes \omega_S)$. The push-forward π_* sends such a pair (F, θ) to F.

Let $E \in \mathrm{Coh}_{\leq 1}(X)$ be a σ-semistable sheaf which corresponds to a pair (F, θ) such that $[F] = (\beta, n)$. Suppose that F is not σ-semistable. Then there exists an exact sequence $0 \to F' \to F \to F'' \to 0$ in $\mathrm{Coh}(S)$ such that F', F'' are pure one dimensional sheaves and $\mu_\sigma(F') > \mu_\sigma(F'')$. We consider the following diagram

$$
\begin{array}{ccccccccc}
0 & \longrightarrow & F' & \longrightarrow & F & \longrightarrow & F'' & \longrightarrow & 0 \\
 & & \downarrow & & \downarrow{\scriptstyle\theta} & & \downarrow & & \\
0 & \longrightarrow & F' \otimes \omega_S & \longrightarrow & F \otimes \omega_S & \longrightarrow & F'' \otimes \omega_S & \longrightarrow & 0.
\end{array}
\tag{6.4.13}
$$

By the assumption, the sheaf F is scheme theoretically supported on a reduced divisor on S. Therefore the supports of F' and F'' do not have common irreducible components, hence we have $\mathrm{Hom}(F', F'' \otimes \omega_S) = 0$. Then there exist dotted arrows in the diagram (6.4.13) which makes the diagram (6.4.13) commutative. However then the diagram (6.4.13) destabilizes E, which is a contradiction. $\square$

By the above lemma, in the reduced case the condition (6.4.9) is satisfied for any $\sigma \in A(S)_{\mathbb{C}}$. Therefore by Corollary 6.4.8, we have the following:

Corollary 6.4.10 *Let S be a smooth projective surface, and take $v = (\beta, n) \in N_{\leq 1}(S)$ such that β is a reduced class. Then for any $\sigma_{\pm} \in A(S)_{\mathbb{C}}$ which do not lie walls, we have an equivalence*

$$ D^b_{\mathrm{coh}}(\mathfrak{M}^{\sigma_+}_S(v)^{\mathbb{C}^*\text{-rig}}) \xrightarrow{\sim} D^b_{\mathrm{coh}}(\mathfrak{M}^{\sigma_-}_S(v)^{\mathbb{C}^*\text{-rig}}). $$

We also have the following lemma for the conditions (6.4.9) and (6.4.10) to hold:

Lemma 6.4.11 *Suppose that $c_1(\omega_S) \in \mathbb{R}_{\leq 0} \cdot H$ for some $H \in A(S)_{\mathbb{R}}$. Then for $\sigma = iH$, both of the conditions (6.4.9) and (6.4.10) hold for any primitive v.*

Proof We follow the proof of Lemma 6.4.9. As for the condition (6.4.9), let $E \in \mathrm{Coh}_{\leq 1}(X)$ be a σ-semistable sheaf which corresponds to a pair (F, θ) such that $[F] = (\beta, n)$. Suppose that F is not σ-semistable on S. Then by taking Harder-Narasimhan filtration in σ-stability, there exists an exact sequence

$$ 0 \to F' \to F \to F'' \to 0 \qquad\qquad (6.4.14) $$

in $\mathrm{Coh}(S)$ such that $\mu^{\min}_{\sigma}(F') > \mu^{\max}_{\sigma}(F'')$. Here $\mu^{\max}_{\sigma}(-)$ (resp. $\mu^{\min}_{\sigma}(-)$) is the maximal (resp. minimal) slope among the Harder-Narasimhan factors of $(-)$. By the assumption $c_1(\omega_S) \in \mathbb{R}_{\leq 0} \cdot H$, taking the tensor product with ω_S preserves σ-stability. Therefore we have

$$ \mu^{\min}_{\sigma}(F') > \mu^{\max}_{\sigma}(F'') \geq \mu^{\max}_{\sigma}(F'' \otimes \omega_S), $$

hence the vanishing $\mathrm{Hom}(F', F'' \otimes \omega_S) = 0$ holds. Then we obtain the diagram (6.4.13), and a contradiction.

As for the condition (6.4.10), the same argument as above applies to the case of $c_1(\omega_S) = 0$. So we assume that $c_1(\omega_S) \in \mathbb{R}_{<0} \cdot H$. Suppose that $E = (F, \theta)$ is $\sigma_{\pm}$-semistable but F is not $\sigma_{\pm}$-semistable. Then similarly to above, we have an exact sequence (6.4.14) such that $\mu^{\min}_{\sigma_{\pm}}(F') > \mu^{\max}_{\sigma_{\pm}}(F'')$. Since E is σ-semistable, F is also σ-semistable by the above argument, therefore F' and F'' are also σ-semistable with the same μ_{σ}-slope. Then $F'' \otimes \omega_S$ is also σ-semistable with

$$ \mu_{\sigma}(F') \geq \mu_{\sigma}(F'') > \mu_{\sigma}(F'' \otimes \omega_S), $$

hence $\mathrm{Hom}(F', F'' \otimes \omega_S) = 0$ holds. Similarly to above, we obtain a contradiction. $\square$

A surface S with $c_1(\omega_S) = 0$ is classified into four types: K3 surface, abelian surface, Enriques surface and bielliptic surface. In the above case, by Corollary 6.4.8 and Lemma 6.4.11 we have the following:

Corollary 6.4.12 *Suppose that S is a smooth projective surface satisfying $c_1(\omega_S) = 0$ in $H^2(S, \mathbb{R})$. Then for any primitive $v \in N_{\leq 1}(S)$ and $\sigma_\pm \in A(S)_\mathbb{C}$ which do not lie on walls, there exists an equivalence*

$$D^b_{\mathrm{coh}}(\mathfrak{M}^{\sigma_+}_S(v)^{\mathbb{C}^*\text{-rig}}) \xrightarrow{\sim} D^b_{\mathrm{coh}}(\mathfrak{M}^{\sigma_-}_S(v)^{\mathbb{C}^*\text{-rig}}). \tag{6.4.15}$$

Remark 6.4.13 In the case that S is a K3 surface, the classical truncations

$$M^{\sigma_\pm}_S(v) := t_0(\mathfrak{M}^{\sigma_\pm}_S(v))$$

are holomorphic symplectic manifolds [13, 92]. A derived equivalence

$$D^b_{\mathrm{coh}}(M^{\sigma_+}_S(v)) \xrightarrow{\sim} D^b_{\mathrm{coh}}(M^{\sigma_-}_S(v)) \tag{6.4.16}$$

is a special case of Halpern-Leistner's result [46]. Since the closed immersion $M^{\sigma_\pm}_S(v) \hookrightarrow \mathfrak{M}^{\sigma_\pm}_S(v)$ is not an equivalence of derived stacks, the equivalence (6.4.15) does not directly imply (6.4.16). This is caused by the existence of surjections $\mathrm{Ext}^2_S(E, E) \twoheadrightarrow \mathbb{C}$ for any coherent sheaf E on S. In order to obtain an equivalence (6.4.16), we replace $\mathfrak{M}^{\mathbb{C}^*\text{-rig}}_S(v)$ with another derived stack $\mathfrak{M}^{\mathrm{rig}}_S(v)$ obtained from $\mathfrak{M}^{\mathbb{C}^*\text{-rig}}_S(v)$ by getting rid of the above one dimensional obstruction space from $\mathfrak{M}^{\mathbb{C}^*\text{-rig}}_S(v)$ as in [46, Proposition 3.4.7]. Then we can recover the Halpern-Leistner's equivalence (6.4.16) from the argument of Theorem 6.4.7.

In the case of Enriques surface, Sacca [120] proved the following:

Theorem 6.4.14 ([120]) *Let S be a general Enriques surface and $|C|$ be a linear system on it which contains an irreducible divisor $C \subset S$ with arithmetic genus $g \geq 2$. We take $v = ([C], n) \in \mathrm{NS}(S) \oplus \mathbb{Z}$ such that $n \neq 0$ and $([C], 2n)$ is coprime. Then for generic $H_\pm \in A(S)_\mathbb{R}$, the moduli spaces $M^{H_\pm}_S(v)$ are smooth $(\beta^2 + 1)$-dimensional birational Calabi-Yau manifolds.*

By Corollary 6.4.12, we have an equivalence of Sacca's Calabi-Yau manifolds for different polarizations:

Corollary 6.4.15 *For a general Enriques surface S, let $v \in N_{\leq 1}(S)$ be as in Theorem 6.4.14. Then for generic $H_\pm \in A(S)_\mathbb{R}$, there exists an equivalence*

$$D^b_{\mathrm{coh}}(M^{H_+}_S(v)) \xrightarrow{\sim} D^b_{\mathrm{coh}}(M^{H_-}_S(v)). \tag{6.4.17}$$

Proof Let $H \in A(S)_\mathbb{R}$ lies on a wall, and $H_\pm$ lie on its adjacent chambers. It is enough to show an equivalence (6.4.17) for such $H_\pm$. It is proved in [120, Lemma 2.5] that there is no obstruction space for $H_\pm$-stable sheaf E such that

$[E] \in M_S^{H_\pm}(v)$, i.e. $\mathrm{Ext}^2(E, E) = 0$. Therefore for $\sigma_\pm = iH_\pm$, the closed immersions

$$M_S^{H_\pm}(v) = t_0(\mathfrak{M}_S^{\sigma_\pm}(v)^{\mathbb{C}^*\text{-rig}}) \hookrightarrow \mathfrak{M}_S^{\sigma_\pm}(v)^{\mathbb{C}^*\text{-rig}}$$

are equivalences. Therefore the corollary follows from Corollary 6.4.12. □

Let S be a del-Pezzo surface, i.e. $-K_S$ is ample. In this case, we have the following:

Corollary 6.4.16 *Let S be a del-Pezzo surface. Then for any primitive $v \in N_{\leq 1}(S)$ and generic perturbations $\sigma_\pm$ of $-iK_S$, we have a derived equivalence of smooth projective varieties*

$$D_{\mathrm{coh}}^b(M_S^{\sigma_+}(v)) \xrightarrow{\sim} D_{\mathrm{coh}}^b(M_S^{\sigma_-}(v)).$$

Proof The moduli stack $\mathcal{M}_S^\sigma(v)$ is a smooth stack without obstruction space, i.e. for any Gieseker $-K_S$-semistable sheaf E on S with $[E] = v$, we have

$$\mathrm{Ext}^2(E, E) = \mathrm{Hom}(E, E \otimes \omega_S)^\vee = 0.$$

Therefore for generic perturbations $\sigma_\pm$ of $-iK_S$, the moduli spaces $M_S^{\sigma_\pm}(v)$ are smooth projective varieties such that the closed immersions

$$M_S^{\sigma_\pm}(v) = t_0(\mathfrak{M}_S^{\sigma_\pm}(v)^{\mathbb{C}^*\text{-rig}}) \hookrightarrow \mathfrak{M}_S^{\sigma_\pm}(v)^{\mathbb{C}^*\text{-rig}}$$

are equivalences. Therefore the corollary follows from Corollary 6.4.8 and Lemma 6.4.11. □

6.5 Application to Categorical MNOP/PT Correspondence

Here we apply the result of Corollary 6.2.8 to prove Conjecture 4.3.3 for reduced curve classes. Below we use the notation of Sect. 4.3.

6.5.1 The Moduli Stack of Semistable Objects on MNOP/PT Wall

Let S be a smooth projective surface, and consider the moduli stacks of pairs $\mathcal{M}_S^\dagger(v)$, moduli stacks of D0-D2-D6 bound states $\mathcal{M}_X^\dagger(v)$ considered in Sects. 4.2.1 and 4.2.2 respectively. For $v = (\beta, n) \in N_{\leq 1}(S)$, we consider open substacks

$$\mathcal{T}_n(S, \beta) \subset \mathcal{M}_S^\dagger(\beta, n), \quad \mathcal{T}_n(X, \beta) \subset \mathcal{M}_X^\dagger(\beta, n)$$

corresponding to pairs $(\mathcal{O}_S \overset{\xi}{\to} F)$ such that $\mathrm{Cok}(\xi)$ is at most zero dimensional, objects $\mathcal{E} \in \mathcal{A}_X$ such that $\mathcal{H}^1(\mathcal{E})$ is at most zero dimensional, respectively. Note that the latter moduli stack $\mathcal{T}_n(X, \beta)$ is considered in (4.3.2) as the stack of semistable objects on the MNOP/PT wall. Similarly to Lemma 4.3.13, we have the open immersion

$$(\pi^{\dagger}_*)^{-1}(\mathcal{T}_n(S, \beta)) \subset \mathcal{T}_n(X, \beta). \tag{6.5.1}$$

We define some line bundles on the above moduli stacks.

Definition 6.5.1   We define $l_I, l_P \in \mathrm{Pic}(\mathcal{M}^{\dagger}_S(v))$ to be

$$l_I := \det(\mathbf{R}p_{\mathcal{M}^{\dagger}*}\mathcal{F}), \ \ l_P := \det(\mathbf{R}p_{\mathcal{M}^{\dagger}*}\mathcal{F})^{-1}.$$

Here $\mathcal{O}_{S \times \mathcal{M}^{\dagger}_S(v)} \to \mathcal{F}$ be the universal pair. The restrictions of l_I, l_P to $\mathcal{T}_n(S, \beta)$ and pull-backs via $\pi^{\dagger}_*\colon \mathcal{T}_n(X, \beta) \to \mathcal{M}^{\dagger}_S(v)$ are also denoted by l_I, l_P.

We have the following proposition:

Proposition 6.5.2   *For $l_I, l_P \in \mathrm{Pic}(\mathcal{T}_n(X, \beta))$, we have the identities*

$$\mathcal{I}_n(X, \beta) = \mathcal{T}_n(X, \beta)^{l_I\text{-ss}}, \ \ \mathcal{P}_n(X, \beta) = \mathcal{T}_n(X, \beta)^{l_P\text{-ss}}. \tag{6.5.2}$$

Proof   It is enough to show the identities (6.5.2) on the fibers of good moduli space morphisms $\pi_{\mathcal{T}}\colon \mathcal{T}_n(X, \beta) \to T_n(X, \beta)$. A closed point $y \in T_n(X, \beta)$ corresponds to a polystable object on the MNOP/PT wall (see [137, Appendix B])

$$I_C \oplus \bigoplus_{i=1}^{m} V_i \otimes \mathcal{O}_{p_i}[-1]. \tag{6.5.3}$$

Here $I_C \subset \mathcal{O}_{\overline{X}}$ is the ideal sheaf of a Cohen-Macaulay curve $C \subset X$, $p_1, \ldots, p_m \in X$ are distinct points and V_i is a finite dimensional vector space. Let Q be the Ext-quiver associated with the collection $\{I_C, \mathcal{O}_{p_1}[-1], \ldots, \mathcal{O}_{p_m}[-1]\}$. Then the vertex set $V(Q)$ is $\{0, 1, \ldots, m\}$, and the edge set $E(Q)$ is given by

$$\sharp(0 \to i) = \hom(I_C, \mathcal{O}_{p_i}), \ \ \sharp(i \to 0) = \mathrm{ext}^2(\mathcal{O}_{p_i}, I_C), \ \ \sharp(0 \to 0) = \mathrm{ext}^1(I_C, I_C)$$

and also for $1 \leq i, j \leq m$

$$\sharp(i \to j) = \begin{cases} 3, \ (i = j) \\ 0, \ (i \neq j). \end{cases}$$

By the argument in Sect. 8.4.2, we have the closed immersion

$$\pi_{\mathcal{T}}^{-1}(y) \hookrightarrow \left[\left\{ (V_i)^{\sharp(0\to i)} \oplus (V_i^{\vee})^{\sharp(i\to 0)} \oplus \bigoplus_{i=1}^{m} \text{End}(V_i)^{\oplus 3} \oplus \mathbb{C}^{\sharp(0\to 0)} \right\}^{\text{nil}} / G \right].$$
(6.5.4)

Here the RHS is the moduli stack of nilpotent Q-representations with dimension vector $(1, \{\dim V_i\}_{1 \le i \le k})$. The algebraic group $G = \prod_{i=1}^{m} \text{GL}(V_i)$ acts on V_i, $V_i^{\vee}$ in a standard way, on $\text{End}(V_i)$ by the conjugation and on $\mathbb{C}^{\sharp(0\to 0)}$ trivially.

Let $x \in \mathcal{T}_n(X, \beta)$ be a closed point corresponding to the object (6.5.3). Then we have $G = \text{Aut}(x)$ and

$$(l_I)_x = \det \mathbf{R}\Gamma \left(\mathcal{O}_C \oplus \bigoplus_{i=1}^{m} V_i \otimes \mathcal{O}_{p_i} \right) \cong \bigotimes_{i=1}^{m} \det V_i$$

as an element of $\text{Pic}(BG)$, i.e. it is the determinant character

$$\chi_0 \colon G \to \mathbb{C}^*, \ (g_i)_{1 \le i \le m} \mapsto \prod_{i=1}^{m} \det(g_i).$$
(6.5.5)

Let e_i for $1 \le i \le m$ be the simple Q-representation corresponding to the vertex i, and $Q_0 \subset Q$ be the full subquiver whose vertex set is $\{1, \ldots, m\}$. By Lemma 6.1.9, a Q-representation R with dimension vector $(1, \{\dim V_i\}_{1 \le i \le m})$ is χ_0-(semi)stable if and only if the images of $\mathbb{C} \to V_i$ for $(0 \to i)$ generate $\oplus_{i=1}^{m} V_i$ as a $\mathbb{C}[Q_0]$-module. This is equivalent to that $\text{Hom}(R, e_i) = 0$ for all $1 \le i \le m$. On the other hand, the fiber $\pi_{\mathcal{T}}^{-1}(y)$ parametrizes objects in the extension closure

$$\langle I_C, \mathcal{O}_{p_1}[-1], \ldots, \mathcal{O}_{p_m}[-1] \rangle_{\text{ex}}.$$
(6.5.6)

Under the embedding (6.5.4), an object $\mathcal{E}$ in (6.5.6) corresponds to a Q-representation R with $\text{Hom}(R, e_i) = 0$ if and only if $\text{Hom}(\mathcal{E}, \mathcal{O}_{p_i}[-1]) = 0$. The object $\mathcal{E}$ is represented by a pair $(\mathcal{O}_{\overline{X}} \xrightarrow{\xi} E)$ for a generically surjective ξ by Lemma 4.3.2, and the condition $\text{Hom}(\mathcal{E}, \mathcal{O}_{p_i}[-1]) = 0$ for all i is equivalent to that ξ is surjective. Therefore the identity for $I_n(X, \beta)$ holds on $\pi_{\mathcal{T}}^{-1}(y)$.

The identity for $P_n(X, \beta)$ is similar. By Lemma 6.1.9, a Q-representation R with dimension vector $(1, \{\dim V_i\}_{1 \le i \le m})$ is χ_0^{-1}-(semi)stable if and only if the images of duals of $V_i \to \mathbb{C}$ for $(i \to 0)$ generate $\oplus_{i=1}^{m} V_i^{\vee}$ as a $\mathbb{C}[Q_0]$-module, which is equivalent to that $\text{Hom}(e_i, R) = 0$ for all $1 \le i \le m$. Under the embedding (6.5.4), an object $\mathcal{E}$ in (6.5.6) corresponds to a Q-representation R with $\text{Hom}(e_i, R) = 0$ if and only if $\text{Hom}(\mathcal{O}_{p_i}[-1], \mathcal{E}) = 0$. As $\mathcal{E}$ is represented by a pair $(\mathcal{O}_{\overline{X}} \xrightarrow{\xi} E)$ for a generically surjective ξ, the above condition is equivalent to that E is pure. Therefore the identity for $P_n(X, \beta)$ holds on $\pi_{\mathcal{T}}^{-1}(y)$. $\square$

We have the derived open substack

$$\mathfrak{T}_n(S, \beta) \subset \mathfrak{M}_S^{\dagger}(\beta, n)$$

whose classical truncation is $\mathcal{T}_n(S, \beta)$. In the following proposition, we show the existence of a symmetric structure on $\mathfrak{T}_n(S, \beta)$ such that l_P is compatible with it.

Proposition 6.5.3 *There exists a symmetric structure* $\mathbb{S}$ *on* $\mathfrak{T}_n(S, \beta)$ *such that* l_I, l_P *are* $\mathbb{S}$*-generic and* l_P *is compatible with* $\mathbb{S}$.

Proof A closed point x of the stack $\mathcal{T}_n(S, \beta)$ corresponds to a direct sum of pairs

$$(\mathcal{O}_S \to F) = (\mathcal{O}_S \twoheadrightarrow \mathcal{O}_C) \oplus \bigoplus_{i=1}^{m} V_i \otimes (0 \to \mathcal{O}_{p_i}).$$

Here $C \subset S$ is a Cohen-Macaulay curve, $p_1, \dots, p_m \in S$ are distinct points and V_i is a finite dimensional vector space. Then $\mathrm{Aut}(x) = \prod_{i=1}^{m} \mathrm{GL}(V_i)$. From the description of the cotangent complex (4.2.2), we have

$$\mathbb{T}_{\mathfrak{T}_n(S,\beta)}|_x = \mathbf{RHom}\left(\mathcal{O}_S(-C) \oplus \bigoplus_{i=1}^{m}(V_i \otimes \mathcal{O}_{p_i})[-1], \mathcal{O}_C \oplus \bigoplus_{i=1}^{m} V_i \otimes \mathcal{O}_{p_i}\right).$$

By the above, an easy computation shows that

$$\mathcal{H}^0(\mathbb{T}_{\mathfrak{T}_n(S,\beta)}|_x) \oplus \mathcal{H}^1(\mathbb{T}_{\mathfrak{T}_n(S,\beta)}|_x)^{\vee} \tag{6.5.7}$$

$$= \left(\bigoplus_{i=1}^{m} \mathrm{Ext}_S^2(\mathcal{O}_{p_i}, \mathcal{O}_C)^{\vee} \oplus H^0(\mathcal{O}_S(C)|_{p_i})\right) \otimes V_i \oplus \left(\bigoplus_{i=1}^{m} \mathrm{Ext}_S^1(\mathcal{O}_{p_i}, \mathcal{O}_C)\right) \otimes V_i^{\vee}$$

$$\oplus \bigoplus_{i=1}^{m} \mathrm{End}(V_i)^{\oplus 3} \oplus H^0(\mathcal{O}_C(C)) \oplus H^1(\mathcal{O}_C(C))^{\vee}.$$

We take the the decomposition $\mathbb{S}_x \oplus \mathbb{U}_x$ of (6.5.7) to be

$$\mathbb{S}_x = \left(\bigoplus_{i=1}^{m} \mathrm{Ext}_S^2(\mathcal{O}_{p_i}, \mathcal{O}_C)^{\vee}\right) \otimes V_i \oplus \left(\bigoplus_{i=1}^{m} \mathrm{Ext}_S^1(\mathcal{O}_{p_i}, \mathcal{O}_C)\right) \otimes V_i^{\vee}$$

$$\oplus \bigoplus_{i=1}^{m} \mathrm{End}(V_i)^{\oplus 3} \oplus H^0(\mathcal{O}_C(C)) \oplus H^1(\mathcal{O}_C(C))^{\vee},$$

$$\mathbb{U}_x = \bigoplus_{i=1}^{m} H^0(\mathcal{O}_S(C)|_{p_i}) \otimes V_i.$$

As C is Cohen-Macaulay we have $\text{Hom}(\mathcal{O}_{p_i}, \mathcal{O}_C) = 0$, so the Riemann-Roch theorem shows that $\text{ext}_S^1(\mathcal{O}_{p_i}, \mathcal{O}_C) = \text{ext}_S^2(\mathcal{O}_{p_i}, \mathcal{O}_C)$. Therefore $\mathbb{S}_x$ is a symmetric $\text{Aut}(x)$-representation, and the decomposition $\mathbb{S}_x \oplus \mathbb{U}_x$ of (6.5.7) gives its symmetric structure.

Note that, similarly to Remark 6.4.2, the $\text{Aut}(x)$-representation (6.5.7) is obtained as the space of Q-representations with dimension vector $\{1, \{\dim V_i\}_{1 \leq i \leq m}\}$, where Q is the Ext-quiver associated with the collection

$$\{I_{i(C)}, i_* \mathcal{O}_{p_1}[-1], \ldots, i_* \mathcal{O}_{p_m}[-1]\}.$$

Here $i: S \hookrightarrow X$ is the zero section. From the proof of Proposition 6.5.2, the $(l_I)_x$-stability on (6.5.7) is given by the determinant character χ_0 in (6.5.5), and $(l_P)_x$ is given by χ_0^{-1}. By Lemma 6.1.12, these characters are $\mathbb{S}_x$-generic, and $(l_P)_x$ is compatible with $\mathbb{S}_x$. Therefore l_I, l_P are $\mathbb{S}$-generic, and l_P is compatible with $\mathbb{S}$. □

6.5.2 Proof of Categorical MNOP/PT Correspondence

In the case of reduced curve class, we have the following lemma:

Lemma 6.5.4 *If β is a reduced class, then the inclusion (6.5.1) is the identity. So we have*

$$\mathcal{T}_n(X, \beta) = t_0(\Omega_{\mathfrak{T}_n(S,\beta)}[-1]). \tag{6.5.8}$$

Proof By Lemma 4.3.2, a $\mathbb{C}$-valued point in the right hand side of (6.5.1) corresponds to a pair $(s: \mathcal{O}_{\overline{X}} \to F)$ such that $\text{Cok}(s)$ is at most zero dimensional. If β is reduced, then s is non-zero at each generic point of the support of F. Therefore $\pi_* s: \mathcal{O}_S \to \pi_* F$ is non-zero at each generic point of the support of $\pi_* F$. Since the fundamental one cycle of $\pi_* F$ is reduced, this implies that $\text{Cok}(\pi_* s)$ is at most zero dimensional. □

As an application of Corollary 6.2.8, we give the following application to Conjecture 4.3.3:

Theorem 6.5.5 *Suppose that β is a reduced class. Then there exists a fully-faithful functor*

$$\mathcal{DT}^{\mathbb{C}^*}(P_n(X, \beta)) \hookrightarrow \mathcal{DT}^{\mathbb{C}^*}(I_n(X, \beta)).$$

Proof By Lemma 6.5.4 and Lemma 3.2.9, we can take a quasi-compact derived open substack in Definition 4.3.1 to be

$$\mathfrak{M}_S^\dagger(\beta, n)_{\text{qc}} = \mathfrak{T}_n(S, \beta).$$

By (6.5.8) and Proposition 6.5.2, we have

$$I_n(X, \beta) = t_0(\Omega_{\mathfrak{T}_n(S,\beta)}[-1])^{l_I\text{-ss}}, \quad P_n(X, \beta) = t_0(\Omega_{\mathfrak{T}_n(S,\beta)}[-1])^{l_P\text{-ss}}.$$

As we mentioned before $\mathcal{T}_n(S, \beta)$ admits a good moduli space, and $\mathfrak{T}_n(S, \beta)$ satisfies formal neighborhood theorem by Lemma 8.4.3 (by applying it for $t = \infty$). Then by Proposition 6.5.3, we can apply Corollary 6.2.8 to conclude the theorem.

□

Remark 6.5.6 For a stable pair $(\mathcal{O}_X \to E)$ on X, if E has a reduced support then it push-forwards to a stable pair $(\mathcal{O}_S \to \pi_* E)$ on S. Therefore $P_n(X, \beta) = t_0(\Omega_{\mathfrak{P}_n(S,\beta)}[-1])$ for a reduced class β, where $\mathfrak{P}_n(S, \beta)$ is the derived moduli space of stable pairs on S. Therefore we have an equivalence by Lemma 3.2.9

$$\mathcal{DT}^{\mathbb{C}^*}(P_n(X, \beta)) \xrightarrow{\sim} D^b_{\text{coh}}(\mathfrak{P}_n(S, \beta)).$$

On the other hand, a surjection $\mathcal{O}_X \twoheadrightarrow F$ does not push-forward to a surjection $\mathcal{O}_S \twoheadrightarrow \pi_* F$ in general, even if the support of F is reduced. So for a reduced class β, the $\mathbb{C}^*$-equivariant MNOP category $\mathcal{DT}^{\mathbb{C}^*}(I_n(X, \beta))$ is not necessary equivalent to $D^b_{\text{coh}}(\mathfrak{I}_n(S, \beta))$, where $\mathfrak{I}_n(S, \beta)$ is the derived moduli space of closed subschemes in S.

Remark 6.5.7 In [115], the fully-faithful functor in Theorem 6.5.5 is refined to a semiorthogonal decomposition, which categorifies MNOP/PT correspondence (1.3.2). See Sect. 1.5.3.

Chapter 7
Categorified Hall Products on DT Categories

In this chapter, we use categorified Hall products for DT categories to give an approach toward Conjectures 4.3.3 and 4.3.14. The two dimensional categorified Hall algebras are introduced by Porta-Sala [109] in order to categorify cohomological Hall algebras for surfaces introduced by Kapranov-Vasserot [66]. For a smooth projective surface S, the Porta-Sala categorified Hall product is defined by pull-back/push-forward diagram associated with the derived moduli stack $\mathfrak{M}_S^{\text{ext}}(v_\bullet)$ of short exact sequences

$$0 \to F_1 \to F_3 \to F_2 \to 0, \ [F_i] = v_i.$$

It is equipped with evaluation morphisms $\text{ev}_i \colon \mathfrak{M}_S^{\text{ext}}(v_\bullet) \to \mathfrak{M}_S(v_i)$, so we have a correspondence diagram

$$
\begin{array}{ccc}
\mathfrak{M}_S^{\text{ext}}(v_\bullet) & \xrightarrow{\ \text{ev}_3\ } & \mathfrak{M}_S(v_3) \\[4pt]
{\scriptstyle(\text{ev}_1,\text{ev}_2)}\Big\downarrow & & \\[8pt]
\mathfrak{M}_S(v_1) \times \mathfrak{M}_S(v_2). & &
\end{array}
$$

The Porta-Sala categorified Hall product is given by the functor

$$\text{ev}_{3*}(\text{ev}_1, \text{ev}_2)^* \colon D_{\text{coh}}^b(\mathfrak{M}_S(v_1)) \times D_{\text{coh}}^b(\mathfrak{M}_S(v_2)) \to D_{\text{coh}}^b(\mathfrak{M}_S(v_3)).$$

In a separate paper [134], we proved that Porta-Sala categorified Hall product is compatible with singular supports, so that it descends to the categorified Hall product on DT categories for the local surface $X = \text{Tot}_S(\omega_S)$

$$\mathcal{DT}^{\mathbb{C}^*}(\mathcal{M}_X^\sigma(v_1)) \times \mathcal{DT}^{\mathbb{C}^*}(\mathcal{M}_X^\sigma(v_2)) \to \mathcal{DT}^{\mathbb{C}^*}(\mathcal{M}_X^\sigma(v_3)).$$

A similar construction yields an action of DT categories for one or zero dimensional semistable sheaves on X to those for stable D0-D2-D6 bound states on X. Here we

Y. Toda, *Categorical Donaldson-Thomas Theory for Local Surfaces*, Lecture Notes in Mathematics 2350, https://doi.org/10.1007/978-3-031-61705-8_7

explain these actions specializing to the MNOP/PT wall. Let $\mathcal{T}_n(X, \beta)$ be the moduli stack of semistable objects on the MNOP/PT wall, and $\mathcal{M}_i(X)$ be the moduli stack of zero dimensional sheaves on X with length i. Then we will see that there is a left/right action of DT categories for $\mathcal{M}_i(X)$ to those for $\mathcal{T}_n(X, \beta)$:

$$\bigoplus_{i \geq 0} \mathcal{DT}^{\mathbb{C}^*}(\mathcal{M}_i(X)) \curvearrowright \bigoplus_{n \in \mathbb{Z}} \mathcal{DT}^{\mathbb{C}^*}(\mathcal{T}_n(X, \beta)) \curvearrowleft \bigoplus_{i \geq 0} \mathcal{DT}^{\mathbb{C}^*}(\mathcal{M}_i(X)).$$

The above actions descend to those on left/right actions on MNOP/PT categories, and in particular the right actions on PT categories are studied in [134] in detail. On the other hand, the above actions do not descend to right/left actions on MNOP/PT categories, which we focus in this chapter.

Our main conjecture in Conjecture 7.2.4 specialized to the MNOP/PT wall yields a conjectural semiorthogonal decomposition of $\mathcal{DT}^{\mathbb{C}^*}(\mathcal{T}_n(X, \beta))$ described in Theorem 1.4.8, which we will prove when β is reduced. Although the proof works only for reduced β, this approach may give an ansatz toward Conjecture 4.3.3 for any curve class. In particular it gives another proof of Theorem 6.5.5. Our main conjecture in Conjecture 7.2.4 extends the above construction to further wall-crossing of D0-D2-D6 bound states, which gives an ansatz toward Conjecture 4.3.14. By proving the above conjecture, we will settle Conjecture 4.3.14 when β is a reduced class and $t > m(\beta)$ where $m(\beta)$ only depends on β.

The organization of this chapter is as follows. In Sect. 7.1, we recall categorified Hall products on surfaces and local surfaces. In Sect. 7.2, we formulate the main conjecture on semiorthogonal decompositions via categorified Hall products. In Sect. 7.3, we prove our conjectures under the assumption that the semistability on X is preserved under push-forward to S at the wall.

In this chapter, we will use the following notation. Let $T_i = \operatorname{Spec} A_i$ for complete local $\mathbb{C}$-algebras A_i for $i = 1, 2$. We denote by $A_1 \widehat{\otimes} A_2$ the complete tensor product over $\mathbb{C}$, and write $T_1 \widehat{\times}_R T_2 := \operatorname{Spec}(A_1 \widehat{\otimes} A_2)$. For derived stacks $\mathcal{X}_i \to T_i$ over T_i, we denote by $\mathcal{X}_1 \widehat{\times} \mathcal{X}_2 \to T_1 \widehat{\times} T_2$ the pull-back of $\mathcal{X}_1 \times \mathcal{X}_2 \to T_1 \times T_2$ via $T_1 \widehat{\times} T_2 \to T_1 \times T_2$. We denote by $D^b(\mathcal{X}_1) \widehat{\boxtimes} D^b(\mathcal{X}_2) = D^b(\mathcal{X}_1 \widehat{\times} \mathcal{X}_2)$. The triangulated subcategory $\mathcal{C}_1 \widehat{\boxtimes} \mathcal{C}_2$ in $D^b(\mathcal{X}_1) \widehat{\boxtimes} D^b(\mathcal{X}_2)$ is defined to be split generated by $(C_1 \boxtimes C_2)|_{\mathcal{X}_1 \widehat{\times} \mathcal{X}_2}$ for $C_i \in \mathcal{C}_i$. These notation also apply to categories of ($\mathbb{C}^*$-equivariant) factorizations in an obvious way.

7.1 Categorified Hall Products

7.1.1 Derived Moduli Stacks of Extensions

Let S be a smooth projective surface. We define the derived moduli stack of exact sequences of coherent sheaves on S, following [109, Section 3]. It is given by the derived Artin stack

$$\mathfrak{M}_S^{\text{ext}} : dAff^{op} \to SSets$$

which sends an affine derived scheme T to the ∞-groupoid of fiber sequences of perfect complexes on $T \times S$,

$$\mathfrak{F}_1 \to \mathfrak{F}_3 \to \mathfrak{F}_2 \tag{7.1.1}$$

whose restrictions to $t_0(T) \times S$ are flat families of exact sequences of coherent sheaves on S over $t_0(T)$. The classical truncation of $\mathfrak{M}_S^{\text{ext}}$ is denoted by $\mathcal{M}_S^{\text{ext}} := t_0(\mathfrak{M}_S^{\text{ext}})$. We have the decompositions into open and closed substacks

$$\mathfrak{M}_S^{\text{ext}} = \coprod_{v_\bullet = (v_1, v_2)} \mathfrak{M}_S^{\text{ext}}(v_\bullet), \ \ \mathcal{M}_S^{\text{ext}} = \coprod_{v_\bullet = (v_1, v_2)} \mathcal{M}_S^{\text{ext}}(v_\bullet)$$

where each component corresponds to exact sequences $0 \to F_1 \to F_3 \to F_2 \to 0$ on S with $[F_i] = v_i$ for $i = 1, 2$.

By sending a sequence (7.1.1) to $\mathfrak{F}_i$, we have the evaluation morphisms

$$\mathrm{ev}_i \colon \mathfrak{M}_S^{\text{ext}}(v_\bullet) \to \mathfrak{M}_S(v_i), \ 1 \le i \le 3$$

where we set $v_3 = v_1 + v_2$. Below we use the following diagram

$$
\begin{array}{ccc}
\mathfrak{M}_S^{\text{ext}}(v_\bullet) & \xrightarrow{\ \mathrm{ev}_3\ } & \mathfrak{M}_S(v_3) \\[1mm]
{\scriptstyle (\mathrm{ev}_1, \mathrm{ev}_2)} \Big\downarrow & & \\[2mm]
\mathfrak{M}_S(v_1) \times \mathfrak{M}_S(v_2). & &
\end{array}
\tag{7.1.2}
$$

The morphism $(\mathrm{ev}_1, \mathrm{ev}_2)$ is quasi-smooth (see [109, Corollary 3.10]), so we have the well-defined functor

$$(\mathrm{ev}_1, \mathrm{ev}_2)^* \colon D_{\mathrm{coh}}^b(\mathfrak{M}_S(v_1) \times \mathfrak{M}_S(v_2)) \to D_{\mathrm{coh}}^b(\mathfrak{M}_S^{\text{ext}}(v_\bullet)).$$

The horizontal morphism in (7.1.2) is a proper morphism, so we have the well-defined functor

$$\mathrm{ev}_{3*} \colon D_{\mathrm{coh}}^b(\mathfrak{M}_S^{\text{ext}}(v_\bullet)) \to D_{\mathrm{coh}}^b(\mathfrak{M}_S(v_3)).$$

The Porta-Sala categorified Hall product is given by the composition [109]

$$D_{\mathrm{coh}}^b(\mathfrak{M}_S(v_1)) \times D_{\mathrm{coh}}^b(\mathfrak{M}_S(v_2)) \xrightarrow{\boxtimes} D_{\mathrm{coh}}^b(\mathfrak{M}_S(v_1) \times \mathfrak{M}_S(v_2)) \tag{7.1.3}$$

$$\xrightarrow{\mathrm{ev}_{3*}(\mathrm{ev}_1, \mathrm{ev}_2)^*} D_{\mathrm{coh}}^b(\mathfrak{M}_S(v_3))$$

which categorifies cohomological Hall algebra on surfaces by Kapranov-Vasserot [66].

Similarly we consider the classical Artin stack of short exact sequences of compactly supported coherent sheaves on X. It is given by the 2-functor

$$\mathcal{M}_X^{\text{ext}} \colon Aff^{\text{op}} \to Groupoid$$

whose T-valued points for $T \in Aff$ form the groupoid of exact sequences of coherent sheaves on $X \times T$,

$$0 \to \mathcal{E}_1 \to \mathcal{E}_3 \to \mathcal{E}_2 \to 0, \ \mathcal{E}_i \in \mathcal{M}_X(T).$$

We have the decomposition into open and closed substacks

$$\mathcal{M}_X^{\text{ext}} = \coprod_{v_\bullet = (v_1, v_2)} \mathcal{M}_X^{\text{ext}}(v_\bullet)$$

where each component corresponds to exact sequences $0 \to E_1 \to E_3 \to E_2 \to 0$ on X with $[\pi_* E_i] = v_i$. We have the evaluation morphisms

$$\text{ev}_i^X \colon \mathcal{M}_X^{\text{ext}}(v_\bullet) \to \mathcal{M}_X(v_i), \ E_\bullet \mapsto E_i$$

and obtain the diagram

$$
\begin{array}{ccc}
\mathcal{M}_X^{\text{ext}}(v_\bullet) & \xrightarrow{\ \text{ev}_3^X\ } & \mathcal{M}_X(v_3) \\
{\scriptstyle (\text{ev}_1^X, \text{ev}_2^X)} \Big\downarrow & & \\
\mathcal{M}_X(v_1) \times \mathcal{M}_X(v_2). & &
\end{array}
\tag{7.1.4}
$$

We also have the morphism given by the push-forward along $\pi \colon X \to S$

$$\pi_* \colon \mathcal{M}_X^{\text{ext}}(v_\bullet) \to \mathcal{M}_S^{\text{ext}}(v_\bullet), \ E_\bullet \mapsto \pi_* E_\bullet.$$

Here note that π_* preserves the exact sequences of coherent sheaves as π is affine.

7.1.2   Categorified Hall Algebras for Local Surfaces

Below we fix an ample divisor H on S. For $(\beta, n) \in N_{\leq 1}(S)$, we use the following notation for moduli stacks of semistable sheaves (see Sect. 3.4)

$$\mathcal{M}_n^H(X, \beta) := \mathcal{M}_X^{\sigma = iH}(\beta, n), \ \mathfrak{M}_n^H(S, \beta) := \mathfrak{M}_S^{\sigma = iH}(\beta, n).$$

For each $t \in \mathbb{Q} \cup \{\infty\}$, we set

$$N_{\leq 1}(S)_t := \{v \in N_{\leq 1}(S) : \mu_H(v) = t\}.$$

Here $\mu_H(v) = n/(H \cdot \beta)$ for $v = (\beta, n)$, which is ∞ if $H \cdot \beta = 0$. The following result in [134] shows that the categorified Hall products (7.1.3) descend to those on DT categories.

Theorem 7.1.1 ([134, Theorem 3.5]) *For* $v_\bullet = (v_1, v_2) \in N_{\leq 1}(S)_t^{\times 2}$ *with* $v_3 = v_1 + v_2$, $v_i = (\beta_i, n_i)$, *the functor (7.1.3) descends to the functor*

$$\mathcal{DT}^{\mathbb{C}^*}(\mathcal{M}_{n_1}^H(X, \beta_1)) \times \mathcal{DT}^{\mathbb{C}^*}(\mathcal{M}_{n_2}^H(X, \beta_2)) \to \mathcal{DT}^{\mathbb{C}^*}(\mathcal{M}_{n_3}^H(X, \beta_3)).$$

Proof For the reader's convenience, we give an outline of the proof in [134, Theorem 3.5]. Let ev be the morphism

$$\text{ev} := (\text{ev}_1, \text{ev}_2, \text{ev}_3) \colon \mathfrak{M}_S^{\text{ext}}(v_\bullet) \to \mathfrak{M}_S(v_1) \times \mathfrak{M}_S(v_2) \times \mathfrak{M}_S(v_3).$$

Then we have the following diagram by (3.2.19)

$$
\begin{array}{ccc}
t_0(\Omega_{\text{ev}}[-2]) & \longrightarrow & t_0(\Omega_{\mathfrak{M}_S(v_3)}[-1]) \\
\downarrow & & \\
\end{array}
$$
$$t_0(\Omega_{\mathfrak{M}_S(v_1)}[-1]) \times t_0(\Omega_{\mathfrak{M}_S(v_2)}[-1]).$$

Then one can show (see [134, Proposition 3.1]) that the above diagram is isomorphic to the diagram (7.1.4). Moreover from the definition of H-stability, the diagram (7.1.4) restricts to the diagram

$$
\begin{array}{ccc}
(\text{ev}_3^X)^{-1}(\mathcal{M}_{n_3}^H(X, \beta_3)) & \xrightarrow{\text{ev}_3^X} & \mathcal{M}_{n_3}^H(X, \beta_3) \\
{\scriptstyle (\text{ev}_1^X, \text{ev}_2^X)} \downarrow & & \\
\end{array}
$$
$$\mathcal{M}_{n_1}^H(X, \beta_1) \times \mathcal{M}_{n_2}^H(X, \beta_2).$$

Therefore the claim follows from Proposition 3.2.13.   □

7.1.3 Derived Moduli Stacks of Extensions of Pairs

Let $\mathfrak{M}_S^\dagger$ be the derived moduli stack of pairs in Sect. 4.2.1. We define the derived stack $\mathfrak{M}_S^{\text{ext}, \dagger}$ by the Cartesian square

$$\begin{array}{ccc}
\mathfrak{M}_S^{\mathrm{ext},\dagger} & \longrightarrow & \mathfrak{M}_S^{\mathrm{ext}} \\
{\scriptstyle (\mathrm{ev}_1^\dagger, \mathrm{ev}_2^\dagger)}\downarrow & \quad\square & \downarrow{\scriptstyle (\mathrm{ev}_1, \mathrm{ev}_2)} \\
\mathfrak{M}_S^\dagger \times \mathfrak{M}_S & \xrightarrow{\;(\rho^\dagger,\mathrm{id})\;} & \mathfrak{M}_S \times \mathfrak{M}_S .
\end{array}$$

For $T \in dAff$, the T-valued points of $\mathfrak{M}_S^{\mathrm{ext},\dagger}$ form the ∞-groupoid of diagrams

$$\begin{array}{c}
\mathcal{O}_{S\times T} \\
\xi\downarrow \\
\mathfrak{F}_1 \longrightarrow \mathfrak{F}_3 \longrightarrow \mathfrak{F}_2 .
\end{array} \qquad (7.1.5)$$

Here the bottom sequence is a T-valued point of $\mathfrak{M}_S^{\mathrm{ext}}$. Let us take $v_\bullet = (v_1, v_2) \in N_{\leq 1}(S)^{\times 2}$ and $v_3 = v_1 + v_2$. We have the open and closed derived substack

$$\mathfrak{M}_S^{\mathrm{ext},\dagger}(v_\bullet) \subset \mathfrak{M}_S^{\mathrm{ext},\dagger}$$

corresponding to the diagram (7.1.5) such that the bottom sequence is a T-valued point of $\mathfrak{M}_S^{\mathrm{ext}}(v_\bullet)$. By [134, Lemma 4.3], the derived stack $\mathfrak{M}_S^{\mathrm{ext},\dagger}(v_\bullet)$ is quasi-smooth. We have the diagram

$$\begin{array}{ccc}
\mathfrak{M}_S^{\mathrm{ext},\dagger}(v_\bullet) & \xrightarrow{\;\mathrm{ev}_3^\dagger\;} & \mathfrak{M}_S^\dagger(v_3) \\
{\scriptstyle (\mathrm{ev}_1^\dagger, \mathrm{ev}_2^\dagger)}\downarrow & & \\
\mathfrak{M}_S^\dagger(v_1) \times \mathfrak{M}_S(v_2) . & &
\end{array} \qquad (7.1.6)$$

Here $\mathrm{ev}_3^\dagger$ is obtained by sending a diagram (7.1.5) to the composition $\mathcal{O}_{S\times T} \xrightarrow{\xi} \mathfrak{F}_1 \to \mathfrak{F}_3$. Note that the vertical arrow is quasi-smooth and the horizontal arrow is proper (see [134, Lemma 4.4]). Therefore the diagram (7.1.6) induces the functor

$$\mathrm{ev}_{3*}^\dagger(\mathrm{ev}_1^\dagger, \mathrm{ev}_2^\dagger)^* : D_{\mathrm{coh}}^b(\mathfrak{M}_S^\dagger(v_1) \times \mathfrak{M}_S(v_2)) \to D_{\mathrm{coh}}^b(\mathfrak{M}_S^\dagger(v_3)). \qquad (7.1.7)$$

Similarly let $v_\bullet' = (v_2, v_1)$ and define the derived stack $\mathfrak{M}_S^{\mathrm{ext},\ddagger}(v_\bullet')$ by the Cartesian square

$$\begin{array}{ccc}
\mathfrak{M}_S^{\mathrm{ext},\ddagger}(v_\bullet') & \xrightarrow{\;\mathrm{ev}_3^\ddagger\;} & \mathfrak{M}_S^\dagger(v_3) \\
\downarrow & \quad\square & \downarrow{\scriptstyle \rho^\dagger} \\
\mathfrak{M}_S^{\mathrm{ext}}(v_\bullet') & \xrightarrow{\;\mathrm{ev}_3\;} & \mathfrak{M}_S(v_3) .
\end{array}$$

For $T \in dAff$, the T-valued points of $\mathfrak{M}_S^{\text{ext},\ddagger}(v_\bullet')$ form the ∞-groupoid of diagrams

$$
\begin{array}{ccccc}
 & & \mathcal{O}_{S \times T} & & \\
 & & \downarrow{\scriptstyle \xi} & & \\
\mathfrak{F}_2 & \longrightarrow & \mathfrak{F}_3 & \longrightarrow & \mathfrak{F}_1.
\end{array}
\tag{7.1.8}
$$

Here the bottom sequence is a T-valued point of $\mathfrak{M}_S^{\text{ext}}(v_\bullet)$. We have the following diagram

$$
\begin{array}{ccc}
\mathfrak{M}_S^{\text{ext},\ddagger}(v_\bullet) & \xrightarrow{\ \text{ev}_3^{\ddagger}\ } & \mathfrak{M}_S^{\dagger}(v_3) \\
{\scriptstyle (\text{ev}_2^{\ddagger}, \text{ev}_1^{\ddagger})} \downarrow & & \\
\mathfrak{M}_S(v_2) \times \mathfrak{M}_S^{\dagger}(v_1). & &
\end{array}
\tag{7.1.9}
$$

Here $\text{ev}_2^{\ddagger}$ sends a diagram (7.1.8) to $\mathfrak{F}_2$, and $\text{ev}_1^{\ddagger}$ sends a diagram (7.1.8) to the composition $\mathcal{O}_{S \times T} \xrightarrow{\xi} \mathfrak{F}_3 \to \mathfrak{F}_1$. Note that $\text{ev}_3^{\ddagger}$ is proper by definition. By Toda [134, Lemma 7.1], the morphism $(\text{ev}_2^{\ddagger}, \text{ev}_1^{\ddagger})$ is quasi-smooth, in particular the derived stack $\mathfrak{M}_S^{\text{ext},\ddagger}(v_\bullet')$ is quasi-smooth. Therefore the diagram (7.1.9) induces the functor

$$
\text{ev}_{3*}^{\ddagger}(\text{ev}_2^{\ddagger}, \text{ev}_1^{\ddagger})^* \colon D_{\text{coh}}^b(\mathfrak{M}_S(v_2) \times \mathfrak{M}_S^{\dagger}(v_1)) \to D_{\text{coh}}^b(\mathfrak{M}_S^{\dagger}(v_3)).
\tag{7.1.10}
$$

7.1.4 Moduli Stacks of Extensions in $\mathcal{A}_X$

Let $\mathcal{M}_X^{\dagger}$ be the moduli stack of D0-D2-D6 bound states in Sect. 4.2.2. We take $v_\bullet = (v_1, v_2) \in N_{\leq 1}(S)^{\times 2}$ with $v_3 = v_1 + v_2$, and define the classical stack

$$
\mathcal{M}_X^{\text{ext},\dagger}(v_\bullet) \colon Aff^{op} \to Groupoid
$$

by sending $T \in Aff$ to the groupoid of distinguished triangles

$$
\mathcal{E}_1 \xrightarrow{j} \mathcal{E}_3 \to \mathcal{E}_2[-1]
$$

where $(\mathcal{E}_i, \lambda_i)$ for $i = 1, 3$ are T-valued points of $\mathcal{M}_X^\dagger(v_i)$, $\mathcal{E}_2$ is a T-valued point of $\mathcal{M}_X(v_2)$ and j commutes with trivializations λ_i at S_∞. We also have the evaluation morphisms

$$
\begin{array}{ccc}
\mathcal{M}_X^{\text{ext},\dagger}(v_\bullet) & \xrightarrow{\text{ev}_3^{X,\dagger}} & \mathcal{M}_X^\dagger(v_3) \\
{\scriptstyle (\text{ev}_1^{X,\dagger}, \text{ev}_2^{X,\dagger})} \downarrow & & \\
\mathcal{M}_X^\dagger(v_1) \times \mathcal{M}_X(v_2) & &
\end{array}
\tag{7.1.11}
$$

where $\text{ev}_i^{X,\dagger}$ sends $\mathcal{E}_\bullet$ to $\mathcal{E}_i$.

For $v_\bullet' = (v_2, v_1)$, we also define the classical stack

$$
\mathcal{M}_X^{\text{ext},\ddagger}(v_\bullet') \colon Aff^{op} \to Groupoid
$$

by sending $T \in Aff$ to the groupoid of distinguished triangles

$$
\mathcal{E}_2[-1] \to \mathcal{E}_3 \xrightarrow{j'} \mathcal{E}_1
$$

where $(\mathcal{E}_i, \lambda_i)$ for $i = 1, 3$ are T-valued points of $\mathcal{M}_X^\dagger(v_i)$, $\mathcal{E}_2$ is a T-valued point of $\mathcal{M}_X(v_2)$ and j' commutes with trivializations λ_i at S_∞. We also have the evaluation morphisms

$$
\begin{array}{ccc}
\mathcal{M}_X^{\text{ext},\ddagger}(v_\bullet') & \xrightarrow{\text{ev}_3^{X,\ddagger}} & \mathcal{M}_X^\dagger(v_3) \\
{\scriptstyle (\text{ev}_2^{X,\ddagger}, \text{ev}_1^{X,\ddagger})} \downarrow & & \\
\mathcal{M}_X(v_2) \times \mathcal{M}_X^\dagger(v_1) & &
\end{array}
\tag{7.1.12}
$$

where $\text{ev}_i^{X,\ddagger}$ sends $\mathcal{E}_\bullet$ to $\mathcal{E}_i$. For substacks $\mathcal{Z}_1 \subset \mathcal{M}_X^\dagger(v_1)$ and $\mathcal{Z}_2 \subset \mathcal{M}_X(v_2)$, from the diagrams (7.1.11), (7.1.12) we define

$$
\mathcal{Z}_1 * \mathcal{Z}_2 := \text{ev}_3^{X,\dagger}(\text{ev}_1^{X,\dagger}, \text{ev}_2^{X,\dagger})^{-1}(\mathcal{Z}_1 \times \mathcal{Z}_2),
\tag{7.1.13}
$$
$$
\mathcal{Z}_2 * \mathcal{Z}_1 := \text{ev}_3^{X,\ddagger}(\text{ev}_2^{X,\ddagger}, \text{ev}_1^{X,\ddagger})^{-1}(\mathcal{Z}_2 \times \mathcal{Z}_1).
$$

Note that they are closed substacks in $\mathcal{M}_X^\dagger(v_3)$, $\mathcal{M}_X^\ddagger(v_3)$ if $\mathcal{Z}_i$ are closed substacks, since $\text{ev}_3^{X,\dagger}, \text{ev}_3^{X,\ddagger}$ are proper.

The above stacks of extensions in $\mathcal{A}_X$ are realized as (-2)-shifted conormal stacks as follows. Let $\mathrm{ev}^\dagger$, $\mathrm{ev}^\ddagger$ be the following morphisms in the diagrams (7.1.6) and (7.1.9)

$$\mathrm{ev}^\dagger = (\mathrm{ev}_1^\dagger, \mathrm{ev}_2^\dagger, \mathrm{ev}_3^\dagger)\colon \mathfrak{M}_S^{\mathrm{ext},\dagger}(v_\bullet) \to \mathfrak{M}_S^\dagger(v_1) \times \mathfrak{M}_S(v_2) \times \mathfrak{M}_S^\dagger(v_3),$$

$$\mathrm{ev}^\ddagger = (\mathrm{ev}_1^\ddagger, \mathrm{ev}_2^\ddagger, \mathrm{ev}_3^\ddagger)\colon \mathfrak{M}_S^{\mathrm{ext},\ddagger}(v_\bullet) \to \mathfrak{M}_S^\dagger(v_1) \times \mathfrak{M}_S(v_2) \times \mathfrak{M}_S^\dagger(v_3).$$

Then similarly to Theorem 7.1.1, the diagrams (7.1.11) and (7.1.12) are isomorphic to the diagrams

$$
\begin{array}{ccc}
t_0(\Omega_{\mathrm{ev}^\dagger}[-2]) & \longrightarrow & t_0(\Omega_{\mathfrak{M}_S^\dagger(v_3)}[-1]) \\
\downarrow & & \\
\end{array}
$$
$$t_0(\Omega_{\mathfrak{M}_S^\dagger(v_1)}[-1]) \times t_0(\Omega_{\mathfrak{M}_S(v_2)}[-1]),$$

$$
\begin{array}{ccc}
t_0(\Omega_{\mathrm{ev}^\ddagger}[-2]) & \longrightarrow & t_0(\Omega_{\mathfrak{M}_S^\dagger(v_3)}[-1]) \\
\downarrow & & \\
\end{array}
$$
$$t_0(\Omega_{\mathfrak{M}_S(v_2)}[-1]) \times t_0(\Omega_{\mathfrak{M}_S^\dagger(v_1)}[-1]),$$

respectively (see [134, Proposition 4.7]). Therefore similarly to Theorem 7.1.1, Proposition 3.2.13 yields the following:

Proposition 7.1.2 ([134, Corollary 4.8]) *Let $\mathcal{M}_X^\dagger(v_i)^\circ \subset \mathcal{M}_X(v_i)$ for $i = 1, 3$ and $\mathcal{M}_X(v_2)^\circ \subset \mathcal{M}_X(v_2)$ be open substacks of finite type. We regard them as open substacks in $t_0(\Omega_{\mathfrak{M}_S^\dagger(v_i)}[-1])$, $t_0(\Omega_{\mathfrak{M}_S(v_2)}[-1])$ by the isomorphisms (4.2.6), (3.4.6) respectively.*

(i) Suppose that the diagram (7.1.11) restricts to the diagram

$$
\begin{array}{ccc}
(\mathrm{ev}_3^{X,\dagger})^{-1}(\mathcal{M}_X^\dagger(v_3)^\circ) & \xrightarrow{\mathrm{ev}_3^{X,\dagger}} & \mathcal{M}_X^\dagger(v_3)^\circ \\
{\scriptstyle (\mathrm{ev}_1^{X,\dagger},\mathrm{ev}_2^{X,\dagger})}\downarrow & & \\
\mathcal{M}_X^\dagger(v_1)^\circ \times \mathcal{M}_X(v_2)^\circ. & &
\end{array}
\qquad (7.1.14)
$$

Then the functor (7.1.7) descends to the functor

$$\mathcal{DT}^{\mathbb{C}^*}(\mathcal{M}_X^\dagger(v_1)^\circ) \otimes \mathcal{DT}^{\mathbb{C}^*}(\mathcal{M}_X(v_2)^\circ) \to \mathcal{DT}^{\mathbb{C}^*}(\mathcal{M}_X^\dagger(v_3)^\circ).$$

(ii) Suppose that the diagram (7.1.12) restricts to the diagram

$$(\mathrm{ev}_3^{X,\ddagger})^{-1}(\mathcal{M}_X^{\ddagger}(v_3)^{\circ}) \xrightarrow{\mathrm{ev}_3^{X,\ddagger}} \mathcal{M}_X^{\ddagger}(v_3)^{\circ}$$

$$\Big\downarrow {\scriptstyle (\mathrm{ev}_2^{X,\ddagger},\mathrm{ev}_1^{X,\ddagger})}$$

$$\mathcal{M}_X(v_2)^{\circ} \times \mathcal{M}_X^{\dagger}(v_1)^{\circ}.$$

Then the functor (7.1.10) descends to the functor

$$\mathcal{DT}^{\mathbb{C}^*}(\mathcal{M}_X(v_2)^{\circ}) \otimes \mathcal{DT}^{\mathbb{C}^*}(\mathcal{M}_X^{\dagger}(v_1)^{\circ}) \to \mathcal{DT}^{\mathbb{C}^*}(\mathcal{M}_X^{\dagger}(v_3)^{\circ}).$$

The following is an analogue of Theorem 7.1.1 for moduli stacks of pairs, which is essentially proved in [134].

Theorem 7.1.3 *We take $(v_1, v_2) \in N_{\leq 1}(S)^{\times 2}$ for $v_i = (\beta_i, n_i)$ such that $v_2 \in N_{\leq 1}(S)_t$.*

(i) The functor (7.1.7) descends to the functor

$$\mathcal{DT}^{\mathbb{C}^*}(\mathcal{P}_{n_1}^t(X, \beta_1)) \otimes \mathcal{DT}^{\mathbb{C}^*}(\mathcal{M}_{n_2}^H(X, \beta_2)) \to \mathcal{DT}^{\mathbb{C}^*}(\mathcal{P}_{n_1+n_2}^t(X, \beta_1 + \beta_2)).$$
$$(7.1.15)$$

(ii) The functor (7.1.10) descends to the functor

$$\mathcal{DT}^{\mathbb{C}^*}(\mathcal{M}_{n_2}^H(X, \beta_2)) \otimes \mathcal{DT}^{\mathbb{C}^*}(\mathcal{P}_{n_1}^t(X, \beta_1)) \to \mathcal{DT}^{\mathbb{C}^*}(\mathcal{P}_{n_1+n_2}^t(X, \beta_1 + \beta_2)).$$
$$(7.1.16)$$

Proof By Lemma 7.1.4 below, the diagrams (7.1.11) and (7.1.14) restrict to the diagrams

$$(\mathrm{ev}_3^{X,\dagger})^{-1}(\mathcal{P}_{n_1+n_2}^t(X, \beta_1 + \beta_2)) \xrightarrow{\mathrm{ev}_3^{X,\dagger}} \mathcal{P}_{n_1+n_2}^t(X, \beta_1 + \beta_2)$$

$$\Big\downarrow {\scriptstyle (\mathrm{ev}_1^{X,\dagger},\mathrm{ev}_2^{X,\dagger})}$$

$$\mathcal{P}_{n_1}^t(X, \beta_1) \times \mathcal{M}_{n_2}^H(X, \beta_2),$$

$$(\mathrm{ev}_3^{X,\ddagger})^{-1}(\mathcal{P}_{n_1+n_2}^t(X, \beta_1 + \beta_2)) \xrightarrow{\mathrm{ev}_3^{X,\ddagger}} \mathcal{P}_{n_1+n_2}^t(X, \beta_1 + \beta_2)$$

$$\Big\downarrow {\scriptstyle (\mathrm{ev}_2^{X,\ddagger},\mathrm{ev}_1^{X,\ddagger})}$$

$$\mathcal{M}_{n_2}^H(X, \beta_2) \times \mathcal{P}_{n_1}^t(X, \beta_1).$$
$$(7.1.17)$$

Then we have the desired functors by Proposition 7.1.2.   □

We have used the following lemma, which is obvious from the definition of $\mu_t^\dagger$-stability

Lemma 7.1.4

(i) Let $0 \to E_1 \to E_3 \to F_2[-1] \to 0$ be an exact sequence in $\mathcal{A}_X$ such that $\mathrm{rank}(E_1) = \mathrm{rank}(E_3) = 1$ and $F_2 \in \mathrm{Coh}_{\leq 1}(X)$ satisfies $\mu_H(F_2) = t$. Then E_3 is $\mu_t^\dagger$-semistable if and only if E_1 is $\mu_t^\dagger$-semistable and F_2 is μ_H-semistable.

(ii) Let $0 \to F_2[-1] \to E_3 \to E_1 \to 0$ be an exact sequence in $\mathcal{A}_X$ such that $\mathrm{rank}(E_1) = \mathrm{rank}(E_3) = 1$ and $F_2 \in \mathrm{Coh}_{\leq 1}(X)$ satisfies $\mu_H(F_2) = t$. Then E_3 is $\mu_t^\dagger$-semistable if and only if E_1 is $\mu_t^\dagger$-semistable and F_2 is μ_H-semistable.

7.2 Conjectural SOD via Categorified Hall Products

7.2.1 Stratifications of $\mathcal{P}_n^t(X, \beta)$

In the situation of Sect. 4.3.2, let us take $t \in \mathbb{R}_{>0} \cup \{\infty\}$. Then for $t_\pm = t \pm \varepsilon$ for $0 < \varepsilon \ll 1$, Conjectures 4.3.3 and 4.3.14 claim the existence of a fully-faithful functor

$$\mathcal{DT}^{\mathbb{C}^*}(P_n^{t_-}(X, \beta)) \hookrightarrow \mathcal{DT}^{\mathbb{C}^*}(P_n^{t_+}(X, \beta)).$$

We refine the above conjecture using the categorified Hall products for DT categories. For $(\beta, n) \in N_{\leq 1}(S)$ and $t \in \mathbb{R} \cup \{\infty\}$, we define $\mathbf{D}_{(\beta,n)}^t$ to be the set of decompositions of (β, n)

$$\mathbf{D}_{(\beta,n)}^t :=$$

$$\left\{ (\beta, n) = (\beta_1, n_1) + (\beta_2, n_2) : (\beta_2, n_2) \in N_{\leq 1}(S)_t, \, \mathcal{P}_{n_1}^t(X, \beta_1) \right.$$

$$\left. \times \mathcal{M}_{n_2}^H(X, \beta_2) \neq \emptyset \right\}.$$

There is a partial order on $\mathbf{D}_{(\beta,n)}^t$ given by

$$(\beta_1', n_1') + (\beta_2', n_2') > (\beta_1, n_1) + (\beta_2, n_2) \qquad (7.2.1)$$

if there is a decomposition $(\beta_1, n_1) = (\beta_1'', n_1'') + (\beta_2'', n_2'')$ in $\mathbf{D}_{(\beta_1, n_1)}^t$ such that

$$(\beta_1', n_1') = (\beta_1'', n_1''), \; (\beta_2', n_2') = (\beta_2, n_2) + (\beta_2'', n_2'').$$

It defines the decomposition of $\mathbf{D}^t_{(\beta,n)}$

$$\mathbf{D}^t_{(\beta,n)} = \mathbf{D}^{t,1}_{(\beta,n)} \sqcup \cdots \sqcup \mathbf{D}^{t,N}_{(\beta,n)}$$

defined inductive so that

$$\mathbf{D}^{t,i}_{(\beta,n)} \subset \mathbf{D}^t_{(\beta,n)} \setminus \bigcup_{j<i} \mathbf{D}^{t,j}_{(\beta,n)}$$

is the set of maximal elements with respect to the partial order (7.2.1). In the following, we set

$$\mathbf{D}^{t,\leq l}_{(\beta,n)} := \mathbf{D}^{t,1}_{(\beta,n)} \sqcup \mathbf{D}^{t,2}_{(\beta,n)} \sqcup \cdots \sqcup \mathbf{D}^{t,l}_{(\beta,n)}.$$

For each $1 \leq l \leq N$, we define

$$\mathcal{Z}^{t_+\text{-us}}_l := \bigcup_{\substack{(\beta,n)=(\beta_1,n_1)+(\beta_2,n_2) \\ \in \mathbf{D}^{t,\leq l}_{(\beta,n)}}} \mathcal{P}^t_{n_1}(X,\beta_1) * \mathcal{M}^H_{n_2}(X,\beta_2),$$

$$\mathcal{Z}^{t_-\text{-us}}_l := \bigcup_{\substack{(\beta,n)=(\beta_1,n_1)+(\beta_2,n_2) \\ \in \mathbf{D}^{t,\leq l}_{(\beta,n)}}} \mathcal{M}^H_{n_2}(X,\beta_2) * \mathcal{P}^t_{n_1}(X,\beta_1).$$

Here $*$ is defined in (7.1.13). Note that the substacks $\mathcal{Z}^{t_\pm\text{-us}}_l$ are closed substacks in $\mathcal{P}^t_n(X,\beta)$. We have the following lemma.

Lemma 7.2.1 *We have the following identities*

$$\mathcal{Z}^{t_+\text{-us}}_l = \bigsqcup_{\substack{(\beta,n)=(\beta_1,n_1)+(\beta_2,n_2) \\ \in \mathbf{D}^{t,\leq l}_{(\beta,n)}}} \mathcal{P}^{t_+}_{n_1}(X,\beta_1) * \mathcal{M}^H_{n_2}(X,\beta_2), \qquad (7.2.2)$$

$$\mathcal{Z}^{t_-\text{-us}}_l = \bigsqcup_{\substack{(\beta,n)=(\beta_1,n_1)+(\beta_2,n_2) \\ \in \mathbf{D}^{t,\leq l}_{(\beta,n)}}} \mathcal{M}^H_{n_2}(X,\beta_2) * \mathcal{P}^{t_-}_{n_1}(X,\beta_1).$$

Proof By the Harder-Narasimhan filtrations in $\mu^\dagger_{t_\pm}$-stability, we have

$$\mathcal{P}^t_n(X,\beta) = \bigsqcup_{\substack{(\beta,n)=(\beta_1,n_1)+(\beta_2,n_2) \\ \in \mathbf{D}^t_{(\beta,n)}}} \mathcal{P}^{t_+}_{n_1}(X,\beta_1) * \mathcal{M}^H_{n_2}(X,\beta_2) \qquad (7.2.3)$$

$$= \bigsqcup_{\substack{(\beta,n)=(\beta_1,n_1)+(\beta_2,n_2) \\ \in \mathbf{D}^t_{(\beta,n)}}} \mathcal{M}^H_{n_2}(X,\beta_2) * \mathcal{P}^{t_-}_{n_1}(X,\beta_1).$$

In particular the RHS in (7.2.2) are disjoint unions. Since $\mathcal{P}_n^{t\pm}(X, \beta) \subset \mathcal{P}_n^t(X, \beta)$, the RHS in (7.2.2) are contained in the LHS. On the other hand applying (7.2.3) for (β_1, n_1), we see that the LHS is contained in the RHS.   □

Lemma 7.2.2 *Let $(\beta, n) = (\beta_1, n_1) + (\beta_2, n_2)$ be a decomposition in $\mathbf{D}_{(\beta,n)}^{t,l}$. Then in the diagrams (7.1.17), we have*

$$(\mathrm{ev}_3^{X,\dagger})^{-1}\left(\mathcal{P}_n^t(X, \beta) \setminus \mathcal{Z}_{l-1}^{t_+\text{-us}}\right) = (\mathrm{ev}_1^{X,\dagger}, \mathrm{ev}_2^{X,\dagger})^{-1}\left(\mathcal{P}_{n_1}^{t+}(X, \beta_1) \times \mathcal{M}_{n_2}^H(X, \beta_2)\right),$$
(7.2.4)

$$(\mathrm{ev}_3^{X,\ddagger})^{-1}\left(\mathcal{P}_n^t(X, \beta) \setminus \mathcal{Z}_{l-1}^{t_-\text{-us}}\right) = (\mathrm{ev}_2^{X,\ddagger}, \mathrm{ev}_1^{X,\ddagger})^{-1}\left(\mathcal{M}_{n_2}^H(X, \beta_2) \times \mathcal{P}_{n_1}^{t-}(X, \beta_1)\right).$$

Proof We only prove the first identity. Let us take an exact sequence in $\mathcal{A}_X$

$$0 \to E_1 \to E \to F_2[-1] \to 0 \qquad\qquad (7.2.5)$$

such that $\mathrm{cl}(E_1) = (1, \beta_1, n_1)$ and $F_2 \in \mathrm{Coh}_{\leq 1}(X)$ with $\mathrm{cl}(F_2[-1]) = (0, \beta_2, n_2)$. If E corresponds to a point in $\mathcal{P}_n^t(X, \beta) \setminus \mathcal{Z}_{l-1}^{t_+\text{-us}}$, then E_1 is $\mu_{t_+}^\dagger$-semistable. Indeed if otherwise, there is an exact sequence

$$0 \to E_1'' \to E_1 \to F_2''[-1] \to 0 \qquad\qquad (7.2.6)$$

in $\mathcal{A}_X$ such that E_1'' is $\mu_{t_+}^\dagger$-semistable and $0 \neq F_2'' \in \mathrm{Coh}_{\leq 1}(X)$ satisfies $\mathrm{cl}(F_2''[-1]) = (0, \beta_2'', n_2'')$, where $(\beta_2'', n_2'') \in N_{\leq 1}(S)_t$. Let $(\beta_1', n_1') = (\beta_1'', n_1'')$ and $(\beta_2', n_2') = (\beta_2, n_2) + (\beta_2'', n_2'')$. Then we have

$$(\beta_1', n_1') + (\beta_2', n_2') > (\beta_1, n_1) + (\beta_2, n_2),$$

so the LHS is contained in $\mathbf{D}_{(\beta,n)}^{t, \leq l-1}$. By combining exact sequences (7.2.5) and (7.2.6), we obtain the exact sequences in $\mathcal{A}_X$ and $\mathrm{Coh}_{\leq 1}(X)$

$$0 \to E_1' \to E \to F_2'[-1] \to 0, \ 0 \to F_2 \to F_2' \to F_2'' \to 0$$

where $E_1' = E_1''$ so that $\mathrm{cl}(E_1') = (1, \beta_1', n_1')$ and $\mathrm{cl}(F_2'[-1]) = (0, \beta_2', n_2')$. This contradicts to the assumption that E does not correspond to a point in $\mathcal{Z}_{l-1}^{t_+\text{-us}}$. Therefore the LHS of (7.2.4) is contained in the RHS.

Conversely suppose that an exact sequence (7.2.5) is given such that E_1 is μ_{t_+}-semistable and F_2 is μ_H-semistable. Suppose by contradiction that $E \in \mathcal{Z}_{l-1}^{t_+\text{-us}}$. Since (7.2.5) is a Hadar-Narasimhan filtration in μ_{t_+}-stability, the uniqueness of Harder-Narasimhan filtration together with Lemma 7.2.1 imply that $(\beta, n) = (\beta_1, n_1) + (\beta_2, n_2)$ is a decomposition in $\mathbf{D}_{(\beta,n)}^{t, \leq l-1}$. This is a contradiction, so the RHS of (7.2.4) is contained in the LHS.   □

7.2.2 Conjectures

For each $1 \leq l \leq N$, let $(\beta, n) = (\beta_1, n_1) + (\beta_2, n_2)$ be a decomposition in $\mathbf{D}_{(\beta,n)}^{t,l}$. By Proposition 7.1.2 and Lemma 7.2.2, the functors (7.1.15), (7.1.16) descend to functors

$$*\colon \mathcal{DT}^{\mathbb{C}^*}(\mathcal{P}_{n_1}^{t+}(X, \beta_1)) \otimes \mathcal{DT}^{\mathbb{C}^*}(\mathcal{M}_{n_2}^{H}(X, \beta_2)) \to \mathcal{DT}^{\mathbb{C}^*}\left(\mathcal{P}_n^{t}(X, \beta) \setminus \mathcal{Z}_{l-1}^{t+,-\mathrm{us}}\right),$$
(7.2.7)

$$*\colon \mathcal{DT}^{\mathbb{C}^*}(\mathcal{M}_{n_2}^{H}(X, \beta_2)) \otimes \mathcal{DT}^{\mathbb{C}^*}(\mathcal{P}_{n_1}^{t-}(X, \beta_1)) \to \mathcal{DT}^{\mathbb{C}^*}\left(\mathcal{P}_n^{t}(X, \beta) \setminus \mathcal{Z}_{l-1}^{t-,-\mathrm{us}}\right).$$

By Remark 3.4.6, we have the decomposition into j-twisted parts

$$\mathcal{DT}^{\mathbb{C}^*}(\mathcal{M}_n^{H}(X, \beta)) = \bigoplus_{j \in \mathbb{Z}} \mathcal{DT}^{\mathbb{C}^*}(\mathcal{M}_n^{H}(X, \beta))_j.$$

We denote by i_j, pr_j the inclusion, projection with respect to the above decomposition

$$\mathcal{DT}^{\mathbb{C}^*}(\mathcal{M}_n^{H}(X, \beta))_j \underset{\mathrm{pr}_j}{\overset{i_j}{\rightleftarrows}} \mathcal{DT}^{\mathbb{C}^*}(\mathcal{M}_n^{H}(X, \beta)).$$

We propose the following conjecture.

Conjecture 7.2.3 For each $1 \leq l \leq N$ let $\mathbf{d} \in \mathbf{D}_{(\beta,n)}^{t,l}$ be a decomposition $(\beta, n) = (\beta_1, n_1) + (\beta_2, n_2)$, and take $k^{\pm}(\mathbf{d}) \in \mathbb{R}$. Then for $j \in \mathbb{Z}$, the functors (7.2.7) composed with i_j

$$*_j := * \circ i_j \colon \mathcal{DT}^{\mathbb{C}^*}(\mathcal{P}_{n_1}^{t+}(X, \beta_1)) \otimes \mathcal{DT}^{\mathbb{C}^*}(\mathcal{M}_{n_2}^{H}(X, \beta_2))_j \qquad (7.2.8)$$
$$\to \mathcal{DT}^{\mathbb{C}^*}\left(\mathcal{P}_n^{t}(X, \beta) \setminus \mathcal{Z}_{l-1}^{t+,-\mathrm{us}}\right),$$

$$*_j := * \circ i_j \colon \mathcal{DT}^{\mathbb{C}^*}(\mathcal{M}_{n_2}^{H}(X, \beta_2))_j \otimes \mathcal{DT}^{\mathbb{C}^*}(\mathcal{P}_{n_1}^{t-}(X, \beta_1))$$
$$\to \mathcal{DT}^{\mathbb{C}^*}\left(\mathcal{P}_n^{t}(X, \beta) \setminus \mathcal{Z}_{l-1}^{t-,-\mathrm{us}}\right)$$

are fully-faithful. By setting $\Upsilon_j^{\pm,\mathbf{d}}$ to be the essential images of the above functors, we have the semiorthogonal decompositions

$$\mathcal{DT}^{\mathbb{C}^*}\left(\mathcal{P}_n^t(X,\beta)\setminus\mathcal{Z}_{l-1}^{t_+\text{-us}}\right) = \left\langle \bigoplus_{\mathbf{d}\in\mathbf{D}_{(\beta,n)}^{t,l}} \Upsilon_{>k^+(\mathbf{d})}^{+,\mathbf{d}}, \mathcal{W}_{l,k^+}^+, \bigoplus_{\mathbf{d}\in\mathbf{D}_{(\beta,n)}^{t,l}} \Upsilon_{\leq k^+(\mathbf{d})}^{+,\mathbf{d}} \right\rangle,$$
(7.2.9)

$$\mathcal{DT}^{\mathbb{C}^*}\left(\mathcal{P}_n^t(X,\beta)\setminus\mathcal{Z}_{l-1}^{t_-\text{-us}}\right) = \left\langle \bigoplus_{\mathbf{d}\in\mathbf{D}_{(\beta,n)}^{t,l}} \Upsilon_{<k^-(\mathbf{d})}^{-,\mathbf{d}}, \mathcal{W}_{l,k^-}^-, \bigoplus_{\mathbf{d}\in\mathbf{D}_{(\beta,n)}^{t,l}} \Upsilon_{\geq k^-(\mathbf{d})}^{-,\mathbf{d}} \right\rangle$$

together with semiorthogonal decompositions

$$\Upsilon_{>k^+(\mathbf{d})}^{+,\mathbf{d}} = \langle \ldots, \Upsilon_{\lfloor k^+(\mathbf{d})\rfloor+2}^{+,\mathbf{d}}, \Upsilon_{\lfloor k^+(\mathbf{d})\rfloor+1}^{+,\mathbf{d}} \rangle, \quad \Upsilon_{\leq k^+(\mathbf{d})}^{+,\mathbf{d}} = \langle \Upsilon_{\lfloor k^+(\mathbf{d})\rfloor}^{+,\mathbf{d}}, \Upsilon_{\lfloor k^+(\mathbf{d})\rfloor-1}^{+,\mathbf{d}}, \ldots \rangle,$$
(7.2.10)

$$\Upsilon_{<k^-(\mathbf{d})}^{-,\mathbf{d}} = \langle \ldots, \Upsilon_{\lceil k^-(\mathbf{d})\rceil-2}^{-,\mathbf{d}}, \Upsilon_{\lceil k^-(\mathbf{d})\rceil-1}^{-,\mathbf{d}} \rangle, \quad \Upsilon_{\geq k^-(\mathbf{d})}^- = \langle \Upsilon_{\lceil k^-(\mathbf{d})\rceil}^{-,\mathbf{d}}, \Upsilon_{\lceil k^-(\mathbf{d})\rceil+1}^{-,\mathbf{d}}, \ldots \rangle$$

such that the composition functors

$$\mathcal{W}_{l,k^\pm}^\pm \hookrightarrow \mathcal{DT}^{\mathbb{C}^*}\left(\mathcal{P}_n^t(X,\beta)\setminus\mathcal{Z}_{l-1}^{t_\pm\text{-us}}\right) \twoheadrightarrow \mathcal{DT}^{\mathbb{C}^*}\left(\mathcal{P}_n^t(X,\beta)\setminus\mathcal{Z}_l^{t_\pm\text{-us}}\right)$$
(7.2.11)

are equivalences.

The following conjecture holds if Conjecture 7.2.3 holds for all $1\leq l\leq N$.

Conjecture 7.2.4 Let $k^\pm\colon \mathbf{D}_{(\beta,n)}^t \to \mathbb{R}$ be maps. There exist semiorthogonal decompositions

$$\mathcal{DT}^{\mathbb{C}^*}(\mathcal{P}_n^t(X,\beta)) = \left\langle \bigoplus_{\mathbf{d}\in\mathbf{D}_{(\beta,n)}^{t,1}} \Upsilon_{>k^+(\mathbf{d})}^{+,\mathbf{d}}, \bigoplus_{\mathbf{d}\in\mathbf{D}_{(\beta,n)}^{t,2}} \Upsilon_{>k^+(\mathbf{d})}^{+,\mathbf{d}}, \ldots, \bigoplus_{\mathbf{d}\in\mathbf{D}_{(\beta,n)}^{t,N}} \Upsilon_{>k^+(\mathbf{d})}^{+,\mathbf{d}}, \mathcal{W}_{k^+}^+, \right.$$

$$\left. \bigoplus_{\mathbf{d}\in\mathbf{D}_{(\beta,n)}^{t,N}} \Upsilon_{\leq k^+(\mathbf{d})}^{+,\mathbf{d}}, \bigoplus_{\mathbf{d}\in\mathbf{D}_{(\beta,n)}^{t,N-1}} \Upsilon_{\leq k^+(\mathbf{d})}^{+,\mathbf{d}}, \ldots, \bigoplus_{\mathbf{d}\in\mathbf{D}_{(\beta,n)}^{t,1}} \Upsilon_{\leq k^+(\mathbf{d})}^{+,\mathbf{d}} \right\rangle$$

$$= \left\langle \bigoplus_{\mathbf{d}\in\mathbf{D}_{(\beta,n)}^{t,1}} \Upsilon_{<k^-(\mathbf{d})}^{-,\mathbf{d}}, \bigoplus_{\mathbf{d}\in\mathbf{D}_{(\beta,n)}^{t,2}} \Upsilon_{<k^-(\mathbf{d})}^{-,\mathbf{d}}, \ldots, \bigoplus_{\mathbf{d}\in\mathbf{D}_{(\beta,n)}^{t,N}} \Upsilon_{<k^-(\mathbf{d})}^{-,\mathbf{d}}, \mathcal{W}_{k^-}^-, \right.$$

$$\left. \bigoplus_{\mathbf{d}\in\mathbf{D}_{(\beta,n)}^{t,N}} \Upsilon_{\geq k^-(\mathbf{d})}^{-,\mathbf{d}}, \bigoplus_{\mathbf{d}\in\mathbf{D}_{(\beta,n)}^{t,N-1}} \Upsilon_{\geq k^-(\mathbf{d})}^{-,\mathbf{d}}, \ldots, \bigoplus_{\mathbf{d}\in\mathbf{D}_{(\beta,n)}^{t,1}} \Upsilon_{\geq k^-(\mathbf{d})}^{-,\mathbf{d}} \right\rangle$$

with semiorthogonal decomposition as in (7.2.10) and equivalences induced by categorified Hall products

$$*_j \colon \mathcal{DT}^{\mathbb{C}^*}(P_{n_1}^{t+}(X, \beta_1)) \otimes \mathcal{DT}^{\mathbb{C}^*}(\mathcal{M}_{n_2}^H(X, \beta_2))_j \xrightarrow{\sim} \Upsilon_j^{+,\mathbf{d}},$$

$$*_j \colon \mathcal{DT}^{\mathbb{C}^*}(\mathcal{M}_{n_2}^H(X, \beta_2))_j \otimes \mathcal{DT}^{\mathbb{C}^*}(P_{n_1}^{t-}(X, \beta_1)) \xrightarrow{\sim} \Upsilon_j^{-,\mathbf{d}}$$

such that the composition functors

$$\mathcal{W}_{k\pm}^{\pm} \hookrightarrow \mathcal{DT}^{\mathbb{C}^*}(P_n^t(X, \beta)) \to \mathcal{DT}^{\mathbb{C}^*}(P_n^{t\pm}(X, \beta))$$

are equivalences.

In addition to the above conjecture, we also propose the following conjecture which in particular implies Conjecture 4.3.14.

Conjecture 7.2.5 In the situation of Conjecture 7.2.4, we set

$$k^+(\mathbf{d}) = \frac{1}{2}(n_2 + \beta_1\beta_2), \ k^-(\mathbf{d}) = -\frac{1}{2}\beta_1\beta_2 \qquad (7.2.12)$$

where $\mathbf{d}$ is a decomposition $(\beta, n) = (\beta_1, n_1) + (\beta_2, n_2)$. Then we have $\mathcal{W}_{k^-}^- \subset \mathcal{W}_{k^+}^+$. In particular, we have a fully-faithful functor

$$\mathcal{DT}^{\mathbb{C}^*}(P_n^{t-}(X, \beta)) \hookrightarrow \mathcal{DT}^{\mathbb{C}^*}(P_n^{t+}(X, \beta)).$$

We say that $t \in \mathbb{R}$ is a *simple wall* if $N = 1$. In this case, we also propose the description of semiorthogonal complements of the fully-faithful embedding in Conjecture 7.2.5.

Conjecture 7.2.6 If $t \in \mathbb{R}_{>0}$ is a simple wall, there exists a semiorthogonal decomposition

$$\mathcal{DT}^{\mathbb{C}^*}(P_n^{t+}(X, \beta)) \qquad (7.2.13)$$

$$= \left\langle \bigoplus_{\mathbf{d} \in \mathbf{D}_{(\beta,n)}^{t,1}} \Upsilon_{[-\frac{1}{2}\beta_1\beta_2 - \frac{1}{2}n_2, -\frac{1}{2}\beta_1\beta_2)}^{-,\mathbf{d}}, \mathcal{DT}^{\mathbb{C}^*}(P_n^{t-}(X, \beta)),\right.$$

$$\left. \bigoplus_{\mathbf{d} \in \mathbf{D}_{(\beta,n)}^{t,1}} \Upsilon_{[-\frac{1}{2}\beta_1\beta_2, -\frac{1}{2}\beta_1\beta_2 + \frac{1}{2}n_2)}^{-,\mathbf{d}} \right\rangle.$$

Here **d** is a decomposition $(\beta, n) = (\beta_1, n_1) + (\beta_2, n_2)$ in $\mathbf{D}^{t,1}_{(\beta,n)}$, and $\Upsilon^{-,\mathbf{d}}_{[a,b)}$ admits semiorthogonal decomposition

$$\Upsilon^{-,\mathbf{d}}_{[a,b)} = \left\langle \Upsilon^{-,\mathbf{d}}_{\lceil a \rceil}, \Upsilon^{-,\mathbf{d}}_{\lceil a \rceil + 1}, \ldots, \Upsilon^{-,\mathbf{d}}_{\lceil b \rceil - 1} \right\rangle.$$

Remark 7.2.7 The above Conjectures 7.2.4 and 7.2.5 for $t = \infty$ implies the semiorthogonal decomposition in Theorem 1.4.8 in the introduction. Indeed let $n(\beta) \in \mathbb{Z}$ be as in (4.3.4). Then $\mathbf{D}^{\infty,l}_{(\beta,n)}$ consists of one element

$$\mathbf{D}^{\infty,l}_{(\beta,n)} = \{(\beta, n) = (\beta, n(\beta) + l - 1) + (0, n - n(\beta) - l + 1)\}.$$

In particular $t = \infty$ is a simple wall if and only if $n = n(\beta) + 1$. Note that

$$P_n(X, \beta) = I_n(X, \beta),\ n \le n(\beta),$$

indeed they are empty for $n < n(\beta)$ by the definition of $n(\beta)$. For $t = \infty$ and $n = n(\beta) + 1$, the semiorthogonal decomposition (7.2.13) is

$$\mathcal{DT}^{\mathbb{C}^*}(I_n(X, \beta)) = \langle \mathcal{DT}^{\mathbb{C}^*}(P_n(X, \beta)), \mathcal{DT}^{\mathbb{C}^*}(\mathcal{M}_1(X))_0 \otimes \mathcal{DT}^{\mathbb{C}^*}(P_{n-1}(X, \beta)) \rangle,$$

where $\mathcal{DT}^{\mathbb{C}^*}(\mathcal{M}_1(X))_0$ is equivalent to $D^b_{\mathrm{coh}}(S)$.

7.2.3   The Formally Local Descriptions of $\mathcal{P}^t_n(X, \beta)$

Let $\mathcal{S}^{t\pm}_l$ be substacks in $\mathcal{P}^t_n(X, \beta)$ defined by

$$\mathcal{S}^{t\pm}_l := \mathcal{Z}^{t\pm\text{-us}}_l \setminus \mathcal{Z}^{t\pm\text{-us}}_{l-1} \subset \mathcal{P}^t_n(X, \beta).$$

Then we have the stratifications

$$\mathcal{P}^t_n(X, \beta) = \mathcal{S}^{t\pm}_1 \sqcup \mathcal{S}^{t\pm}_2 \sqcup \cdots \sqcup \mathcal{S}^{t\pm}_N \sqcup \mathcal{P}^{t\pm}_n(X, \beta). \qquad (7.2.14)$$

Here we interpret the above stratification in terms of KN stratifications formally locally on good moduli spaces of $\mathcal{P}^t_n(X, \beta)$.

Let us consider the good moduli space morphism

$$\pi_{\mathcal{P}} \colon \mathcal{P}^t_n(X, \beta) \to P^t_n(X, \beta).$$

For a closed point $p \in P_n^t(X, \beta)$, we consider its formal fiber

$$
\begin{CD}
\widehat{\mathcal{P}}_n^t(X, \beta)_p @>>> \mathcal{P}_n^t(X, \beta) \\
@VVV @. \square @VVV \\
\operatorname{Spec} \widehat{\mathcal{O}}_{P_n(X,\beta),p} @>>> P_n^t(X, \beta).
\end{CD}
$$

As in [132] (also see [137, Theorem 9.11] and Sect. 8.4) we can describe the above formal fiber in terms of Ext-quivers with super-potentials associated with polystable objects as follows. A closed point p is represented by a $\mu_t^\dagger$-polystable object of the form

$$
\mathcal{E} = \mathcal{E}_0 \oplus \bigoplus_{i=1}^{k} W_i \otimes E_i[-1] \tag{7.2.15}
$$

where $\mathcal{E}_0$ is a rank one $\mu_t^\dagger$-stable object in $\mathcal{A}_X$, $(E_1, \ldots, E_k)$ are mutually non-isomorphic μ_H-stable sheaves in $\operatorname{Coh}_{\leq 1}(X)$ with $\mu_H(E_i) = t$ and W_i is a finite dimensional vector space. Let $Q_{\mathcal{E}}$ be the Ext-quiver associated with the collection

$$
(\mathcal{E}_0, E_1[-1], \ldots, E_k[-1]),
$$

i.e. the set of vertices is $\{0, 1, \ldots, k\}$ and the number of arrows from i to j is $\operatorname{ext}^1(E_i, E_j)$ where we have set $E_0 = \mathcal{E}_0[1]$. Let $Q_{\mathcal{E},0} \subset Q_{\mathcal{E}}$ be the full sub quiver as in Sect. 6.1.5 whose vertex set is $\{1, \ldots, k\}$, which is the Ext-quiver for the collection $(E_1, \ldots, E_k)$. Let $\vec{w} = (\dim W_i)_{1 \leq i \leq k}$ be the dimension vector for $Q_{\mathcal{E},0}$, and set $G_{\mathcal{E}} = \prod_{i=1}^{k} \operatorname{GL}(W_i)$. As in Sect. 6.1.5, we have the moduli stack of $Q_{\mathcal{E}}$-representations with dimension vector $(1, \vec{w})$, given by the quotient stack together with its good moduli space

$$
\left[R^\dagger(Q_{\mathcal{E},0}, \vec{w}) / G_{\mathcal{E}} \right] \to R^\dagger(Q_{\mathcal{E},0}, \vec{w}) /\!\!/ G_{\mathcal{E}}. \tag{7.2.16}
$$

Then by Toda [137, Theorem 9.11], there is a $G_{\mathcal{E}}$-equivariant function on the formal fiber of (7.2.16) at 0

$$
w_{\mathcal{E}} \colon \widehat{R}^\dagger(Q_{\mathcal{E},0}, \vec{w})_0 \to \mathbb{C}
$$

such that we have the following commutative isomorphisms

$$
\begin{CD}
[\operatorname{Crit}(w_{\mathcal{E}}) / G_{\mathcal{E}}] @>\cong>> \widehat{\mathcal{P}}_n^t(X, \beta)_p \\
@VVV @VVV \\
\operatorname{Crit}(w_{\mathcal{E}}) /\!\!/ G_{\mathcal{E}} @>\cong>> \operatorname{Spec} \widehat{\mathcal{O}}_{P_n(X,\beta),p}.
\end{CD}
\tag{7.2.17}
$$

We have the following lemma:

Lemma 7.2.8 *The stratifications (7.2.14) pulled back via the top arrow in (7.2.17) is the KN stratification on $R^\dagger(Q_{\mathcal{E},0}, \vec{w})$ restricted to $\mathrm{Crit}(w_{\mathcal{E}})$ with respect to characters $\chi_0^\pm$, where χ_0 is the determinant character*

$$\chi_0 \colon G_{\mathcal{E}} = \prod_{i=1}^{k} \mathrm{GL}(W_i) \to \mathbb{C}^*, \ (g_i)_{1 \le i \le k} \mapsto \prod_{i=1}^{k} \det(g_i).$$

Proof We only prove the lemma for t_+. We first note that, on the formal fiber at $p \in P_n^t(X, \beta)$ we can restrict decompositions $(\beta, n) = (\beta_1, n_1) + (\beta_2, n_2)$ in $\mathbf{D}_{(\beta,n)}^t$ such that (β_2, n_2) is of the form

$$(\beta_2, n_2) = \sum_{i=1}^{k} w_i' \cdot [\pi_* E_i], \ 0 < w_i' < \dim W_i. \tag{7.2.18}$$

Therefore after discarding strata which do not intersect with $\widehat{\mathcal{P}}_n^t(X, \beta)_p$ and renumbering, the stratification (7.2.14) restricts to the filtration

$$\widehat{\mathcal{P}}_n^l(X, \beta)_p = \mathcal{S}_{0,p}^{t+} \sqcup \mathcal{S}_{1,p}^{t+} \sqcup \cdots \sqcup \mathcal{S}_{|\vec{w}|-1,p}^{t+} \sqcup \widehat{\mathcal{P}}_n^{t+}(X, \beta)_p.$$

Here $\cup_{i \le l} \mathcal{S}_{i,p}^{t\pm}$ correspond to $\mu_i^\dagger$-semistable object $\mathcal{E}'$ which admits an exact sequence

$$0 \to \mathcal{E}_1' \to \mathcal{E}' \to E_2'[-1] \to 0, \ E_2' \in \mathrm{Coh}_{\le 1}(X)$$

where $(\beta_2, n_2) = [\pi_* E_2']$ is of the form (7.2.18) such that $\sum_{i=1}^{k} w_i' \ge |\vec{w}| - l$.

Under the top isomorphism (7.2.17), the above $\mathcal{E}'$ corresponds to a $Q_{\mathcal{E}}$-representation T which admits an exact sequence

$$0 \to T_1 \to T \to T_2 \to 0$$

such that T_2 is a $Q_{\mathcal{E},0}$-representation with $|\mathbf{dim}(T_2)| \ge |\vec{w}| - l$. A $Q_{\mathcal{E}}$-representation T admits an exact sequence as above if and only if it lies in $\cup_{i \le l} S_i^+$, where $S_i^+ \subset R^\dagger(Q_{\mathcal{E},0}, \vec{w})$ is a KN strata with respect to χ_0 given in Lemma 6.1.9. Therefore the lemma holds. $\square$

7.3   Proofs of Conjectures 7.2.3, 7.2.4, 7.2.5, and 7.2.6

In this section, we prove the conjectures in the previous section under Assumption 7.3.1.

7.3.1 Assumption

Let us take $t \in \mathbb{R}_{>0} \cup \{\infty\}$ and $(\beta, n) \in N_{\leq 1}(S)$. Below we impose the following assumption:

Assumption 7.3.1 For each decomposition $(\beta, n) = (\beta_1, n_1) + (\beta_2, n_2)$ in $\mathbf{D}^t_{(\beta, n)}$, we assume that the morphisms (4.2.5) and (3.4.5) restrict to the morphisms

$$\pi^\dagger_* \colon \mathcal{P}^t_{n_1}(X, \beta_1) \to \mathcal{P}^t_{n_1}(S, \beta_1), \ \pi_* \colon \mathcal{M}^{H\text{-st}}_{n_2}(X, \beta_2) \to \mathcal{M}^{H\text{-st}}_{n_2}(S, \beta_2). \qquad (7.3.1)$$

By Lemmas 4.3.13 and 3.4.7, the above assumption implies that

$$\mathcal{P}^t_{n_1}(X, \beta) = t_0(\Omega_{\mathfrak{P}^t_{n_1}(S, \beta_1)}[-1]), \ \mathcal{M}^H_{n_2}(X, \beta_2) = t_0(\Omega_{\mathfrak{M}^H_{n_2}(S, \beta_2)}).$$

Therefore we have equivalences

$$\mathcal{DT}^{\mathbb{C}^*}(\mathcal{P}^t_{n_1}(X, \beta_1)) \xrightarrow{\sim} D^b_{\mathrm{coh}}(\mathfrak{P}^t_{n_1}(S, \beta_1)), \qquad (7.3.2)$$

$$\mathcal{DT}^{\mathbb{C}^*}(\mathcal{M}^H_{n_2}(X, \beta_2)) \xrightarrow{\sim} D^b_{\mathrm{coh}}(\mathfrak{M}^H_{n_2}(S, \beta_2)).$$

Moreover as we have open immersions $P^{t\pm}_{n_1}(X, \beta_2) \subset \mathcal{P}^t_{n_1}(X, \beta_1)$, we have

$$\mathcal{DT}^{\mathbb{C}^*}(P^{t\pm}_{n_1}(X, \beta_1)) \xrightarrow{\sim} D^b_{\mathrm{coh}}(\mathfrak{P}^t_{n_1}(S, \beta_1))/\mathcal{C}_{\mathcal{Z}^{t\pm}\text{-us}}, \qquad (7.3.3)$$

where $\mathcal{Z}^{t\pm}\text{-us}$ is the complement of $P^{t\pm}_{n_1}(X, \beta_1)$ in $\mathcal{P}^t_{n_1}(X, \beta_1)$.

Remark 7.3.2 In Assumption 7.3.1, we don't impose the first condition in (7.3.1) for $t = t_\pm$, so (7.3.3) are not necessary equivalent to $\mathcal{DT}^{\mathbb{C}^*}(\mathfrak{P}^{t\pm}_{n_1}(S, \beta_1))$.

Note that the second condition in (7.3.1) is satisfied if β is a reduced class by Lemma 6.4.9. We discuss the cases where the first condition in (7.3.1) is satisfied. We prepare the following two lemmas:

Lemma 7.3.3 *For each effective class $\beta \in \mathrm{NS}(S)$, there exist $m(\beta) \in \mathbb{Q}$ such that for any H-semistable sheaf $E \in \mathrm{Coh}_{\leq 1}(X)$ with $l(\pi_* E) \leq \beta$ and $\mu_H(E) > m(\beta)$, we have $H^{>0}(X, E) = 0$.*

Proof For each $0 \leq i < \beta \cdot H$ and $k \in \mathbb{Z}$, we have the isomorphism of stacks

$$\otimes \mathcal{O}_X(k\pi^*H) \colon \mathcal{M}^H_i(X, \beta) \xrightarrow{\cong} \mathcal{M}^H_{i+kH\cdot\beta}(X, \beta).$$

Since $\mathcal{M}_i^H(X, \beta)$ is of finite type, there exist $k(i) \in \mathbb{Z}$ such that $H^{>0}(X, E \otimes \mathcal{O}_X(k\pi^*H)) = 0$ for $k \geq k(i)$ and $[E] \in \mathcal{M}_i^H(X, \beta)$. By setting

$$m'(\beta) := \max\left\{k(i) + \frac{i}{\beta \cdot H} : 0 \leq i < \beta \cdot H\right\},$$

$$m(\beta) := \max\left\{m'(\beta') : 0 < \beta' \leq \beta\right\}$$

the lemma holds.   □

Lemma 7.3.4 *For an effective reduced class $\beta \in \mathrm{NS}(S)$, let $m(\beta) \in \mathbb{Q}$ be defined in Lemma 7.3.3. Then for $t \in (m(\beta), \infty]$, the morphism (4.2.5) restricts to the morphism*

$$\pi_*^\dagger : \mathcal{P}_n^t(X, \beta) \to \mathcal{P}_n^t(S, \beta).$$

Proof Let $\mathcal{E} \in \mathcal{A}_X$ be a rank one $\mu_t^\dagger$-semistable object with $\mathrm{cl}(\mathcal{E}) = (1, \beta, n)$. Then we have the exact sequence

$$0 \to \mathcal{H}^0(\mathcal{E}) \to \mathcal{E} \to \mathcal{H}^1(\mathcal{E})[-1] \to 0$$

in $\mathcal{A}_X$, where $\mathcal{H}^1(\mathcal{E}) \in \mathrm{Coh}_{\leq 1}(X)$. The $\mu_t^\dagger$-stability of $\mathcal{E}$ implies that any Harder-Narasimhan factor T of $\mathcal{H}^1(\mathcal{E})$ with respect to μ_H-stability satisfies $\mu_H(T) \geq t$. From the definition of $m(\beta)$, we have $H^1(X, \mathcal{H}^1(\mathcal{E})) = 0$. On the other hand as $\mathcal{H}^0(\mathcal{E})$ is a rank one torsion free sheaf, it is the ideal sheaf I_C for a compactly supported closed subscheme $C \subset X$ with $\dim C \leq 1$. The composition

$$\mathcal{H}^1(\mathcal{E})[-2] \to \mathcal{H}^0(\mathcal{E}) = I_C \hookrightarrow \mathcal{O}_{\overline{X}}$$

vanishes by the Serre duality and the vanishing $H^1(X, \mathcal{H}^1(\mathcal{E})) = 0$. So the morphism $\mathcal{H}^0(\mathcal{E}) \hookrightarrow \mathcal{O}_{\overline{X}}$ factors through $\mathcal{E} \to \mathcal{O}_{\overline{X}}$. By taking the cones, we obtain exact sequences in $\mathcal{A}_X$ and $\mathrm{Coh}_{\leq 1}(X)$

$$0 \to F[-1] \to \mathcal{E} \to \mathcal{O}_{\overline{X}} \to 0, \quad 0 \to \mathcal{O}_C \to F \to \mathcal{H}^1(\mathcal{E}) \to 0.$$

Therefore $\mathcal{E}$ is isomorphic to the two term complex $(s : \mathcal{O}_{\overline{X}} \to F)$, and the morphism $\pi_*^\dagger$ in (4.2.5) sends it to $(\pi_*s : \mathcal{O}_S \to \pi_*F)$.

Below we show that $(\pi_*s : \mathcal{O}_S \to \pi_*F)$ is $\mu_t^\dagger$-semistable in $\mathcal{A}_S$. Namely for any exact sequence in $\mathcal{A}_S$

$$0 \to (W_1 \otimes \mathcal{O}_S \to F_1) \to (\mathcal{O}_S \to \pi_*F) \to (W_2 \otimes \mathcal{O}_S \to F_2) \to 0$$

we show the inequality

$$\mu_t^\dagger(I_1) \le t \le \mu_t^\dagger(I_2), \ I_i := (W_i \otimes \mathcal{O}_S \to F_i). \tag{7.3.4}$$

Note that (W_1, W_2) is either $(0, \mathbb{C})$ or $(\mathbb{C}, 0)$. By the $\mu_t^\dagger$-stability of $\mathcal{E}$, we see that any Harder-Narasimhan factor F' of F with respect to μ_H-stability satisfies $\mu_H(F') \le t$. As β is reduced, $\pi_* F'$ is also μ_H-semistable with $\mu_H(\pi_* F') \le t$ by Lemma 6.4.9. Therefore (7.3.4) holds when $(W_1, W_2) = (0, \mathbb{C})$. Also by pushing forward the exact sequence $\mathcal{O}_{\overline{X}} \to F \to \mathrm{Cok}(s) \to 0$, we obtain the exact sequence

$$\mathcal{O}_S \to \pi_* F \to \pi_* \mathrm{Cok}(s) \to 0.$$

Therefore we have the surjection $\mathrm{Cok}(\pi_* s) \twoheadrightarrow \pi_* \mathrm{Cok}(s)$. We take the exact sequence

$$0 \to Q \to \mathrm{Cok}(\pi_* s) \to \pi_* \mathrm{Cok}(s) \to 0.$$

Since β is reduced, both of $\pi_* \mathrm{Cok}(s)$ and $\mathrm{Cok}(\pi_* s)$ has the same one dimensional reduced supports, so Q must be zero dimensional. As $\mathrm{Cok}(s) = \mathcal{H}^1(\mathcal{E})$, it follows that any Harder-Narasimhan factor T' of $\mathrm{Cok}(\pi_* s)$ with respect to μ_H-stability satisfies $\mu_H(T') \ge t$. Therefore (7.3.4) holds when $(W_1, W_2) = (\mathbb{C}, 0)$. □

For example, Assumption 7.3.1 holds in the following cases.

Example 7.3.5 Let β be a reduced curve class and take $t \in (m(\beta), \infty)$. Then the assumption (7.3.1) is satisfied by Lemma 7.3.4. In this case, the first condition (7.3.1) also holds for $t = t_\pm$ so we have equivalences

$$\mathcal{DT}^{\mathbb{C}^*}(P_{n_1}^{t_\pm}(X, \beta)) \xrightarrow{\sim} D^b_{\mathrm{coh}}(\mathfrak{P}_{n_1}^{t_\pm}(S, \beta_1)).$$

Example 7.3.6 Let β be a reduced curve class and take $t = \infty$. Then the assumption (7.3.1) is satisfied by Lemma 7.3.4. In this case, the first condition (7.3.1) is satisfied for $t = t_-$, but not satisfied for $t = t_+$. So we have

$$\mathcal{DT}^{\mathbb{C}^*}(I_n(X, \beta)) \ne D^b_{\mathrm{coh}}(\mathfrak{I}_n(S, \beta)), \ \mathcal{DT}^{\mathbb{C}^*}(P_n(X, \beta)) \xrightarrow{\sim} D^b_{\mathrm{coh}}(\mathfrak{P}_n(S, \beta))$$

where $\mathfrak{I}_n(S, \beta)$ is the derived Hilbert scheme as in Remark 6.5.6.

Example 7.3.7 Let β be an irreducible curve class. In this case, there is only one wall $t = n/(H \cdot \beta)$, and the assumption (7.3.1) is satisfied at this wall. In this case, the first condition (7.3.1) is satisfied for $t = t_+$, but not satisfied for $t = t_-$. So we have

$$\mathcal{DT}^{\mathbb{C}^*}(P_n^{t+}(X, \beta)) \xrightarrow{\sim} D^b_{\mathrm{coh}}(\mathfrak{P}_n(S, \beta)), \ \mathcal{DT}^{\mathbb{C}^*}(P_n^{t-}(X, \beta)) \ne D^b_{\mathrm{coh}}(\mathfrak{P}_n^{t-}(S, \beta)).$$

Indeed we have $\mathfrak{P}_n^{t-}(S, \beta) = \emptyset$, but $\mathcal{DT}^{\mathbb{C}^*}(P_n^{t-}(X, \beta))$ is not necessary zero (see Sect. 5.1).

7.3.2   The Formal Local Descriptions of $\mathcal{P}_n^t(X, \beta)$ Over $P_n^t(S, \beta)$

Below we suppose that Assumption 7.3.1 holds. Then by Lemma 4.3.13, we have the following commutative diagram

$$
\begin{array}{ccc}
t_0(\Omega_{\mathfrak{P}_n^t(S,\beta)}[-1]) \xrightarrow{\;\cong\;} & \mathcal{P}_n^t(X, \beta) \longrightarrow & P_n^t(X, \beta) \\
\downarrow \qquad\qquad & \pi_*^\dagger \downarrow \qquad & \pi_*^\dagger \downarrow \\
\mathcal{P}_n^t(S, \beta) =\!=\!=\!=\!= & \mathcal{P}_n^t(S, \beta) \longrightarrow & P_n^t(S, \beta).
\end{array}
\qquad (7.3.5)
$$

Here the right horizontal arrows are good moduli space morphisms, and the right vertical arrow is the induced morphism on good moduli spaces. By Lemma 8.4.3, the derived stack $\mathfrak{P}_n^t(S, \beta)$ satisfies the formal neighborhood theorem. Together with Lemma 6.2.5, we have the formal local description of $\mathfrak{P}_n^t(S, \beta)$ over the good moduli space as we discuss below.

Let us take a closed point $p \in P_n^t(S, \beta)$, which corresponds to a $\mu_t^\dagger$-polystable object in $\mathcal{A}_S$

$$
I = (\mathcal{O}_S \to F) = (\mathcal{O}_S \to F_0) \oplus \bigoplus_{i=1}^k V_i \otimes (0 \to F_i) \qquad (7.3.6)
$$

where V_i is a finite dimensional vector space, $I_0 = (\mathcal{O}_S \to F_0)$ is $\mu_t^\dagger$-stable and $F_1, \ldots, F_k$ are mutually non-isomorphic μ_H-stable sheaves in $\mathrm{Coh}_{\leq 1}(S)$ with $\mu_H(F_i) = t$ for $1 \leq i \leq k$. The automorphism group of I is

$$
G := \mathrm{Aut}(I) = \prod_{i=1}^k \mathrm{GL}(V_i).
$$

By (4.2.2), the tangent complex of $\mathfrak{P}_n^t(S, \beta)$ at I is given by $\mathbf{RHom}(I, F)$, which is a complex of G-representations. The G-action is determined by the decomposition

$$
\mathbf{RHom}(I, F) = \mathbf{RHom}(I_0, F_0) \oplus \bigoplus_{i=1}^k \mathbf{RHom}(I_0, F_i) \otimes V_i
$$

$$
\oplus \bigoplus_{i=1}^k \mathbf{RHom}(F_i, F_0)[1] \otimes V_i^\vee
$$

$$
\oplus \bigoplus_{i,j} \mathbf{RHom}(F_i, F_j)[1] \otimes \mathrm{Hom}(V_i, V_j)
$$

and the natural actions of G to V_i. Let $\widehat{\mathrm{Hom}(I, F)}_0$ be the formal fiber of $\mathrm{Hom}(I, F) \to \mathrm{Hom}(I, F) /\!\!/ G$ at 0 (see Sect. 6.1.4 for the notation of the formal fiber). Let κ be a G-equivariant Kuranishi map

$$\kappa: \widehat{\mathrm{Hom}(I, F)}_0 \to \mathrm{Hom}^1(I, F) \tag{7.3.7}$$

and $\widehat{\mathfrak{N}}_0$ its derived zero locus. Then the derived stack $\mathfrak{P}_n^t(S, \beta)$ along the good moduli space morphism $\mathcal{P}_n^t(S, \beta) \to P_n^t(S, \beta)$ is equivalent to the quotient derived stack

$$[\widehat{\mathfrak{N}}_0 / G] \hookrightarrow [\widehat{\mathrm{Hom}(I, F)}_0 / G].$$

Let w be the function

$$w: \widehat{\mathrm{Hom}(I, F)}_0 \oplus \mathrm{Hom}^1(I, F)^\vee \to \mathbb{C}$$

determined from κ by the construction (2.1.3). Then from the diagram (7.3.5), we see that

$$t_0 \left(\Omega_{[\widehat{\mathfrak{N}}_0 / G]}[-1] \right) = [\mathrm{Crit}(w)/G] \tag{7.3.8}$$

is isomorphic to the formal fiber of the morphism $\mathcal{P}_n^t(X, \beta) \to P_n^t(S, \beta)$ at p in the diagram (7.3.5).

Remark 7.3.8 The G-representation $\mathrm{Hom}(I, F) \oplus \mathrm{Hom}^1(I, F)^\vee$ is explicitly written as

$$\mathrm{Hom}(I_0, F_0) \oplus \mathrm{Hom}^1(I_0, F_0)^\vee \oplus \bigoplus_{i=1}^k \left(\mathrm{Hom}(I_0, F_i) \oplus \mathrm{Ext}^2(F_i, F_0)^\vee \right) \otimes V_i$$

$$\oplus \bigoplus_{i=1}^k \left(\mathrm{Hom}^1(I_0, F_i)^\vee \oplus \mathrm{Ext}^1(F_i, F_0) \right) \otimes V_i^\vee$$

$$\oplus \bigoplus_{i,j} \left(\mathrm{Ext}^1(F_i, F_j) \oplus \mathrm{Ext}^1(F_j, F_i)^\vee \right) \otimes \mathrm{Hom}(V_i, V_j).$$

Similarly to Remark 6.4.2, the above G-representation is the space of representations of the Ext-quiver associated with the collection

$$\{ (\mathcal{O}_X \to i_* F_0), i_* F_1[-1], \ldots, i_* F_k[-1] \}$$

on the three-fold $X = \mathrm{Tot}_S(\omega_S)$, with dimension vector $(1, \{\dim V_i\}_{1 \le i \le k})$. Here $i: S \hookrightarrow X$ is the zero section.

By restricting $P_n^{t\pm}(X, \beta)$ and $\mathcal{Z}_l^{t\pm\text{-us}}$ to the formal fiber (7.3.8), we have open/closed substacks

$$[\mathrm{Crit}(w)^{t\pm\text{-ss}}/G] \subset [\mathrm{Crit}(w)/G] \supset \widehat{\mathcal{Z}}_l^{t\pm\text{-us}}.$$

We also set $\widehat{\mathcal{S}}_l^{t\pm} := \widehat{\mathcal{Z}}_l^{t\pm\text{-us}} \setminus \widehat{\mathcal{Z}}_{l-1}^{t\pm\text{-us}}$.

Lemma 7.3.9 *The stratifications*

$$[\mathrm{Crit}(w)/G] = \widehat{\mathcal{S}}_1^{t\pm} \sqcup \widehat{\mathcal{S}}_2^{t\pm} \sqcup \cdots \sqcup \widehat{\mathcal{S}}_N^{t\pm} \sqcup [\mathrm{Crit}(w)^{t\pm\text{-ss}}/G]$$

are KN stratifications with respect to $\chi_0^{\pm} : G \to \mathbb{C}^*$, where χ_0 is the determinant character

$$\chi_0 : G = \prod_{i=1}^{k} \mathrm{GL}(V_i) \to \mathbb{C}^*, \quad (g_i)_{1 \le i \le k} \mapsto \prod_{i=1}^{k} \det(g_i).$$

Proof A closed point of the stack $[\mathrm{Crit}(w)/G]$ corresponds to an object $\mathcal{E}$ of the form (7.2.15) such that $\pi_* \mathcal{E}$ is S-equivalent to (7.3.6). By the first condition in (7.3.1), the object $\pi_* \mathcal{E}_0$ is $\mu_t^{\dagger}$-semistable, so it is S-equivalent to the object in $\mathcal{A}_S$ of the form

$$(\mathcal{O}_S \to F_0') \oplus \bigoplus_{i=1}^{k} W_i' \otimes (0 \to F_i),$$

where $I_0' = (\mathcal{O}_S \to F_0')$ is $\mu_t^{\dagger}$-stable. The second condition in (7.3.1) implies that each $\pi_* E_i$ is μ_H-stable. Therefore by comparing the multiplicity of F_i, we have

$$V_i = W_i' \oplus \bigoplus_{\pi_* E_j \cong F_i} W_j.$$

The map $\mathrm{Aut}(\mathcal{E}) \overset{\pi_*}{\to} \mathrm{Aut}(\pi_* \mathcal{E}) \subset \mathrm{Aut}(I)$ is the inclusion $G_{\mathcal{E}} \hookrightarrow G$ given by

$$\prod \mathrm{GL}(W_i) \to \prod \mathrm{GL}(V_i), \quad (g_i) \mapsto \left(\mathrm{id}_{W_i'} \oplus \bigoplus_{\pi_* E_j \cong F_i} g_j \right).$$

Therefore the composition $G_{\mathcal{E}} \hookrightarrow G \overset{\chi_0}{\to} \mathbb{C}^*$ coincides with the determinant character for $G_{\mathcal{E}}$ in Lemma 7.2.8. It follows that the claim holds for each fiber of the map $[\mathrm{Crit}(w)/G] \to \mathrm{Crit}(w)/\!/G$ by Lemma 7.2.8, so the lemma follows. □

7.3.3 The Formal Local Descriptions of the Stack of Exact Sequences

Let $(\beta, n) = (\beta_1, n_1) + (\beta_2, n_2)$ be a decomposition in $\mathbf{D}^t_{(\beta,n)}$. For $v_i = (\beta_i, n_i)$, we have the open substacks in the diagrams (7.1.6) and (7.1.9)

$$\mathfrak{M}^{\mathrm{ext},\dagger,t}_S(v_\bullet) := (\mathrm{ev}_3^\dagger)^{-1}(\mathfrak{P}^t_n(S, \beta)) \subset \mathfrak{M}^{\mathrm{ext},\dagger}_S(v_\bullet),$$

$$\mathfrak{M}^{\mathrm{ext},\ddagger,t}_S(v_\bullet) := (\mathrm{ev}_3^\ddagger)^{-1}(\mathfrak{P}^t_n(S, \beta)) \subset \mathfrak{M}^{\mathrm{ext},\ddagger}_S(v_\bullet).$$

Similarly to (7.1.17), the diagrams (7.1.6) and (7.1.9) restrict to the diagrams

$$
\begin{array}{ccc}
\mathfrak{M}^{\mathrm{ext},\dagger,t}_S(v_\bullet) & \xrightarrow{\mathrm{ev}_3^\dagger} & \mathfrak{P}^t_n(S, \beta) \\
{\scriptstyle (\mathrm{ev}_1^\dagger, \mathrm{ev}_2^\dagger)}\downarrow & & \\
\mathfrak{P}^t_{n_1}(S, \beta_1) \times \mathfrak{M}^H_{n_2}(S, \beta_2), & &
\end{array}
\tag{7.3.9}
$$

$$
\begin{array}{ccc}
\mathfrak{M}^{\mathrm{ext},\dagger,t}_S(v_\bullet) & \xrightarrow{\mathrm{ev}_3^\ddagger} & \mathfrak{P}^t_n(S, \beta) \\
{\scriptstyle (\mathrm{ev}_2^\ddagger, \mathrm{ev}_1^\ddagger)}\downarrow & & \\
\mathfrak{M}^H_{n_2}(S, \beta_2) \times \mathfrak{P}^t_{n_1}(S, \beta_1). & &
\end{array}
$$

By taking the classical truncations and good moduli spaces, we obtain the following commutative diagrams

$$
\begin{array}{ccc}
\mathcal{M}^{\mathrm{ext},\dagger,t}_S(v_\bullet) & \xrightarrow{\mathrm{ev}_3^\dagger} & \mathcal{P}^t_n(S, \beta) \\
{\scriptstyle (\mathrm{ev}_1^\dagger, \mathrm{ev}_2^\dagger)}\downarrow & & \downarrow \\
\mathcal{P}^t_{n_1}(S, \beta_1) \times \mathcal{M}^H_{n_2}(S, \beta_2) & & \\
\downarrow & & \downarrow \\
P^t_{n_1}(S, \beta_1) \times M^H_{n_2}(S, \beta_2) & \xrightarrow{\oplus} & P^t_n(S, \beta),
\end{array}
\tag{7.3.10}
$$

$$\mathcal{M}_S^{\mathrm{ext},\ddagger,t}(v_\bullet) \xrightarrow{\quad \mathrm{ev}_3^{\ddagger} \quad} \mathcal{P}_n^t(S, \beta)$$

$$(\mathrm{ev}_2^{\ddagger}, \mathrm{ev}_1^{\ddagger}) \downarrow \qquad\qquad\qquad \downarrow$$

$$\mathcal{M}_{n_2}^H(S, \beta_2) \times \mathcal{P}_{n_1}^t(S, \beta_1)$$

$$\downarrow \qquad\qquad\qquad\qquad \downarrow$$

$$M_{n_2}^H(S, \beta_2) \times P_{n_1}^t(S, \beta_1) \xrightarrow{\quad \oplus \quad} P_n^t(S, \beta).$$

Here the bottom arrows are given by taking the direct sums of polystable objects, and the left bottom vertical arrow, the right vertical arrows are good moduli space morphisms.

Any point $(p^{(1)}, p^{(2)}) \in P_{n_1}^t(S, \beta_1) \times M_{n_2}^H(S, \beta_2)$ of the bottom arrows in (7.3.10) over the point $p \in P_n^t(S, \beta)$ corresponds to a direct sum decomposition of the object I in (7.3.6)

$$I = I_0 \oplus \bigoplus_{i=1}^k V_i^{(1)} \otimes (0 \to F_i) \oplus \bigoplus_{i=1}^k V_i^{(2)} \otimes (0 \to F_i). \tag{7.3.11}$$

Here $V_i = V_i^{(1)} \oplus V_i^{(2)}$ and

$$I^{(1)} = (\mathcal{O}_S \to F^{(1)}) := I_0 \oplus \bigoplus_{i=1}^k V_i^{(1)} \otimes (0 \to F_i) \in P_{n_1}^t(S, \beta_1),$$

$$F^{(2)} := \bigoplus_{i=1}^k V_i^{(2)} \otimes F_i \in M_{n_2}^H(S, \beta_2),$$

such that $p^{(1)}$ corresponds to $I^{(1)}$ and $p^{(2)}$ corresponds to $F^{(2)}$. We have the decomposition

$$\mathbf{R}\mathrm{Hom}(I, F) = \mathbf{R}\mathrm{Hom}(I^{(1)}, F^{(1)})$$

$$\oplus \mathbf{R}\mathrm{Hom}(I^{(1)}, F^{(2)})$$

$$\oplus \mathbf{R}\mathrm{Hom}(F^{(2)}, F^{(1)})[1] \oplus \mathbf{R}\mathrm{Hom}(F^{(2)}, F^{(2)})[1].$$

Let $G^{(1)}, G^{(2)}$ be

$$G^{(1)} = \mathrm{Aut}(I^{(1)}) = \prod_{i=1}^k \mathrm{GL}(V_i^{(1)}), \quad G^{(2)} = \mathrm{Aut}(F^{(2)}) = \prod_{i=1}^k \mathrm{GL}(V_i^{(2)}).$$

Let $T^{(ij)}$ and $O^{(ij)}$ be defined by

$$T^{(11)} = \operatorname{Hom}(I^{(1)}, F^{(1)}), \quad T^{(12)} = \operatorname{Hom}(I^{(1)}, F^{(2)}),$$

$$T^{(21)} = \operatorname{Ext}^1(F^{(2)}, F^{(1)}), \quad T^{(22)} = \operatorname{Ext}^1(F^{(2)}, F^{(2)}),$$

$$O^{(11)} = \operatorname{Hom}^1(I^{(1)}, F^{(1)}), \quad O^{(12)} = \operatorname{Hom}^1(I^{(1)}, F^{(2)}),$$

$$O^{(21)} = \operatorname{Ext}^2(F^{(2)}, F^{(1)}), \quad O^{(22)} = \operatorname{Ext}^2(F^{(2)}, F^{(2)}).$$

We also write $T^{(i)} = T^{(ii)}$, $O^{(i)} = O^{(ii)}$ for simplicity. We have $G^{(i)}$-equivariant Kuranishi maps for $i = 1, 2$

$$\kappa^{(i)} \colon \widehat{T}_0^{(i)} \to O^{(i)}.$$

Let $\widehat{\mathfrak{N}}_0^{(1)}$, $\widehat{\mathfrak{N}}_0^{(2)}$ be derived zero loci of $\kappa^{(1)}$, $\kappa^{(2)}$ respectively. By Lemma 6.2.5, the derived stacks

$$[\widehat{\mathfrak{N}}_0^{(1)}/G^{(1)}]\widehat{\times}[\widehat{\mathfrak{N}}_0^{(2)}/G^{(2)}]$$

is equivalent to the derived stack $\mathfrak{P}_{n_1}^t(S, \beta_1) \times \mathfrak{M}_{n_2}^H(S, \beta_2)$ along the formal fiber of the left bottom arrow of (7.3.10) at $(p^{(1)}, p^{(2)})$. We have the function

$$w^{(i)} \colon \widehat{T}_0^{(i)} \times O^{(i)\vee} \to \mathbb{C}$$

determined by $\kappa^{(i)}$ by the construction (2.1.3).

We next give formal local description of the stack of exact sequences.

Lemma 7.3.10 *We have the following*

$$\widehat{\operatorname{Hom}(I, F)}_0 \times_{\operatorname{Hom}(I,F)} \left(\bigoplus_{i \geq j} T^{(ij)} \right) = (\widehat{T}_0^{(1)}\widehat{\times}\widehat{T}_0^{(2)}) \times T^{(21)}, \qquad (7.3.12)$$

$$\widehat{\operatorname{Hom}(I, F)}_0 \times_{\operatorname{Hom}(I,F)} \left(\bigoplus_{i \leq j} T^{(ij)} \right) = (\widehat{T}_0^{(2)}\widehat{\times}\widehat{T}_0^{(1)}) \times T^{(12)}.$$

Proof We only prove the first identity. Let us consider the composition

$$\bigoplus_{i \geq j} T^{(ij)} \hookrightarrow \operatorname{Hom}(I, F) \twoheadrightarrow \operatorname{Hom}(I, F)/\!\!/G. \qquad (7.3.13)$$

Since any point in the LHS is specialized to a point in $T^{(1)} \oplus T^{(2)}$ by the action of the one parameter subgroup

$$\mathbb{C}^* \to G^{(1)} \times G^{(2)} \subset G, \quad t \mapsto (\operatorname{id}, t^{-1}\operatorname{id})$$

the composition (7.3.13) factors through

$$\bigoplus_{i \geq j} T^{(ij)} \twoheadrightarrow \bigoplus_{i=1}^{2} T^{(i)} \to \mathrm{Hom}(I, F) /\!\!/ G. \tag{7.3.14}$$

The right arrow further factors as

$$\bigoplus_{i=1}^{2} T^{(i)} \twoheadrightarrow T^{(1)} /\!\!/ G^{(1)} \times T^{(2)} /\!\!/ G^{(2)} \to \mathrm{Hom}(I, F) /\!\!/ G \tag{7.3.15}$$

The right morphism is finite by Meinhardt and Reineke [89, Lemma 2.1], whose preimage of 0 is $(0, 0)$. Therefore the formal fiber of the composition (7.3.15) at 0 is $\widehat{T}_0^{(1)} \widehat{\times} \widehat{T}_0^{(2)}$. By taking the formal fiber of (7.3.14) at 0, we conclude that the formal fiber of (7.3.13) is given by the RHS of (7.3.12). □

By Lemma 7.3.10 and the fact that the Kuranishi map κ in (7.3.7) is G-equivariant, it restricts to morphisms

$$\kappa^{\mathrm{ext}} : (\widehat{T}_0^{(1)} \widehat{\times} \widehat{T}_0^{(2)}) \times T^{(21)} \to \bigoplus_{i \geq j} O^{(ij)},$$

$$\kappa'^{\mathrm{ext}} : (\widehat{T}_0^{(2)} \widehat{\times} \widehat{T}_0^{(1)}) \times T^{(12)} \to \bigoplus_{i \leq j} O^{(ij)}.$$

Let $\widehat{\mathfrak{N}}_0^{\mathrm{ext}}$, $\widehat{\mathfrak{N}}_0'^{\mathrm{ext}}$ be derived zero locus of κ^{ext}, κ'^{ext} respectively. We have closed immersions

$$\widehat{\mathfrak{N}}_0^{\mathrm{ext}} \hookrightarrow \widehat{\mathfrak{N}}_0, \quad \widehat{\mathfrak{N}}_0'^{\mathrm{ext}} \hookrightarrow \widehat{\mathfrak{N}}_0$$

which correspond to deformations of I which preserve subobjects $I^{(1)} \subset I$, $(0 \to F^{(2)}) \subset I$ respectively. The deformations preserving these subobjects are identified by the actions of the subgroups

$$G^{\mathrm{ext}} := \{g \in G : g(I^{(1)}) \subset I^{(1)}\}, \quad G'^{\mathrm{ext}} := \{g \in G : g(0 \to F^{(2)}) \subset (0 \to F^{(2)})\}$$

respectively. They are explicitly written as

$$G^{\mathrm{ext}} = \prod_{i=1}^{k} \mathrm{GL}(V_i^{(1)}) \times \prod_{i=1}^{k} \mathrm{GL}(V_i^{(2)}) \times \prod_{i=1}^{k} \mathrm{Hom}(V_i^{(2)}, V_i^{(1)}),$$

$$G'^{\mathrm{ext}} = \prod_{i=1}^{k} \mathrm{GL}(V_i^{(2)}) \times \prod_{i=1}^{k} \mathrm{GL}(V_i^{(1)}) \times \prod_{i=1}^{k} \mathrm{Hom}(V_i^{(1)}, V_i^{(2)}).$$

Therefore the derived stacks $\mathfrak{M}_S^{\text{ext},\dagger,t}(v_\bullet)$, $\mathfrak{M}_S^{\text{ext},\ddagger,t}(v_\bullet)$ along the formal fiber of the composition of the left arrows in (7.3.10) at $(p^{(1)}, p^{(2)})$ are equivalent to the derived stacks $[\widehat{\mathfrak{N}}_0^{\text{ext}}/G^{\text{ext}}]$, $[\widehat{\mathfrak{N}}_0^{\text{ext}}/G'^{\text{ext}}]$ respectively.

By the above arguments, the diagrams (7.3.9) along the formal fibers at $p \in P_n^t(S, \beta)$ are equivalent to the following diagrams

$$
\begin{array}{ccc}
[\widehat{\mathfrak{N}}_0^{\text{ext}}/G^{\text{ext}}] & \xrightarrow{\;\text{ev}_3^\dagger\;} & [\widehat{\mathfrak{N}}_0/G] \\
{\scriptstyle (\text{ev}_1^\dagger,\text{ev}_2^\dagger)}\big\downarrow & & \\
[\widehat{\mathfrak{N}}_0^{(1)}/G^{(1)}]\,\widehat{\times}\,[\widehat{\mathfrak{N}}_0^{(2)}/G^{(2)}], & &
\end{array}
\qquad
\begin{array}{ccc}
[\widehat{\mathfrak{N}}_0'^{\text{ext}}/G'^{\text{ext}}] & \xrightarrow{\;\text{ev}_3^\ddagger\;} & [\widehat{\mathfrak{N}}_0/G] \\
{\scriptstyle (\text{ev}_2^\ddagger,\text{ev}_1^\ddagger)}\big\downarrow & & \\
[\widehat{\mathfrak{N}}_0^{(2)}/G^{(2)}]\,\widehat{\times}\,[\widehat{\mathfrak{N}}_0^{(1)}/G^{(1)}]. & &
\end{array}
$$

$$(7.3.16)$$

Here the horizontal arrows are induced by the inclusions of (7.3.12) into $\widehat{\text{Hom}}(I, F)_0$, and the vertical arrows are induced by the projections from $\oplus_{i \geq j} T^{(ij)}$, $\oplus_{i \leq j} T^{(ij)}$ onto $T^{(1)} \oplus T^{(2)}$, $T^{(2)} \oplus T^{(1)}$ respectively. Let w^{ext} and w'^{ext} be the functions

$$
w^{\text{ext}} \colon (\widehat{T}_0^{(1)} \widehat{\times} \widehat{T}_0^{(2)}) \times T^{(21)} \times \bigoplus_{i \geq j} O^{(ij)\vee} \to \mathbb{C},
$$

$$
w'^{\text{ext}} \colon (\widehat{T}_0^{(2)} \widehat{\times} \widehat{T}_0^{(1)}) \times T^{(12)} \times \bigoplus_{i \leq j} O^{(ij)\vee} \to \mathbb{C}
$$

determined from κ^{ext}, κ'^{ext} by the construction (2.1.3). We have the following commutative diagrams

$$(7.3.17)$$

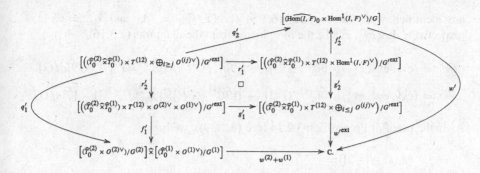

Here $g_1, r_1, f_2, q_2, g_1', r_1', f_2', q_2'$ are induced by embeddings into direct summands, and $f_1, g_2, r_2, q_1, f_1', g_2', r_2', q_1'$ are induced by projections. In order to simplify the notation, we write the above diagrams as

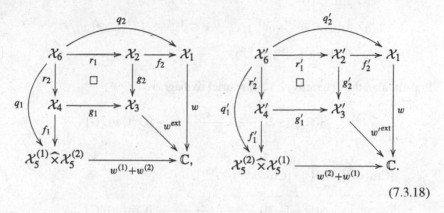

$$(7.3.18)$$

e.g. $\mathcal{X}_1 = [(\widehat{\mathrm{Hom}}(I, F)_0 \times \mathrm{Hom}^1(I, F)^\vee)/G]$, etc. We will also write $w_i : \mathcal{X}_i \to \mathbb{C}$, $w_i' : \mathcal{X}_i' \to \mathbb{C}$ for the induced functions on the above diagrams for $i = 2, 4, 6$.

Let λ be the one parameter subgroup

$$\lambda \colon \mathbb{C}^* \to G^{(1)} \times G^{(2)}, \quad t \mapsto (\mathrm{id}, t \cdot \mathrm{id}). \qquad (7.3.19)$$

Below we also regard λ as one parameter subgroups for G^{ext}, G'^{ext} and G as they contain $G^{(1)} \times G^{(2)}$ as a subgroup. Note that the diagrams

$$
\begin{array}{ccc}
\mathcal{X}_6 & \xrightarrow{\ q_2\ } & \mathcal{X}_1 \\
{\scriptstyle q_1}\big\downarrow & & \\
\mathcal{X}_5^{(1)} \widehat{\times} \mathcal{X}_5^{(2)}, & &
\end{array}
\qquad
\begin{array}{ccc}
\mathcal{X}_6' & \xrightarrow{\ q_2'\ } & \mathcal{X}_1 \\
{\scriptstyle q_1'}\big\downarrow & & \\
\mathcal{X}_5^{(2)} \widehat{\times} \mathcal{X}_5^{(1)} & &
\end{array}
\qquad (7.3.20)
$$

are identified with the diagram (6.1.8) for $[Y/G] = \mathcal{X}_1$ and $\lambda_\alpha = \lambda^{-1}, \lambda$ respectively. Let $\widehat{\mathrm{ev}}^\dagger$, $\widehat{\mathrm{ev}}^\ddagger$ be the morphisms from the diagrams (7.3.16),

$$\widehat{\mathrm{ev}}^\dagger := (\mathrm{ev}_1^\dagger, \mathrm{ev}_2^\dagger, \mathrm{ev}_3^\dagger) \colon [\widehat{\mathfrak{N}}_0^{\mathrm{ext}}/G^{\mathrm{ext}}] \to ([\widehat{\mathfrak{N}}_0^{(1)}/G^{(1)}]\widehat{\times}[\widehat{\mathfrak{N}}_0^{(2)}/G^{(2)}]) \times [\widehat{\mathfrak{N}}_0/G],$$

$$\widehat{\mathrm{ev}}^\ddagger := (\mathrm{ev}_1^\ddagger, \mathrm{ev}_2^\ddagger, \mathrm{ev}_3^\ddagger) \colon [\widehat{\mathfrak{N}}_0'^{\mathrm{ext}}/G^{\mathrm{ext}}] \to ([\widehat{\mathfrak{N}}_0^{(1)}/G^{(1)}]\widehat{\times}[\widehat{\mathfrak{N}}_0^{(2)}/G^{(2)}]) \times [\widehat{\mathfrak{N}}_0/G].$$

As in the proof of Proposition 3.2.14 (see (3.2.29)), we have

$$t_0(\Omega_{\widehat{\mathrm{ev}}^\dagger}[-2]) =$$
$$(q_1^{-1}([\mathrm{Crit}(w^{(1)}+w^{(2)})/G^{(1)} \times G^{(2)}]) \cap q_2^{-1}([\mathrm{Crit}(w)/G]))$$
$$\times_{\mathcal{X}_3} [\mathrm{Crit}(w^{\mathrm{ext}})/G^{\mathrm{ext}}],$$

$$t_0(\Omega_{\widehat{\mathrm{ev}}^\ddagger}[-2]) =$$
$$(q_1'^{-1}([\mathrm{Crit}(w^{(2)}+w^{(1)})/G^{(2)} \times G^{(1)}]) \cap q_2'^{-1}([\mathrm{Crit}(w)/G]))$$
$$\times_{\mathcal{X}_3'} [\mathrm{Crit}(w'^{\mathrm{ext}})/G'^{\mathrm{ext}}].$$

In particular, the diagrams (7.3.20) restrict to diagrams

$$
\begin{array}{ccc}
t_0(\Omega_{\widehat{\mathrm{ev}}^\dagger}[-2]) & \xrightarrow{q_2} & [\mathrm{Crit}(w)/G] \\
{\scriptstyle q_1}\downarrow & & \\
[\mathrm{Crit}(w^{(1)})/G^{(1)}]\widehat{\times}[\mathrm{Crit}(w^{(2)})/G^{(2)}], & &
\end{array}
$$

$$
\begin{array}{ccc}
t_0(\Omega_{\widehat{\mathrm{ev}}^\ddagger}[-2]) & \xrightarrow{q_2'} & [\mathrm{Crit}(w)/G] \\
{\scriptstyle q_1'}\downarrow & & \\
[\mathrm{Crit}(w^{(2)})/G^{(2)}]\widehat{\times}[\mathrm{Crit}(w^{(1)})/G^{(1)}], & &
\end{array}
\tag{7.3.21}
$$

which give formal local descriptions of the diagrams (7.1.17) at $p \in P_n^t(S, \beta)$.

7.3.4 Adjoint Functors of Categorified Hall Products

Let us consider the ind-completions of the functors (7.2.7)

$$* \colon \mathrm{Ind}\left(\mathcal{DT}^{\mathbb{C}^*}(\mathcal{P}_{n_1}^{t_+}(X, \beta_1)) \otimes \mathcal{DT}^{\mathbb{C}^*}(\mathcal{M}_{n_2}^H(X, \beta_2))\right)$$
$$\to \mathrm{Ind}\,\mathcal{DT}^{\mathbb{C}^*}\left(\mathcal{P}_n^t(X, \beta) \setminus \mathcal{Z}_{l-1}^{t_+,-\mathrm{us}}\right),$$

$$*: \mathrm{Ind}\left(\mathcal{DT}^{\mathbb{C}^*}(\mathcal{M}_{n_2}^H(X,\beta_2)) \otimes \mathcal{DT}^{\mathbb{C}^*}(\mathcal{P}_{n_1}^{t-}(X,\beta_1))\right)$$

$$\rightarrow \mathrm{Ind}\,\mathcal{DT}^{\mathbb{C}^*}\left(\mathcal{P}_n^t(X,\beta) \setminus \mathcal{Z}_{l-1}^{t--\mathrm{us}}\right). \tag{7.3.22}$$

We show that they admit right adjoints under Assumption 7.3.1.

Lemma 7.3.11 *Under Assumption 7.3.1, the functors (7.3.22) admit right adjoints*

$$*^R: \mathrm{Ind}\,\mathcal{DT}^{\mathbb{C}^*}\left(\mathcal{P}_n^t(X,\beta) \setminus \mathcal{Z}_{l-1}^{t+-\mathrm{us}}\right)$$

$$\rightarrow \mathrm{Ind}\left(\mathcal{DT}^{\mathbb{C}^*}(\mathcal{P}_{n_1}^{t+}(X,\beta_1)) \otimes \mathcal{DT}^{\mathbb{C}^*}(\mathcal{M}_{n_2}^H(X,\beta_2))\right),$$

$$*^R: \mathrm{Ind}\,\mathcal{DT}^{\mathbb{C}^*}\left(\mathcal{P}_n^t(X,\beta) \setminus \mathcal{Z}_{l-1}^{t--\mathrm{us}}\right)$$

$$\rightarrow \mathrm{Ind}\left(\mathcal{DT}^{\mathbb{C}^*}(\mathcal{M}_{n_2}^H(X,\beta_2)) \otimes \mathcal{DT}^{\mathbb{C}^*}(\mathcal{P}_{n_1}^{t-}(X,\beta_1))\right).$$

Proof We only prove the statement for t_+. By the equivalences (7.3.2), under the assumption the functor (7.2.7) is a descendant of the functor from the diagram (7.3.9)

$$\mathrm{ev}_{3*}^\dagger(\mathrm{ev}_1^\dagger,\mathrm{ev}_2^\dagger)^*: D_{\mathrm{coh}}^b(\mathfrak{P}_{n_1}^t(S,\beta_1)) \otimes D_{\mathrm{coh}}^b(\mathfrak{M}_{n_2}^H(S,\beta_2)) \tag{7.3.23}$$

$$\rightarrow D_{\mathrm{coh}}^b(\mathfrak{P}_n(S,\beta)).$$

Its ind-completion admits a right adjoint functor (see Sect. 3.1.2)

$$(\mathrm{ev}_1^\dagger,\mathrm{ev}_2^\dagger)_*^{\mathrm{ind}}(\mathrm{ev}_3^\dagger)^!: \mathrm{Ind}\,D_{\mathrm{coh}}^b(\mathfrak{P}_n(S,\beta))$$

$$\rightarrow \mathrm{Ind}\left(D_{\mathrm{coh}}^b(\mathfrak{P}_{n_1}^t(S,\beta_1)) \otimes D_{\mathrm{coh}}^b(\mathfrak{M}_{n_2}^H(S,\beta_2))\right).$$

We see that the above functor restricts to the functor

$$(\mathrm{ev}_1^\dagger,\mathrm{ev}_2^\dagger)_*^{\mathrm{ind}}(\mathrm{ev}_3^\dagger)^!: \mathrm{Ind}\,\mathcal{C}_{\mathcal{Z}_{l-1}^{t+-\mathrm{us}}} \rightarrow \mathrm{Ind}\left(\mathcal{C}_{\mathcal{Z}^{t+-\mathrm{us}(1)} \times \mathcal{M}_{n_2}^H(X,\beta_2)}\right). \tag{7.3.24}$$

Here we have set

$$\mathcal{Z}^{t+-\mathrm{us}(1)} = \mathcal{P}_{n_1}^t(X,\beta_1) \setminus \mathcal{P}_{n_1}^{t+}(X,\beta_1).$$

Indeed it is enough to show the above claim formally locally on $\mathcal{P}_n^t(S,\beta)$ in the diagram (7.3.10), i.e. the functor in the diagram (7.3.16)

$$(\mathrm{ev}_1^\dagger,\mathrm{ev}_2^\dagger)_*^{\mathrm{ind}}(\mathrm{ev}_3^\dagger)^!: \mathrm{Ind}\,D_{\mathrm{coh}}^b([\widehat{\mathfrak{N}}_0/G]) \tag{7.3.25}$$

$$\rightarrow \mathrm{Ind}\,D_{\mathrm{coh}}^b([\widehat{\mathfrak{N}}_0^{(1)}/G^{(1)}]\widehat{\times}[\widehat{\mathfrak{N}}_0^{(2)}/G^{(2)}])$$

restricts to the functor

$$(\mathrm{ev}_1^\dagger, \mathrm{ev}_2^\dagger)_*^{\mathrm{ind}}(\mathrm{ev}_3^\dagger)^! \colon \operatorname{Ind} \mathcal{C}_{\widehat{\mathcal{Z}}_{l-1}^{t+\text{-us}}} \to \operatorname{Ind}\left(\mathcal{C}_{\widehat{\mathcal{Z}}^{t+\text{-us}(1)} \widehat{\times} [\mathrm{Crit}(w^{(2)})/G^{(2)}]}\right). \qquad (7.3.26)$$

Here $\widehat{\mathcal{Z}}^{t+\text{-us}(1)} \subset [\mathrm{Crit}(w^{(1)})/G^{(1)}]$ is the pull-back of $\mathcal{Z}^{t+\text{-us}(1)}$ to $[\mathrm{Crit}(w^{(1)})/G^{(1)}]$. By Lemma 7.2.2, we have the identity in the diagram (7.3.21)

$$q_1^{-1}(\widehat{\mathcal{Z}}^{t+\text{-us}(1)} \widehat{\times} [\mathrm{Crit}(w^{(2)})/G^{(2)}]) = q_2^{-1}(\widehat{\mathcal{Z}}_{l-1}^{t+\text{-us}}).$$

Since the Cartesian diagram in (7.3.17) is derived Cartesian and f_2 is proper, we can apply Proposition 3.2.14 to conclude that the functor (7.3.25) restricts to the functor (7.3.26).

By Theorem 8.2.2, both sides in (7.3.24) are compactly generated with compact objects $\mathcal{C}_{\mathcal{Z}_{l-1}^{t+\text{-us}}}$, $\mathcal{C}_{\mathcal{Z}^{t+\text{-us}(1)} \times \mathcal{M}_{n_2}^H(X, \beta_2)}$ respectively. Therefore by the equivalences in Proposition 3.2.7, the desired right adjoint $*^R$ is obtained by taking the Verdier quotients of both sides in (7.3.23) by the subcategories in (7.3.24). $\qquad \square$

Lemma 7.3.12 *In the setting of Lemma 7.3.11, the functors $*^R$ in Lemma 7.3.11 composed with the projection*

$$\mathrm{pr}_j \colon \operatorname{Ind} \mathcal{DT}^{\mathbb{C}^*}(\mathcal{M}_{n_2}^H(X, \beta_2)) \to \operatorname{Ind} \mathcal{DT}^{\mathbb{C}^*}(\mathcal{M}_{n_2}^H(X, \beta_2))_j$$

restrict to the functors

$$*_j^R = \mathrm{pr}_j \circ *^R \colon \mathcal{DT}^{\mathbb{C}^*}\left(\mathcal{P}_n^t(X, \beta) \setminus \mathcal{Z}_{l-1}^{t+\text{-us}}\right)$$

$$\to \mathcal{DT}^{\mathbb{C}^*}(\mathcal{P}_{n_1}^{t+}(X, \beta_1)) \otimes \mathcal{DT}^{\mathbb{C}^*}(\mathcal{M}_{n_2}^H(X, \beta_2))_j,$$

$$*_j^R = \mathrm{pr}_j \circ *^R \colon \mathcal{DT}^{\mathbb{C}^*}\left(\mathcal{P}_n^t(X, \beta) \setminus \mathcal{Z}_{l-1}^{t-\text{-us}}\right)$$

$$\to \mathcal{DT}^{\mathbb{C}^*}(\mathcal{M}_{n_2}^H(X, \beta_2))_j \otimes \mathcal{DT}^{\mathbb{C}^*}(\mathcal{P}_{n_1}^{t-}(X, \beta_1))$$

which give right adjoint functors of the functors (7.2.8).

Proof We only prove the lemma for t_+. For an object $(-)$ in the LHS, we need to prove that the object $*_j^R(-) := \mathrm{pr}_j *^R (-)$, which is a priori an object in the ind-completion of the RHS by Lemma 7.3.11, is indeed an object in the RHS. It is enough to prove this formally locally on each point of $P_n^t(S, \beta)$ in the diagram (7.3.10). Let us take a closed point $p \in P_n^t(S, \beta)$ corresponding to the polystable object (7.3.6). Then formally locally around p, the functor $*_j^R$ is a descendant of the functor (7.3.25) composed with the projection

$$\mathrm{pr}_j \colon \operatorname{Ind} D_{\mathrm{coh}}^b([\widehat{\mathfrak{M}}_0^{(1)}/G^{(1)}] \widehat{\times} [\widehat{\mathfrak{M}}_0^{(2)}/G^{(2)}])$$

$$\to \operatorname{Ind} D_{\mathrm{coh}}^b([\widehat{\mathfrak{M}}_0^{(1)}/G^{(1)}] \widehat{\times} [\widehat{\mathfrak{M}}_0^{(2)}/G^{(2)}])_{\lambda\text{-wt}=j}.$$

Here λ is the one parameter subgroup (7.3.19). Therefore it is enough to show that the composition of (7.3.25) with the above projection restricts to the functor

$$\mathrm{pr}_j(\mathrm{ev}_1^\dagger, \mathrm{ev}_2^\dagger)_*^{\mathrm{ind}}(\mathrm{ev}_3^\dagger)^!\colon D_{\mathrm{coh}}^b([\widehat{\mathfrak{N}}_0/G])$$
$$\to D_{\mathrm{coh}}^b([\widehat{\mathfrak{N}}_0^{(1)}/G^{(1)}]\widehat{\times}[\widehat{\mathfrak{N}}_0^{(2)}/G^{(2)}])_{\lambda\text{-wt}=j}.$$

Below we use the notation of the diagram (7.3.18). By Lemmas 2.4.4 and 2.4.6, we are reduced to showing that the composition of the functors

$$\mathrm{MF}_{\mathrm{qcoh}}^{\mathbb{C}^*}(\mathcal{X}_1, w) \overset{f_2^!}{\to} \mathrm{MF}_{\mathrm{qcoh}}^{\mathbb{C}^*}(\mathcal{X}_2, w_2) \overset{g_{2*}}{\to} \mathrm{MF}_{\mathrm{qcoh}}^{\mathbb{C}^*}(\mathcal{X}_3, w^{\mathrm{ext}}) \overset{g_1^*}{\to} \mathrm{MF}_{\mathrm{qcoh}}^{\mathbb{C}^*}(\mathcal{X}_4, w_4)$$
$$\overset{f_{1*}}{\to} \mathrm{MF}_{\mathrm{qcoh}}^{\mathbb{C}^*}(\mathcal{X}_5^{(1)}\widehat{\times}\mathcal{X}_5^{(2)}, w^{(1)}+w^{(2)}) \overset{\mathrm{pr}_j}{\to} \mathrm{MF}_{\mathrm{qcoh}}^{\mathbb{C}^*}(\mathcal{X}_5^{(1)}\widehat{\times}\mathcal{X}_5^{(2)}, w^{(1)}+w^{(2)})_{\lambda\text{-wt}=j}$$

restricts to the functor

$$\mathrm{MF}_{\mathrm{coh}}^{\mathbb{C}^*}(\mathcal{X}_1, w) \to \mathrm{MF}_{\mathrm{coh}}^{\mathbb{C}^*}(\mathcal{X}_5^{(1)}\widehat{\times}\mathcal{X}_5^{(2)}, w^{(1)}+w^{(2)})_{\lambda\text{-wt}=j}.$$

By the derived base change, the above composition functor is equivalent to the following composition

$$\mathrm{MF}_{\mathrm{qcoh}}^{\mathbb{C}^*}(\mathcal{X}_1, w) \overset{f_2^!}{\to} \mathrm{MF}_{\mathrm{qcoh}}^{\mathbb{C}^*}(\mathcal{X}_2, w_2) \overset{r_1^*}{\to} \mathrm{MF}_{\mathrm{qcoh}}^{\mathbb{C}^*}(\mathcal{X}_6, w_6)$$
$$\overset{q_{1*}}{\to} \mathrm{MF}_{\mathrm{qcoh}}^{\mathbb{C}^*}(\mathcal{X}_5^{(1)}\widehat{\times}\mathcal{X}_5^{(2)}, w^{(1)}+w^{(2)}) \overset{\mathrm{pr}_j}{\to} \mathrm{MF}_{\mathrm{qcoh}}^{\mathbb{C}^*}(\mathcal{X}_5^{(1)}\widehat{\times}\mathcal{X}_5^{(2)}, w^{(1)}+w^{(2)})_{\lambda\text{-wt}=j}.$$

Since f_2 is a representable morphism of smooth stacks, it is quasi-smooth and $f_2^!$ is given by $f_2^!(-) = f_2^*(-) \otimes \omega_{f_2}$. Therefore $r_1^* f_2^!$ gives the functor

$$r_1^* f_2^!\colon \mathrm{MF}_{\mathrm{coh}}^{\mathbb{C}^*}(\mathcal{X}_1, w) \to \mathrm{MF}_{\mathrm{coh}}^{\mathbb{C}^*}(\mathcal{X}_6, w_6).$$

It is enough to show that the functor $\mathrm{pr}_j q_{1*}$ gives the functor

$$\mathrm{pr}_j q_{1*}\colon \mathrm{MF}_{\mathrm{coh}}^{\mathbb{C}^*}(\mathcal{X}_6, w_6) \to \mathrm{MF}_{\mathrm{coh}}^{\mathbb{C}^*}(\mathcal{X}_5^{(1)}\widehat{\times}\mathcal{X}_5^{(2)}, w^{(1)}+w^{(2)})_{\lambda\text{-wt}=j}. \qquad (7.3.27)$$

The morphism q_1 factors as

$$q_1\colon \mathcal{X}_6 \overset{q_1'}{\to} \mathcal{X}_7 := \left[\left((\widehat{T}_0^{(1)}\widehat{\times}\widehat{T}_0^{(2)}) \times O^{(1)\vee} \times O^{(2)\vee}\right)/G^{\mathrm{ext}}\right] \overset{q_1''}{\to} \mathcal{X}_5^{(1)}\widehat{\times}\mathcal{X}_5^{(2)}$$

where G^{ext} acts on $((\widehat{T}_0^{(1)} \widehat{\times} \widehat{T}_0^{(2)}) \times O^{(1)\vee} \times O^{(2)\vee})$ through the projection $G^{\mathrm{ext}} \to G^{(1)} \times G^{(2)}$. Since $T^{(21)} \oplus O^{(12)\vee}$ is of weight -1 with respect to λ, by Lemma 2.2.3 the push-forward q'_{1*} restricts to the functor

$$q'_{1*} \colon \mathrm{MF}_{\mathrm{coh}}^{\mathbb{C}^*}(\mathcal{X}_6, w_6) \to \mathrm{MF}_{\mathrm{coh}}^{\mathbb{C}^*}(\mathcal{X}_7, w_7)_{\lambda\text{-below}},$$

where $w_7 = q_1''^*(w^{(1)} + w^{(2)})$. Since q_{1*}'' is a gerbe over a finite dimensional linear space $\oplus_i \mathrm{Hom}(V_i^{(2)}, V_i^{(1)})$, the functor q_{1*}'' gives

$$q_{1*}'' \colon \mathrm{MF}_{\mathrm{coh}}^{\mathbb{C}^*}(\mathcal{X}_7, w_7) \to \mathrm{MF}_{\mathrm{coh}}^{\mathbb{C}^*}(\mathcal{X}_5^{(1)} \widehat{\times} \mathcal{X}_5^{(2)}, w^{(1)} + w^{(2)}).$$

Therefore $q_{1*} = q_{1*}'' \circ q'_{1*}$ restricts to the functor

$$q_{1*} \colon \mathrm{MF}_{\mathrm{coh}}^{\mathbb{C}^*}(\mathcal{X}_6, w_6) \to \mathrm{MF}_{\mathrm{coh}}^{\mathbb{C}^*}(\mathcal{X}_5^{(1)} \widehat{\times} \mathcal{X}_5^{(2)}, w^{(1)} + w^{(2)})_{\lambda\text{-below}},$$

which concludes that $\mathrm{pr}_j q_{1*}$ gives the functor (7.3.27). $\qquad\qquad\qquad\square$

Lemma 7.3.13 *Under Assumption 7.3.1, the functors (7.2.8) are fully-faithful.*

Proof We only prove the lemma for t_+. By Lemma 7.3.12, it is enough to show that the natural transform

$$\mathrm{id} \to *_j^R \circ *_j$$

is an isomorphism. It is enough to this formally locally on $P_n^t(S, \beta)$. So we are reduced to showing that the functor

$$(\mathrm{ev}_3^\dagger)_* (\mathrm{ev}_1^\dagger, \mathrm{ev}_2^\dagger)^* \colon$$
$$D_{\mathrm{coh}}^b([\widehat{\mathfrak{N}}_0^{(1)} \widehat{\times} \widehat{\mathfrak{N}}_0^{(2)} / G^{(1)} \times G^{(2)}])_{\lambda\text{-wt}=j} / \mathcal{C}_{\widehat{\mathcal{Z}}_1^{t_+ - \mathrm{us}(1)}} \widehat{\times} [\mathrm{Crit}(w^{(2)}) / G^{(2)}]$$
$$\to D_{\mathrm{coh}}^b([\widehat{\mathfrak{N}}_0 / G]) / \mathcal{C}_{\widehat{\mathcal{Z}}_{l-1}^{t_+ - \mathrm{us}}}$$

from the diagram (7.3.16) is fully-faithful. Let us consider the following composition functor

$$\mathrm{MF}_{\mathrm{coh}}^{\mathbb{C}^*}(\mathcal{X}_5^{(1)} \widehat{\times} \mathcal{X}_5^{(2)}, w^{(1)} + w^{(2)})_{\lambda\text{-wt}=j} \xrightarrow{f_1^*} \mathrm{MF}_{\mathrm{coh}}^{\mathbb{C}^*}(\mathcal{X}_4, w_4) \xrightarrow{g_{1!}} \mathrm{MF}_{\mathrm{coh}}^{\mathbb{C}^*}(\mathcal{X}_3, w^{\mathrm{ext}})$$

$$\xrightarrow{g_2^*} \mathrm{MF}_{\mathrm{coh}}^{\mathbb{C}^*}(\mathcal{X}_2, w_2) \xrightarrow{f_{2*}} \mathrm{MF}_{\mathrm{coh}}^{\mathbb{C}^*}(\mathcal{X}_1, w).$$

By Lemmas 2.4.4 and 2.4.7, it is enough to show that the descendant of the above composition functor

$$f_{2*}g_2^*g_{1!}f_1^* \colon \mathrm{MF}_{\mathrm{coh}}^{\mathbb{C}^*}((\mathcal{X}_5^{(1)} \setminus \widehat{\mathcal{Z}}^{t_+ - \mathrm{us}(1)}) \widehat{\times} \mathcal{X}_5^{(2)}, w^{(1)} + w^{(2)})_{\lambda\text{-wt}=j} \qquad (7.3.28)$$
$$\to \mathrm{MF}_{\mathrm{coh}}^{\mathbb{C}^*}(\mathcal{X}_1 \setminus \widehat{\mathcal{Z}}_{l-1}^{t_+ - \mathrm{us}}, w)$$

is fully-faithful. Note that we have

$$g_{1!}(-) = g_{1*}(- \otimes \det O^{(21)\vee}[-\dim O^{(21)}]).$$

Since the G^{ext}-action on $O^{(21)}$ factors through $G^{\mathrm{ext}} \twoheadrightarrow G^{(1)} \times G^{(2)}$, the G^{ext}-character $\det O^{(21)\vee}$ is written as $f_1^* \det O^{(21)\vee}$ where we regard $\det O^{(21)\vee}$ as a $(G^{(1)} \times G^{(2)})$-character. Therefore we have

$$f_{2*}g_2^*g_{1!}f_1^*(-) \cong q_{2*}q_1^*(- \otimes \det O^{(21)\vee}[-\dim O^{(21)}]).$$

Here the first functor is an equivalence

$$\otimes \det O^{(21)\vee}[-\dim O^{(21)}] \colon \mathrm{MF}_{\mathrm{coh}}^{\mathbb{C}^*}((\mathcal{X}_5^{(1)} \setminus \widehat{\mathcal{Z}}^{t_+ - \mathrm{us}(1)}) \widehat{\times} \mathcal{X}_5^{(2)}, w^{(1)} + w^{(2)})_{\lambda\text{-wt}=j}$$
$$(7.3.29)$$
$$\xrightarrow{\sim} \mathrm{MF}_{\mathrm{coh}}^{\mathbb{C}^*}((\mathcal{X}_5^{(1)} \setminus \widehat{\mathcal{Z}}^{t_+ - \mathrm{us}(1)}) \widehat{\times} \mathcal{X}_5^{(2)}, w^{(1)} + w^{(2)})_{\lambda\text{-wt}=j+\dim O^{(21)}}$$

By Lemma 7.3.9, the substack $\widehat{\mathcal{S}}_l^{t_+} \subset [\mathrm{Crit}(w)/G] \setminus \widehat{\mathcal{Z}}_{l-1}^{t_+ - \mathrm{us}}$ is a KN strata for the determinant character $\chi_0 \colon G \to \mathbb{C}^*$, with associated one parameter subgroup λ^{-1}. The center is given by

$$[\mathrm{Crit}(w)^\lambda/G^\lambda] \setminus \widehat{\mathcal{Z}}_{l-1}^{t_+ - \mathrm{us}} = \left([\mathrm{Crit}(w^{(1)})/G^{(1)}] \setminus \widehat{\mathcal{Z}}^{t_+ - \mathrm{us}(1)}\right) \widehat{\times} [\mathrm{Crit}(w^{(2)})/G^{(2)}]$$

by Lemma 7.2.2. Moreover as we already mentioned, the left diagram in (7.3.20) is identified with the diagram (6.1.8) for $\lambda_\alpha = \lambda^{-1}$. Therefore from the equivalence (6.1.9), we conclude that the functor

$$q_{2*}q_1^* \colon \mathrm{MF}_{\mathrm{coh}}^{\mathbb{C}^*}((\mathcal{X}_5^{(1)} \setminus \widehat{\mathcal{Z}}^{t_+ - \mathrm{us}(1)}) \widehat{\times} \mathcal{X}_5^{(2)}, w^{(1)} + w^{(2)})_{\lambda\text{-wt}=j+\dim O^{(21)}} \qquad (7.3.30)$$
$$\to \mathrm{MF}_{\mathrm{coh}}^{\mathbb{C}^*}(\mathcal{X}_1 \setminus \widehat{\mathcal{Z}}_{l-1}^{t_+ - \mathrm{us}})$$

is fully-faithful. Therefore the functor (7.3.28) is fully-faithful.   □

We denote by

$$\Upsilon_j^{\pm,\mathbf{d}} \subset \mathcal{DT}^{\mathbb{C}^*}(\mathcal{P}_n^t(X, \beta) \setminus \mathcal{Z}_{l-1}^{l \pm - \mathrm{us}})$$

the essential images of the fully-faithful functors (7.2.8).

Lemma 7.3.14 *We have*

$$\text{Hom}(\Upsilon_j^{+,\mathbf{d}}, \Upsilon_{j'}^{+,\mathbf{d}}) = 0 \ (j < j'), \ \text{Hom}(\Upsilon_j^{-,\mathbf{d}}, \Upsilon_{j'}^{-,\mathbf{d}}) = 0 \ (j > j').$$

Proof We only prove the first case. It is enough to show that $*_j^R \circ *_{j'} \cong 0$ for $j < j'$. As in the proof of Lemma 7.3.12, it is enough to prove this formally locally on $P_n^t(S, \beta)$. Note that the LHS of (7.3.30) have weight $-j - \dim O^{(21)}$ with respect to λ^{-1}. Therefore by Theorem 6.1.2, the essential images of the functors (7.3.30) are semiorthogonal for $j < j'$, so the lemma follows.   □

Lemma 7.3.15 *The functors (7.2.8) admit left adjoints*

$$*_j^L : \mathcal{DT}^{\mathbb{C}^*}\left(\mathcal{P}_n^t(X, \beta) \setminus \mathcal{Z}_{l-1}^{t+,-\text{us}}\right) \to \left(\mathcal{DT}^{\mathbb{C}^*}(\mathcal{P}_{n_1}^{t+}(X, \beta_1)) \otimes \mathcal{DT}^{\mathbb{C}^*}(\mathcal{M}_{n_2}^H(X, \beta_2))\right),$$

$$*_j^L : \mathcal{DT}^{\mathbb{C}^*}\left(\mathcal{P}_n^t(X, \beta) \setminus \mathcal{Z}_{l-1}^{t-,-\text{us}}\right) \to \left(\mathcal{DT}^{\mathbb{C}^*}(\mathcal{M}_{n_2}^H(X, \beta_2)) \otimes \mathcal{DT}^{\mathbb{C}^*}(\mathcal{P}_{n_1}^{t-}(X, \beta_1))\right).$$

Proof We only prove the lemma for t_+. In order to simplify the notation, we write the left diagram of (7.3.9) as

$$\mathfrak{M}_3 \xleftarrow{(\text{ev}_1^\dagger, \text{ev}_2^\dagger)} \mathfrak{M}_2 \xrightarrow{\text{ev}_3^\dagger} \mathfrak{M}_1.$$

We denote by $\mathbb{D}_i$ the Serre duality equivalence for $D_{\text{coh}}^b(\mathfrak{M}_i)$ in (3.2.4), and denote by $\omega_{(\text{ev}_1^\dagger, \text{ev}_2^\dagger)}$ the relative dualizing complex of $(\text{ev}_1^\dagger, \text{ev}_2^\dagger)$. It is of the form $(\text{ev}_1^\dagger, \text{ev}_2^\dagger)^* \mathcal{L}[k]$ for a line bundle $\mathcal{L}$ on $\mathfrak{M}_3$ and $k \in \mathbb{Z}$. Let

$$i_j : D_{\text{coh}}^b(\mathfrak{M}_3)_j \hookrightarrow D_{\text{coh}}^b(\mathfrak{M}_3), \ \text{pr}_j : D_{\text{coh}}^b(\mathfrak{M}_3) \twoheadrightarrow D_{\text{coh}}^b(\mathfrak{M}_3)_j$$

the inclusion from the weight j-part, projection to the weight j-part, respectively. Then for $A \in D_{\text{coh}}^b(\mathfrak{M}_1)$ and $B \in D_{\text{coh}}^b(\mathfrak{M}_3)_j$, using [41, Corollary 9.5.9] we have

$$\text{Hom}(A, \text{ev}_{3*}^\dagger(\text{ev}_1^\dagger, \text{ev}_2^\dagger)^* i_j(B))$$

$$= \text{Hom}(\mathbb{D}_1 \, \text{ev}_{3*}^\dagger(\text{ev}_1^\dagger, \text{ev}_2^\dagger)^* i_j(B), \mathbb{D}_1(A))$$

$$= \text{Hom}(\text{ev}_{3*}^\dagger \mathbb{D}_2(\text{ev}_1^\dagger, \text{ev}_2^\dagger)^* i_j(B), \mathbb{D}_1(A))$$

$$= \text{Hom}(\text{ev}_{3*}^\dagger(\text{ev}_1^\dagger, \text{ev}_2^\dagger)^*((\mathbb{D}_3 \circ i_j(B)) \otimes \mathcal{L}[k]), \mathbb{D}_1(A))$$

$$= \text{Hom}((\mathbb{D}_3 i_j(B)) \otimes \mathcal{L}[k], (\text{ev}_1^\dagger, \text{ev}_2^\dagger)_*(\text{ev}_3^\dagger)^! \mathbb{D}_1(A))$$

$$= \text{Hom}(i_{-j}\mathbb{D}_3(B), ((\text{ev}_1^\dagger, \text{ev}_2^\dagger)_*(\text{ev}_3^\dagger)^! \mathbb{D}_1(A)) \otimes \mathcal{L}^\vee[-k])$$

$$= \text{Hom}(\mathbb{D}_3(B), \text{pr}_{-j}(((\text{ev}_1^\dagger, \text{ev}_2^\dagger)_*(\text{ev}_3^\dagger)^! \mathbb{D}_1(A)) \otimes \mathcal{L}^\vee[-k])).$$

Here the last functor is

$$\mathrm{pr}_{-j}(((\mathrm{ev}_1^\dagger, \mathrm{ev}_2^\dagger)_*(\mathrm{ev}_3^\dagger)^! \mathbb{D}_1(-)) \otimes \mathcal{L}^\vee[-k]) \colon D^b_{\mathrm{coh}}(\mathfrak{M}_1)^{\mathrm{op}} \to \mathrm{Ind}\, D^b_{\mathrm{coh}}(\mathfrak{M}_3)_{-j}.$$

Since the dualizing functors and tensor products with line bundle preserve singular supports and compact objects, as in the proof of Lemma 7.3.11 the above functor induces the functor

$$\phi \colon \mathcal{DT}^{\mathbb{C}^*}\!\left(\mathcal{P}^t_n(X, \beta) \setminus \mathcal{Z}^{t_+\text{-us}}_{l-1}\right)^{\mathrm{op}} \to \mathcal{DT}^{\mathbb{C}^*}(\mathcal{P}^{t_+}_{n_1}(X, \beta_1)) \otimes \mathcal{DT}^{\mathbb{C}^*}(\mathcal{M}^H_{n_2}(X, \beta_2))_{-j}.$$

Then the functor

$$\mathbb{D}_3 \circ \phi^{\mathrm{op}} \colon \mathcal{DT}^{\mathbb{C}^*}\!\left(\mathcal{P}^t_n(X, \beta) \setminus \mathcal{Z}^{t_+\text{-us}}_{l-1}\right)$$
$$\to \mathcal{DT}^{\mathbb{C}^*}(\mathcal{P}^{t_+}_{n_1}(X, \beta_1)) \otimes \mathcal{DT}^{\mathbb{C}^*}(\mathcal{M}^H_{n_2}(X, \beta_2))_j$$

gives a desired left adjoint.   □

7.3.5 Proofs of Conjectures

Let $*^R_j$, $*^L_j$ be the right and left adjoints functors of $*_j$ given in Lemma 7.3.12, Lemma 7.3.15. We define the following subcategories

$$\mathcal{W}^+_{k^+(\mathbf{d})} := \bigcap_{j \le k^+(\mathbf{d})} \mathrm{Ker}(*^R_j) \cap \bigcap_{j > k^+(\mathbf{d})} \mathrm{Ker}(*^L_j) \subset \mathcal{DT}^{\mathbb{C}^*}\!\left(\mathcal{P}^t_n(X, \beta) \setminus \mathcal{Z}^{t_+\text{-us}}_{l-1}\right),$$

$$\mathcal{W}^-_{k^-(\mathbf{d})} := \bigcap_{j \ge k^-(\mathbf{d})} \mathrm{Ker}(*^R_j) \cap \bigcap_{j < k^-(\mathbf{d})} \mathrm{Ker}(*^L_j) \subset \mathcal{DT}^{\mathbb{C}^*}\!\left(\mathcal{P}^t_n(X, \beta) \setminus \mathcal{Z}^{t_-\text{-us}}_{l-1}\right).$$

Lemma 7.3.16 *There exist semiorthogonal decompositions of the form*

$$\mathcal{DT}^{\mathbb{C}^*}\!\left(\mathcal{P}^t_n(X, \beta) \setminus \mathcal{Z}^{t_+\text{-us}}_{l-1}\right)$$
$$= \left\langle \ldots, \Upsilon^{+,\mathbf{d}}_{\lfloor k^+(\mathbf{d})\rfloor+2}, \Upsilon^{+,\mathbf{d}}_{\lfloor k^+(\mathbf{d})\rfloor+1}, \mathcal{W}^+_{k^+(\mathbf{d})}, \Upsilon^{+,\mathbf{d}}_{\lfloor k^+(\mathbf{d})\rfloor}, \Upsilon^{+,\mathbf{d}}_{\lfloor k^+(\mathbf{d})\rfloor-1}, \ldots \right\rangle, \qquad (7.3.31)$$

$$\mathcal{DT}^{\mathbb{C}^*}\!\left(\mathcal{P}^t_n(X, \beta) \setminus \mathcal{Z}^{t_-\text{-us}}_{l-1}\right)$$
$$= \left\langle \ldots, \Upsilon^{-,\mathbf{d}}_{\lceil k^-(\mathbf{d})\rceil-2}, \Upsilon^{-,\mathbf{d}}_{\lceil k^-(\mathbf{d})\rceil-1}, \mathcal{W}^-_{k^-(\mathbf{d})}, \Upsilon^{-,\mathbf{d}}_{\lceil k^-(\mathbf{d})\rceil}, \Upsilon^{-,\mathbf{d}}_{\lceil k^-(\mathbf{d})\rceil+1}, \ldots \right\rangle.$$

Proof We only prove the lemma for t_+. Note that the RHS is semiorthogonal by Lemma 7.3.14 and the definition of $\mathcal{W}^+_{k^+(\mathbf{d})}$. It is enough to show that any object in the LHS is obtained as successive extensions of objects in the RHS.

For an object $\mathcal{E}$ in the LHS, the proof of Lemma 7.3.12 shows that $*_j^R(\mathcal{E}) = 0$ for $j \ll 0$ formally locally at $p \in P_n^t(S, \beta)$. Since $\mathcal{P}_n^t(S, \beta) \to P_n^t(S, \beta)$ is universally closed, there exists a Zariski open neighborhood of $p \in P_n^t(S, \beta)$ on which $*_j^R(\mathcal{E}) = 0$ for $j \ll 0$. As $P_n^t(S, \beta)$ is of finite type, we conclude that $*_j^R(\mathcal{E}) = 0$ for $j \ll 0$. A similar argument shows that $*_j^L(\mathcal{E}) = 0$ for $j \gg 0$.

Suppose that $\mathcal{E}$ is not an object in $\mathcal{W}_{k^+(\mathbf{d})}^+$. Then there is $j_1 \leq k^+(\mathbf{d})$ such that $*_j^R(\mathcal{E}) = 0$ for $j < j_1$ and $*_{j_1}^R(\mathcal{E}) \neq 0$, or there is $j_1' > k^+(\mathbf{d})$ such that $*_j^L(\mathcal{E}) = 0$ for $j > j_1'$ and $*_{j_1'}^L(\mathcal{E}) \neq 0$. Below we assume the former case. The latter case is similarly discussed. We have the distinguished triangle $*_{j_1} *_{j_1}^R (\mathcal{E}) \to \mathcal{E} \to \mathcal{E}_1$, where $*_j^R(\mathcal{E}_1) = 0$ for $j \leq j_1$. Repeating the above constructions for $\mathcal{E}_1$, we have a distinguished triangle

$$\mathcal{E}_2 \to \mathcal{E} \to \mathcal{E}_3, \quad \mathcal{E}_2 \in \langle \Upsilon_{\lfloor k^+(\mathbf{d}) \rfloor}^{+,\mathbf{d}}, \dots, \Upsilon_{j_1}^{+,\mathbf{d}} \rangle, \quad \mathcal{E}_3 \in \bigcap_{j \leq k^+(\mathbf{d})} \mathrm{Ker}(*_j^R).$$

If $*_j^L(\mathcal{E}_3) = 0$ for all $j > k^+(\mathbf{d})$, then $\mathcal{E}_3 \in \mathcal{W}_{k^+(\mathbf{d})}^+$. Otherwise there is $j_1' > k^+(\mathbf{d})$ such that $*_j^L(\mathcal{E}_3) \neq 0$ for $j > j_1'$ and $*_{j_1'}^L(\mathcal{E}_3) \neq 0$. Similarly to above, we have the distinguished triangle $\mathcal{E}_4 \to \mathcal{E}_3 \to *_{j_1'} *_{j_1'}^L (\mathcal{E})$ such that $*_j^L(\mathcal{E}_4) = 0$ for $j \geq j_1'$. We also have $*_j^R(\mathcal{E}_4) = 0$ for $j \leq k^+(\mathbf{d})$ by applying $*_j^R$ to the above triangle and using Lemma 7.3.14. By repeating the above construction for $\mathcal{E}_4$, we obtain the distinguished triangle

$$\mathcal{E}_5 \to \mathcal{E}_3 \to \mathcal{E}_6, \quad \mathcal{E}_6 \in \langle \Upsilon_{j_1'}^{+,\mathbf{d}}, \dots, \Upsilon_{\lfloor k^+(\mathbf{d}) \rfloor + 1}^{+,\mathbf{d}} \rangle, \quad \mathcal{E}_5 \in \mathcal{W}_{k^+(\mathbf{d})}^+.$$

Therefore we obtain the desired semiorthogonal decomposition. $\qquad\square$

Using the above preparations, we prove Conjectures 7.2.3 and 7.2.4.

Theorem 7.3.17 *Under Assumption 7.3.1, Conjectures 7.2.3 and 7.2.4 hold.*

Proof It is enough to prove Conjecture 7.2.3. Below we only prove Conjecture 7.2.3 for t_+ under Assumption 7.3.1. For a decomposition $\mathbf{d} \in \mathbf{D}_{(\beta,n)}^{t,l}$ denoted as $(\beta, n) = (\beta_1, n_1) + (\beta_2, n_2)$, we already proved that the functors (7.2.8) are fully-faithful in Lemma 7.3.13, whose essential images fit into the semiorthogonal decomposition in Lemma 7.3.16. Let $\mathbf{d}' \in \mathbf{D}_{(\beta,n)}^{t,l}$ be another decomposition $(\beta, n) = (\beta_1', n_1') + (\beta_2', n_2')$. Note that objects in $\Upsilon_j^{+,\mathbf{d}}$, $\Upsilon_j^{+,\mathbf{d}'}$ have singular supports contained in

$$(\mathcal{P}_{n_1}^{t+}(X, \beta_1) * \mathcal{M}_{n_2}^H(X, \beta_2)) \cup \mathcal{Z}_{l-1}^{t+\text{-us}}, \quad (\mathcal{P}_{n_1'}^{t+}(X, \beta_1') * \mathcal{M}_{n_2'}^H(X, \beta_2')) \cup \mathcal{Z}_{l-1}^{t+\text{-us}}$$

respectively. Since they are disjoint outside $\mathcal{Z}_{l-1}^{t+-\mathrm{us}}$, by Lemma 3.2.8 $\Upsilon_j^{\pm,\mathbf{d}}$ is orthogonal to $\Upsilon_{j'}^{\pm,\mathbf{d}'}$. Therefore from Lemma 7.3.16, we obtain the semiorthogonal decomposition (7.2.9).

Below we show that the composition functor (7.2.11) is an equivalence. In order to simplify the argument, we assume that $\mathbf{D}_{(\beta,n)}^{t,l}$ consists of only $\mathbf{d}$. So we show that the composition

$$
\mathcal{W}_{k^+(\mathbf{d})}^+ \hookrightarrow \mathcal{DT}^{\mathbb{C}^*}\left(\mathcal{P}_n^t(X,\beta)\setminus\mathcal{Z}_{l-1}^{t+-\mathrm{us}}\right) \twoheadrightarrow \mathcal{DT}^{\mathbb{C}^*}\left(\mathcal{P}_n^t(X,\beta)\setminus\mathcal{Z}_l^{t+-\mathrm{us}}\right)
$$

$$(7.3.32)$$

is an equivalence. As objects in $\Upsilon_j^{+,\mathbf{d}}$ have singular supports contained in $\mathcal{Z}_l^{t+-\mathrm{us}}$, the composition

$$
\Upsilon_j^{+,\mathbf{d}} \hookrightarrow \mathcal{DT}^{\mathbb{C}^*}\left(\mathcal{P}_n^t(X,\beta)\setminus\mathcal{Z}_{l-1}^{t+-\mathrm{us}}\right) \twoheadrightarrow \mathcal{DT}^{\mathbb{C}^*}\left(\mathcal{P}_n^t(X,\beta)\setminus\mathcal{Z}_l^{t+-\mathrm{us}}\right)
$$

is zero. Therefore the functor (7.3.32) is essentially surjective from the semiorthogonal decomposition in Lemma 7.3.16.

It remains to show that the functor (7.3.32) is fully-faithful. Let $\iota\colon U \to P_n^t(S,\beta)$ be an etale morphism as in Theorem 3.1.4, and take the following Cartesian diagrams as in (3.1.12)

$$
\begin{array}{ccccc}
\mathfrak{P}_U & \longleftarrow & \mathcal{P}_U & \longrightarrow & U_\cdot \\
{\scriptstyle \iota\mathfrak{P}}\downarrow & \square & \downarrow & \square & \downarrow{\scriptstyle \iota} \\
\mathfrak{P}_n^t(S,\beta) & \longleftarrow & \mathcal{P}_n^t(S,\beta) & \longrightarrow & P_n^t(S,\beta).
\end{array}
$$

By Lemma 8.2.3, we have the fully-faithful functor

$$
\mathcal{DT}^{\mathbb{C}^*}\left(\mathcal{P}_n^t(X,\beta)\setminus\mathcal{Z}_{l-1}^{t+-\mathrm{us}}\right) \hookrightarrow \lim_{U \xrightarrow{\iota} P_n^t(S,\beta)}\left(D_{\mathrm{coh}}^b(\mathfrak{P}_U)/\mathcal{C}_{\iota_{\mathfrak{P}}^*\mathcal{Z}_{l-1}^{t+-\mathrm{us}}}\right).
$$

Let $\mathcal{E}_1,\mathcal{E}_2$ be an object in $\mathcal{W}_{k^+(\mathbf{d})}^+$. By the above fully-faithful functor, it is enough to show that the natural morphism

$$
\mathrm{Hom}_{D_{\mathrm{coh}}^b(\mathfrak{P}_U)/\mathcal{C}_{\iota_{\mathfrak{P}}^*\mathcal{Z}_{l-1}^{t+-\mathrm{us}}}}(\mathcal{E}_1|_{\mathfrak{P}_U},\mathcal{E}_2|_{\mathfrak{P}_U}) \to \mathrm{Hom}_{D_{\mathrm{coh}}^b(\mathfrak{P}_U)/\mathcal{C}_{\iota_{\mathfrak{P}}^*\mathcal{Z}_l^{t+-\mathrm{us}}}}(\mathcal{E}_1|_{\mathfrak{P}_U},\mathcal{E}_2|_{\mathfrak{P}_U})
$$

$$(7.3.33)$$

is an isomorphism. The above morphism is regarded as a morphism in $D_{\mathrm{qcoh}}(U)$, so it is enough to show the above isomorphism formally locally at any point in U. Similarly to Lemmas 7.3.12 and 7.3.13, we prove the corresponding claim for derived categories of factorizations via Koszul duality.

For each $p \in P_n^t(S, \beta)$ corresponding to the object (7.3.6) and a decomposition (7.3.11), we use the notation of the diagram (7.3.18). Let $\widehat{\Upsilon}_j^+$ be the essential image of the functor (7.3.28). Since the functor (7.3.28) is the composition of (7.3.29) and (7.3.30), by Theorem 6.1.2 we have the semiorthogonal decomposition

$$\mathrm{MF}_{\mathrm{coh}}^{\mathbb{C}^*}(\mathcal{X}_1 \setminus \widehat{\mathcal{Z}}_{l-1}^{t_+\text{-us}}, w) = \tag{7.3.34}$$

$$\langle \ldots, \widehat{\Upsilon}_{\lfloor k^+(\mathbf{d})\rfloor+2}^+, \widehat{\Upsilon}_{\lfloor k^+(\mathbf{d})\rfloor+1}^+, \widehat{\mathcal{W}}_{k^+(\mathbf{d})}^+, \widehat{\Upsilon}_{\lfloor k^+(\mathbf{d})\rfloor}^+, \widehat{\Upsilon}_{\lfloor k^+(\mathbf{d})\rfloor-1}^+, \ldots \rangle.$$

such that the composition functor

$$\widehat{\mathcal{W}}_{k^+(\mathbf{d})}^+ \hookrightarrow \mathrm{MF}_{\mathrm{coh}}^{\mathbb{C}^*}(\mathcal{X}_1 \setminus \widehat{\mathcal{Z}}_{l-1}^{t_+\text{-us}}, w) \to \mathrm{MF}_{\mathrm{coh}}^{\mathbb{C}^*}(\mathcal{X}_1 \setminus \widehat{\mathcal{Z}}_{l}^{t_+\text{-us}}, w)$$

is an equivalence. Since $*_j^R$ and $*_j^L$ commute with base change, each object $\mathcal{E}_i$ restricted to the formal fiber at $p \in P_n^t(S, \beta)$ corresponds to an object in $\widehat{\mathcal{W}}_{k^+(\mathbf{d})}^+$ under the equivalence (2.3.9)

$$D_{\mathrm{coh}}^b([\widehat{\mathfrak{M}}_0 / G])/\mathcal{C}_{\widehat{\mathcal{Z}}_{l-1}^{t_+\text{-us}}} \xrightarrow{\sim} \mathrm{MF}_{\mathrm{coh}}^{\mathbb{C}^*}(\mathcal{X}_1 \setminus \widehat{\mathcal{Z}}_{l-1}^{t_+\text{-us}}, w).$$

Therefore we conclude the formal local isomorphism of (7.3.33).   □

We next prove Conjecture 7.2.5

Theorem 7.3.18 *Under Assumption 7.3.1, Conjecture 7.2.5 holds.*

Proof For each point $p \in P_n^t(S, \beta)$ corresponding to a polystable object I as in (7.3.6) and a decomposition (7.3.11), we consider the diagram (7.3.18) as before. Let λ be the one parameter subgroup (7.3.19). We set

$$\eta^+ := \mathrm{wt}_{\lambda^{-1}}(\det \mathbb{L}_{q_2}^{\vee}|_0) = \hom(I^{(1)}, F^{(2)}) + \mathrm{ext}^2(F^{(2)}, F^{(1)}) - \hom^{-1}(I^{(1)}, F^{(2)}),$$

$$\eta^- := \mathrm{wt}_{\lambda}(\det \mathbb{L}_{q_2'}^{\vee}|_0) = \mathrm{ext}^1(F^{(2)}, F^{(1)}) + \hom^1(I^{(1)}, F^{(2)}) - \hom(F^{(2)}, F^{(1)}).$$

We also denote by

$$i^+ : (\mathcal{X}_5^{(1)} \setminus \widehat{\mathcal{Z}}^{t_+\text{-us}(1)}) \widehat{\times} \mathcal{X}_5^{(2)} \to \mathcal{X}_1, \quad i^- : \mathcal{X}_5^{(2)} \widehat{\times} (\mathcal{X}_5^{(1)} \setminus \widehat{\mathcal{Z}}^{t_-\text{-us}(1)}) \to \mathcal{X}_1$$

the inclusions into the λ-fixed parts. We denote by

$$\widehat{\mathcal{W}}_{k^+}^+ \subset \mathrm{MF}_{\mathrm{coh}}^{\mathbb{C}^*}(\mathcal{X}_1, w), \quad \widehat{\mathcal{W}}_{k^-}^- \subset \mathrm{MF}_{\mathrm{coh}}^{\mathbb{C}^*}(\mathcal{X}_1, w) \tag{7.3.35}$$

the subcategories of objects $\mathcal{P}$ such that for any decomposition (7.3.11) of I we have

$$\mathrm{wt}_{\lambda^{-1}}(i^{+*}\mathcal{P}) \subset [-k^+(\mathbf{d}) - \dim O^{(21)}, -k^+(\mathbf{d}) - \dim O^{(21)} + \eta^+), \qquad (7.3.36)$$

$$\mathrm{wt}_{\lambda}(i^{-*}\mathcal{P}) \subset [k^-(\mathbf{d}) - \dim O^{(12)}, k^-(\mathbf{d}) - \dim O^{(12)} + \eta^-)$$

respectively. Here $\mathbf{d}$ denotes the decomposition $(\beta, n) = (\beta_1, n_1) + (\beta_2, n_2)$ with $\mathrm{cl}(I^{(1)}) = (1, \beta_1, n_1)$ and $[F^{(2)}] = (\beta_2, n_2)$. We set $k^{\pm}(\mathbf{d})$ to be

$$k^+(\mathbf{d}) = \frac{1}{2}\chi(I^{(1)}, F^{(2)}) = \frac{1}{2}(n_2 + \beta_1\beta_2),$$

$$k^-(\mathbf{d}) = \frac{1}{2}\chi(F^{(2)}, F^{(1)}) = -\frac{1}{2}\beta_1\beta_2.$$

Then we have

$$
\begin{aligned}
k^+(\mathbf{d}) + \dim O^{(21)} &= \frac{1}{2}\hom(I^{(1)}, F^{(2)}) - \frac{1}{2}\hom^{-1}(I^{(1)}, F^{(2)}) \\
&\quad - \frac{1}{2}\hom^1(I^{(1)}, F^{(2)}) + \mathrm{ext}^2(F^{(2)}, F^{(1)}) \\
&= \frac{1}{2}\eta^+ + \frac{1}{2}(\mathrm{ext}^2(F^{(2)}, F^{(1)}) - \hom^1(I^{(1)}, F^{(2)})) \\
&= \frac{1}{2}\eta^+ + \frac{1}{2}\langle \delta_0, \lambda^{-1} \rangle.
\end{aligned}
$$

Here δ_0 is the following G-character

$$\delta_0 := \det \mathrm{Hom}^1(I, F) = \bigotimes_{i,j} \det O^{(ij)}$$

which satisfies that

$$\langle \delta_0, \lambda \rangle = \hom^1(I^{(1)}, F^{(2)}) - \mathrm{ext}^2(F^{(2)}, F^{(1)}).$$

Therefore the first condition in (7.3.36) is equivalent to

$$\mathrm{wt}_{\lambda^{-1}}(i^{+*}\mathcal{P}) \subset \left[-\frac{1}{2}\eta^+ - \frac{1}{2}\langle \delta_0, \lambda^{-1} \rangle, \frac{1}{2}\eta^+ - \frac{1}{2}\langle \delta_0, \lambda^{-1} \rangle \right). \qquad (7.3.37)$$

A similar computation shows that the second condition in (7.3.36) is equivalent to

$$\mathrm{wt}_{\lambda}(i^{-*}\mathcal{P}) \subset \left[-\frac{1}{2}\eta^- - \frac{1}{2}\langle \delta_0, \lambda \rangle, \frac{1}{2}\eta^- - \frac{1}{2}\langle \delta_0, \lambda \rangle \right). \qquad (7.3.38)$$

By Theorem 6.1.2 and an argument of Theorem 7.3.17, an object $\mathcal{E} \in$ $\mathcal{DT}^{\mathbb{C}^*}(\mathcal{P}_n^t(X, \beta))$ lies in $\mathcal{W}_{k\pm}^{\pm}$ if and only if for any point $p \in P_n^t(S, \beta)$ the restriction of $\mathcal{E}$ to the formal fiber at p corresponds to an object in $\widehat{\mathcal{W}}_{k\pm}^{\pm}$ under the Koszul duality in Theorem 2.3.3,

$$D_{\mathrm{coh}}^b([\widehat{\mathfrak{N}}_0/G]) \xrightarrow{\sim} \mathrm{MF}_{\mathrm{coh}}^{\mathbb{C}^*}(\mathcal{X}_1, w).$$

Therefore it is enough to show the inclusion

$$\widehat{\mathcal{W}}_{k-}^- \subset \widehat{\mathcal{W}}_{k+}^+. \tag{7.3.39}$$

Note that we have

$$\mathrm{hom}(I_0, F_i) + \mathrm{ext}^2(F_i, F_0) - \mathrm{hom}^1(I_0, F_i) - \mathrm{ext}^1(F_i, F_0) = \chi(I_0, F_i) + \chi(F_i, F_0)$$
$$= \chi(F_i) > 0.$$

Here the last inequality follows from $\chi(F_i) = t \cdot (H \cdot [F_i]) > 0$. The above inequality implies that, by Remark 7.3.8, the multiplicity of V_i in the G-representation $\mathrm{Hom}(I, F)$ is strictly bigger than that of $V_i^\vee$. So $\mathrm{Hom}(I, F)$ is the space of quiver representations satisfying the condition in (6.1.21). Moreover by Lemma 7.3.9, the subcategories in (7.3.39) are window subcategories with respect to $\chi_0^\pm$. Therefore we can try to apply Proposition 6.1.13 to conclude the inclusion (7.3.37).

However there is a slight issue, as δ_0 is not necessary χ_0-generic. Instead we discuss by perturbing δ_0 as follows. Let $0 < \varepsilon \ll 1$ be an irrational number, and set $\delta_0' = \delta_0 + \varepsilon \chi_0$. Then δ_0' is χ_0-generic. Since $\langle \chi_0, \lambda \rangle > 0$, the condition (7.3.38) is equivalent to

$$\mathrm{wt}_\lambda(i^{-*}\mathcal{P}) \subset \left[-\frac{1}{2}\eta^- - \frac{1}{2}\langle \delta_0', \lambda \rangle, \frac{1}{2}\eta^- - \frac{1}{2}\langle \delta_0', \lambda \rangle \right].$$

Moreover since we have

$$\eta^+ - \eta^- = \chi(I^{(1)}, F^{(2)}) + \chi(F^{(2)}, F^{(1)}) \tag{7.3.40}$$
$$= \chi(F^{(2)}) = n_2 > 0,$$

we have

$$\left[-\frac{1}{2}\eta^- - \frac{1}{2}\langle \delta_0', \lambda^{-1} \rangle, \frac{1}{2}\eta^- - \frac{1}{2}\langle \delta_0', \lambda^{-1} \rangle \right]$$
$$\subset \left[-\frac{1}{2}\eta^+ - \frac{1}{2}\langle \delta_0, \lambda^{-1} \rangle, \frac{1}{2}\eta^+ - \frac{1}{2}\langle \delta_0, \lambda^{-1} \rangle \right).$$

Therefore we can directly apply the argument of Proposition 6.1.7 (also see Corollary 6.1.8) to show that the LHS of (7.3.39) is the magic window over $\mathbb{S}$ with respect to δ_0', where $\mathbb{S}$ coincides with the symmetric structure of $\mathrm{Hom}(I, F)$ defined as in (6.1.22), and the RHS of (7.3.39) contains this magic window so that the inclusion (7.3.39) holds. $\square$

Finally we prove Conjecture 7.2.6.

Theorem 7.3.19 *Under Assumption 7.3.1, Conjecture 7.2.6 holds.*

Proof For simplicity, we assume that $\mathbf{D}_{(\beta, n)}^{t, 1}$ consists of one element $\mathbf{d}$, denoted as $(\beta, n) = (\beta_1, n_1) + (\beta_2, n_2)$. We take $k^{\pm}(\mathbf{d})$ as in (7.2.12). It is enough to show the following identity in $\mathcal{DT}^{\mathbb{C}^*}(\mathcal{P}_n^t(X, \beta))$

$$
\mathcal{W}_{k^+}^+ = \left\langle \Upsilon_{[-\frac{1}{2}\beta_1\beta_2 - \frac{1}{2}n_2, -\frac{1}{2}\beta_1\beta_2)}^{-, \mathbf{d}}, \mathcal{W}_{k^-}^-, \Upsilon_{[-\frac{1}{2}\beta_1\beta_2, -\frac{1}{2}\beta_1\beta_2 + \frac{1}{2}n_2)}^{, \mathbf{d}} \right\rangle. \tag{7.3.41}
$$

Let us take a point $p \in P_n^t(S, \beta)$ corresponding to a polystable object (7.3.6). If it is not stable, then the simple wall condition implies that $k = 1$ and $\dim V_1 = 1$, so that $G = \mathbb{C}^*$. The semiorthogonal decomposition (7.3.34) in this case is

$$
\mathrm{MF}_{\mathrm{coh}}^{\mathbb{C}^*}(\mathcal{X}_1, w) = \langle \dots, \widehat{\Upsilon}_{\lfloor k^+(\mathbf{d})\rfloor + 2}^+, \widehat{\Upsilon}_{\lfloor k^+(\mathbf{d})\rfloor + 1}^+, \widehat{\mathcal{W}}_{k^+}^+, \widehat{\Upsilon}_{\lfloor k^+(\mathbf{d})\rfloor}^+, \widehat{\Upsilon}_{\lfloor k^+(\mathbf{d})\rfloor - 1}^+, \dots \rangle
$$

where $\widehat{\mathcal{W}}_{k^+}^+$ is the same as (7.3.35). We have the similar semiorthogonal decomposition for t_--part

$$
\mathrm{MF}_{\mathrm{coh}}^{\mathbb{C}^*}(\mathcal{X}_1, w) = \langle \dots, \widehat{\Upsilon}_{\lceil k^-(\mathbf{d})\rceil - 2}^-, \widehat{\Upsilon}_{\lceil k^-(\mathbf{d})\rceil - 1}^-, \widehat{\mathcal{W}}_{k^-}^-, \widehat{\Upsilon}_{\lceil k^-(\mathbf{d})\rceil}^-, \widehat{\Upsilon}_{\lceil k^-(\mathbf{d})\rceil + 1}^-, \dots \rangle.
$$

Here $\widehat{\Upsilon}_j^-$ is the essential image of the fully-faithful functor from the right diagram of (7.3.18)

$$
f_{2*}'g_2'^* g_{1!}' f_1'^* : \mathrm{MF}_{\mathrm{coh}}^{\mathbb{C}^*}(\mathcal{X}_5^{(2)} \widehat{\times} \mathcal{X}_5^{(1)}, w^{(2)} + w^{(1)})_{\lambda\text{-wt}=j} \to \mathrm{MF}_{\mathrm{coh}}^{\mathbb{C}^*}(\mathcal{X}_1, w),
$$

where similarly to the proof Lemma 7.3.12 the above functor is written as

$$
f_{2*}'g_2'^* g_{1!}' f_1'^*(-) \cong q_{2*}'q_1'^*(- \otimes \det O^{(12)\vee}[-\dim O^{(12)}]).
$$

Note that the condition (7.3.37) is equivalent to

$$
\mathrm{wt}_\lambda(i^{+*}\mathcal{P}) \subset \left(-\frac{1}{2}\eta^+ - \frac{1}{2}\langle \delta_0, \lambda \rangle, \frac{1}{2}\eta^+ - \frac{1}{2}\langle \delta_0, \lambda \rangle \right].
$$

By (7.3.40), comparing with the condition (7.3.38) the following identity is well-known in [11, 45] (also see [78, Theorem 5.3]) for variation of GIT quotients for $\mathbb{C}^*$-actions

$$\widehat{\mathcal{W}}_{k^+}^+ = \langle \widehat{\Upsilon}_{\lceil k^-(\mathbf{d})-n_2/2 \rceil}^-, \cdots \widehat{\Upsilon}_{\lceil k^-(\mathbf{d})\rceil-1}^-, \widehat{\mathcal{W}}_{k^-}^-, \widehat{\Upsilon}_{\lfloor k^-(\mathbf{d})\rfloor}^-, \cdots, \widehat{\Upsilon}_{\lfloor k^-(\mathbf{d})+n_2/2 \rfloor}^- \rangle.$$

This means that the identity (7.3.41) holds formally locally at $p \in P_n^t(S, \beta)$. Since this holds for any p, the identity (7.3.41) holds.   □

Chapter 8
Some Auxiliary Results

In this chapter, we prove several technical results which were postponed in previous sections.

8.1 Comparisons of DT Categories

Here we give comparisons of DT categories in Definitions 3.2.2 and 3.2.4. Below we use the following:

Theorem 8.1.1 ([79, Theorem 7.2.1]) *Let $\mathcal{D}$ be a triangulated category and $\mathcal{D}' \subset \mathcal{D}$ a thick triangulated subcategory closed under taking small coproducts. Suppose that $\mathcal{D}$ and $\mathcal{D}'$ are compactly generated. Then $\mathcal{D}/\mathcal{D}'$ is also compactly generated, and we have the fully-faithful functor $\mathcal{D}^{\mathrm{cp}}/\mathcal{D}'^{\mathrm{cp}} \hookrightarrow (\mathcal{D}/\mathcal{D}')^{\mathrm{cp}}$ with dense image. Here $\mathcal{D}^{\mathrm{cp}} \subset \mathcal{D}$ is the subcategory of compact objects.*

8.1.1 Proof of Proposition 3.2.7

Proof Applying Theorem 8.1.1 for the subcategory $\operatorname{Ind}\mathcal{C}_{\mathcal{Z}} \subset \operatorname{Ind} D^b_{\mathrm{coh}}(\mathfrak{M})$ and using the assumption on the compact generation of $\operatorname{Ind}\mathcal{C}_{\mathcal{Z}}$ and Theorem 3.1.1, we have the fully-faithful functor

$$D^b_{\mathrm{coh}}(\mathfrak{M})/\mathcal{C}_{\mathcal{Z}} \hookrightarrow \left(\operatorname{Ind} D^b_{\mathrm{coh}}(\mathfrak{M})/\operatorname{Ind}\mathcal{C}_{\mathcal{Z}} \right)^{\mathrm{cp}}$$

with dense image. Moreover $\operatorname{Ind} D^b_{\mathrm{coh}}(\mathfrak{M})/\operatorname{Ind} \mathcal{C}_\mathcal{Z}$ is compactly generated, so we have the equivalence

$$\operatorname{Ind}\left(D^b_{\mathrm{coh}}(\mathfrak{M})/\mathcal{C}_\mathcal{Z}\right) \xrightarrow{\sim} \operatorname{Ind} D^b_{\mathrm{coh}}(\mathfrak{M})/\operatorname{Ind} \mathcal{C}_\mathcal{Z}. \qquad (8.1.1)$$

Applying the equivalence (8.1.1) for each smooth morphism $\alpha \colon \mathfrak{U} \to \mathfrak{M}$, we have the localization sequence of triangulated categories

$$\operatorname{Ind}(\mathcal{C}_{\alpha^*\mathcal{Z}}) \xrightarrow{i_\mathfrak{U}} \operatorname{Ind}(D^b_{\mathrm{coh}}(\mathfrak{U})) \xrightarrow{j_\mathfrak{U}} \operatorname{Ind}\left(D^b_{\mathrm{coh}}(\mathfrak{U})/\mathcal{C}_{\alpha^*\mathcal{Z}}\right). \qquad (8.1.2)$$

By [8, Section 4.3], the functor $i_\mathfrak{U}$ admits a right adjoint $i^R_\mathfrak{U}$. Therefore there also exists a right adjoint $j^R_\mathfrak{U}$ of $j_\mathfrak{U}$ such that for each $\mathcal{E} \in \operatorname{Ind}(D^b_{\mathrm{coh}}(\mathfrak{U}))$ we have the exact triangle

$$i_\mathfrak{U} \circ i^R_\mathfrak{U}(\mathcal{E}) \to \mathcal{E} \to j^R_\mathfrak{U} \circ j_\mathfrak{U}(\mathcal{E}).$$

By the above triangle, we have $j_\mathfrak{U} \circ j^R_\mathfrak{U} \cong \mathrm{id}$. On the other hand by taking the limit of the sequence (8.1.2), we obtain the sequence

$$\operatorname{Ind} \mathcal{C}_\mathcal{Z} \to \operatorname{Ind} D^b_{\mathrm{coh}}(\mathfrak{M}) \xrightarrow{j} \varprojlim_{\mathfrak{U} \xrightarrow{\alpha} \mathfrak{M}} \operatorname{Ind}\left(D^b_{\mathrm{coh}}(\mathfrak{U})/\mathcal{C}_{\alpha^*\mathcal{Z}}\right). \qquad (8.1.3)$$

Since $j^R_\mathfrak{U}$ is functorial for $\mathfrak{U}$, the functor j also admits a right adjoint j^R by taking the limit of $j^R_\mathfrak{U}$ such that $j \circ j^R \cong \mathrm{id}$. Also by the construction of j, we have that $\operatorname{Ker}(j) = \operatorname{Ind} \mathcal{C}_\mathcal{Z}$. By [51, Lemma 3.4], the above properties of the sequence (8.1.3) implies that the sequence (8.1.3) is a localization sequence, i.e. we have the equivalence

$$\operatorname{Ind} D^b_{\mathrm{coh}}(\mathfrak{M})/\operatorname{Ind} \mathcal{C}_\mathcal{Z} \xrightarrow{\sim} \varprojlim_{\mathfrak{U} \xrightarrow{\alpha} \mathfrak{M}} \operatorname{Ind}\left(D^b_{\mathrm{coh}}(\mathfrak{U})/\mathcal{C}_{\alpha^*\mathcal{Z}}\right).$$

We have the commutative diagram

$$
\begin{array}{ccc}
D^b_{\mathrm{coh}}(\mathfrak{M})/\mathcal{C}_\mathcal{Z} & \longrightarrow & \varprojlim_{\mathfrak{U} \xrightarrow{\alpha} \mathfrak{M}} \left(D^b_{\mathrm{coh}}(\mathfrak{U})/\mathcal{C}_{\alpha^*\mathcal{Z}}\right) \\
\downarrow & & \downarrow \\
\left(\operatorname{Ind} D^b_{\mathrm{coh}}(\mathfrak{M})/\operatorname{Ind} \mathcal{C}_\mathcal{Z}\right)^{\mathrm{cp}} & \xrightarrow{\sim} & \left(\varprojlim_{\mathfrak{U} \xrightarrow{\alpha} \mathfrak{M}} \operatorname{Ind}\left(D^b_{\mathrm{coh}}(\mathfrak{U})/\mathcal{C}_{\alpha^*\mathcal{Z}}\right)\right)^{\mathrm{cp}}.
\end{array}
$$

Here the left vertical arrow is fully-faithful with dense image, the right vertical arrow exists by Lemma 8.1.2 below and it is fully-faithful. Therefore the top horizontal arrow is also full-faithful with dense image. $\square$

We have used the following lemma:

Lemma 8.1.2 *Any object in the subcategory*

$$\lim_{\mathfrak{U} \xrightarrow{\alpha} \mathfrak{M}} \left(D^b_{\mathrm{coh}}(\mathfrak{U}) / \mathcal{C}_{\alpha^* \mathcal{Z}} \right) \subset \lim_{\mathfrak{U} \xrightarrow{\alpha} \mathfrak{M}} \mathrm{Ind} \left(D^b_{\mathrm{coh}}(\mathfrak{U}) / \mathcal{C}_{\alpha^* \mathcal{Z}} \right) \qquad (8.1.4)$$

is a compact object.

Proof The lemma is proved for $\mathcal{Z} = \emptyset$ in [38, Proposition 3.4.2 (b)]. Since the action of $D_{\mathrm{qcoh}}(\mathfrak{U})$ on $\mathrm{Ind}(D^b_{\mathrm{coh}}(\mathfrak{U}))$ by taking tensor products preserves $\mathrm{Ind}(\mathcal{C}_{\alpha^* \mathcal{Z}})$ (see [8, Lemma 4.2.2]), the quotient category

$$\mathrm{Ind}(D^b_{\mathrm{coh}}(\mathfrak{U})) / \mathrm{Ind}(\mathcal{C}_{\alpha^* \mathcal{Z}}) = \mathrm{Ind}(D^b_{\mathrm{coh}}(\mathfrak{U}) / \mathcal{C}_{\alpha^* \mathcal{Z}})$$

is a module over $D_{\mathrm{qcoh}}(\mathfrak{U})$. Therefore the proof of [38, Proposition 3.4.2 (b)] applies verbatim. Here we give the argument for the reader's convenience.

Let $\mathcal{E}_1$ be an object in the LHS of (8.1.4), and $\mathcal{E}_2$ an object in the RHS of (8.1.4). For each smooth morphism $\alpha \colon \mathfrak{U} \to \mathfrak{M}$, we have the objects $\alpha^* \mathcal{E}_1 \in D^b_{\mathrm{coh}}(\mathfrak{U}) / \mathcal{C}_{\alpha^* \mathcal{Z}}$ and $\alpha^* \mathcal{E}_2 \in \mathrm{Ind}\left(D^b_{\mathrm{coh}}(\mathfrak{U}) / \mathcal{C}_{\alpha^* \mathcal{Z}} \right)$. We have its inner Hom

$$\mathcal{H}om(\alpha^* \mathcal{E}_1, \alpha^* \mathcal{E}_2) \in D_{\mathrm{qcoh}}(\mathfrak{U}),$$

determined by the property that for any $\mathcal{F} \in D_{\mathrm{qcoh}}(\mathfrak{U})$ we have

$$\mathrm{Hom}_{D_{\mathrm{qcoh}}(\mathfrak{U})}(\mathcal{F}, \mathcal{H}om(\alpha^* \mathcal{E}_1, \alpha^* \mathcal{E}_2)) \cong \mathrm{Hom}_{\mathrm{Ind}(D^b_{\mathrm{coh}}(\mathfrak{U}) / \mathcal{C}_{\alpha^* \mathcal{Z}})}(\mathcal{F} \otimes \alpha^* \mathcal{E}_1, \alpha^* \mathcal{E}_2).$$

Since $\alpha^* \mathcal{E}_1$ is a compact object in $\mathrm{Ind}(D^b_{\mathrm{coh}}(\mathfrak{U}) / \mathcal{C}_{\alpha^* \mathcal{Z}})$, the assignment $\mathcal{E}_2 \mapsto \mathcal{H}om(\alpha^* \mathcal{E}_1, \alpha^* \mathcal{E}_2)$ preserves colimit. Moreover for each smooth morphism $\rho \colon \mathfrak{U}' \to \mathfrak{U}$ as in the diagram (3.1.2), we have the natural morphism

$$\rho^* \mathcal{H}om(\alpha^* \mathcal{E}_1, \alpha^* \mathcal{E}_2) \to \mathcal{H}om(\alpha'^* \mathcal{E}_1, \alpha'^* \mathcal{E}_2)$$

which is an isomorphism by the argument of [38, Lemma 3.4.4]. Therefore we have the object

$$\mathcal{H}om(\mathcal{E}_1, \mathcal{E}_2) = \{\mathcal{H}om(\alpha^* \mathcal{E}_1, \alpha^* \mathcal{E}_2)\}_{\mathfrak{U} \xrightarrow{\alpha} \mathfrak{M}} \in D_{\mathrm{qcoh}}(\mathfrak{M}).$$

Moreover the assignment $\mathcal{E}_2 \mapsto \mathcal{H}om(\mathcal{E}_1, \mathcal{E}_2)$ preserves colimit. We have

$$\mathrm{Hom}(\mathcal{E}_1, \mathcal{E}_2) = \Gamma(\mathfrak{M}, \mathcal{H}om(\mathcal{E}_1, \mathcal{E}_2)).$$

As $\mathfrak{M}$ is QCA, the functor $\Gamma(\mathfrak{M}, -)$ preserves colimit by [38, Theorem 1.4.2]. Therefore the assignment $\mathcal{E}_2 \mapsto \mathrm{Hom}(\mathcal{E}_1, \mathcal{E}_2)$ preserves colimit, which implies that $\mathcal{E}_1$ is a compact object in the RHS of (8.1.4). $\square$

In general, it is not known whether $\operatorname{Ind} \mathcal{C}_{\mathcal{Z}}$ is compactly generated or not (see [8, Remark 8.12]). This is the case for a derived stack in Definition 2.3.1 by Lemma 2.3.2, so we obtain the following:

Corollary 8.1.3 *Let $\mathfrak{M} = [\mathfrak{U}/G]$ be a derived stack as in Definition 2.3.1, and $\mathcal{Z} \subset t_0(\Omega_{\mathfrak{M}}[-1])$ a conical closed substack. Then the natural functor*

$$D^b_{\mathrm{coh}}([\mathfrak{U}/G])/\mathcal{C}_{\mathcal{Z}} \to \lim_{\mathfrak{U}' \xrightarrow{\alpha'} [\mathfrak{U}/G]} \left(D^b_{\mathrm{coh}}(\mathfrak{U}')/\mathcal{C}_{\alpha'^* \mathcal{Z}} \right)$$

is fully-faithful with dense image. Here $\mathfrak{U}'$ is an affine derived scheme of the form (2.1.1), and α' is a smooth morphism.

Proof The corollary follows from Lemma 2.3.2 and Proposition 3.2.7. ☐

8.1.2 Restrictions to Open Substacks

Let $\mathfrak{M}$ be a quasi-smooth and QCA derived stack with classical truncation $\mathcal{M} = t_0(\mathfrak{M})$. Let $\mathcal{W} \subset \mathcal{M}$ be a closed substack, and take the derived open substack $\mathfrak{M}_\circ \subset \mathfrak{M}$ whose truncation is $\mathcal{M} \setminus \mathcal{W}$. We define $\operatorname{Ind} D^b_{\mathrm{coh}}(\mathfrak{M})_{\mathcal{W}}$ to be the kernel of the restriction functor

$$\operatorname{Ind} D^b_{\mathrm{coh}}(\mathfrak{M})_{\mathcal{W}} := \operatorname{Ker} \left(\operatorname{Ind} D^b_{\mathrm{coh}}(\mathfrak{M}) \to \operatorname{Ind} D^b_{\mathrm{coh}}(\mathfrak{M}_\circ) \right),$$

and set $D^b_{\mathrm{coh}}(\mathfrak{M})_{\mathcal{W}} := \operatorname{Ind} D^b_{\mathrm{coh}}(\mathfrak{M})_{\mathcal{W}} \cap D^b_{\mathrm{coh}}(\mathfrak{M})$. Let $p_0 \colon \mathcal{N} = t_0(\Omega_{\mathfrak{M}}[-1]) \to \mathcal{M}$ be the projection and set $\mathcal{Z} = p_0^{-1}(\mathcal{W})$. By [8, Corollary 4.1.2], we have

$$\operatorname{Ind} \mathcal{C}_{\mathcal{Z}} = \operatorname{Ind} D^b_{\mathrm{coh}}(\mathfrak{M})_{\mathcal{W}}, \ \mathcal{C}_{\mathcal{Z}} = D^b_{\mathrm{coh}}(\mathfrak{M})_{\mathcal{W}}, \tag{8.1.5}$$

as subcategories in $\operatorname{Ind} D^b_{\mathrm{coh}}(\mathfrak{M})$. We have the following generalization of Theorem 3.1.1:

Lemma 8.1.4 *The category $\operatorname{Ind} D^b_{\mathrm{coh}}(\mathfrak{M})_{\mathcal{W}}$ is compactly generated with compact objects $D^b_{\mathrm{coh}}(\mathfrak{M})_{\mathcal{W}}$.*

Proof Using the fact that $\operatorname{Coh}(\mathcal{M})_{\mathcal{W}} := \operatorname{Ind} D^b_{\mathrm{coh}}(\mathfrak{M})_{\mathcal{W}} \cap \operatorname{Coh}(\mathcal{M})$ is the heart of a t-structure on $D^b_{\mathrm{coh}}(\mathfrak{M})_{\mathcal{W}}$, the same argument of [38, Proposition 3.5.1] shows that $\operatorname{Coh}(\mathcal{M})_{\mathcal{W}}$ generates $\operatorname{Ind} D^b_{\mathrm{coh}}(\mathfrak{M})_{\mathcal{W}}$. Then the arguments of [38, Theorem 3.3.5] apply verbatim to show that $\operatorname{Ind} D^b_{\mathrm{coh}}(\mathfrak{M})_{\mathcal{W}}$ is compactly generated with compact objects $D^b_{\mathrm{coh}}(\mathfrak{M})_{\mathcal{W}}$. We give the argument for the completeness.

We show that $\operatorname{Coh}(\mathcal{M})_{\mathcal{W}}$ generates $\operatorname{Ind} D^b_{\mathrm{coh}}(\mathfrak{M})_{\mathcal{W}}$. Since $\operatorname{Ind} D^b_{\mathrm{coh}}(\mathfrak{M})_{\mathcal{W}}$ is generated by the image of $\iota_* \colon \operatorname{Ind} D^b_{\mathrm{coh}}(\mathcal{M})_{\mathcal{W}} \to \operatorname{Ind} D^b_{\mathrm{coh}}(\mathfrak{M})_{\mathcal{W}}$ where $\iota \colon \mathcal{M} \hookrightarrow \mathfrak{M}$ is the natural closed immersion, we can assume that $\mathfrak{M} = \mathcal{M}$. By the similar argument, we can also assume that $\mathcal{M}$ is reduced. Then as $\mathcal{M}$ is QCA, it admits a finite stratification $\mathcal{M}_i$ such that each strata $\mathcal{M}_i$ is smooth. Let $\mathcal{W}_i = \mathcal{W} \cap \mathcal{M}_i$. For each strata $\mathcal{M}_i$, the category $\operatorname{Ind} D^b_{\mathrm{coh}}(\mathcal{M}_i)_{\mathcal{W}_i}$ is compactly generated with compact objects $D^b_{\mathrm{coh}}(\mathcal{M}_i)_{\mathcal{W}_i}$ by [8, Corollary 9.2.7, 9.2.8]. Since any object in the latter category is a successive extension of objects in $\operatorname{Coh}(\mathcal{M}_i)_{\mathcal{W}_i}$ up to shift, the assertion holds on $\mathcal{M}_i$. Therefore by induction, it is enough to show the following: for a closed substack $\mathcal{M}_1 \subset \mathcal{M}$ and $\mathcal{M}_2 := \mathcal{M} \setminus \mathcal{M}_1$, if $\operatorname{Ind} D^b_{\mathrm{coh}}(\mathcal{M}_i)_{\mathcal{W}_i}$ is generated by $\operatorname{Coh}(\mathcal{M}_i)_{\mathcal{W}_i}$, then $\operatorname{Ind} D^b_{\mathrm{coh}}(\mathcal{M})_{\mathcal{W}}$ is generated by $\operatorname{Coh}(\mathcal{M})_{\mathcal{W}}$.

For $\mathcal{F} \in \operatorname{Ind} D^b_{\mathrm{coh}}(\mathcal{M})_{\mathcal{W}}$, suppose that $\operatorname{Hom}(\mathcal{E}, \mathcal{F}) = 0$ for all $\mathcal{E} \in \operatorname{Coh}(\mathcal{M})_{\mathcal{W}}$. We need to show that $\mathcal{F} = 0$. Let $j_i \colon \mathcal{M}_i \to \mathcal{M}$ be the inclusion. We have the distinguished triangle

$$j^{\mathrm{ind}}_{1*} j^!_1(\mathcal{F}) \to \mathcal{F} \to j^{\mathrm{ind}}_{2*} j^{\mathrm{ind}*}_2 \mathcal{F}.$$

For any $\mathcal{F}' \in \operatorname{Coh}(\mathcal{M}_1)_{\mathcal{W}_1}$, we have

$$\operatorname{Hom}_{\mathcal{M}_1}(\mathcal{F}', j^!_1(\mathcal{F})) = \operatorname{Hom}_{\mathcal{M}}(j^{\mathrm{ind}}_{1*} \mathcal{F}', \mathcal{F}) = 0$$

since $j^{\mathrm{ind}}_{1*} \mathcal{F}' \in \operatorname{Coh}(\mathcal{M})_{\mathcal{W}}$. Therefore $j^!_1(\mathcal{F}) = 0$, so the natural map $\mathcal{F} \to j^{\mathrm{ind}}_{2*} j^{\mathrm{ind}*}_2 \mathcal{F}$ is an isomorphism. In particular for every $\mathcal{E} \in \operatorname{Ind} D^b_{\mathrm{coh}}(\mathfrak{M})_{\mathcal{W}}$, we have the isomorphism

$$\operatorname{Hom}_{\mathcal{M}}(\mathcal{E}, \mathcal{F}) \xrightarrow{\cong} \operatorname{Hom}_{\mathcal{M}_2}(j^{\mathrm{ind}*}_2 \mathcal{E}, j^{\mathrm{ind}*}_2 \mathcal{F}).$$

Since $j^*_2 \colon \operatorname{Coh}(\mathcal{M})_{\mathcal{W}} \to \operatorname{Coh}(\mathcal{M}_2)_{\mathcal{W}_2}$ is essentially surjective (see the argument of [82, Corollary 15.5]), we also conclude that $j^{\mathrm{ind}*}_2 \mathcal{F} = 0$, therefore $\mathcal{F} = 0$.

It remains to show that compact objects of $\operatorname{Ind} D^b_{\mathrm{coh}}(\mathfrak{M})_{\mathcal{W}}$ coincide with $D^b_{\mathrm{coh}}(\mathfrak{M})_{\mathcal{W}}$. Since the inclusion $\operatorname{Ind} D^b_{\mathrm{coh}}(\mathfrak{M})_{\mathcal{W}} \hookrightarrow \operatorname{Ind} D^b_{\mathrm{coh}}(\mathfrak{M})$ admits a continuous right adjoint (see (3.2.2)), we have

$$(\operatorname{Ind} D^b_{\mathrm{coh}}(\mathfrak{M})_{\mathcal{W}})^{\mathrm{cp}} = (\operatorname{Ind} D^b_{\mathrm{coh}}(\mathfrak{M}))^{\mathrm{cp}} \cap \operatorname{Ind} D^b_{\mathrm{coh}}(\mathfrak{M})_{\mathcal{W}}.$$

Therefore the claim holds from Theorem 3.1.1.   □

Remark 8.1.5 *For a conical closed substack $\mathcal{Z} \subset t_0(\Omega_{\mathfrak{M}}[-1])$, the subcategory $\operatorname{Ind} \mathcal{C}_{\mathcal{Z}} \cap \operatorname{Coh}(\mathcal{M}) \subset \mathcal{C}_{\mathcal{Z}}$ is not necessary the heart of a t-structure on $\mathcal{C}_{\mathcal{Z}}$ in general, so the above argument does not apply to the compact generation of $\operatorname{Ind} \mathcal{C}_{\mathcal{Z}}$. In Sect. 8.2.2 we will show the compact generation of $\operatorname{Ind} \mathcal{C}_{\mathcal{Z}}$ when $\mathcal{M}$ admits a good moduli space.*

Lemma 8.1.6 *The sequence of triangulated categories*

$$D^b_{\mathrm{coh}}(\mathfrak{M})_{\mathcal{W}} \to D^b_{\mathrm{coh}}(\mathfrak{M}) \to D^b_{\mathrm{coh}}(\mathfrak{M}_{\circ})$$

is a localization sequence.

Proof By [41, Section 4.1], the sequence of triangulated categories

$$\text{Ind } D^b_{\text{coh}}(\mathfrak{M})_{\mathcal{W}} \to \text{Ind } D^b_{\text{coh}}(\mathfrak{M}) \to \text{Ind } D^b_{\text{coh}}(\mathfrak{M}_{\circ})$$

is a localization sequence. Therefore by Theorems 8.1.1, 3.1.1, and Lemma 8.1.4, we have the induced fully-faithful functor

$$D^b_{\text{coh}}(\mathfrak{M})/D^b_{\text{coh}}(\mathfrak{M})_{\mathcal{W}} \hookrightarrow D^b_{\text{coh}}(\mathfrak{M}_{\circ}). \qquad (8.1.6)$$

It is enough to show that the above functor is essentially surjective. Note that cohomology sheaves of an object in the RHS of (8.1.6) is a coherent sheaf on the classical stack $\mathcal{M}\backslash\mathcal{W}$. Also any coherent sheaf on the classical stack $\mathcal{M}\backslash\mathcal{W}$ extends to a coherent sheaf on $\mathcal{M}$ (see [82, Corollary 15.5]). Since the LHS of (8.1.6) is a triangulated subcategory, it follows that the restriction functor in (8.1.6) is essentially surjective. $\square$

Let us take a conical closed substack $\mathcal{Z} \subset \mathcal{N} = t_0(\Omega_{\mathfrak{M}}[-1])$, and use notation in Sect. 3.2.3. We have the following lemma:

Lemma 8.1.7 *We have the localization sequence*

$$D^b_{\text{coh}}(\mathfrak{M})_{\mathcal{W}}/\mathcal{C}_{\mathcal{Z}\cap p_0^{-1}(\mathcal{W})} \to \mathcal{DT}^{\mathbb{C}^*}(\mathcal{N}^{\text{ss}}) \to \mathcal{DT}^{\mathbb{C}^*}(\mathcal{N}^{\text{ss}}_{\circ}).$$

Proof We have the following commutative diagram

$$
\begin{array}{ccccc}
D^b_{\text{coh}}(\mathfrak{M})_{\mathcal{W}} & \hookrightarrow & D^b_{\text{coh}}(\mathfrak{M}) & \twoheadrightarrow & D^b_{\text{coh}}(\mathfrak{M}_{\circ}) \\
\uparrow & & \uparrow & & \uparrow \\
\mathcal{C}_{\mathcal{Z}\cap p_0^{-1}(\mathcal{W})} & \hookrightarrow & \mathcal{C}_{\mathcal{Z}} & \twoheadrightarrow & \mathcal{C}_{\mathcal{Z}_{\circ}}.
\end{array}
$$

Here each horizontal sequence is a localization sequence by Lemma 8.1.6. By taking quotients, we obtain the desired localization sequence. $\square$

When $\mathcal{Z} = p_0^{-1}(\mathcal{W})$ for the projection $p_0 \colon \mathcal{N} \to \mathcal{M}$, we have a simpler description of the quotient category (3.2.3) by the following lemma:

Lemma 8.1.8 *Suppose that $\mathcal{Z} = p_0^{-1}(\mathcal{W})$. Then the restriction functors give equivalences*

$$\mathcal{DT}^{\mathbb{C}^*}(\mathcal{N}^{\text{ss}}) \xrightarrow{\sim} \widehat{\mathcal{DT}}^{\mathbb{C}^*}(\mathcal{N}^{\text{ss}}) \xrightarrow{\sim} D^b_{\text{coh}}(\mathfrak{M}_{\circ}).$$

Proof The equivalence $\mathcal{DT}^{\mathbb{C}^*}(\mathcal{N}^{\text{ss}}) \xrightarrow{\sim} D^b_{\text{coh}}(\mathfrak{M}_{\circ})$ follows from the identity $\mathcal{C}_{\mathcal{Z}} = D^b_{\text{coh}}(\mathfrak{M})_{\mathcal{W}}$ in (8.1.5) together with Lemma 8.1.6. In particular $\mathcal{DT}^{\mathbb{C}^*}(\mathcal{N}^{\text{ss}})$ is idempotent complete. Then the equivalence $\mathcal{DT}^{\mathbb{C}^*}(\mathcal{N}^{\text{ss}}) \xrightarrow{\sim} \widehat{\mathcal{DT}}^{\mathbb{C}^*}(\mathcal{N}^{\text{ss}})$ follows from Proposition 3.2.7 and Lemma 8.1.4. $\square$

We next consider the case that $p_0^{-1}(\mathcal{W}) \subset \mathcal{Z}$. In this case, the open immersion (3.2.6) is an isomorphism.

Lemma 8.1.9 *When $p_0^{-1}(\mathcal{W}) \subset \mathcal{Z}$, the restriction functors give equivalences*

$$\mathcal{DT}^{\mathbb{C}^*}(\mathcal{N}^{\mathrm{ss}}) \xrightarrow{\sim} \mathcal{DT}^{\mathbb{C}^*}(\mathcal{N}_\circ^{\mathrm{ss}}), \ \widehat{\mathcal{DT}}^{\mathbb{C}^*}(\mathcal{N}^{\mathrm{ss}}) \xrightarrow{\sim} \widehat{\mathcal{DT}}^{\mathbb{C}^*}(\mathcal{N}_\circ^{\mathrm{ss}})$$

Proof The assumption together with Lemma 8.1.8 imply that

$$\mathcal{C}_{\mathcal{Z} \cap p_0^{-1}(\mathcal{W})} = \mathcal{C}_{p_0^{-1}(\mathcal{W})} = D_{\mathrm{coh}}^b(\mathfrak{M})_{\mathcal{W}}.$$

Therefore the first equivalence holds from Lemma 8.1.7.

For a smooth morphism $\alpha \colon \mathfrak{U} \to \mathfrak{M}$, by setting $\mathfrak{U}_\circ := \mathfrak{U} \setminus \alpha^{-1}(\mathcal{W})$, the first equivalence gives an equivalence

$$D_{\mathrm{coh}}^b(\mathfrak{U})/\mathcal{C}_{\alpha^* \mathcal{Z}} \xrightarrow{\sim} D_{\mathrm{coh}}^b(\mathfrak{U}_\circ)/\mathcal{C}_{\alpha^* \mathcal{Z}_\circ}.$$

Therefore the second equivalence holds by taking the limit. □

8.2 Compact Generation of Ind $\mathcal{C}_{\mathcal{Z}}$

Let $\mathfrak{M}$ be a quasi-smooth QCA derived stack over $\mathbb{C}$ whose classical truncation $\mathcal{M} = t_0(\mathfrak{M})$ admits a good moduli space $\pi_{\mathcal{M}} \colon \mathcal{M} \to M$. Here we prove the compact generation of Ind $\mathcal{C}_{\mathcal{Z}}$ for a conical closed substack $\mathcal{Z} \subset t_0(\Omega_{\mathfrak{M}}[-1])$. We note that when $\mathcal{Z} = \mathcal{M} \subset t_0(\Omega_{\mathfrak{M}}[-1])$ is the zero section, then

$$\mathrm{Ind}\,\mathcal{C}_{\mathcal{M}} = D_{\mathrm{qcoh}}(\mathfrak{M})$$

and its compact generation is proved in [51, Theorem A], as $\mathfrak{M}$ is s-global type by Theorem 3.1.4. Indeed we will see that the same argument applies almost verbatim to show the compact generating of Ind $\mathcal{C}_{\mathcal{Z}}$.

8.2.1 *Presheaves of Triangulated Categories*

Let $\mathcal{M} = t_0(\mathfrak{M})$ be the classical truncation of $\mathfrak{M}$, and assume that it admits a good moduli space

$$\pi_{\mathcal{M}} \colon \mathcal{M} \to M.$$

Let $\mathcal{D}_{\text{ét}}/M$ be the category of étale morphisms $\iota\colon U \to M$ as in Sect. 3.1.4. Then we have the associated diagram (3.1.12); so in particular a derived stack $\mathfrak{M}_U$ together with a morphism $\iota_{\mathfrak{M}}\colon \mathfrak{M}_U \to \mathfrak{M}$. For a conical closed substack $\mathcal{Z} \subset t_0(\Omega_{\mathfrak{M}}[-1])$ and $(\iota\colon U \to M) \in \mathcal{D}_{\text{ét}}/M$, we set

$$\mathcal{T}_{\mathcal{Z}}(U) := \operatorname{Ind} \mathcal{C}_{\iota_{\mathfrak{M}}^*\mathcal{Z}} \subset \operatorname{Ind} D_{\text{coh}}^b(\mathfrak{M}_U).$$

For a morphism $\rho\colon U' \to U$ in $\mathcal{D}_{\text{ét}}/M$, from the diagram (3.1.13) we have adjoint pairs

$$\mathcal{T}_{\mathcal{Z}}(U) \underset{\rho_*}{\overset{\rho^*}{\rightleftarrows}} \mathcal{T}_{\mathcal{Z}}(U')\,, \quad \rho^* := \rho_{\mathfrak{M}}^{\text{ind}\,*}, \rho_* := \rho_{\mathfrak{M}*}^{\text{ind}}, \quad \rho^* \dashv \rho_*.$$

Therefore $U \mapsto \mathcal{T}_{\mathcal{Z}}(U)$ is a $\mathcal{D}_{\text{ét}}/M$-presheaf of triangulated categories with adjoints in the sense of [51, Section 5]. For an open immersion $U_\circ \subset U$, we set

$$\mathcal{T}_{\mathcal{Z}}(U)_{U \setminus U^\circ} := \operatorname{Ker}(\mathcal{T}_{\mathcal{Z}}(U) \to \mathcal{T}_{\mathcal{Z}}(U_\circ))$$

$$= \operatorname{Ind} \mathcal{C}_{\iota_{\mathfrak{M}}^*\mathcal{Z} \cap p_U^*(\mathcal{M}_U \setminus \mathcal{M}_{U_\circ})}$$

where $p_U\colon t_0(\Omega_{\mathfrak{M}_U}[-1]) \to \mathcal{M}_U$ is the projection.

Proposition 8.2.1 *The presheaf of triangulated categories $\mathcal{T}_{\mathcal{Z}}$ satisfies the following properties.*

 (i) *For $(\iota\colon U \to M) \in \mathcal{D}_{\text{ét}}/M$, the category $\mathcal{T}_{\mathcal{Z}}(U)$ is closed under small coproduct.*
 (ii) *For any morphism $\rho\colon U' \to U$ in $\mathcal{D}_{\text{ét}}/M$, the right adjoint ρ_* preserves small coproducts.*
 (iii) *For every Cartesian square in $\mathcal{D}_{\text{ét}}/M$*

$$\begin{array}{ccc} V' & \overset{g'}{\longrightarrow} & U' \\ \rho_V \downarrow & \square & \downarrow \rho \\ V & \overset{g}{\longrightarrow} & U \end{array}$$

the natural transformation $g^*\rho_* \to \rho_{V*}(g')^*$ is an isomorphism.
 (iv) *For any open immersion $U_\circ \subset U$ and an étale neighborhood $\rho\colon U' \to U$ of $U \setminus U_\circ$, the pull-back ρ^* induces an equivalence $\rho^*\colon \mathcal{T}_{\mathcal{Z}}(U)_{U \setminus U_\circ} \overset{\sim}{\to} \mathcal{T}_{\mathcal{Z}}(U')_{U' \setminus U'_\circ}$.*
 (v) *For every finite faithfully flat morphism $\rho\colon U' \to U$, the functor $\rho_*\colon \mathcal{T}_{\mathcal{Z}}(U') \to \mathcal{T}_{\mathcal{Z}}(U)$ admits a right adjoint $\rho^!$ that preserves small coproducts, conservative, and commutes with pull-back along open immersions.*

Proof

(i) For any quasi-smooth affine derived scheme $\mathfrak{U}$, the category Ind $D^b_{\mathrm{coh}}(\mathfrak{U})$ is co-complete, and the subcategory of fixed singular supports are closed under direct sums as it is a left orthogonal of some subcategory in Ind $D^b_{\mathrm{coh}}(\mathfrak{U})$ (see [8, Section 3.1.4]). Then (i) follows from the definition of Ind $\mathcal{C}_{\mathcal{Z}}$ as a limit (3.1.4).

(ii) Since $\rho_{\mathfrak{M}}$ is representable, the functor ρ_* is continuous by [41, Proposition 3.1.1].

(iii) The desired isomorphism is proved in [42, Proposition 3.2.2].

(iv) Let $\mathcal{W} = \mathcal{M}_U \setminus \mathcal{M}_{U_\circ}$ and $\mathcal{W}' = \mathcal{M}_{U'} \setminus \mathcal{M}_{U'_\circ}$. It is enough to show that the adjoint pairs

$$\text{Ind } D^b_{\mathrm{coh}}(\mathfrak{M}_U)_{\mathcal{W}} \underset{\rho^{\mathrm{ind}}_{\mathfrak{M}*}}{\overset{\rho^{\mathrm{ind}\,*}_{\mathfrak{M}}}{\rightleftarrows}} \text{Ind } D^b_{\mathrm{coh}}(\mathfrak{M}_{U'})_{\mathcal{W}'} \tag{8.2.1}$$

are equivalences, since both functors preserve subcategories $\mathcal{T}_{\mathcal{Z}}(U)_{U \setminus U_\circ}$ and $\mathcal{T}_{\mathcal{Z}}(U')_{U' \setminus U'_\circ}$. It is proved in [53, Proposition 4.2, (1) $\Rightarrow$ (4)] that the adjoint pairs

$$D_{\mathrm{qcoh}}(\mathfrak{M}_U)_{\mathcal{W}} \underset{\rho_{\mathfrak{M}*}}{\overset{\rho^*_{\mathfrak{M}}}{\rightleftarrows}} D_{\mathrm{qcoh}}(\mathfrak{M}_{U'})_{\mathcal{W}'}$$

are equivalences. As ρ is an étale neighborhood of $U \setminus U_\circ$, the above functors restrict to equivalences between $D^b_{\mathrm{coh}}(\mathfrak{M}_U)_{\mathcal{W}}$ and $D^b_{\mathrm{coh}}(\mathfrak{M}_{U'})_{\mathcal{W}'}$. As both sides in (8.2.1) are obtained as ind-completions of $D^b_{\mathrm{coh}}(\mathfrak{M}_U)_{\mathcal{W}}$ and $D^b_{\mathrm{coh}}(\mathfrak{M}_{U'})_{\mathcal{W}'}$ by Lemma 8.1.4, the functors in (8.2.1) are equivalences.

(v) As ρ is proper, the right adjoint $\rho^!_{\mathfrak{M}}$ is continuous (see [41, Section 3.3.7]). Since ρ is étale, we have $\rho^!_{\mathfrak{M}}(-) = \rho^*_{\mathfrak{M}}(-) \otimes \omega_\rho$, which is obviously conservative and commutes with pull-back along open immersions.

$\square$

8.2.2 Compact Generation of Ind $\mathcal{C}_{\mathcal{Z}}$

In the setting of the previous subsection, we have the following:

Theorem 8.2.2 *Let $\mathfrak{M}$ be a quasi-smooth and QCA derived stack such that $M = t_0(\mathfrak{M})$ admits a good moduli space $\pi_{\mathcal{M}} \colon M \to M$. Then the category Ind $\mathcal{C}_{\mathcal{Z}}$ is compactly generated with compact objects $\mathcal{C}_{\mathcal{Z}}$.*

Proof We note that the conditions in Proposition 8.2.1 are exactly the assumptions imposed in [51, Theorem 6.9], so the compact generation of Ind $\mathcal{C}_{\mathcal{Z}}$ follows from the argument of *loc. cit.* almost verbatim. More precisely, let $\mathcal{D}_{\mathrm{ét}}/M^{\mathrm{cp}} \subset \mathcal{D}_{\mathrm{ét}}/M$ be the subcategory of $\iota \colon U \to M$ such that $\mathcal{T}_{\mathcal{Z}}(U)$ is compactly generated. If $\iota \colon U \to$

M satisfies the condition in Theorem 3.1.4, then $\mathfrak{M}_U$ is of the form $[\mathfrak{U}/G]$ as in Proposition 3.1.5, so $\mathcal{T}_{\mathcal{Z}}(U)$ is compactly generated by Lemma 2.3.2. Therefore there is a separated, quasi-finite and faithfully-flat morphism $\iota\colon U \to M$ such that $(\iota\colon U \to M) \in \mathcal{D}_{\text{ét}}/M^{\text{cp}}$. By applying [53, Theorem 6.1], we have $(\mathrm{id}\colon M \to M) \in \mathcal{D}_{\text{ét}}/M^{\text{cp}}$ if we have the followings:

(D1) For any morphism $\rho\colon U' \to U$ in $\mathcal{D}_{\text{ét}}/M$, if $(U \to M) \in \mathcal{D}_{\text{ét}}/M^{\text{cp}}$, then $(U' \to M) \in \mathcal{D}_{\text{ét}}/M^{\text{cp}}$.
(D2) For any finite morphism $\rho\colon U' \to U$ in $\mathcal{D}_{\text{ét}}/M$, if $(U' \to M) \in \mathcal{D}_{\text{ét}}/M^{\text{cp}}$, then $(U \to M) \in \mathcal{D}_{\text{ét}}/M^{\text{cp}}$.
(D3) For any open immersion $U_\circ \subset U$ and an étale neighborhood $\rho\colon U' \to U$ of $U \setminus U_\circ$, if $(U_\circ \to M)$, $(U'_\circ \to M)$ and $(U' \to M)$ are objects in $\mathcal{D}_{\text{ét}}/M^{\text{cp}}$, then $(U \to M) \in \mathcal{D}_{\text{ét}}/M^{\text{cp}}$.

The claim for (D1) holds from [51, Example 3.11], since $\rho_*\colon \mathcal{T}_{\mathcal{Z}}(U') \to \mathcal{T}_{\mathcal{Z}}(U)$ is continuous and conservative. Indeed $\rho_{\mathfrak{M}}^* \mathcal{T}_{\mathcal{Z}}(U)^{\text{cp}}$ are compact objects and generate $\mathcal{T}_{\mathcal{Z}}(U')$. The claim for (D2) follows from [51, Proposition 6.9], as we have properties (i)–(v) in Proposition 8.2.1. Indeed $\rho_*(\mathcal{T}_{\mathcal{Z}}(U'))$ are compact objects and generate $\mathcal{T}_{\mathcal{Z}}(U)$. The claim for (D3) follows from [51, Proposition 6.8] by setting $V = \emptyset$ in *loc. cit.*

The above argument shows that $\mathrm{Ind}\,\mathcal{C}_{\mathcal{Z}}$ is compactly generated. It remains to show that the subcategory of compact objects $(\mathrm{Ind}\,\mathcal{C}_{\mathcal{Z}})^{\text{cp}}$ coincide with $\mathcal{C}_{\mathcal{Z}}$. Since $D_{\text{coh}}^b(\mathfrak{M}) = (\mathrm{Ind}\,D_{\text{coh}}^b(\mathfrak{M}))^{\text{cp}}$ by Theorem 3.1.1, we have

$$\mathcal{C}_{\mathcal{Z}} \subset (\mathrm{Ind}\,D_{\text{coh}}^b(\mathfrak{M}))^{\text{cp}} \cap \mathrm{Ind}\,\mathcal{C}_{\mathcal{Z}} \subset (\mathrm{Ind}\,\mathcal{C}_{\mathcal{Z}})^{\text{cp}}.$$

As for the converse direction, let us take $\mathcal{E} \in (\mathrm{Ind}\,\mathcal{C}_{\mathcal{Z}})^{\text{cp}}$ and a smooth morphism $\alpha\colon \mathfrak{U} \to \mathfrak{M}$ for a quasi-smooth affine derived scheme $\mathfrak{U}$. It is enough to show that $\alpha^*\mathcal{E} \in \mathcal{C}_{\alpha^*\mathcal{Z}}$. Since $\alpha_*^{\text{ind}}\colon \mathrm{Ind}\,D_{\text{coh}}^b(\mathfrak{U}) \to \mathrm{Ind}\,D_{\text{coh}}^b(\mathfrak{M})$ is continuous, its restriction $\alpha_*^{\text{ind}}\colon \mathrm{Ind}\,\mathcal{C}_{\alpha^*\mathcal{Z}} \to \mathrm{Ind}\,\mathcal{C}_{\mathcal{Z}}$ is also continuous. Therefore $\alpha^*\mathcal{E} \in (\mathrm{Ind}\,\mathcal{C}_{\alpha^*\mathcal{Z}})^{\text{cp}} = \mathcal{C}_{\alpha^*\mathcal{Z}}$, where the latter identity follows from Lemma 2.3.2. □

We keep the setting of Theorem 8.2.2. As in the proof of Theorem 6.3.13, we denote by $\mathcal{D}'_{\text{ét}}/M \subset \mathcal{D}_{\text{ét}}/M$ the subcategory of étale morphisms $\iota\colon U \to M$ satisfying the condition in Theorem 3.1.4. For a morphism $\rho\colon U' \to U$ in $\mathcal{D}'_{\text{ét}}/M$, we have the Cartesian diagrams (3.1.13). In particular, we have the pull-back functor

$$\rho_{\mathfrak{M}}^*\colon D_{\text{coh}}^b(\mathfrak{M}_U)/\mathcal{C}_{\iota_{\mathfrak{M}}^*\mathcal{Z}} \to D_{\text{coh}}^b(\mathfrak{M}_{U'})/\mathcal{C}_{\iota'^*_{\mathfrak{M}}\mathcal{Z}}$$

so that we can take their limit.

Lemma 8.2.3 *We have a fully-faithful functor,*

$$\mathcal{DT}^{\mathbb{C}^*}(\mathcal{N}^{\text{ss}}) \hookrightarrow \lim_{(U \xrightarrow{\iota} M) \in \mathcal{D}'_{\text{ét}}/M} \left(D_{\text{coh}}^b(\mathfrak{M}_U)/\mathcal{C}_{\iota_{\mathfrak{M}}^*\mathcal{Z}} \right). \tag{8.2.2}$$

Proof Using the result of Theorem 8.2.2, the proof is parallel to that of Proposition 3.2.7. As in the proof of Proposition 3.2.7, for each $(U \xrightarrow{\iota} M) \in \mathcal{D}'_{\text{ét}}/M$ the sequence

$$\operatorname{Ind} \mathcal{C}_{\iota_{\mathfrak{M}}^*\mathcal{Z}} \to \operatorname{Ind} D^b_{\text{coh}}(\mathfrak{M}_U) \to \operatorname{Ind}\left(D^b_{\text{coh}}(\mathfrak{M}_U)/\mathcal{C}_{\iota_{\mathfrak{M}}^*\mathcal{Z}}\right) \qquad (8.2.3)$$

is a localization sequence such that the right arrow admits a functorial right adjoint. As the union of $\iota_{\mathfrak{M}} : \mathfrak{M}_U \to \mathfrak{M}$ is an étale cover of $\mathfrak{M}$, we have the equivalences

$$\operatorname{Ind} D^b_{\text{coh}}(\mathfrak{M}) \xrightarrow{\sim} \lim_{(U \xrightarrow{\iota} M)\in\mathcal{D}'_{\text{ét}}/M} \operatorname{Ind} D^b_{\text{coh}}(\mathfrak{M}_U), \quad \operatorname{Ind}\mathcal{C}_{\mathcal{Z}} \xrightarrow{\sim} \lim_{(U \xrightarrow{\iota} M)\in\mathcal{D}'_{\text{ét}}/M} \operatorname{Ind}\mathcal{C}_{\iota_{\mathfrak{M}}^*\mathcal{Z}}.$$

Therefore by taking the limit of (8.2.3) for all $U \to M$ in $\mathcal{D}'_{\text{ét}}/M$, we obtain the localizing sequence

$$\operatorname{Ind}\mathcal{C}_{\mathcal{Z}} \to \operatorname{Ind} D^b_{\text{coh}}(\mathfrak{M}) \to \lim_{(U \xrightarrow{\iota} M)\in\mathcal{D}'_{\text{ét}}/M} \operatorname{Ind}\left(D^b_{\text{coh}}(\mathfrak{M}_U)/\mathcal{C}_{\iota_{\mathfrak{M}}^*\mathcal{Z}}\right).$$

By Theorems 8.1.1 and 8.2.2, we have the fully-faithful functor

$$D^b_{\text{coh}}(\mathfrak{M})/\mathcal{C}_{\mathcal{Z}} \hookrightarrow \lim_{(U \xrightarrow{\iota} M)\in\mathcal{D}'_{\text{ét}}/M} \operatorname{Ind}\left(D^b_{\text{coh}}(\mathfrak{M}_U)/\mathcal{C}_{\iota_{\mathfrak{M}}^*\mathcal{Z}}\right).$$

We have the natural functors

$$D^b_{\text{coh}}(\mathfrak{M})/\mathcal{C}_{\mathcal{Z}} \to \lim_{(U \xrightarrow{\iota} M)\in\mathcal{D}'_{\text{ét}}/M} (D^b_{\text{coh}}(\mathfrak{M}_U)/\mathcal{C}_{\iota_{\mathfrak{M}}^*\mathcal{Z}})$$

$$\to \lim_{(U \xrightarrow{\iota} M)\in\mathcal{D}'_{\text{ét}}/M} \operatorname{Ind}\left(D^b_{\text{coh}}(\mathfrak{M}_U)/\mathcal{C}_{\iota_{\mathfrak{M}}^*\mathcal{Z}}\right).$$

Since the above composition and the second arrow are fully-faithful, the first arrow is also fully-faithful. □

8.3 Some Lemmas in Derived Algebraic Geometry

8.3.1 Equivariant Affine Derived Schemes

Let G be a reductive algebraic group. We denote by $cdga^G$ the ∞-category of G-equivariant cdga's consisting of non-positive degrees. Here a G-equivariant cdga is a cdga which admits a G-action and satisfies obvious compatibility with differentials and products. The ∞-category of G-equivariant affine derived schemes is defined by

$$dAff^G := W^{-1}(cdga^G)^{\mathrm{op}}.$$

Here $W^{-1}(-)$ means ∞-categorical localization by weak homotopy equivalences (cf. [145, Section 2.2]). Let dSt/BG be the ∞-category of derived stacks over BG. We have the natural ∞-functor

$$\Pi\colon dAff^G \to dSt/BG, \quad \mathfrak{U} \mapsto [\mathfrak{U}/G].$$

We use the following two propositions:

Proposition 8.3.1 *Let $\mathfrak{M}$ be a quasi-smooth derived stack over BG such that $t_0(\mathfrak{M}) = [\mathcal{U}/G]$ for an affine scheme $\mathcal{U}$ with G-action. Then there exists a G-equivariant tuple (Y, V, s) as in Definition 2.3.1 and an equivalence $\mathfrak{M} \sim \Pi(\mathfrak{U})$ where $\mathfrak{U} = \operatorname{Spec} \mathcal{R}(V \to Y, s)$ is an object in $dAff^G$.*

Proof Note that we have a sequence of square-zero extensions

$$\mathcal{M} = t_0(\mathfrak{M}) \hookrightarrow t_{\leq 1}(\mathfrak{M}) \hookrightarrow t_{\leq 2}(\mathfrak{M}) \hookrightarrow \cdots$$

such that $t_{\leq n}(\mathfrak{M}) \hookrightarrow \mathfrak{M}$ is an equivalence for $n \gg 0$. Let us take distinguished triangles

$$\mathcal{I}_n \to \mathcal{O}_{t_{\leq n+1}(\mathfrak{M})} \to \mathcal{O}_{t_{\leq n}(\mathfrak{M})}.$$

Then the square zero extension $t_{\leq n}(\mathfrak{M}) \hookrightarrow t_{\leq n+1}(\mathfrak{M})$ in dSt/BG corresponds to an element in (see [43, Section 1])

$$\operatorname{Hom}_{t_{\leq n}(\mathfrak{M})}(\mathbb{L}_{t_{\leq n}(\mathfrak{M})/BG}, \mathcal{I}_n[1]). \tag{8.3.1}$$

Suppose that $t_{\leq n}(\mathfrak{M})$ is equivalent to $\Pi(\mathfrak{U}_{\leq n})$ for some $\mathfrak{U}_{\leq n} \in dAff^G$. Then as G is reductive, the Hom space in (8.3.1) is isomorphic to $\operatorname{Hom}_{\mathfrak{U}_{\leq n}}(\mathbb{L}_{\mathfrak{U}_{\leq n}}, \mathcal{I}_n[1])^G$. An element of the above corresponds to a square zero extension $\mathfrak{U}_{\leq n} \hookrightarrow \mathfrak{U}_{\leq n+1}$ in $dAff^G$. Therefore $t_{\leq n+1}(\mathfrak{M})$ is equivalent to $\Pi(\mathfrak{U}_{\leq n+1})$ for some $\mathfrak{U}_{\leq n+1} \in dAff^G$. By the induction, we see that $\mathfrak{M}$ is equivalent to $[\mathfrak{U}/G]$ for some $\mathfrak{U} \in dAff^G$.

We are left to prove that $\mathfrak{U}$ is equivalent in $dAff^G$ to an affine derived scheme $\operatorname{Spec} \mathcal{R}(V \to Y, s)$ associated with a G-equivariant tuple (Y, V, s) as in Definition 2.3.1. We prove this by following an argument of [22, Theorem 4.1]. Similarly to *loc. cit.*, below we will not fix a particular model for the G-equivariant cdga $\mathcal{O}_{\mathfrak{U}}$, and regard $\mathfrak{U}$ as an object in the ∞-category $dAff^G$. Also any map $\mathfrak{U} \to \mathfrak{U}'$ is regarded as a morphism in the ∞-category $dAff^G$.

As G is reductive, we can find a G-equivariant closed embedding $\mathcal{U} \hookrightarrow Y$ for some smooth affine scheme Y with a G-action, and let $I \subset \mathcal{O}_Y$ be the ideal sheaf which defines $\mathcal{U}$. As $\mathfrak{U}$ is quasi-smooth, the natural morphism of cotangent complexes $\phi: \mathbb{L}_{\mathfrak{U}}|_{\mathcal{U}} \to \tau_{\geq -1}\mathbb{L}_{\mathcal{U}}$ is a G-equivariant perfect obstruction theory on $\mathcal{U}$ (see [15]), i.e. ϕ is a morphism in the derived category of G-equivariant coherent sheaves on $\mathcal{U}$, $\mathcal{H}^0(\phi)$ is an isomorphism and $\mathcal{H}^{-1}(\phi)$ is a surjection. Since $\mathcal{U}$ and Y are affine, the morphism ϕ is represented by a morphism of complexes of G-equivariant sheaves on $\mathcal{U}$

$$
\begin{array}{ccc}
V^\vee|_{\mathcal{U}} & \longrightarrow & \Omega_Y|_{\mathcal{U}} \\
\downarrow & & \| \\
I/I^2 & \longrightarrow & \Omega_Y|_{\mathcal{U}}.
\end{array}
$$

Here $V \to Y$ is a G-equivariant vector bundle, and the left arrow $V^\vee|_{\mathcal{U}} \to I/I^2$ is a surjection. One can lift the surjection $V^\vee|_{\mathcal{U}} \to I/I^2$ to a G-equivariant map $s: V^\vee \to I$, which we can assume to be surjective by shrinking Y if necessary. Since Y is smooth, we can lift the closed immersion $\mathcal{U} \hookrightarrow Y$ to a map $j: \mathfrak{U} \to Y$ in $dAff^G$. Then the diagram

$$
\begin{array}{ccc}
\mathfrak{U} & \xrightarrow{\ j\ } & Y \\
{\scriptstyle j}\downarrow & & \downarrow{\scriptstyle 0} \\
Y & \xrightarrow{\ s\ } & V
\end{array}
$$

is commutative in $dAff^G$ up to equivalence. Therefore the above diagram induces a map $\mathfrak{U} \to \operatorname{Spec} \mathcal{R}(V \to Y, s)$ in $dAff^G$. The above map induces the isomorphism on the classical truncation, and the induced map on cotangent complexes is a quasi-isomorphism by the construction. Therefore $\mathfrak{U}$ is equivalent to $\operatorname{Spec} \mathcal{R}(V \to Y, s)$ in $dAff^G$. $\square$

Proposition 8.3.2 *For $\mathfrak{U}, \mathfrak{U}' \in dAff^G$, let $f: \Pi(\mathfrak{U}) \to \Pi(\mathfrak{U}')$ be a morphism in dSt/BG. Then there exists a morphism $\widetilde{f}: \mathfrak{U} \to \mathfrak{U}'$ in $dAff^G$ such that $f \sim \Pi(\widetilde{f})$.*

Proof The proof is similar to Proposition 8.3.1 using deformation argument. We set

$$
f_{\leq n} := f|_{t_{\leq n}(\Pi(\mathfrak{U}))} : t_{\leq n}(\Pi(\mathfrak{U})) \to \Pi(\mathfrak{U}').
$$

For $n = 0$, $f_{\leq 0}$ factors through a morphism of classical Artin stacks $[t_0(\mathfrak{U})/G] \to [t_0(\mathfrak{U}')/G]$ over BG. By pulling back via $\operatorname{Spec}\mathbb{C} \to BG$, it lifts to a G-equivariant morphisms of affine schemes $t_0(\mathfrak{U}) \to t_0(\mathfrak{U}')$. By setting $\widetilde{f}_{\leq 0}$ to be the composition of the above morphism with $t_0(\mathfrak{U}') \hookrightarrow \mathfrak{U}'$, we have $f_{\leq 0} \sim \Pi(\widetilde{f}_{\leq 0})$.

Suppose that there exists $\widetilde{f}_{\leq n} : t_{\leq n}(\mathfrak{U}) \to \mathfrak{U}'$ in $dAff^G$ such that $f_{\leq n} \sim \Pi(\widetilde{f}_{\leq n})$. We take the distinguished triangle

$$\mathcal{J}_n \to \mathcal{O}_{t_{\leq n+1}(\Pi(\mathfrak{U}))} \to \mathcal{O}_{t_{\leq n}(\Pi(\mathfrak{U}))}.$$

Note that $\mathcal{J}_n$ is an object in $D^b_{\mathrm{coh}}(\Pi(\mathcal{U}))$ for $\mathcal{U} = t_0(\mathfrak{U})$ push-forward along with the closed immersion $\Pi(\mathcal{U}) \hookrightarrow \Pi(\mathfrak{U}_{\leq n+1})$. We set S^i_n, T^i_n to be

$$S^i_n := \mathrm{Hom}_{\mathcal{U}}(\widetilde{f}^*_{\leq 0}(\mathbb{L}_{\mathfrak{U}'}|_{\mathcal{U}'}), p^* \mathcal{J}_n[i])^G,$$

$$T^i_n := \mathrm{Hom}_{[\mathcal{U}/G]}(f^*_{\leq 0}(\mathbb{L}_{\Pi(\mathfrak{U}')/BG}|_{\Pi(\mathcal{U}')}), \mathcal{J}_n[i]).$$

Here $\mathcal{U}' = t_0(\mathfrak{U}')$ and $p \colon \mathcal{U} \to [\mathcal{U}/G]$. Since G is reductive, we have natural isomorphisms

$$S^i_n \xrightarrow{\cong} T^i_n. \tag{8.3.2}$$

The obstruction of extending $\widetilde{f}_{\leq n}$ to $\widetilde{f}_{\leq n+1}$ in $dAff^G$ lies in S^1_n, which corresponds to the obstruction in T^1_n of extending $f_{\leq n}$ to $f_{\leq n+1}$ in dSt/BG under the isomorphism (8.3.2). As $f_{\leq n}$ is extended to $f_{\leq n+1}$ by the assumption, the above obstruction vanishes so that there exists an extension $\widetilde{f}_{\leq n+1}$ of $\widetilde{f}_{\leq n}$ in $dAff^G$. An extension of $\widetilde{f}_{\leq n}$ to $\widetilde{f}_{\leq n+1}$ in $dAff^G$ is classified by S^0_n, while an extension of $f_{\leq n}$ to $f_{\leq n+1}$ in dSt/BG is classified by T^0_n. Therefore by the isomorphism (8.3.2) there exists an extension $\widetilde{f}_{\leq n+1}$ such that $\Pi(\widetilde{f}_{\leq n+1}) \sim f_{\leq n+1}$. By the induction, we obtain $\widetilde{f} \colon \mathfrak{U} \to \mathfrak{U}'$ in $dAff^G$ such that $\Pi(\widetilde{f}) \sim f$. □

8.3.2 Proof of Proposition 3.1.3

Proof We use the theory of good moduli spaces for derived Artin stacks developed in [3]. By [3, Theorem 2.12], there is a derived algebraic space $\widetilde{M}$ with $t_0(\widetilde{M}) = M$ and a commutative diagram

$$
\begin{array}{ccc}
M & \longrightarrow & \mathfrak{M} \\
\downarrow & & \downarrow \\
M & \longrightarrow & \widetilde{M}.
\end{array}
$$

Here the right vertical arrow is a good moduli space for $\mathfrak{M}$ defined in [3, Definition 2.1] and the horizontal arrows are closed immersions.

Let $\mathcal{D}_{\text{ét}/\widetilde{M}}$ be the category of étale morphisms $(\widetilde{U} \to \widetilde{M})$ for a derived algebraic space $\widetilde{U}$. We have the truncation functor

$$t_0: \mathcal{D}_{\text{ét}/\widetilde{M}} \to \mathcal{D}_{\text{ét}}/M. \tag{8.3.3}$$

The above functor is an equivalence. Indeed suppose that M is affine and let

$$\mathcal{D}_{\text{ét,aff}/M} \subset \mathcal{D}_{\text{ét}}/M, \quad \mathcal{D}_{\text{ét,aff}/\widetilde{M}} \subset \mathcal{D}_{\text{ét}/\widetilde{M}}$$

be the subcategories of $U \to M, \widetilde{U} \to \widetilde{M}$ such that $U, \widetilde{U}$ are affine. Then by [148, Corollary 2.2.2.9], the truncation functor induces the equivalence

$$t_0: \mathcal{D}_{\text{ét,aff}/\widetilde{M}} \xrightarrow{\sim} \mathcal{D}_{\text{ét,aff}/M}. \tag{8.3.4}$$

The equivalence (8.3.3) follows from the equivalence (8.3.4) and étale descent.

Then by the equivalence (8.3.4), for each object $(U \to M)$ in $\mathcal{D}_{\text{ét}}/M$, it uniquely lifts to $(\widetilde{U} \to \widetilde{M})$ in $\mathcal{D}_{\text{ét}/\widetilde{M}}$. By taking the fiber products with $\mathfrak{M} \to \widetilde{M}$, we obtain the derived stack $\mathfrak{M}_U$ which fits into the diagram (3.1.12). Moreover for each morphism $U \to U'$ in $\mathcal{D}_{\text{ét}}/M$, it uniquely lifts to a morphism $\widetilde{U} \to \widetilde{U}'$ which induces the diagram (3.1.13). □

8.3.3 Proof of Proposition 3.1.5

Proof By Proposition 3.1.3, there is a unique derived stack $\mathfrak{M}_U$ which fits into the Cartesian diagram (3.1.12). Then by Proposition 8.3.1, the derived stack $\mathfrak{M}_U$ is equivalent to $[\mathfrak{U}/G]$ associated with a G-equivariant tuple (Y, V, s).

The final statement follows from Lemma 6.1.1. Indeed by shrinking U if necessary, the $\mathbb{R}$-line bundle $\iota_{\mathfrak{M}}^*(l)$ on $[\mathcal{U}/G]$ is pulled back via $[\mathcal{U}/G] \to BG$, so can be extended to $[Y/G]$ by taking the pull-back via $[Y/G] \to BG$. □

8.3.4 Proof of Lemma 6.2.15

Proof We take a G-equivariant closed immersion $\iota: Y \hookrightarrow W$ for a smooth affine $\mathbb{C}$-scheme with a G-action. We choose such W which admits a G-equivariant vector bundle $W' \to W$ with a G-invariant regular section $t: W \to W'$ such that its zero locus is a single reduced point $\{w_0\}$. For example as Y is of finite presentation, we can take W to be a finite dimensional G-representation and $W' = W \times W \to W$ with section given by the diagonal. We set

$$Y'' = W \times Y', \quad V'' = W' \times V'$$

and regard V'' as a G-equivariant vector bundle on Y'' by the projection $V'' \to Y''$. We have the following G-equivariant commutative diagram

$$
\begin{array}{c}
\xleftarrow{\hspace{2cm}g\hspace{2cm}} \\
V \dashrightarrow^{g''} V'' \longrightarrow V' \hookleftarrow^{g'} V'' \\
s \downarrow \quad\uparrow \qquad \downarrow s'' \qquad \downarrow s' \qquad \downarrow s'' \\
Y \hookrightarrow^{f''} Y'' \xrightarrow{f'''} Y' \hookleftarrow Y''. \\
\xleftarrow{\hspace{2cm}f\hspace{2cm}}
\end{array}
\tag{8.3.5}
$$

Here the middle horizontal arrows are projections, and

$$f''(y) = (\iota(y), f(y)), \ s''(w, y') = (t(w), s'(y')),$$
$$f'(y') = (w_0, y'), \ g'(v') = (t(w_0), v').$$

Note that f', f'' are closed immersions.

Let $\mathfrak{U}''$ be the derived zero locus of s''. By the constructions of (Y'', V'', s''), the projection f''' and the closed immersion f' in the diagram (8.3.5) induce equivalences

$$\mathbf{f'''} \colon [\mathfrak{U}''/G] \xrightarrow{\sim} [\mathfrak{U}'/G], \ \mathbf{f'} \colon [\mathfrak{U}'/G] \xrightarrow{\sim} [\mathfrak{U}''/G] \tag{8.3.6}$$

which are inverse each other in the ∞-category dSt/BG. Since we assumed that the diagram (6.2.9) induces an equivalence $[\mathfrak{U}/G] \xrightarrow{\sim} [\mathfrak{U}'/G]$, by composing with (8.3.6) we obtain an equivalence $[\mathfrak{U}/G] \sim [\mathfrak{U}''/G]$ in dSt/BG. By Proposition 8.3.2, the equivalence $[\mathfrak{U}/G] \sim [\mathfrak{U}''/G]$ lifts to an equivalence $\mathfrak{U} \sim \mathfrak{U}''$ in $dAff^G$. Now we have the morphisms $\mathfrak{U} \hookrightarrow Y \xrightarrow{f''} Y''$ in $dAff^G$, and as $\mathcal{O}_{\mathfrak{U}''}$ is cofibrant over $\mathcal{O}_{Y''}$ in $cdga^G$, the equivalence $\mathfrak{U} \sim \mathfrak{U}''$ is given by an actual morphism of G-equivariant cdga's, i.e. we have a G-equivariant vector bundle morphism g'' in the dotted arrow in (8.3.5) which induces an equivalence $\mathfrak{U} \xrightarrow{\sim} \mathfrak{U}''$. Then it induces an equivalence $\mathbf{f''} \colon [\mathfrak{U}/G] \xrightarrow{\sim} [\mathfrak{U}''/G]$ as desired. $\square$

8.3.5 Proof of Lemma 6.2.17

Proof By Proposition 8.3.2, the equivalence $\mathbf{f}$ lifts to an équivalence $\mathbf{f}^{\sharp} \colon \mathfrak{U} \xrightarrow{\sim} \mathfrak{U}'$ in $dAff^G$. We then construct the diagram (6.2.17) following the argument of [22, Theorem 4.2]. By the equivalence $\mathbf{f}^{\sharp} \colon \mathfrak{U} \xrightarrow{\sim} \mathfrak{U}'$ together with closed immersions $\mathfrak{U} \hookrightarrow Y, \mathfrak{U}' \hookrightarrow Y'$, we have the morphism in $dAff^G$ closed immersion $\mathfrak{U} \hookrightarrow Y \times Y'$ which is a closed immersion. Then similarly to the proof Proposition 8.3.1, there exist a G-invariant affine open subset $\widetilde{Y} \subset Y \times Y'$ which contains $\mathcal{U}$, a G-equivariant vector bundle $\widetilde{V} \to \widetilde{Y}$ with a G-invariant section $\widetilde{s}$, such that $\mathfrak{U} \hookrightarrow \widetilde{Y}$ factors through an equivalence $\mathfrak{U} \xrightarrow{\sim} \widetilde{\mathfrak{U}}$ in $dAff^G$, where $\widetilde{\mathfrak{U}}$ is the derived zero locus of $\widetilde{s}$. Let us consider the composition $\widetilde{\mathfrak{U}} \hookrightarrow \widetilde{Y} \to Y$, where the latter map is the projection. Since $\mathcal{O}_{\mathfrak{U}}$ is cofibrant over $\mathcal{O}_Y$ in $cdga^G$, the above map factors through an equivalence $\widetilde{\mathfrak{U}} \xrightarrow{\sim} \mathfrak{U}$ in $dAff^G$, which is induced by an actual morphism of G-equivariant cdga's. So we obtain a left diagram in (6.2.17) and an equivalence $\widetilde{\mathbf{f}}$. The existence of a right diagram in (6.2.17) and $\widetilde{\mathbf{f}}'$ also follow similarly by taking the projection $\widetilde{Y} \to Y'$. $\qquad\qquad\square$

8.3.6 The Case of Formal Fibers

Here we discuss the formal fiber version of Lemmas 6.2.15 and 6.2.17, which is used in Lemmas 6.3.5 and 6.3.8. We consider the setting of Lemma 6.3.5, so that we have the equivalence of derived stacks $\widehat{\mathbf{f}}$ in the diagram (6.3.12). We assume that $\widehat{\mathbf{f}}$ commutes with maps to BG given by canonical G-torsors. We have the formal fiber version of Lemma 6.2.17:

Lemma 8.3.3 *There exist G-equivariant commutative diagram*

$$
\begin{array}{ccccc}
\widehat{V}_y & \longleftarrow & \widetilde{V} & \longrightarrow & \widehat{V}'_{y'} \\
\widehat{s}_y \Big\uparrow\Big\downarrow & & \widetilde{s}\Big\uparrow\Big\downarrow & & \Big\uparrow\Big\downarrow \widehat{s}_{y'} \\
\widehat{Y}_y & \xleftarrow{\ \widetilde{f}\ } & \widetilde{Y} & \xrightarrow{\ \widetilde{f}'\ } & \widehat{Y}'_{y'}
\end{array}
\tag{8.3.7}
$$

such that, by setting $\widetilde{\mathfrak{U}}$ to be the derived zero locus of $\widetilde{s}$, the above diagram induces equivalences $\widetilde{\mathbf{f}} \colon [\widetilde{\mathfrak{U}}/G] \xrightarrow{\sim} [\widehat{\mathfrak{U}}_y/G]$, $\widetilde{\mathbf{f}}' \colon [\widetilde{\mathfrak{U}}/G] \xrightarrow{\sim} [\widehat{\mathfrak{U}}'_{y'}/G]$ which commute with $\widehat{\mathbf{f}}$, i.e. $\mathbf{f} \circ \widetilde{\mathbf{f}} \sim \widetilde{\mathbf{f}}'$. Here $\widetilde{Y}$ is a G-invariant open subset of $\widehat{Y}_y \widehat{\times} \widehat{Y}'_{y'}$ which contains (x, x'), $\widetilde{V} \to \widetilde{Y}$ is a G-equivariant vector bundle and $\widetilde{s}$ is a G-invariant section.

Proof The proof is obvious from the proof of Lemma 6.2.17, by replacing $Y \times Y'$ in *loc. cit.* with $\widehat{Y}_y \widehat{\times} \widehat{Y}'_{y'}$. $\qquad\qquad\square$

We also have the formal fiber version Lemma 6.2.15:

Lemma 8.3.4 *In the setting of Lemma 8.3.3, we have G-equivariant commutative diagrams*

$$
\begin{array}{ccc}
\widetilde{V} & \xrightarrow{\ g''\ } & V'' \\
s\Big\uparrow\Big\downarrow & & \Big\uparrow\Big\downarrow s'' \\
\widetilde{Y} & \xhookrightarrow{\ f''\ } & Y'',
\end{array}
\qquad\qquad
\begin{array}{ccc}
\widehat{V}_y & \xrightarrow{\ g'\ } & V'' \\
s'\Big\uparrow\Big\downarrow & & \Big\uparrow\Big\downarrow s'' \\
\widehat{Y}_y & \xhookrightarrow{\ f'\ } & Y'',
\end{array}
\tag{8.3.8}
$$

where $Y'' = \widetilde{Y} \times \widehat{Y}_y$, $V'' \to Y''$ is a G-equivariant vector bundle, s'' is a G-invariant section, satisfying the followings:

(i) The diagrams (8.3.8) induce equivalences of derived stacks

$$\mathbf{f}' : [\widetilde{\mathfrak{U}}/G] \xrightarrow{\sim} [\mathfrak{U}''/G], \quad \widehat{\mathbf{f}} : [\widehat{\mathfrak{U}}_y/G] \xrightarrow{\sim} [\mathfrak{U}''/G]$$

such that $\mathbf{f}' \circ \widehat{\mathbf{f}} \sim \mathbf{f}''$. Here $[\mathfrak{U}''/G]$ is the derived zero locus of s'': $[Y''/G] \to [V''/G]$.

(ii) The morphisms f', f'' are closed immersions.

Moreover the same statement holds for the right square of (8.3.7).

Proof The same argument of the proof of Lemma 6.2.15 applies. In this case, we can also take $Y'' = \widetilde{Y} \times \widehat{Y}_y$ as we show in the following. From the proof of Lemma 6.2.15, we first take a G-equivariant closed immersion $\widetilde{Y} \hookrightarrow W$ with a vector bundle $W' \to W$ and a regular section t whose zero locus is a reduced single point. In the situation of the lemma, we can take $W = \widetilde{Y}$. Indeed the condition $G = \mathrm{Aut}(x)$ and étale slice theorem implies that $\widehat{Y}_y$ is isomorphic to the formal fiber of $T_x Y \to T_x Y /\!\!/ G$ at 0 (see the proof of Proposition 6.3.7), so we can assume that Y is a G-representation and $x = 0$. The same also applies to Y'. Then we have the trivial vector bundle $(Y \times Y') \times (Y \times Y') \to (Y \times Y')$ by the second projection, with a regular section given by the diagonal whose zero locus is $(0, 0)$. By restricting the above vector bundle to the formal fiber $\widehat{Y}_y \times \widehat{Y}'_{y'} \to Y \times Y'$ and its G-invariant open subset $\widetilde{Y}$, we obtain the G-equivariant vector bundle $W' \to \widetilde{Y}$ with a regular section whose zero locus is a reduced point $(0, 0)$. So we can take $W = \widetilde{Y}$. Then the argument of the proof of Lemma 6.2.15 implies that we can take $Y'' = \widetilde{Y} \times \widehat{Y}_y$. $\qquad\square$

8.3.7 Proof of Lemma 6.3.6

Proof Let $\mathfrak{m} \subset \mathcal{O}_{\widehat{\mathcal{U}}_y}$ be the maximal ideal of x and $\mathcal{U}^{[n]} \subset \widehat{\mathcal{U}}_y$ the closed subscheme defined by $\mathfrak{m}^n$. We claim that the compatible isomorphisms

$$f|_{[\mathcal{U}^{[n]}/G]} \cong f'|_{[\mathcal{U}^{[n]}/G]} \tag{8.3.9}$$

exist by the induction on n. The $n = 1$ case follows from $f \circ \mu \cong f' \circ \mu$. Suppose that the claim holds for n. By a standard deformation theory of morphisms of stacks, the set of possible extensions of $f|_{[\mathfrak{U}^{[n]}/G]}$ to $[\mathfrak{U}^{[n+1]}/G]$ is a torsor over

$$\mathrm{Hom}_{[\mathcal{U}^{[n]}/G]}(f|^*_{[\mathcal{U}^{[n]}/G]}\mathbb{L}_{BG}, \mathfrak{m}^n/\mathfrak{m}^{n+1}) = 0$$

since $\mathbb{L}_{BG} = \mathfrak{g}^\vee[-1]$, $\mathcal{U}^{[n]}$ is affine and G is reductive. Therefore the isomorphism (8.3.9) lifts to an isomorphism for $n + 1$, so the claim holds.

Since $\widehat{\mathcal{U}}_y /\!\!/ G$ is complete local, by [5, Corollary 3.6] a morphism $[\widehat{\mathcal{U}}_y/G] \to BG$ is determined by its restriction to $\varinjlim [\mathcal{U}^{[n]}/G]$. Therefore we conclude that $f|_{[\widehat{\mathcal{U}}_y/G]} \cong f'|_{[\widehat{\mathcal{U}}_y/G]}$. We then apply the same deformation argument above for the square zero extensions

$$[\widehat{\mathcal{U}}_y/G] = t_0([\widehat{\mathcal{U}}_y/G]) \hookrightarrow t_{\leq 1}([\widehat{\mathcal{U}}_y/G]) \hookrightarrow t_{\leq 2}([\widehat{\mathcal{U}}_y/G]) \hookrightarrow \cdots$$

and conclude that $f \sim f'$. $\qquad\qquad\qquad\qquad\qquad\qquad\qquad\qquad\qquad\qquad$ □

8.4 Formal Neighborhood Theorem

In this section, we prove the formal neighborhood theorem for moduli stacks of semistable sheaves on smooth projective varieties. A similar result is proved in [132] analytic locally on the good moduli spaces, using gauge theory argument. Instead of gauge theory, we use formal GAGA theorem [7, 44] to give a purely algebraic proof for the formal neighborhood theorem.

8.4.1 Formal GAGA for Good Moduli Spaces

Let $\mathcal{X}$ be a classical Artin stack with affine diagonal, and

$$\pi : \mathcal{X} \to \operatorname{Spec} R$$

its good moduli space. We assume that R is a complete local Noetherian ring with maximal ideal $\mathfrak{m} \subset R$, and π is of finite type. Let $\mathcal{I} \subset \mathcal{O}_{\mathcal{X}}$ be the ideal sheaf generated by the pull-back of $\mathfrak{m}$. We define the closed substack $\mathcal{X}_n \subset \mathcal{X}$ to be defined by the ideal $\mathcal{I}^n$. We have the formal stack $\varinjlim \mathcal{X}_n$. The formal GAGA theorem for $\mathcal{X}$ is stated as follows:

Theorem 8.4.1 ([44, Theorem 1.1], [7, Corollary 1.7]) *The restriction functor*

$$\operatorname{Coh}(\mathcal{X}) \to \operatorname{Coh}(\varinjlim \mathcal{X}_n)$$

is an equivalence of categories.

8.4.2 Ext-Quivers Associated with Simple Collections

Let Z be a smooth projective variety over $\mathbb{C}$. A collection of coherent sheaves $(E_1, \ldots, E_m)$ on Z is called a *simple collection* if $\operatorname{Hom}(E_i, E_j) = \mathbb{C} \cdot \delta_{ij}$. The

Ext-quiver $Q_{E_\bullet}$ associated with the collection $(E_1, \ldots, E_m)$ is defined as follows: the vertex set $V(Q_{E_\bullet})$ and the edge set $E(Q_{E_\bullet})$ are given by

$$V(Q_{E_\bullet}) = \{1, \ldots, m\}, \quad E(Q_{E_\bullet}) = \bigcup_{1 \le i,j \le m} E_{i,j}.$$

Here $E_{i,j}$ the set of edges from i to j with $\sharp E_{i,j} = \mathrm{ext}^1(E_i, E_j)$. Let $\mathbf{E}_{i,j}$ be the $\mathbb{C}$-vector space spanned by $E_{i,j}$. We set

$$E_{i,j}^\vee := \{e^\vee : e \in E_{i,j}\} \subset \mathbf{E}_{i,j}^\vee.$$

Here for $e \in E_{i,j}$, the element $e^\vee \in \mathbf{E}_{i,j}^\vee$ is defined by the condition $e^\vee(e) = 1$ and $e^\vee(e') = 0$ for any $e \ne e' \in E_{i,j}$, i.e. $E_{i,j}^\vee$ is the dual basis of $E_{i,j}$.

By setting $\overline{E} = \oplus_{i=1}^m E_i$, we have linear maps

$$m_n \colon \mathrm{Ext}^1(\overline{E}, \overline{E})^{\otimes n} \to \mathrm{Ext}^2(\overline{E}, \overline{E}) \tag{8.4.1}$$

given by a minimal A_∞-structure of the dg-algebra $\mathbf{RHom}(\overline{E}, \overline{E})$. The map (8.4.1) only consists of the direct sum factors of the form (see [132, Section 5.1])

$$m_n \colon \mathrm{Ext}^1(E_{\psi(1)}, E_{\psi(2)}) \otimes \mathrm{Ext}^1(E_{\psi(2)}, E_{\psi(3)}) \otimes \cdots$$
$$\cdots \otimes \mathrm{Ext}^1(E_{\psi(n)}, E_{\psi(n+1)}) \to \mathrm{Ext}^2(E_{\psi(1)}, E_{\psi(n+1)}). \tag{8.4.2}$$

Here ψ is a map $\psi \colon \{1, \ldots, n+1\} \to \{1, \ldots, m\}$, and the above $\{m_n\}_{n \ge 2}$ give a minimal A_∞-category structure on the dg-category generated by $(E_1, \ldots, E_m)$. By taking the dual and the products of (8.4.2) for all $n \ge 2$, we obtain the linear map

$$\mathbf{m}^\vee := \prod_{n \ge 2} m_n^\vee \colon \mathrm{Ext}^2(\overline{E}, \overline{E})^\vee \to \prod_{\substack{n \ge 2 \\ \{1,\ldots,n+1\} \\ \overset{\psi}{\to} \{1,\ldots,m\}}} \bigoplus \mathrm{Ext}^1(E_{\psi(1)}, E_{\psi(2)})^\vee \otimes \cdots$$

$$\cdots \otimes \mathrm{Ext}^1(E_{\psi(n)}, E_{\psi(n+1)})^\vee.$$

Note that an element of the RHS is an element of the completed path algebra $\mathbb{C}[\![Q_{E_\bullet}]\!]$. Let $\mathbf{I} \subset \mathbb{C}[\![Q_{E_\bullet}]\!]$ be the topological closure of the ideal generated by the image of $\mathbf{m}^\vee$. By [132, Section 6.4], the quotient algebra

$$A := \mathbb{C}[\![Q_{E_\bullet}]\!]/\mathbf{I}$$

is a pro-representable hull for the NC deformation functor associated with the collection $(E_1, \ldots, E_m)$.

Let $E \in \mathrm{Coh}(X)$ be given by

$$E = \bigoplus_{i=1}^{m} V_i \otimes E_i$$

for finite dimensional vector spaces V_i. Then

$$\left[\mathrm{Ext}^1(E, E) / \mathrm{Aut}(E) \right] = \left[\bigoplus_{(i \to j) \in Q_{E_\bullet}} \mathrm{Hom}(V_i, V_j) / \prod_{i=1}^{m} \mathrm{GL}(V_i) \right]$$

is the moduli stack of $Q_{E_\bullet}$-representations with dimension vector $\vec{v} = (\dim V_i)_{1 \le i \le m}$. Here $\mathrm{Aut}(E)$ acts on $\mathrm{Ext}^1(E, E)$ by the conjugation. The fiber of the morphism to the good moduli space

$$\left[\mathrm{Ext}^1(E, E) / \mathrm{Aut}(E) \right] \to \mathrm{Ext}^1(E, E) /\!\!/ \mathrm{Aut}(E) \tag{8.4.3}$$

at the origin consists of nilpotent $Q_{E_\bullet}$-representations. More precisely, let $I \subset \mathcal{O}_{\mathrm{Ext}^1(E, E)}$ be the ideal sheaf which defines the fiber (8.4.3) at the origin, and $T_n \hookrightarrow \mathrm{Ext}^1(E, E)$ the closed subscheme defined by I^n. Then the formal stack $\varinjlim [T_n / \mathrm{Aut}(E)]$ represents the 2-functor

$$\mathcal{M}_{Q_{E_\bullet}}^{\mathrm{nil}}(\vec{v}) \colon Aff^{op} \to Groupoid \tag{8.4.4}$$

which sends an affine $\mathbb{C}$-scheme T to the groupoid of data

$$\{(\mathcal{V}_i, \phi_e)\}_{i \in V(Q_{E_\bullet}), e \in E(Q_{E_\bullet})}, \ \phi_e \colon \mathcal{V}_{s(e)} \to \mathcal{V}_{t(e)}. \tag{8.4.5}$$

Here $\mathcal{V}_i$ is a vector bundle on T with rank $\dim V_i$, $(s(e), t(e)) = (i, j)$ for $e \in E_{i,j}$, and ϕ_e are $\mathcal{O}_T$-module homomorphisms whose sufficiently large number of compositions are zero.

For an element

$$u = (u_e)_{e \in E(Q_{E_\bullet})} \in \mathrm{Ext}^1(E, E), \ u_e \colon V_{s(e)} \to V_{t(e)},$$

we consider the following $\mathrm{Ext}^2(E, E)$-valued $\mathrm{Aut}(E)$-equivariant formal function

$$\kappa(u) = \sum_{\substack{n \ge 2, \\ \{1, \ldots, n+1\} \overset{\psi}{\to} \{1, \ldots, m\}}} \ \sum_{e_i \in E_{\psi(i), \psi(i+1)}} m_n(e_1^\vee, \ldots, e_n^\vee) \cdot u_{e_n} \circ \cdots \circ u_{e_2} \circ u_{e_1}.$$

The above sum is a finite sum if u corresponds to a nilpotent Q_E-representation. Therefore the restriction $\kappa|_{T_n}$ determines the $\mathrm{Aut}(E)$-equivariant algebraic map

$$\kappa_n := \kappa|_{T_n} \colon T_n \to \mathrm{Ext}^2(E, E).$$

By the above arguments, we obtain sections of vector bundles over T_n

$$\left[(\mathrm{Ext}^2(E, E) \times T_n)/\mathrm{Aut}(E) \right] \xrightarrow{\quad\kappa_n\quad} \left[T_n/\mathrm{Aut}(E) \right].$$

Let $\mathcal{N}_n \hookrightarrow T_n$ be the closed subscheme defined by $\kappa_n = 0$. Then the formal stack $\varinjlim [\mathcal{N}_n/\mathrm{Aut}(E)]$ represents the sub 2-functor of (8.4.4),

$$\mathcal{M}^{\mathrm{nil}}_{(Q_{E_\bullet}, \mathbf{I})}(\vec{v}) \subset \mathcal{M}^{\mathrm{nil}}_{Q_{E_\bullet}}(\vec{v})$$

consisting of groupoids (8.4.5) such that $\{\phi_e\}_{e \in E(Q_{E_\bullet})}$ satisfy the relation $\mathbf{I} \subset \mathbb{C}[\![Q_{E_\bullet}]\!]$.

By Theorem 8.4.1, the system of sections $\{\kappa_n\}_{n \geq 1}$ above uniquely lifts to a section of a vector bundle on $\left[\widehat{\mathrm{Ext}^1(E, E)_0}/\mathrm{Aut}(E) \right]$

$$\left[(\mathrm{Ext}^2(E, E) \times \widehat{\mathrm{Ext}^1(E, E)_0})/\mathrm{Aut}(E) \right] \xrightarrow{\quad\kappa\quad} \left[\widehat{\mathrm{Ext}^1(E, E)_0}/\mathrm{Aut}(E) \right].$$

Here as in Sect. 6.1.4, $\widehat{\mathrm{Ext}^1(E, E)_0}$ is the formal fiber at the origin $0 \in \mathrm{Ext}^1(E, E)/\!\!/\mathrm{Aut}(E)$. Let $\widehat{\mathcal{N}_0} \subset \widehat{\mathrm{Ext}^1(E, E)_0}$ be the closed subscheme defined by $\kappa = 0$. We have the quotient stack with good moduli space

$$[\widehat{\mathcal{N}_0}/\mathrm{Aut}(E)] \to \widehat{\mathcal{N}_0}/\!\!/\mathrm{Aut}(E). \qquad\qquad (8.4.6)$$

The good moduli space $\widehat{\mathcal{N}_0}/\!\!/\mathrm{Aut}(E)$ is a closed subscheme of $\widehat{\mathrm{Ext}^1(E, E)_0}/\!\!/\mathrm{Aut}(E)$, hence written as $\mathrm{Spec}\, R$ for a complete local Noetherian ring R. The formal stack $\varinjlim [\mathcal{N}_n/\mathrm{Aut}(E)]$ is obtained by taking the colimit of thickened fibers of the morphism (8.4.6) at the origin as in Sect. 8.4.1.

8.4.3 Moduli Stacks of Semistable Sheaves

For an ample divisor H on Z and $v \in H^{2*}(Z, \mathbb{Q})$, we denote by $\mathcal{M}_{Z,H}(v)$ the moduli stack of Gieseker H-semistable sheaves E on Z with $\mathrm{ch}(E) = v$. By the

GIT construction of the moduli stack $\mathcal{M}_{Z,H}(v)$ (see [57]), it admits a good moduli space (see [4, Example 8.7])

$$\pi : \mathcal{M}_{Z,H}(v) \twoheadrightarrow M_{Z,H}(v).$$

The good moduli space $M_{Z,H}(v)$ is a projective scheme which parametrizes H-polystable sheaves, i.e. a closed point $y \in M_{Z,H}(v)$ corresponds to a direct sum

$$E = \bigoplus_{i=1}^{m} V_i \otimes E_i \tag{8.4.7}$$

where each E_i is a Gieseker H-stable sheaf, E_i is not isomorphic to E_j for $i \neq j$, and E_i has the same reduced Hilbert polynomial with E_j. Note that $(E_1, \dots, E_m)$ is a simple collection. Let $R = \widehat{\mathcal{O}}_{M_{Z,H}(v),y}$, which is a complete local Noetherian ring, and set

$$\widehat{\mathcal{M}}_{Z,H}(v)_y := \mathcal{M}_{Z,H}(v) \times_{M_{Z,H}(v)} \operatorname{Spec} R \to \operatorname{Spec} R. \tag{8.4.8}$$

Note that the above map is a good moduli space morphism of $\widehat{\mathcal{M}}_{Z,H}(v)_y$, since the good moduli morphism is preserved by the base change (see [4, Proposition 4.7]).

Let $[\widehat{\mathcal{N}}_0/\operatorname{Aut}(E)]$ be the quotient stack constructed in (8.4.6) associated with the above collection $(E_1, \dots, E_m)$ for the polystable sheaf (8.4.7). The following result gives the formal neighborhood theorem for Gieseker moduli spaces.

Theorem 8.4.2 *We have the commutative isomorphisms*

$$
\begin{array}{ccc}
[\widehat{\mathcal{N}}_0/\operatorname{Aut}(E)] & \xrightarrow{\cong} & \widehat{\mathcal{M}}_{Z,H}(v)_y \\
\downarrow & & \downarrow \\
\widehat{\mathcal{N}}_0 /\!\!/ \operatorname{Aut}(E) & \xrightarrow{\cong} & \operatorname{Spec} R.
\end{array} \tag{8.4.9}
$$

Proof Note that the closed fiber of the morphism (8.4.8) corresponds to H-semistable sheaves which are S-equivalent to E, so in particular they are objects in $\langle E_1, \dots, E_m \rangle_{\mathrm{ex}}$. Here $\langle - \rangle_{\mathrm{ex}}$ is the extension closure of $(-)$. Let $\mathcal{I} \subset \mathcal{O}_{\widehat{\mathcal{M}}_{Z,H}(v)_y}$ be the ideal sheaf which is the pull-back of the maximal ideal $\mathfrak{m} \subset R$ by the morphism (8.4.8). Let $\mathcal{M}_n \hookrightarrow \widehat{\mathcal{M}}_{Z,H}(v)_y$ be the closed substack defined by $\mathcal{I}^n$. Then the formal stack $\varinjlim \mathcal{M}_n$ represents the functor

$$\mathcal{M}_{Z,H}^E(v) : Aff^{op} \to Groupoid$$

sending an affine $\mathbb{C}$-scheme T to the groupoid of flat families of coherent sheaves in $\langle E_1, \dots, E_m \rangle_{\mathrm{ex}}$.

By [132, Corollary 6.7], we have an equivalence of categories

$$\Phi \colon \mathrm{mod}_{\mathrm{nil}}(A) \xrightarrow{\sim} \langle E_1, \dots, E_m \rangle_{\mathrm{ex}}.$$

Here the left hand side is the abelian category of finitely generated nilpotent right
A-modules. The above functor is given by

$$\Phi(M) = M \otimes_A \mathcal{E}, \ \mathcal{E} \in \mathrm{Coh}(\mathcal{O}_X \widehat{\otimes} A)$$

where $\mathcal{E}$ is the universal object of NC deformation theory associated with the
collection $(E_1, \dots, E_m)$. Therefore the functor Φ gives an isomorphism of 2-
functors

$$\Phi \colon \mathcal{M}^{\mathrm{nil}}_{(Q_{E_\bullet}, \mathbf{I})}(\vec{v}) \xrightarrow{\cong} \mathcal{M}^E_{Z,H}(v).$$

Therefore the functor Φ also induces the isomorphism of formal stacks

$$\Phi \colon \varinjlim [\mathcal{N}_n / \mathrm{Aut}(E)] \xrightarrow{\cong} \varinjlim \mathcal{M}_n.$$

Then applying Theorem 8.4.1, we have equivalences

$$
\begin{array}{ccc}
\mathrm{Coh}\left(\widehat{\mathcal{M}}_{Z,H}(v)_y\right) & \xrightarrow{\ \sim\ } & \mathrm{Coh}\left([\widehat{\mathcal{N}}_0 / \mathrm{Aut}(E)]\right) \\
\Big\downarrow{\scriptstyle\sim} & & \Big\downarrow{\scriptstyle\sim} \\
\mathrm{Coh}\left(\varinjlim \mathcal{M}_n\right) & \xrightarrow[\ \Phi^*\ ]{\sim} & \mathrm{Coh}\left(\varinjlim [\mathcal{N}_n / \mathrm{Aut}(E)]\right).
\end{array}
$$
(8.4.10)

Here the vertical arrows are restriction functors and the top arrow is defined by the
above commutative diagram. The top arrow sends $\mathcal{O}_{\widehat{\mathcal{M}}_{Z,H}(v)_y}$ to $\mathcal{O}_{[\widehat{\mathcal{N}}_0 / \mathrm{Aut}(E)]}$ and
preserves the tensor products, as these properties are satisfied in the bottom arrow.
Therefore by the Tannaka duality for Artin stacks (see [52, Theorem 1.1]), there
exists an unique isomorphism of stacks

$$\Psi \colon [\widehat{\mathcal{N}}_0 / \mathrm{Aut}(E)] \xrightarrow{\cong} \widehat{\mathcal{M}}_{Z,H}(v)_y$$

such that the top arrow of (8.4.10) is isomorphic to Ψ^*. Therefore we obtain the top
isomorphism in the diagram (8.4.9). The bottom isomorphism in the diagram (8.4.9)
follows from the uniqueness of good moduli spaces.   □

The case of moduli stacks of pairs is similarly proved as follows.

Lemma 8.4.3 *The derived stack $\mathfrak{P}^t_n(S, \beta)$ in (4.3.8) satisfies formal neighborhood
theorem.*

Proof The same argument of Theorem 8.4.2 applies by replacing $\text{Coh}(Z)$ with the abelian category $\mathcal{A}_S$. The only thing to check is that the cotangent complex as deformations of pairs is the same as that as objects in $\mathcal{A}_S$. Namely for a pair $J = (\mathcal{O}_S \to F) \in \mathcal{A}_S$, let $I = (\mathcal{O}_S \to F) \in D_{\text{coh}}^b(S)$ be the associated object where $\mathcal{O}_S$ is located in degree zero. We show that there is a distinguished triangle

$$\mathbf{R}\text{Hom}_S(I, F) \to \mathbf{R}\text{Hom}_{\mathcal{A}_S}(J, J)[1] \to \mathbb{C}[1]. \qquad (8.4.11)$$

So in particular, we have isomorphisms $\text{Hom}_S^i(I, F) \overset{\cong}{\to} \text{Ext}_{\mathcal{A}_S}^{i+1}(J, J)$ for $i = 0, 1$. From the exact sequence $0 \to (0 \to F) \to J \to (\mathcal{O}_S \to 0) \to 0$, in $\mathcal{A}_S$, we have the distinguished triangle

$$\mathbf{R}\text{Hom}_{\mathcal{A}_S}(J, 0 \to F) \to \mathbf{R}\text{Hom}_{\mathcal{A}_S}(J, J) \to \mathbf{R}\text{Hom}_{\mathcal{A}_S}(J, \mathcal{O}_S \to 0). \qquad (8.4.12)$$

We have the adjoint pair of functors

$$D_{\text{coh}}^b(S) \underset{j}{\overset{i}{\rightleftarrows}} D^b(\mathcal{A}_S), \ i(F) = (0 \to F), \ j(W \otimes \mathcal{O}_S \overset{s}{\to} F) = \text{Cone}(s)$$

such that $j \dashv i$. We also have the adjoint pair

$$D_{\text{coh}}^b(\mathbb{C}) \underset{j'}{\overset{i'}{\rightleftarrows}} D^b(\mathcal{A}_S), \ i'(W) = (W \otimes \mathcal{O}_S \to 0), \ j'(W \otimes \mathcal{O}_S \to F) = W$$

such that $j' \dashv i'$. Therefore the distinguished triangle (8.4.12) implies (8.4.11). $\quad\square$

Bibliography

1. Abramovich, D., Olsson, M., Vistoli, A.: Tame stacks in positive characteristic. Ann. Inst. Fourier (Grenoble) **58**(4), 1057–1091 (2008)
2. Addington, N., Donovan, W., Segal, E.: The Pfaffian-Grassmannian equivalence revisited. Algebr. Geom. **2**(3), 332–364 (2015)
3. Ahlqvist, E., Hekking, J., Pernice, M., Savvas, M.: Good moduli spaces in derived algebraic geometry, arXiv:2309.16574
4. Alper, J.: Good moduli spaces for Artin stacks. Ann. Inst. Fourier (Grenoble) **63**(6), 2349–2402 (2013)
5. Alper, J., Hall, J., David, D.: A Luna étale slice theorem for algebraic stacks. Ann. Math. (2) **191**(3), 675–738 (2020)
6. Alper, J., Halpern-Leistner, D., Heinloth, J.: Existence of moduli spaces for algebraic stacks. Invent. Math. **234**(3), 949–1038 (2023)
7. Alper, J., Hall, J., Rydh, D.: The étale local structure of algebraic stacks, arXiv:1912.06162
8. Arinkin, D., Gaitsgory, D.: Singular support of coherent sheaves and the geometric Langlands conjecture. Selecta Math. (N.S.) **21**(1), 1–199 (2015)
9. Ballard, M., Favero, D., Katzarkov, L.: A category of kernels for equivariant factorizations and its implications for Hodge theory. Publ. Math. Inst. Hautes Études Sci. **120**, 1–111 (2014)
10. Ballard, M., Deliu, D., Favero, D., Isik, M.U., Katzarkov, L.: Resolutions in factorization categories. Adv. Math. **295**, 195–249 (2016)
11. Ballard, M., Favero, D., Katzarkov, L.: Variation of geometric invariant theory quotients and derived categories. J. Reine Angew. Math. **746**, 235–303 (2019)
12. Bassat, O.B., Brav, C., Bussi, V., Joyce, D.: A 'Darboux theorem' for shifted symplectic structures on derived Artin stacks, with applications. Geom. Topol. **19**(3), 1287–1359 (2015)
13. Bayer, A., Macrì, E.: Projectivity and birational geometry of Bridgeland moduli spaces. J. Amer. Math. Soc. **27**(3), 707–752. MR 3194493 (2014)
14. Behrend, K.: Donaldson-Thomas type invariants via microlocal geometry. Ann. Math. (2) **170**(3), 1307–1338 (2009)
15. Behrend, K., Fantechi, B.: The intrinsic normal cone. Invent. Math. **128**, 45–88 (1997)
16. Benson, D., Iyengar, S.B., Krause, H.: Local cohomology and support for triangulated categories. Ann. Sci. Éc. Norm. Supér. (4) **41**(4), 573–619 (2008)
17. Berth, D., Schnurer, O.M.: Decompositions of derived categories of gerbes and families of Brauer-Severi varieties. Doc. Math. **26**, 1465–1500 (2021)
18. Blanc, A., Robalo, M., Toën, B., Vezzosi, G.: Motivic realizations of singularity categories and vanishing cycles. J. Éc. polytech. Math. **5**, 651–747 (2018)

Y. Toda, *Categorical Donaldson-Thomas Theory for Local Surfaces*, Lecture Notes in Mathematics 2350, https://doi.org/10.1007/978-3-031-61705-8

19. Brav, C., Bussi, V., Dupont, D., Joyce, D., Szendrői, B.: Symmetries and stabilization for sheaves of vanishing cycles. J. Singul. **11**, 85–151 (2015), With an appendix by Jörg Schürmann. MR 3353002
20. Bondal, A., Orlov, D.: Semiorthogonal decomposition for algebraic varieties, arXiv:9506012
21. Bondal, A.I., Larsen, M., Lunts, V.A.: Grothendieck ring of pretriangulated categories. Int. Math. Res. Not. 2004(29), 1461–1495 (2004)
22. Brav, C., Bussi, V., Joyce, D.: A Darboux theorem for derived schemes with shifted symplectic structure. J. Amer. Math. Soc. **32**, 399–443 (2019)
23. Bridgeland, T.: Flops and derived categories. Invent. Math **147**, 613–632 (2002)
24. Bridgeland, T.: Stability conditions on triangulated categories. Ann. Math. (2) **166**(2), 317–345 (2007)
25. Bridgeland, T.: Hall algebras and curve-counting invariants. J. Amer. Math. Soc. **24**(4), 969–998 (2011)
26. Bridgeland, T., King, A., Reid, M.: The McKay correspondence as an equivalence of derived categories. J. Amer. Math. Soc. **14**, 535–554 (2001)
27. Bussi, V., Joyce, D., Meinhardt, S.: On motivic vanishing cycles of critical loci. J. Algebraic Geom. **28**(3), 405–438 (2019)
28. Calaque, D.: Shifted cotangent stacks are shifted symplectic. Ann. Fac. Sci. Toulouse **28**, 67–90 (2019)
29. Chen, X.W.: Unifying two results of Orlov on singularity categories. Abh. Math. Semin. Univ. Hambg. **80**(2), 207–212 (2010)
30. Căldăraru, A.: The Mukai pairing, I: The Hochschild structure, arXiv:0308079
31. Căldăraru, A.: Derived categories of twisted sheaves on Calabi-Yau manifolds, ProQuest LLC, Ann Arbor, MI, 2000, Thesis (Ph.D.)–Cornell University
32. Davison, B.: Positivity for quantum cluster algebras. Ann. Math. (2) **187**(1), 157–219 (2018)
33. Davison, B., Meinhardt, S.: Cohomological Donaldson-Thomas theory of a quiver with potential and quantum enveloping algebras. Invent. Math. **221**(3), 777–871 (2020)
34. Davison, B., Hennecart, L., Schlegel Mejia, S.: BPS Lie algebras for totally negative 2-Calabi-Yau categories and nonabelian Hodge theory for stacks, arXiv:2212.07668
35. Diaconescu, D.E.: Moduli of ADHM sheaves and the local Donaldson-Thomas theory. J. Geom. Phys. **62**(4), 763–799 (2012)
36. Dimca, A.: Sheaves in Topology. Universitext. Springer, Berlin (2004)
37. Drinfeld, V.: DG quotients of DG categories. J. Algebra **272**(2), 643–691 (2004)
38. Drinfeld, V., Gaitsgory, D.: On some finiteness questions for algebraic stacks. Geom. Funct. Anal. **23**(1), 149–294 (2013)
39. Efimov, A.I.: Cyclic homology of categories of matrix factorizations. Int. Math. Res. Not. IMRN **2018**(12), 3834–3869 (2018)
40. Efimov, A.I., Positselski, L.: Coherent analogues of matrix factorizations and relative singularity categories. Algebra Number Theory **9**(5), 1159–1292 (2015)
41. Gaitsgory, D.: Ind-coherent sheaves. Mosc. Math. J. **13**(3), 399–528, 553 (2013)
42. Gaitsgory, D., Rozenblyum, N.: A Study in Derived Algebraic Geometry. Vol. I. Correspondences and Duality. Mathematical Surveys and Monographs, vol. 221. American Mathematical Society, Providence (2017)
43. Gaitsgory, D., Rozenblyum, N.: A Study in Derived Algebraic Geometry. Vol. II. Deformations, Lie Theory and Formal Geometry. Mathematical Surveys and Monographs, vol. 221. American Mathematical Society, Providence (2017)
44. Geraschenko, A., Zureick-Brown, D.: Formal GAGA for good moduli spaces. Algebraic Geom. **2**, 214–230 (2015)
45. Halpern-Leistner, D.: The derived category of a GIT quotient. J. Amer. Math. Soc. **28**(3), 871–912 (2015)
46. Halpern-Leistner, D.: Derived Θ -stratifications and the D-equivalence conjecture, arXiv:2010.01127
47. Halpern-Leistner, D.: On the structure of instability in moduli theory, arXiv:1411.0627

48. Halpern-Leistner, D.: Remarks on Theta-stratifications and derived categories, arXiv:1502.03083
49. Halpern-Leistner, D.: Θ -statifications, Θ-reductive stacks, and applications, Proc. Sympos. Pure Math. 97.1 Amer. Math. Soc., Providence, RI, 349–379 (2018)
50. Halpern-Leistner, D., Sam, S.V.: Combinatorial constructions of derived equivalences. J. Amer. Math. Soc. **33**(3), 735–773 (2020)
51. Hall, J., Rydh, D.: Perfect complexes on algebraic stacks. Compos. Math. **153**(11), 2318–2367 (2017)
52. Hall, J., Rydh, D.: Coherent Tannaka duality and algebraicity of Hom-stacks. Algebra Number Theory **13**(7), 1633–1675 (2019)
53. Hall, J., Rydh, D.: Mayer-Vietoris squares in algebraic geometry. J. Lond. Math. Soc. (2) **107**(5), 1583–1612 (2023)
54. Hartshorne, R.: Residues and Duality: Lecture Notes of a Seminar on the Work of A. Grothendieck, given at Harvard 1963/1964. Lecture Notes in Mathematics, no. 20. Springer, Berlin (1966)
55. Hennion, B., Holstein, J., Robalo, M.: On the Categorification problem for Motivic Donaldson-Thomas invariants. Slide available in https://marco-robalo.perso.math.cnrs.fr/gluingMF.pdf
56. Hirano, Y.: Derived Knörrer periodicity and Orlov's theorem for gauged Landau-Ginzburg models. Compos. Math. **153**(5), 973–1007 (2017)
57. Huybrechts, D., Lehn, M.: The Geometry of Moduli Spaces of Sheaves. Aspects of Mathematics, E31. Friedr. Vieweg & Sohn, Braunschweig (1997)
58. Isik, M.U.: Equivalence of the derived category of a variety with a singularity category. Int. Math. Res. Not. IMRN **2013**(12), 2787–2808 (2013)
59. Jiang, Q.: Derived categories of Quot schemes of locally free quotients I, arXiv:2107.09193
60. Jiang, Q., Leung, C.: Derived category of projectivization and flops. Adv. Math. **396**, 108–169 (2022)
61. Jiang, Y., Thomas, R.: Virtual signed Euler characteristics. J. Algebraic Geom. **26**(2), 379–397 (2017)
62. Joyce, D.: Shifted symplectic geometry, Calabi-Yau moduli spaces, and generalizations of Donaldson-Thomas theory: our current and future research. In: Talks Given Oxford, October 2013, at a Workshop for EPSRC Programme Grant Research Group (2013). https://people.maths.ox.ac.uk/joyce/PGhandout.pdf
63. Joyce, D.: A classical model for derived critical loci. J. Differential Geom. **101**(2), 289–367 (2015)
64. Joyce, D., Song, Y.: A theory of generalized Donaldson-Thomas invariants. Mem. Amer. Math. Soc. **217**(1020), iv+199 (2012)
65. Joyce, D., Upmeier, M.: Orientation data for moduli spaces of coherent sheaves over Calabi-Yau 3-folds. Adv. Math. **381**, Paper No. 10762, 47 (2021)
66. Kapranov, M., Vasserot, E.: The cohomological Hall algebra of a surface and factorization cohomology, arXiv:1901.07641
67. Kashiwara, M., Schapira, P.: Categories and Sheaves. Grundlehren der Mathematischen Wissenschaften [Fundamental Principles of Mathematical Sciences], vol. 332. Springer, Berlin (2006)
68. Kawamata, Y.: D-equivalence and K-equivalence. J. Differential Geom. **61**(1), 147–171 (2002)
69. Keller, B.: Invariance and localization for cyclic homology of DG algebras. J. Pure Appl. Algebra **123**(1–3), 223–273 (1998)
70. Keller, B.: On the cyclic homology of exact categories. J. Pure Appl. Algebra **136**(1), 1–56 (1999)
71. King, A.: Moduli of representations of finite-dimensional algebras. Q. J. Math. Oxford Ser. (2) **45**, 515–530 (1994)
72. Kinjo, T.: Dimensional reduction in cohomological Donaldson-Thomas theory. Compos. Math. **158**(1), 123–167 (2022)

73. Kollár, J., Mori, S.: Birational Geometry of Algebraic Varieties. Cambridge Tracts in Mathematics, vol. 134. Cambridge University Press, Cambridge (1998), With the collaboration of C. H. Clemens and A. Corti, Translated from the 1998 Japanese original

74. Kontsevich, M.: Homological Algebra of Mirror Symmetry. Proceedings of ICM, vol. 1. Birkhäuser, Basel (1995)

75. Kontsevich, M., Soibelman, Y.: Cohomological Hall algebra, exponential Hodge structures and motivic Donaldson-Thomas invariants. Commun. Number Theory Phys. **5**(2), 231–352 (2011)

76. Kontsevich, M., Soibelman, Y.: Stability structures, motivic Donaldson-Thomas invariants and cluster transformations, arXiv:0811.2435

77. Koseki, N.: Categorical blow-up formula for Hilbert schemes of points, arXiv:2110.08315

78. Koseki, N., Toda, Y.: Derived categories of Thaddeus pair moduli spaces via d-critical flips. Adv. Math. **391**, Paper No. 107965, 55 (2021)

79. Krause, H.: Localization Theory for Triangulated Categories. Triangulated Categories, London Math. Soc. Lecture Note Ser., vol. 375, pp. 161–235. Cambridge University Press, Cambridge (2010). MR 2681709

80. Kuznetsov, A.: Base change for semiorthogonal decompositions. Compos. Math. **147**(3), 852–876 (2011)

81. Kuznetsov, A.: Semiorthogonal decompositions in algebraic geometry. In: Proceedings of the International Congress of Mathematicians—Seoul 2014. Vol. II, Kyung Moon Sa, Seoul, pp. 635–660 (2014)

82. Laumon, G., Moret-Bailly, L.: Champs algébriques, Ergebnisse der Mathematik und ihrer Grenzgebiete. 3. Folge. A Series of Modern Surveys in Mathematics [Results in Mathematics and Related Areas. 3rd Series. A Series of Modern Surveys in Mathematics], vol. 39. Springer, Berlin (2000)

83. Le Potier, J.: Systèmes cohérents et structures de niveau, Astérisque, vol. 214. Société Mathématique de France (1993)

84. Lieblich, M.: Moduli of twisted sheaves. Duke Math. J. **138**(1), 23–118 (2007)

85. Luna, D.: Slices étales. Bull. Soc. Math. France Paris Mém. **33**, 81–105 (1973)

86. Maulik, D., Thomas, R.P.: Sheaf counting on local K3 surfaces. Pure Appl. Math. Q. **2018**, 419–441 (2018)

87. Maulik, D., Toda, Y.: Gopakumar-Vafa invariants via vanishing cycles. Invent. Math. **213**(3), 1017–1097 (2018)

88. Maulik, D., Nekrasov, N., Okounkov, A., Pandharipande, R.: Gromov-Witten theory and Donaldson-Thomas theory. I. Compos. Math. **142**(5), 1263–1285 (2006)

89. Meinhardt, S., Reineke, M.: Donaldson-Thomas invariants versus intersection cohomology of quiver moduli. J. R. Angew. Math. **754**, 143–178 (2019)

90. Mirković, I., Riche, S.: Linear Koszul duality. Compos. Math. **146**(1), 233–258 (2010)

91. Mirković, I., Riche, S.: Linear Koszul duality, II: coherent sheaves on perfect sheaves. J. Lond. Math. Soc. (2) **93**(1), 1–24 (2016). MR 3455779

92. Mukai, S.: On the moduli space of bundles on K3 surfaces I. In: Atiyah, M.F., et al. (eds.) Vector Bundles on Algebraic Varieties, pp. 341–413. Oxford University Press (1987)

93. Nagao, K.: Donaldson-Thomas theory and cluster algebras. Duke Math. J. **162**(7), 1313–1367 (2013)

94. Nagao, K., Nakajima, H.: Counting invariant of perverse coherent sheaves and its wall-crossing. Int. Math. Res. Not. IMRN **2011**(17), 3885–3938 (2011)

95. Narasimhan, M.S., Ramanan, S.: Moduli of vector bundles on a compact Riemann surface. Ann. Math. (2) **89**, 14–51 (1969)

96. Neguţ, A.: Shuffle algebras associated to surfaces. Sel. Math. (N.S.) **25**(3), Art. 36, 57 (2019)

97. Neguţ, A.: Hecke correspondences for smooth moduli spaces of sheaves. Publ. Math. Inst. Hautes Études Sci. **135**, 337–418 (2022)

98. O'Grady, K.: Desingularized moduli spaces of sheaves on a K3. J. R. Angew. Math. **512**, 49–117 (1999)

99. Oblomkov, A., Rozansky, L.: Categorical Chern character and braid groups. Adv. Math. **437**, Paper No. 109436, 66 pp. (2024)

100. Orlov, D.: Derived categories of coherent sheaves and triangulated categories of singularities. In: Algebra, Arithmetic, and Geometry: In Honor of Yu. I. Manin. Vol. II, Progr. Math., vol. 270, pp. 503–531. Birkhäuser Boston, Boston (2009)

101. Pandharipande, R., Pixton, A.: Gromov-Witten/Pairs correspondence for the quintic 3-fold. J. Amer. Math. Soc. **30**(2), 389–449 (2017)

102. Pandharipande, R., Thomas, R.P.: Curve counting via stable pairs in the derived category. Invent. Math. **178**(2), 407–447 (2009)

103. Pandharipande, R., Thomas, R.P.: Stable pairs and BPS invariants. J. Amer. Math. Soc. **23**(1), 267–297 (2010)

104. Pantev, T., Toën, B., Vaquié, M., Vezzosi, G.: Shifted symplectic structures. Publ. Math. Inst. Hautes Études Sci. **117**, 271–328 (2013)

105. Pardon, J.: Universally counting curves in Calabi-Yau threefolds, arXiv:2308.02948

106. Piyaratne, D., Toda, Y.: Moduli of Bridgeland semistable objects on 3-folds and Donaldson Thomas invariants. J. R. Angew. Math. **747**, 175–219 (2019)

107. Polishchuk, A., Positselski, L.: Hochschild (co)homology of the second kind I. Trans. Amer. Math. Soc. **364**(10), 5311–5368 (2012)

108. Polishchuk, A., Vaintrob, A.: Matrix factorizations and singularity categories for stacks. Ann. Inst. Fourier (Grenoble) **61**(7), 2609–2642 (2011)

109. Porta, M., Sala, F.: Two dimensional categorified Hall algebras. J. Eur. Math. Soc. (JEMS) **25**(3), 1113–1205 (2023)

110. Pădurariu, T.: Categorical and K-theoretic Hall algebras for quivers with potential. J. Inst. Math. Jussieu **22**(6), 2717–2747 (2023)

111. Pădurariu, T.: Generators for K-theoretic Hall algebras of quivers with potential. Sel. Math. (N.S.) **30**(1), Paper No. 4, 37 pp. (2024)

112. Pădurariu, T.: K-theoretic Hall algebras for quivers with potential, arXiv:1911.05526

113. Pădurariu, T., Toda, Y.: Categorical and K-theoretic Donaldson-Thomas theory of $\mathbb{C}^3$ (part II). Forum Math. Sigma **11**, Paper No. e108, 47 pp. (2023)

114. Pădurariu, T., Toda, Y.: Categorical and K-theoretic Donaldson-Thomas theory of $\mathbb{C}^3$ (part I). Duke Math. J., to appear, arXiv:2207.01899

115. Pădurariu, T., Toda, Y.: The categorical DT/PT correspondence and quasi-BPS categories for local surfaces, arXiv:2211.12182

116. Pădurariu, T., Toda, Y.: The local categorical DT/PT correspondence, Adv. Math. **442** 109590, 46 p. (2024).

117. Pădurariu, T., Toda, Y.: Quasi-BPS categories for K3 surfaces, arXiv:2309.08437

118. Pădurariu, T., Toda, Y.: Quasi-BPS categories for symmetric quivers with potential, arXiv:2309.08425

119. Pădurariu, T., Toda, Y.: Topological K-theory and uasi-BPS categories for symmetric quivers with potential, arXiv:2309.08432

120. Sacca, G.: Relative compactified Jacobians of linear systems on Enriques surfaces. Trans. AMS. **371**, 7791–7843 (2019)

121. Shipman, I.: A geometric approach to Orlov's theorem. Compos. Math. **148**(5), 1365–1389 (2012)

122. Špenko, Š., Van den Bergh, M.: Non-commutative resolutions of quotient singularities for reductive groups. Invent. Math. **210**(1), 3–67 (2017)

123. Stoppa, J., Thomas, R.P.: Hilbert schemes and stable pairs: GIT and derived category wall crossings. Bull. Soc. Math. France **139**(3), 297–339 (2011)

124. Sumihiro, H.: Equivariant completion. J. Math. Kyoto Univ. **14**, 1–28 (1974)

125. Thomas, R.P.: A holomorphic Casson invariant for Calabi-Yau 3-folds, and bundles on K3 fibrations. J. Differential Geom. **54**(2), 367–438 (2000)

126. Toda, Y.: Limit stable objects on Calabi-Yau 3-folds. Duke Math. J. **149**, 157–208 (2009)

127. Toda, Y.: Curve counting theories via stable objects I: DT/PT correspondence. J. Amer. Math. Soc. **23**, 1119–1157 (2010)

128. Toda, Y.: Generating functions of stable pair invariants via wall-crossings in derived categories. Adv. Stud. Pure Math. **59**, 389–434 (2010). New developments in algebraic geometry, integrable systems and mirror symmetry (RIMS, Kyoto, 2008)

129. Toda, Y.: Stability conditions and curve counting invariants on Calabi-Yau 3-folds. Kyoto J. Math. **52**, 1–50 (2012)

130. Toda, Y.: Stable pairs on local K3 surfaces. J. Differential Geom. **92**, 285–370 (2012)

131. Toda, Y.: Curve counting theories via stable objects II: DT/ncDT flop formula. J. Reine Angew. Math. **675**, 1–51 (2013)

132. Toda, Y.: Moduli stacks of semistable sheaves and representations of Ext-quivers. Geom. Topol. **22**(5), 3083–3144 (2018)

133. Toda, Y.: Hall algebras in the derived category and higher-rank DT invariants. Algebr. Geom. **7**(3), 240–262 (2020)

134. Toda, Y.: Hall-type algebras for categorical Donaldson-Thomas theories on local surfaces. Sel. Math. (N.S.) **26**(4), 64 (2020)

135. Toda, Y.: Semiorthogonal decompositions of stable pair moduli spaces via d-critical flips. J. Eur. Math. Soc. (JEMS) **23**(5), 1675–1725 (2021)

136. Toda, Y.: Recent progress on the Donaldson-Thomas Theory—Wall-crossing and Refined Invariants. Springer Briefs in Mathematical Physics, vol. 43. Springer, Singapore (2021)

137. Toda, Y.: Birational geometry for d-critical loci and wall-crossing in Calabi-Yau 3-folds. Algebr. Geom. **9**(5), 513–573 (2022)

138. Toda, Y.: Categorical Donaldson-Thomas theory for local surfaces: $\mathbb{Z}/2$-periodic version. Int. Math. Res. Not. IMRN **2023**(13), 11172–11216 (2023)

139. Toda, Y.: Derived categories of Quot schemes of locally free quotients via categorified Hall products. Math. Res. Lett. **30**(1), 239–265 (2023)

140. Toda, Y.: Categorical wall-crossing formula for Donaldson-Thomas theory on the resolved conifold. Geom. Topol. **28**(3), 1341–1407 (2024)

141. Toda, Y.: Derived categories of Quot schemes of zero-dimensional quotients on curves, arXiv:2207.09687

142. Toda, Y.: Semiorthogonal decompositions for categorical Donaldson-Thomas theory via Θ-stratifications, arXiv:2106.05496

143. Toën, B.: The homotopy theory of dg-categories and derivedMorita theory. Invent. Math. **167**(3), 615–667 (2007)

144. Toën, B.: Lectures on dg-categories. In: Topics in Algebraic and Topological K-theory. Lecture Notes in Math., vol. 2008, pp. 243–302. Springer, Berlin (2011)

145. Toën, B.: Derived algebraic geometry. EMS Surv. Math. Sci. **1**(2), 153–240 (2014)

146. Toën, B.: Derived algebraic geometry and deformation quantization. In: Proceedings of the International Congress of Mathematicians—Seoul 2014. Vol. II, Kyung Moon Sa, Seoul, pp. 769–792 (2014)

147. Toën, B., Vaquié, M.: Moduli of objects in dg-categories. Ann. Sci. École Norm. Sup. (4) **40**(3), 387–444 (2007)

148. Toën, B., Vezzosi, G.: Homotopical Algebraic Geometry II: Geometric Stacks and Applications. Memoirs of the American Mathematical Society, vol. 193.902, pp. 1–228 (2008)

149. Van den Bergh, M.: Three-dimensional flops and noncommutative rings. Duke Math. J. **122**(3), 423–455 (2004)

150. Van den Bergh, M.: Noncommutative crepant resolutions, an overview. In: ICM—International Congress of Mathematicians. Vol. II. Plenary Lectures, pp. 1354–1391. EMS Press, Berlin (2023)

151. Varagnolo, M., Vasserot, E.: Critical convolution algebras and quantum loop groups, arXiv:2302.01418

152. Varagnolo, M., Vasserot, E.: K-theoretic Hall algebras, quantum groups and super quantum groups. Sel. Math. (N.S.) **28**(1), Paper No. 7, 56 (2022)

153. Zhao, Y.: A categorical quantum toroidal action on Hilbert schemes, arXiv:2009.11267

154. Zhao, Y.: On the K-theoretic Hall algebra of a surface. Int. Math. Res. Not. IMRN **2021**(6), 4445–4486 (2021)

Index

© The Author(s), under exclusive license to Springer Nature Switzerland AG 2024

Y. Toda, *Categorical Donaldson-Thomas Theory for Local Surfaces*, Lecture Notes in Mathematics 2350, https://doi.org/10.1007/978-3-031-61705-8

LECTURE NOTES IN MATHEMATICS  Springer

Editors in Chief: J.-M. Morel, B. Teissier;

Editorial Policy

1. Lecture Notes aim to report new developments in all areas of mathematics and their applications – quickly, informally and at a high level. Mathematical texts analysing new developments in modelling and numerical simulation are welcome.

   Manuscripts should be reasonably self-contained and rounded off. Thus they may, and often will, present not only results of the author but also related work by other people. They may be based on specialised lecture courses. Furthermore, the manuscripts should provide sufficient motivation, examples and applications. This clearly distinguishes Lecture Notes from journal articles or technical reports which normally are very concise. Articles intended for a journal but too long to be accepted by most journals, usually do not have this "lecture notes" character. For similar reasons it is unusual for doctoral theses to be accepted for the Lecture Notes series, though habilitation theses may be appropriate.

2. Besides monographs, multi-author manuscripts resulting from SUMMER SCHOOLS or similar INTENSIVE COURSES are welcome, provided their objective was held to present an active mathematical topic to an audience at the beginning or intermediate graduate level (a list of participants should be provided).

   The resulting manuscript should not be just a collection of course notes, but should require advance planning and coordination among the main lecturers. The subject matter should dictate the structure of the book. This structure should be motivated and explained in a scientific introduction, and the notation, references, index and formulation of results should be, if possible, unified by the editors. Each contribution should have an abstract and an introduction referring to the other contributions. In other words, more preparatory work must go into a multi-authored volume than simply assembling a disparate collection of papers, communicated at the event.

3. Manuscripts should be submitted either online at www.editorialmanager.com/lnm to Springer's mathematics editorial in Heidelberg, or electronically to one of the series editors. Authors should be aware that incomplete or insufficiently close-to-final manuscripts almost always result in longer refereeing times and nevertheless unclear referees' recommendations, making further refereeing of a final draft necessary. The strict minimum amount of material that will be considered should include a detailed outline describing the planned contents of each chapter, a bibliography and several sample chapters. Parallel submission of a manuscript to another publisher while under consideration for LNM is not acceptable and can lead to rejection.

4. In general, **monographs** will be sent out to at least 2 external referees for evaluation.

   A final decision to publish can be made only on the basis of the complete manuscript, however a refereeing process leading to a preliminary decision can be based on a pre-final or incomplete manuscript.

   Volume Editors of **multi-author works** are expected to arrange for the refereeing, to the usual scientific standards, of the individual contributions. If the resulting reports can be forwarded to the LNM Editorial Board, this is very helpful. If no reports are forwarded or if other questions remain unclear in respect of homogeneity etc, the series editors may wish to consult external referees for an overall evaluation of the volume.

5. Manuscripts should in general be submitted in English. Final manuscripts should contain at least 100 pages of mathematical text and should always include

   - a table of contents;
   - an informative introduction, with adequate motivation and perhaps some historical remarks: it should be accessible to a reader not intimately familiar with the topic treated;
   - a subject index: as a rule this is genuinely helpful for the reader.
   - For evaluation purposes, manuscripts should be submitted as pdf files.

6. Careful preparation of the manuscripts will help keep production time short besides ensuring satisfactory appearance of the finished book in print and online. After acceptance of the manuscript authors will be asked to prepare the final LaTeX source files (see LaTeX templates online: https://www.springer.com/gb/authors-editors/book-authors-editors/manuscriptpreparation/5636) plus the corresponding pdf- or zipped ps-file. The LaTeX source files are essential for producing the full-text online version of the book, see http://link.springer.com/bookseries/304 for the existing online volumes of LNM). The technical production of a Lecture Notes volume takes approximately 12 weeks. Additional instructions, if necessary, are available on request from lnm@springer.com.

7. Authors receive a total of 30 free copies of their volume and free access to their book on SpringerLink, but no royalties. They are entitled to a discount of 33.3 % on the price of Springer books purchased for their personal use, if ordering directly from Springer.

8. Commitment to publish is made by a *Publishing Agreement*; contributing authors of multiauthor books are requested to sign a *Consent to Publish form*. Springer-Verlag registers the copyright for each volume. Authors are free to reuse material contained in their LNM volumes in later publications: a brief written (or e-mail) request for formal permission is sufficient.

Addresses:
Professor Jean-Michel Morel, CMLA, École Normale Supérieure de Cachan, France
E-mail: moreljeanmichel@gmail.com

Professor Bernard Teissier, Equipe Géométrie et Dynamique,
Institut de Mathématiques de Jussieu – Paris Rive Gauche, Paris, France
E-mail: bernard.teissier@imj-prg.fr

Springer: Ute McCrory, Mathematics, Heidelberg, Germany,
E-mail: lnm@springer.com

Printed in the United States
by Baker & Taylor Publisher Services